Geology of the United States' Seafloor

Geology of the United States' Seafloor

The View from GLORIA

Edited by

JAMES V. GARDNER, MICHAEL E. FIELD, and DAVID C. TWICHELL

US Geological Survey

CAMBRIDGE
UNIVERSITY PRESS

Published by the Press Syndicate of the University of Cambridge
The Pitt Building, Trumpington Street, Cambridge CB2 1RP
40 West 20th Street, New York, NY 10011-4211, USA
10 Stamford Road, Oakleigh, Melbourne 3166, Australia

First published 1996

Printed in the United States of America

Library of Congress Cataloging-in-Publication Data
Geology of the United States' seafloor : the view from GLORIA / edited
  by James V. Gardner, Michael E. Field, David C. Twichell.
      p.   cm.
    Includes bibliographical references.
    ISBN 0-521-43310-X (hc)
    1. Geology – United States.   2. Continental margins – United States.
  3. Economic zones (Maritime law) – United States.   I. Gardner, James
  V.   II. Field, Michael E.   III. Twichell, David C.
  QE77.G325   1996
  557.3 – dc20   95-46492
                                                              CIP

A catalog record for this book is available from the British Library

ISBN   0-521-43310-X Hardback

Contents

Contributors

Mr. Jeff A. Anderson
U.S. Geological Survey
2255 N. Gemini Dr.
Flagstaff, AZ 86001

Dr. Robert G. Bohannon
U.S. Geological Survey
345 Middlefield Rd.
MS 999
Menlo Park, CA 94025

Ms. JoAnn Bowell
U.S. Geological Survey
2255 N. Gemini Dr.
Flagstaff, AZ 86001

Dr. Terry R. Bruns
U.S. Geological Survey
345 Middlefield Rd.
MS 999
Menlo Park, CA 94025

Dr. David A. Cacchione
U.S. Geological Survey
345 Middlefield Rd.
MS 999
Menlo Park, CA 94025

Dr. Paul R. Carlson
U.S. Geological Survey
345 Middlefield Rd.
MS 999
Menlo Park, CA 94025

Dr. Bobb Carson
Dept. of Earth and Environmental Sciences
Lehigh University
Bethlehem, PA 18015

Mr. T.E. Chase
U.S. Geological Survey
345 Middlefield Rd.
MS 999
Menlo Park, CA 94025

Mr. Pat S. Chavez, Jr.
U.S. Geological Survey
2255 N. Gemini Dr.
Flagstaff, AZ 86001

Mr. Dwight F. Coleman
U.S. Geological Survey
Quisett Campus
Woods Hole, MA 02543

Ms. Catherine M. Delorey
U.S. Geological Survey
Quisett Campus
Woods Hole, MA 02543

Dr. William P. Dillon
U.S. Geological Survey
Quisett Campus
Woods Hole, MA 02543

Dr. Max R. Dobson
Dept. of Geology
University of Aberystwyth
Dyfed SY23 2AX
U.K.

Dr. David E. Drake
U.S. Geological Survey
345 Middlefield Rd.
MS 999
Menlo Park, CA 94025

Dr. N. Terence Edgar
U.S. Geological Survey
12201 Sunrise Valley Dr.
MS 915
Reston, VA 22092

Dr. Brian D. Edwards
U.S. Geological Survey
345 Middlefield Rd.
MS 999
Menlo Park, CA 94025

Dr. Michael E. Field
U.S. Geological Survey
345 Middlefield Rd.
MS 999
Menlo Park, CA 94025

Dr. James V. Gardner
U.S. Geological Survey
345 Middlefield Rd.
MS 999
Menlo Park, CA 94025

Dr. Gary W. Hill
U.S. Geological Survey
12201 Sunrise Valley Dr.
MS 915
Reston, VA 22092

Dr. Mark L. Holmes
University of Washington
School of Oceanography
WB-10
Seattle, WA 98195

Mr. Quentin J. Huggett
Institute of Oceanographic Sciences
Southampton Oceanography Centre
Southampton
U.K.

Dr. Herman A. Karl
U.S. Geological Survey
345 Middlefield Rd.
MS 999
Menlo Park, CA 94025

Dr. Robert E. Kayen
U.S. Geological Survey
345 Middlefield Rd.
MS 999
Menlo Park, CA 94025

Mr. Neil H. Kenyon
Institute of Oceanographic Sciences
Southampton Oceanography Centre
Southampton
U.K.

Dr. Homa J. Lee
U.S. Geological Survey
345 Middlefield Rd.
MS 999
Menlo Park, CA 94025

Mr. Dennis M. Mann
U.S. Geological Survey
345 Middlefield Rd.
MS 999
Menlo Park, CA 94025

Dr. Douglas G. Masson
Institute of Oceanographic Sciences
Southampton Oceanography Centre
Southampton
U.K.

Mr. Dennis W. O'Leary
U.S. Geological Survey
Denver Federal Center
MS 425
Denver, CO 80225

Mr. C.A. Mortera-Gutierrez
Geodynamics Research Institute
Texas A&M University
College Station, TX 77843

Ms. Valerie Paskevich
U.S. Geological Survey
Quisett Campus
Woods Hole, MA 02543

Dr. Charles K. Paull
Univ. of North Carolina
Chapel Hill, NC 27514

Mr. Peter Popenoe
U.S. Geological Survey
Quisett Campus
Woods Hole, MA 02543

Ms. Libby Prueher
Department of Geosciences
University of Michigan
Ann Arbor, MI 48109

Mr. James M. Robb
U.S. Geological Survey
Quisett Campus
Woods Hole, MA 02543

Ms. Kathryn M. Scanlon
U.S. Geological Survey
Quisett Campus
Woods Hole, MA 02543

Mr. Erol Seke
Dept. of Electrical Engineering and Computer Science
Lehigh University
Bethlehem, PA 18015

Mr. Stuart C. Sides
U.S. Geological Survey
2255 N. Gemini Dr.
Flagstaff, AZ 86001

Mr. Michael L. Somers
Churt, Surrey
GU10 2NA
U.K.

Mr. Andrew J. Stevenson
U.S. Geological Survey
345 Middlefield Rd.
MS 999
Menlo Park, CA 94025

Dr. David C. Twichell
U.S. Geological Survey
Quisett Campus
Woods Hole, MA 02543

Dr. Tracy L. Vallier
U.S. Geological Survey
345 Middlefield Rd.
MS 999
Menlo Park, CA 94025

Foreword

In 1879, Congress established the U.S. Geological Survey (USGS) as a scientific research organization and charged it with conducting an "examination of the geological structure, mineral resources, and products of the national domain." In 1962, these examinations were extended to the marine realm when Congress appropriated funds for offshore investigations by the USGS. On March 10, 1983, by proclamation of then-President Ronald Reagan, the United States claimed sovereign rights and jurisdiction within an Exclusive Economic Zone (EEZ). The United States assumed responsibility for the wise management of marine resources and for protection of the marine environment in the EEZ.

The EEZ of the United States is a vast frontier that extends seaward from 3 nautical miles to 200 nautical miles from the coast and brings within the national domain over 3.3 million square nautical miles of submarine lands, an area 30 percent larger than the total onshore area of the United States. With over 50 percent of this country's population clustered along our coasts, the adjacent ocean and ocean floor have become an integral part of the nation's land area to manage, use, and protect.

The rise of the United States to the status of a world power was made possible in part by exploitation of this country's natural resources. As a nation, we have never enjoyed full self-sufficiency in all raw materials that we require, and our growth has been accompanied by the often wasteful use of resources that many believed to be unlimited in quantity. In establishing the EEZ, the United States has gained access to potentially large resources of energy and mineral commodities that may lie on or below the seafloor.

Accordingly, the resources of the EEZ, by allowing the nation to become more self-sufficient, are a possible means for ensuring national economic security. The EEZ contains numerous strategic and critical mineral commodities that this nation relies on politically unstable regions of the world to supply; it also contains other vital natural resources, living and nonliving, in the subsoil and in the water column. Continuing geological studies of marine energy and mineral deposits being formed today also provide insight into the discovery process for land-based mineral and energy deposits.

But, because most of the EEZ has not yet been explored, its resources and their potential benefits to citizens of the United States remain undefined.

Equally important to the economic well-being of the United States is the stability of its large fishing industry that depends on the living resources of the EEZ. Such resources contribute tens of billions of dollars annually to the economy of the United States, and they provide a traditional livelihood for millions of citizens. Although protection of our sovereign interests in the EEZ resources is important, a special concern is to ensure that these living resources remain renewable. Research must continue in order for scientists to determine the relation between the environment and geology.

The fabric of concern woven through the conflicting interests of exploitation and preservation is the protection of marine and nearshore environments. The impact of human activity on the environment can be profound; however, before scientists can determine the extent of these effects, they must first understand the natural processes at work in the environment of the EEZ. Many studies are underway that will yield a road map suitable for future detailed studies, including basic geologic and geodetic framework studies designed to map and chart the seafloor.

The study of this vast underwater land provides unique technological challenges in the effort to visualize a land surface that we cannot see directly with our eyes. This book builds on many of the successes achieved by engineers and research scientists at the USGS and the United Kingdom's Institute of Oceanographic Sciences in carrying out the world's first systematic reconnaissance surveys of the EEZ of the United States. These surveys have collected digital GLORIA sidescan imagery, performed high- and low-resolution seismic-reflection profiling, and conducted gravity and magnetic surveys. Four atlases have been published; two more that display processed images of the seafloor with preliminary interpretations are in preparation. This book presents, for the first time, a series of papers describing the geology and geologic processes of the continental margins of the United States as inferred from GLORIA images.

The book is timely, as many countries are beginning to evaluate how best to survey and inventory their own Exclusive Economic Zones. The United States has more than a decade of experience studying an entire EEZ through a systematic survey. The collective experience of USGS scientists in assembling reconnaissance data and interpreting images, as presented in atlases and in these papers, will be of enormous aid to other countries in showing the techniques and results of such an endeavor.

Equally important for the earth-sciences community are the remarkable scientific results of the first 10 years of USGS surveys. New concepts of geological processes have emerged, including new knowledge about (1) development of submarine fans, which has led to new interpretations about accumulations of hydrocarbons; (2) the origin and history of the Hawaiian Islands, which has led to a new understanding about submarine landslides and the stability of island coastlines; (3) sedimentary processes in the Gulf of Alaska, which has led to new understanding about how fisheries are affected by sediment movement; (4) formation of submarine canyons on a variety of continental margins, which provides new insight into transport and distribution of pollutants; (5) sediment failure along continental slopes, which is critical to submarine communications links; and (6) tectonics of plate margins and continuity of Pacific fracture zones, which has added yet another dimension to our understanding of earthquake and volcanic activity.

This volume is a valuable collection of reports revealing innovative concepts about continental margins, islands and island arcs, and basins, and it was brought about by complete coverage of the U.S. EEZ through reconnaissance sonar imagery. It is analogous to the first books on Landsat imagery of the Earth or early books on planetary imagery. *Geology of the United States' Seafloor: The View from GLORIA* will prove to be a valuable reference document.

Gary W. Hill

Introduction

U.S. Geological Survey, Menlo Park, California

On the fog-shrouded morning of April 26, 1984, the USGS research vessel *Farnella* unceremoniously slipped its berth in San Diego, California, to begin one of the nation's greatest mapping expeditions ever undertaken. The mission was to map the newly declared territories of the United States – the Exclusive Economic Zone (EEZ). The U.S. EEZ was proclaimed just a year earlier by then-President Ronald Reagan. This single stroke of the presidential pen increased the size of the U.S. territory by more than 100%, a much larger increase in area than occurred with the Louisiana Purchase in 1803. Just as the area gained by the Louisiana Purchase was a region little explored, the more than 13 million km² of seafloor represented a vast *terra incognita* (Figure 1). The 1879 Organic Act of the U.S. Geological Survey (USGS) provides a mandate for the USGS to map the territorial lands of the U.S., and mapping the EEZ was a natural extension of that mandate. However, *how* to map this vast region became a problem. It was clear from the beginning that the mapping budget would not be comparable to those set aside to map the Moon, Mars, or Venus. Yet, just a year after the proclamation, and in the tradition of the late nineteenth- and early twentieth-century USGS mapping expeditions of the West, a USGS team ventured forth to map the newly acquired national lands of the EEZ.

Although led by experienced marine geologists, the USGS team was unfamiliar with the principal instrument chosen to map the EEZ. The USGS had selected the British-designed, -built, and -operated GLORIA II long-range sidescan sonar (Figure 2) to map the EEZ because it provided the state-of-the-art technology necessary for mapping large regions of the ocean floor at reconnaissance scale. GLORIA is a digital sidescan sonar system that has the unique capability of mapping swaths 30-, 45-, or 60-km wide, not just the area directly beneath the ship. Other sidescan sonars were available within the United States, but none of them have the capability of mapping such large swaths. Multi-beam-swath bathymetric systems were available in the U.S. in 1983 and were considered for mapping the EEZ, but these

systems, as well as the other available sidescan sonar systems, can map a swath of only 5 to 10 km. Simple calculations showed us that using any of these alternate systems would require decades to complete the EEZ mapping.

So, now a mixed team of USGS marine geologists and U.K. GLORIA experts were plowing through rolling seas off San Diego to begin an eight-year EEZ-SCAN program to map the EEZ. It was well known to the team aboard *Farnella* that morning that they carried with them the only GLORIA II system in existence. The UK Institute of Oceanographic Sciences (IOS) developed the system in the 1970s for their own use but did not have the resources to construct a backup system. If, as often happens on scientific cruises, something were suddenly to go wrong during the cruise and the GLORIA II system were severely damaged or lost, it would mean the abrupt end of the survey and the program, in something just shy of humiliation. Accordingly, this team of scientists, engineers, and ship's crew, and every team that followed, had a large responsibility resting on its collective shoulders. They had to return the GLORIA II system to port in one month, intact and operating, so that the next team could continue the survey. Over the eight years of mapping, this responsibility overrode all other shipboard considerations. In 1986, an improved GLORIA II system was constructed for the USGS, and this system provided for a backup in the event of mishap, but the precautions were never relaxed and GLORIA was never lost.

Anticipating digital images of the seafloor allowed the USGS to develop computer programs to correct distortions in the images and to digitally align the images into quadrangle mosaics. Therefore reconnaissance-view quadrangles of the seafloor could be constructed, similar to Landsat images of the subaerial surface of the Earth. Once again, the USGS was almost uniquely qualified to tackle such a problem because of experience gained in the planetary-exploration programs of the 1970s and 1980s. Consequently, the program began with all the elements in place to be successful.

Eight years later, on the balmy morning of June 15, 1991, with a quarter of a million kilometers of tracks behind them,

This Introduction is not subject to US copyright.

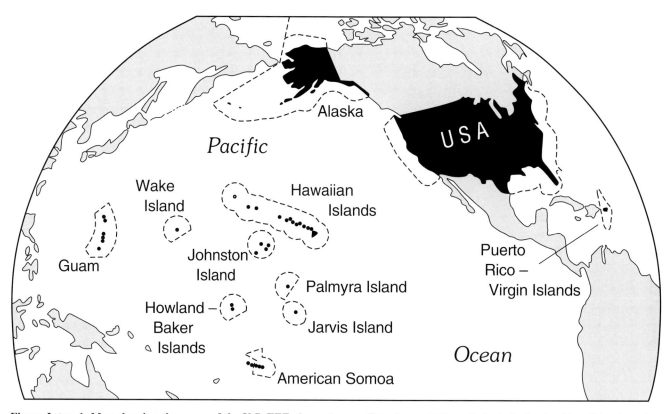

Figure Intro–1. Map showing the areas of the U.S. EEZ. Areas too small to show at this scale include Swains-Rose Islands and Northern Marianas.

Figure Intro–2. The GLORIA II system during a launching operation in the eastern North Pacific.

an experienced team of USGS and IOS scientists brought *Farnella* to the pier in Hilo, Hawaii. This cruise terminated the first and largest stage by mapping more than 6 million km² of the U.S. seafloor. The deep-water portions (deeper than 200 m) off the U.S. East, West, and Gulf Coasts, and the Hawaiian, Aleutian, and Bering Sea EEZs had been mapped with virtually 100 percent coverage. The area was no longer a *terra incognita*; it was now a familiar landscape of seamounts, ridges, fracture zones, canyons, trenches, submarine fans, channels, and plains, all accurately mapped with state-of-the-art navigation and processed digital images. The surveys included more than 100 days at sea per year for eight years, with less than one day per month downtime over the entire period. That remarkable engineering achievement is testimony to the premier GLORIA development team at IOS and those IOS engineers that accompanied the USGS on each cruise. The entire survey cost the U.S. taxpayer only a penny an acre (in 1984 dollars), surely one of the most efficient mapping expeditions of all time.

There is generally no place in a scientific book where one can acknowledge all the people who were essential to make a program of this scope and magnitude a success. I'm taking this opportunity to do so here. A program as large and unconventional as EEZ-SCAN is difficult to get funded, even within a government agency that has experience with large programs. There was a certain urgency to begin a program in the newly formed Exclusive Economic Zone, brought on by the withdrawal of the United States from the UN Law of the Sea proceedings and other political factors. The scientific justifications for the program were easy to formulate, but often scientific justification is not the most important criterion used by those managing the budget. For these reasons, the EEZ-SCAN program was inordinately fortunate in its early stages of development to have the support and leadership of two key science administrators who had vision and perseverance: Dr. David G. Howell, then Branch Chief of the Branch of Pacific Marine Geology, and Dr. Gary W. Hill, then Deputy Chief, Office of Energy and Marine Geology. David enthusiastically encouraged the initiation of the project because he had the vision of road maps of the seafloor to guide generations of more detailed follow-on studies. Gary actively supported and promoted the program because he saw the scientific results as being a vital function of the mandate of the USGS—to map the nation's lands and inventory potential resources and hazards. Without the leadership of these two key people, the goal of mapping the nation's vast new territories would never have been achieved.

Other individuals critical to the successful implementation of the EEZ mapping initiative reside in the United Kingdom. The then-Director of IOS, Sir Anthony Laughton, was a strong supporter of the project from the very beginning, and he was instrumental in the formulation of a strong USGS-IOS collaboration that integrated the collection of data with the interpretations. Our cooperative mapping project enjoyed eight years of success for many reasons, but

none more important than the involvement of the team that designed, built, and improved GLORIA II. This team includes Jack Revie, Mike Somers, Mac Harris, Derek Bishop, Jon Campbell, Chris Flewellen, Eric Darlington, Brian Barrow, and Andy Harris. And then there are the masters, officers, and crew of R/V *Farnella*, managed by one of the truly exceptional individuals of the entire program, Mr. James Hind. These gentlemen made our jobs infinitely easier through their unfailing humor, friendship, and assistance. They also taught us to loathe mushy peas.

In his preface to the first atlas from the EEZ-SCAN Program, David Howell eloquently stated the spirit of EEZ-SCAN with the following words:

> Long before the American Revolution, explorers were venturing westward on missions of discovery, motivated by the lure of the unknown and the quest for riches. The almost continuous westward stream of adventurers ebbed and surged in response to events as varied as the discovery of luxuriant furs in the Northwest, the addition of major land areas such as the Louisiana Purchase and Alaska, and the discovery of gold in California.
>
> The rewards of expeditions extended beyond the bounty of a particular voyage or trek. The more lasting treasures involved new information about the discovered or more fully explored territories. Private or corporate journals were the repositories of the early findings, but when President Thomas Jefferson in 1804 ordered Meriwether Lewis and Captain William Clark on an expedition to explore the unknown country lying between the Mississippi River and the Pacific Ocean, the United States Government assumed a leadership role for all future surveys. It was not until 1879, however, that the Congress formally established the U.S. Geological Survey (USGS) to examine the substance of the national domain. The USGS has remained intact for more than 100 years owing to the importance and lasting relevance of its charge: to provide geologic, topographic, and hydrologic information that contributes to the wise management of the nation's natural resources and that promotes the health, safety, and well-being of the people.
>
> The western limit of exploration for many years was the shoreline of the Pacific, for the ocean beyond was too hostile and the prospect of natural resources seemed minimal. It was ignorance, not information, that temporarily thwarted our manifest destiny.

This book represents the first compilation of derivative studies from the GLORIA surveys of the U.S. undersea territories. The primary sidescan, seismic-reflection, gravity, and magnetic data have been published, or are presently in preparation, in an ongoing series of USGS atlases. CD-ROMs of the data have been produced and distributed by the USGS free of charge to anyone interested in the data. Many studies and areas are not included in this volume because of limitations imposed by time and volume size. For example, interpretative papers on the entire Hawaiian EEZ are planned to appear in a separate volume. More than a hundred derivative studies have already been published in numerous journals, and many more are in various stages of conception, formulation, and draft writing. Like the first cruise in April 1984, this book is just the beginning.

The second stage of mapping America's Exclusive Economic Zone is designed to enhance understanding of the geology represented by the acoustic images of the GLORIA II sidescan sonar. Collecting, processing, and displaying images of the seafloor has been difficult, but in comparison to what follows, it was the easy part. The more difficult job lies ahead. The imagery produced by the GLORIA II sidescan sonar represents the backscatter from the acoustic properties of a variable-thickness volume of seafloor. The more challenging job for marine geologists is to find the Rosetta stone, or stones, that can be used to decipher the backscatter images and produce true geological maps. That task stretches into the 21st century. This stage will require a generation of geoscientists and a major commitment to unravel the physics of the interaction of sound with sediments to provide interpretations equal in detail to those now provided by satellite-mapping systems for the land areas. This second stage is only in its infancy in 1995.

That sunny morning in June 1991 terminated the first stage of EEZ-SCAN mapping, but an additional 3.5 million km^2 of the U.S. Trust Territories remain unmapped, as well as more than a million km^2 of U.S. continental shelf, an area too shallow to be efficiently mapped with GLORIA II. These areas await both the funding base and the technologies only now emerging, so that mapping the nation's frontier can be completed.

References

EEZ-SCAN 84 Scientific Staff 1986. Atlas of the Exclusive Economic Zone, Western Conterminous United States. *U.S. Geological Survey Miscellaneous Investigations Series I-1792,* 152 p., scale 1:500,000, Reston, Va.

EEZ-SCAN 85 Scientific Staff 1987. Atlas of the Exclusive Economic Zone, Gulf of Mexico and Eastern Caribbean areas. *U.S. Geological Survey Miscellaneous Investigations Series I-1864-A,B,* 104 p., scale 1:500,000, Reston, Va.

EEZ-SCAN Scientific Staff 1991. Atlas of the Exclusive Economic Zone, Bering Sea. *U.S. Geological Survey Miscellaneous Investigations Series I-2053,* 145 p., scale 1:500,000, Reston, Va.

EEZ-SCAN 87 Scientific Staff 1991. Atlas of the Exclusive Economic Zone, Atlantic continental margin. *U.S. Geological Survey Miscellaneous Investigations Series I-2054,* 174 p., scale 1:500,000, Reston, Va.

I The GLORIA system and data processing

1 The USGS GLORIA system

Michael L. Somers

Institute of Oceanographic Sciences, Southampton, United Kingdom

The sidescan sonar technique

The sidescan technique is, as the name implies, a scanning process in which attention is directed successively at each point in the target field. It detects and records the energy backscattered from the scanned point onto an image field. The imaging process is a one-to-one mapping from the object field to the image in which the geometry, though containing distortions, is recognizably close to true plan. The technique is commonly applied to radar imaging from the air over land and sea and to sonar imaging of the seabed. GLORIA is an example of the latter. The strength of the technique lies in the fact that with the nearly true geometry, the patterns formed by the recorded backscattered energy give powerful and important clues to the processes at work on the object field surface. In neither radar nor sonar is there much penetration of the radiation below the surface, though there are minor exceptions in both cases, and sidescan is generally classed as a 2D process. The choice of acoustic energy for sidescan mapping of the seabed is dictated by the failure of electromagnetic energy of any wavelength to penetrate useful distances under water. Sidescan differs radically as an imaging technique from the well-known function of a lens in optics where all points on the object field are imaged simultaneously and continuously on the image surface.

The scanning process uses a narrow fan-shaped beam of sound directed at right angles to the survey track, as in Figure 1–1, to illuminate a narrow strip of the seabed. A short pulse of sound is emitted at regular intervals, and the returning echoes are recorded between the transmissions. The transmission interval is the time required for the sound to travel to extreme range and back. Between pulses, the survey vessel advances a known distance along its track. The use of a short pulse ensures that energy reflected from any point on the scan is not confused with that from neighboring points at slightly different ranges. Thus the scanning comprises two mechanisms. Across track, it is the propagation of the sound pulse away from the sonar system, whereas along track it is the physical translation of the array. The former process is essentially continuous, though it is usually sampled in modern systems for digital recording, whereas the pulsing of the transmitter makes the scanning along track inherently discontinuous. The difference is that the designer has much less control over the sampling along track. The minimum size of the patch on the seabed that may be independently mapped, or resolved, on the image is determined by the pulse length across track and, in the limit, by the distance traveled by the survey vessel between pulses along track. In principle this is not rigorously true. Different considerations apply if the so-called near field of the array extends across an appreciable portion of the swath. This condition certainly does not apply to GLORIA, but it sometimes does for very high frequency sonars. In practice the resolution along track is limited for most of the survey swath by the angular spread of the sound beam in the fore and aft direction. The spread is related to the acoustic frequency and the array dimensions, which are crucial considerations in the design of sidescan sonar systems.

Figure 1–2 shows these basic facts of sidescan sonar. There are inherent distortions in the images arising from both the geometry of the technique and the propagation of sound in the sea, but much can be done by suitable processing to alleviate the effects. The real strength of sidescan sonar reveals itself when the sonar energy impinges on the seabed at near-grazing incidence. Not only are the geometrical distortions much less evident, but the variation of backscattering with angle is more rapid at shallow angles, which aids in revealing more subtle processes. The aim of the designer is almost always to have the widest swath possible, and in any case a ratio of maximum range to vehicle altitude of at least 5:1.

This chapter starts with a short resume of the history of sidescan sonar, follows with an expansion of the foregoing material, and ends with a treatment of topics relevant to the USGS GLORIA system.

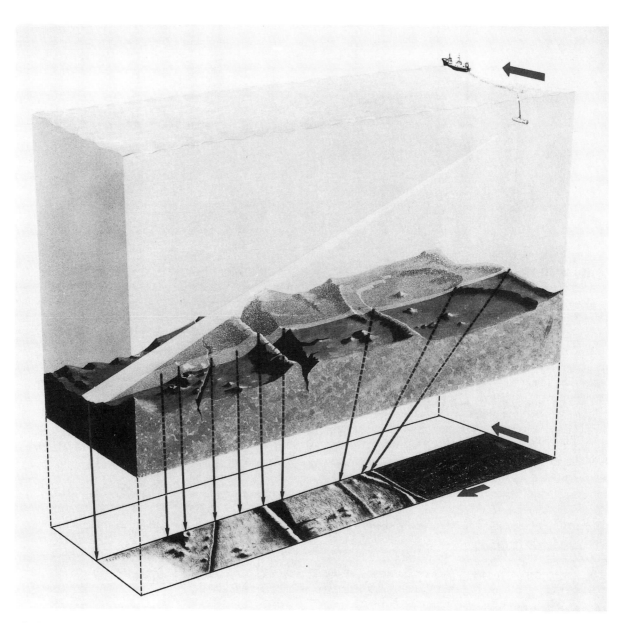

Figure 1–1. The principle of sidescan sonar. The sonar array is shown being towed behind the survey ship in the direction of the arrow. Short pulses of sound are transmitted in the fan-shaped region shown and intercept the seabed along the narrow highlighted strip. The lower part shows the record formed by a large number of recorded pulses laid side-by-side on a facsimile recorder. The correspondence between points on the seabed and features on the image is indicated by the thin arrows. The reader will note that on the image the seabed vertically below the sonar is printed some distance into the record, as though it belonged well off to one side of the track. The delay is the time the sound takes to traverse the water column. Also, features a small distance off the vertical are foreshortened on the image. These two effects, the presence of the water column and the slant-range distortion, are the result of the geometry of the propagation—the recorder simply records events in linear time as they occur. These distortions are normally corrected in a digitally recorded image. Note that, for clarity, only the port-side operation is shown; GLORIA has both port and starboard arrays.

History of sidescan and the GLORIA system

Sidescan sonar in the civilian field dates from the late 1950s. At that time military sonar work was giving information on the backscattering of sound by the seabed, and the geological implications of the patterns observed were soon real-ized, leading shortly to the publication of results in the geological literature (Chesterman, Clynick, and Stride 1958). The first sidescan sonar designed and built in the civilian field was described in 1961 (Tucker and Stubbs) although it was originally intended for fisheries research, and it used military transducer hardware. This fact set the operating frequency at 36 kHz and thereby optimized the design for use

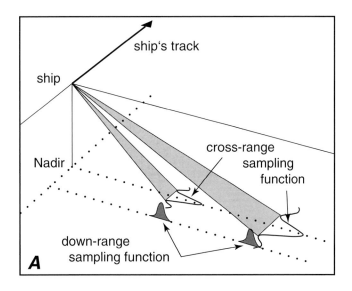

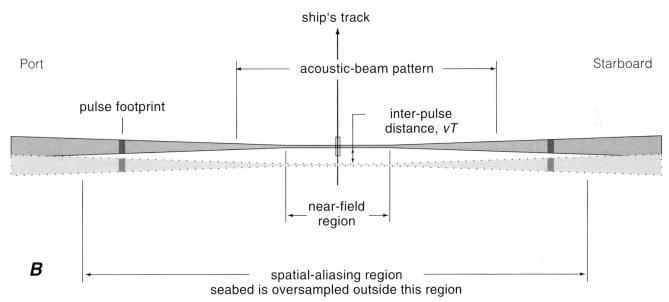

Figure 1–2. A, The basis of scanning and sampling in sidescan. The downrange sampling function is governed by the transmitted pulse shape. Conventionally it is shown, incorrectly, as a rectangular pulse. The bell shaped envelope shown here is a result of the finite system bandwidth. The figure does not show the geometric effect of propagation near the nadir, which smears out the pulse in this region. The cross-range function is the beam pattern of the array, here that of a uniform line array. Note that the sampling functions are not strictly limited in extent. B, A plan view of sidescan geometry showing under- and over-sampling along track. The interpulse distance is the survey velocity times the time required for the sound to complete the round trip to extreme range. There is often a clear economic need to carry out the survey as quickly as possible, which inevitably leaves gaps on the seabed that never get insonified. This is under-sampling and can lead to spatial aliasing (see any good text on communication principles for a discussion of sampling and aliasing for example, Stein and Jones 1967). Beyond the range at which the pulse spread is equal to vT, the seabed is over-sampled in terms of the spatial frequency passed by the sampling pulse. Note that for simplicity this discussion is given in terms of a rectangular beam spread equivalent in width to the more complex pattern shown in Figure 2A.

in continental shelf seas with depths ranging from 50–200 m with a swath of 1.5 km (it was single sided). So far all the work had been done in Britain, and the topic of sidescan sonar could have remained a largely British curiosity for many years but for the tragic loss of the USNS *Thresher*

on April 10, 1963. The subsequent search for the wreckage caused a major upsurge in U.S. activity, mainly in very short range systems with maximum swaths of 500 m and correspondingly high definition. The appearance of these site-specific engineering-scale sonar systems coincided with a major surge in offshore petroleum activity and, with the United States' dominance of that field, leadership in short-range sidescan sonar systems passed to the USA.

Meanwhile in Britain, work continued at IOSDL (then the National Institute of Oceanography) with the original sonar set, where Arthur Stride and his co-workers carried out their pioneering work on geological interpretation of sidescan sonar. However, being primarily a deep water institute, the question was soon asked as to whether the same technique could be extended to apply in the deep oceans, not as deep deployed short-range systems such as had been developed for the USNS *Thresher* search, but as a regional survey tool with a swath of tens of kilometers and speeds of six to eight knots. It was apparent to the potential users that advances in technique to explore such an unknown field must retain some continuity of scale. At this stage all that was known about the detailed morphology of the deep ocean floor was, first, what had been gleaned from vertical sounding lines, which were incapable (except in time-consuming local surveys) of resolving anything smaller than the major features of the ocean floor such as the midocean ridges, and, second, the results of deep sea photography, each frame showing a postage stamp region of not more than 100 sq m. Other techniques (particularly seismic reflection and refraction and the use of the total field magnetometer) were at the time revealing the major features of the earth's crust, but these were not directly related to phenomena on the seabed. The intention with the new instrument was to be able to survey a feature the size of a seamount in considerable detail in a time measured in hours rather than weeks. The existence of seamounts as deep sea features was well known by this time and some had been surveyed in some detail with vertical soundings.

A design study was carried out in 1964 – 5, the project received support in 1966, and after some setbacks, the prototype GLORIA demonstrated the viability of deep ocean sidescan sonar as a technique in 1969 (Rusby 1970). The acronym GLORIA was coined at this time to give the project a recognizable and unique identity. It stands for *Geological LOng Range Inclined ASDIC*, the latter itself being the earlier British acronym for sonar. The acronym is open to the criticism of being somewhat contrived, but it has served its purpose and weathered the passage of time.

In view of the very small proportion of the seabed that lies at depths between 200 and 2,000 meters, and the geometrical requirement of sidescan to maintain range-to-altitude ratios of 5:1 or more, it was decided to step at once to full ocean depths with a range of at least 20 km. It should be remembered that at this time digital recording was not yet a realistic option, so the aim was to maximize the proportion of the swath in which slant-range distortion was not significant. Even today when slant-range correction, at least to first order, is trivial, the range-to-depth ratio needs to be kept at a reasonable level because it is the variations of

backscattering at low grazing angles that are most revealing, and it also maximizes the swath width.

The main achievements of the GLORIA Mk I system were to establish the validity of full ocean depth sidescan sonar as a research technique and to enable the essential parameters of a successful robust system to be identified and quantified. It also turned in a string of useful scientific results in geology and geophysics. However, like many prototypes, it suffered from a number of crippling drawbacks. In particular the towed vehicle was unnecessarily bulky, having to accommodate an array with too large a vertical aperture. The resultant limitations included principally being tied irreversibly to one ship, requiring a large seagoing team in support, requiring good shelter with water at least 300 m deep for deployment and recovery, and being limited to temperate latitudes by the need to use swimmers for launch and recovery.

A new Mark II design was undertaken in 1974 – 5 specifically to address these shortcomings and to add the enhanced capability of having two arrays to survey both sides of the ship's track at the same time. After a design phase the new system went to sea for trials in 1977, and the trials were immediately followed by twelve weeks of operations (Somers et al. 1978). In this new design, because of the advantages (acoustic as well as mechanical) of having an array with a smaller height, and consequently a much broader vertical beam pattern, it was possible to fit port and starboard arrays back-to-back in a vehicle with a diameter of around 0.7 m. The vehicle is neutrally buoyant and is towed from the nose by a heavy cable, giving a simple robust rig. The Mk II GLORIA system has operated from the Arctic Ocean at 79° N through the Tropics to the Antarctic at 60° S and in sea states up to seven or even eight. It can be launched and operational in little over twenty minutes and can be recovered in a full gale in about thirty minutes. The USGS GLORIA III system has also acquitted itself well in areas as widely distributed as Alaska, Hawaii, and both coasts of the continental United States.

The USGS has had a long association with the GLORIA system, starting with a cruise on the eastern seaboard in 1979. The next cruise was in the Gulf of Mexico in January/ February 1982, the results of which were incorporated into the USGS *Atlas of the Exclusive Economic Zone*. This was followed in 1984 by the survey of the western conterminous states EEZ from California to Washington. Further work was carried out the following year in the Gulf of Mexico and Puerto Rico while the USGS acquired their own GLORIA Mk III system. It should be pointed out that the term "Mk III" was used purely to distinguish the USGS system from its IOSDL precursor; the technical performances were closely matched. Operating this in cooperation with IOSDL, the remainder of the U.S. EEZ was surveyed between 1986 and 1991.

Since 1977 the vehicle and launcher have remained almost unchanged, but the laboratory equipment and vehicle instrumentation have been radically altered in a continuing series of upgrades, mainly to the sensor suite and to the signal processing. Image processing has also been subject to major developments, but mainly ashore, and largely by the USGS team at Flagstaff, Arizona (Chavez 1986).

The acoustics of GLORIA

The major strength of sidescan sonar lies in the immediate intuitive acceptance of the images by geologists, and even laymen, who may have only the haziest grasp of ocean acoustics. It is, on the other hand, undeniable that acoustic and technical factors do impinge on the performance of GLORIA (and other sonars), and it is advantageous for the user to have an appreciation of them. This basic knowledge is important to appreciate some of the artifacts.

Sound waves in a fluid are compressional, and the particle motion is aligned with the direction of propagation. The periodic exchange of energy that is characteristic of all wave motions, in an acoustic wave takes place between stored energy in the regions of compression or expansion and kinetic energy in the regions of maximum particle velocity, so that density and compressibility are the physical properties of the medium that govern the speed of sound. The speed of sound in water does not vary with its frequency or pitch, but in the sea it does vary significantly from place to place and to a lesser extent from time to time. The ocean is not by any means ideal as a medium for the transmission of acoustic energy, but it is so much less favorable for the propagation of any sort of electromagnetic radiation (light, radio, radar, etc.) that with all its limitations acoustic energy is the only practical means of probing the ocean. The variations in sound velocity from place to place and from time to time mean that sound energy does not travel in straight lines, nor does it always take the same path from one set point to another. Another feature of sonar is the increase of attenuation, the steady attrition of energy from the sound beam, with frequency, which forces the designer of a sonar towards the use of low frequencies. This course carries penalties arising, in the first place, from the need to deploy large arrays to maintain a high geometric resolution, and in the second, from the fact that acoustic energy penetrates the seabed to increasing depths as the frequency is reduced. In addition there are noise sources in the sea competing for attention in the receiver, and there is the scattering of sound by the surface and midwater objects such as schools of fish.

In the deep ocean (>2 km deep) the water column can on sidescan sonar scales of 15 – 25 km be taken as horizontally stratified with the velocity of sound being purely a function of depth. The principal exception to this generalization is the frequent presence of an internal wave field on the thermocline about 100 m below the surface. The stratification is dominated by temperature and varies as a consequence mainly with latitude. Pressure and salinity also play

a role, though that of the latter is a relatively minor one in the open ocean. Figure 1–3 shows the velocity structure in the open ocean typical of midlatitudes and the paths traveled by rays of sound launched from near the surface at a succession of takeoff angles. The main features are a sharp fall in temperature in the first few hundred meters, lowering the velocity, followed below about 1.2 km by nearly isothermal water in which the effect of pressure is to cause a steady increase in sound velocity. Toward the polar regions the thermocline tapers off and eventually disappears altogether or, more accurately, outcrops at the surface.

Some very complex numerical models of sound propagation in the sea have been published, but for sidescan sonar classical ray theory will suffice, supplemented by extensions that allow the intensity distribution to be computed. The two conditions for the validity of ray theory are always met by a sidescan configuration, namely (1) the water column is many acoustic wavelengths deep and (2) the velocity does not change measurably over distances of a wavelength. Figure 1–3 shows that there are definite limits to the horizontal range of a mapping sonar in the deep ocean. Snell's law, which is the governing law of ray propagation, means that sound rays are refracted toward regions of lower velocity. As a result all rays in deep water will eventually become horizontal and then be refracted upwards toward the surface. This effect means that beyond a certain range no rays will reach the seabed, which therefore lies in an acoustic shadow, the so-called bottom shadow zone. The range at which this occurs is called the half convergence zone from the fact that if allowed to continue, the rays will eventually converge back to roughly a common point at the surface, hence the convergence zone. The other limiting effect of importance is the initial downward refraction of the sound rays in the upper region of falling velocity. Even though they will eventually level out and indeed turn back toward the surface if the water is deep enough, in many places it is so shallow that all rays launched from the surface hit the seabed before they can level out. This initial downward refraction gives rise to a surface shadow zone, and in shallow water the seabed beyond quite modest ranges lies in it. The effect is to place a severe limit on the attainable sonar range in temperate and warm waters unless the water depth exceeds about 4.5 km. Other features of the velocity structure have only minor effects on propagation.

The first requirement of the sonar receiver is to detect the backscattered echo energy above the competing noise mechanisms. The tool used to calculate the excess of target energy over noise is the sonar equation, which takes account of all the mechanisms that affect the outcome, such as transmitted power level, attenuation, spreading, and so forth. It is a well-established equation and widely used, mainly in the logarithmic form employed here. The reader interested in its derivation should refer to Urick (1975). Suffice it to say here that it represents the physics of transmission, propagation, reflection, and detection. The reason for using the

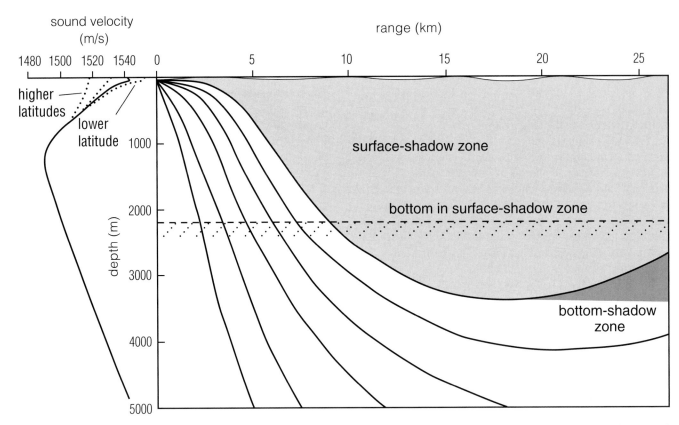

Figure 1–3. Typical midlatitude sound propagation in the ocean with the source near the surface. Note the dominant influence of the surface-shadow zone in shallow and intermediate depths. Because of the combination of downward refraction in the first 1,000 m or so under the influence of the negative velocity gradient and reflection at the surface, there are no direct ray paths from the array into the shadow zone. A minor feature of this profile is the shallow surface channel above the thermocline.

logarithmic form is the usual one of substituting sums for products, and the sonar engineer finds it convenient to convert to decibels by multiplying the logs by ten.

In decibel notation the echo level (dB re 1 μPa) for a transmitted power, P is as follows:

$$EL = 171.61 + \log_{10}P + (DI)_T \\ - 40 \log_{10} R - 2\alpha R + 10 \log_{10} I_R \quad (1)$$

Where I_R is the backscattered intensity referred to a nominal 1 m from the scattering point, and R is the range in meters. The 171.6 is a numerical factor to harmonize the units, and $(DI)_T$ is a term known as the directivity index representing the degree to which the array concentrates the radiated power into preferred directions. It will be apparent that if a sound beam carries a given total power, then the narrower the angular spread, the higher will be the intensity along the axis. Also the directivity index will have contributions from the horizontal and vertical axes.

Let ϕ be the grazing angle, and let I_0 represent the first five terms of Eq. (1); that is, the incident intensity referred to the same nominal distance of 1 meter from the scattering point, actually regarded as a plane wave but corrected for spreading and attenuation. Let dA be an element of area of the (flat) seabed. The total power incident on it is $I_0dA \sin(\phi)$. Suppose that in place of dA we have a small perfectly re-

flecting sphere at the scattering point such that it reflected in the backscattering direction exactly the same intensity as the actual element of seafloor; then the total power scattered by the sphere would be $4\pi I_R$. The backscattering coefficient σ_s is defined as the ratio of these two quantities so that

$$I_R = \frac{I_0 dA \, sin(\phi)}{4\pi} \quad (2)$$

The backscattering strength is conventionally defined as

$$BSS = 10 \log_{10}\left(\frac{\sigma s}{4\pi}\right) \quad (3)$$

Thus we have

$$EL = I_0 + 10 \log_{10}(dA \sin \phi) + BSS \quad (4)$$

And from the geometry

$$dA = \frac{R\delta\theta \, sec(\phi)c\tau}{2} \quad (5)$$

where c is the velocity of sound and τ is the width in seconds of the autocorrelation function of the acoustic pulse. Also $\tau \cong 1/B$ seconds with B being the system bandwidth in Hz. Note that Eq. (5) breaks down at near vertical incidence. So we get

$$EL = I_0 + 10 \log_{10}R + 10 \log_{10}\delta\theta + K(\phi) - 10 \log_{10} B \quad (6)$$

Where $K(\phi)$ is the grazing angle – dependent spreading velocity, which converts inverse bandwidth to length on the ground. Note that the horizontal directivity term (as defined with reference to Eq. [1]) falls out also because of the $\delta\theta$ in the equation for dA. Because the higher power density arising from a narrower beam is compensated by a smaller scattering area; this would not be the case for a target too small to fill the beam. From the echo level we subtract the received noise power (in dB) to get the threshold for the lowest backscattering strength that will allow meaningful numerical estimates to be made, defined arbitrarily as equal backscatter to noise power ratio. For the noise we have a noise spectral density N in Pa^2 per Hz, leading to a noise level of

$$NL = 10\log_{10}N + 10\log_{10}B - (DI)_R \qquad (7)$$

The received directivity can be applied whether the principal source of noise is isotropic or from a fixed point such as the survey ship, provided in the latter case it is well off the acoustic axis. Finally we get the full sonar equation for bottom reverberation at moderate to low grazing angles applicable to sidescan sonar, as follows:

$$EL - NL = 171.6 + 10\log_{10}P + (DI)_T + 10\log_{10}(\delta\theta) - 30\log_{10}R$$
$$- 2\alpha R + BSS - 10\log_{10}N + (DI)_R - 20\log_{10}B + K(\phi) \quad (8)$$

This equation shows two points:
1 The echo level for a given backscattering strength varies with range and thus time, and time-varied gain (TVG) can be applied to compensate, and
2 The excess of echo depends on $-20\log_{10}B$

The second point is very important and not widely appreciated; choice of too high a resolution incurs a surprisingly heavy penalty in terms of S/N ratio. Also the TVG law should reflect any known variation of the vertical beam pattern, though the adjustment would have to assume a flat seabed at a known depth.

To assess the impact of the range-dependent terms of the sonar equation, Eq. (8), note that at close ranges the spreading term ($30\log_{10}R$) dominates the equation but the linearly increasing attenuation, term ($2\alpha R$) overhauls spreading and eventually overwhelms it. It is possible to quantify this effect quite simply by differentiating the range terms, which shows exactly how fast each term is changing, the total rate of increase being the sum of the two. One would expect the maximum range to occur somewhere near the crossover point or where spreading loss and attenuation loss are accruing at equal rates. Remarkably, all published descriptions of sonar imaging systems, regardless of frequency, have been designed close to the point

$$\alpha R_{MAX} = 13 \qquad (9)$$

which is where the attenuation losses are rising at twice the rate of the spreading loss, as described by Somers and Stubbs

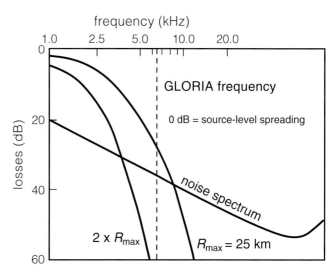

Figure 1–4. Frequency selection curve for GLORIA, targeted at a maximum range of 25 km. This illustrates forcefully the pressures to design at the lowest possible frequency, or from an alternative point of view, the impossibility of designing for 25 km at anything but a low frequency.

(1984). The spreading losses at the full range amount to 130 dB and the attenuation loss is 18 dB, but note that over the outermost 1 km of the swath the spreading losses have increased by only 0.5 dB while the attenuation has done so by very nearly 1 dB at the GLORIA frequency of 6.5 kHz. The transmitted acoustic power is 6 Kw per side for two seconds. With the bandwidth, B, at 100 Hz the processing gain, PG, is 23 dB. The noise power is normally dominated by the ship and so varies from time to time and with the type of propulsion and speed. A fixed pitch propeller is usually to be preferred, but as the EEZ survey showed, the R/V *Farnella* with a CP (controllable pitch) propeller could be made quiet enough by using a high percentage pitch setting with low engine speed. The problem with CP ships is that sonar surveys are usually carried out at less than the optimum speed for the propellor, and the load is shed not by reduction of shaft speed but by flattening the pitch of the propellor, which causes the very noisy phenomenon of cavitation. Occasionally natural sources will dominate, but not for long; for example, heavy precipitation has a lot of energy near the GLORIA band. Figure 1–4 shows the frequency selection curve for GLORIA targeted at R_{MAX} equal to 25 km.

The final term in the sonar equation is the processing gain, PG. So far discussion has concentrated on the idea of a short pulse of a single frequency, though of course such a pulse actually occupies a band of frequencies. However, radar and sonar theory indicate that at the output of an ideal receiver the S/N ratio is proportional to the signal *energy* (Stewart and Westerfield 1959). Thus

$$\left(\frac{S}{N}\right)_{\mathrm{O}} = \frac{E}{N} \qquad (10)$$

Pulse energy is the product of power and time, but peak acoustic power in any transducer is limited at best by cavitation in the water and quite possibly by other engineering limits, so the only recourse is to extend the pulse duration. The apparent loss of resolution is avoided by coding the pulse into a wide band, B, when the receiver can be designed to restore the resolution to approximately $1/B$ seconds by matching the pulse. To see empirically how this works, consider Figure 1–5 where the transmitted pulse is a so-called chirp in which the instantaneous frequency is swept linearly from f_1 to f_2 – that is, $f_2 - f_1 = B$ – in a time T. A point target returns an echo that is a reduced and delayed replica of the outgoing pulse. The echoes are amplified, filtered, and fed into one end of the delay line, DL, which has the property that signals of frequency f_2 travel through it more rapidly than signals of frequency f_1 and is so proportioned that the difference just equals the pulse duration, T. Thus all the pulse energy emerges at about the same time (crowded into a short interval of about $1/B$ seconds, not directly related to the original pulse length, T). The output power is correspondingly increased and because the noise is not given the same privilege, the output S/N ratio is improved (by $10 \log_{10} BT$) whereas range resolution is retained. The GLORIA system employs pulse lengths of 4, 2, or 1 second with a bandwidth, B, of 100 Hz, giving a resolution of 10 mS, or about 10 m. In fact, the whole of the U.S. EEZ survey was carried out with the two-second pulse. To illustrate the power of the long pulse, to achieve the same S/N ratio with a 10 mS pulse would have required a pulse power some 200 times greater, or 1.2 megawatts! The special delay line can be shown to be but one member of a general class of pulse processors with the general title of matched filter, and the GLORIA system uses a different technique. It uses a digital correlator to compare a two-second long section of received signal with a replica of the transmitted pulse. The computation has to be carried out more than $2B$ times per second, and for a point target the match is close only over a short time interval of $1/B$ seconds. More details are given below in the description of the GLORIA instrumentation.

To summarize, the design of a mapping sonar is a multifaceted trade-off between a range of system parameters. The fundamental trade-off is between the acoustic advantages of using the lowest possible frequency and the engineering costs of building and handling the structure to carry the size of transducer array required, with the attendant difficulties of transport, deployment, ship fitting, and so forth. Bandwidth, resolution, power level, processing gain, and ambient noise all come into the equation, but the dominant trade-off is between angular resolution and physical size. A narrow beam pattern in the fore and aft direction requires a large number of wavelengths across the aperture, a good approximation being

$$\text{angle (in degrees)} = \frac{60}{N} \qquad (11)$$

where N is the number of wavelengths across the horizontal aperture. N usually lies in the range 25–50 for a sidescan array.

Technical description of the USGS GLORIA III

Overview

The main components of the system are the towed vehicle with its towcable and launcher; the shipboard equipment, divided roughly into the power amplifiers and associated systems fitted into an ISO container; and the various electronic racks fitted into the ship's laboratory spaces for the signal systems and recording.

The towed vehicle, cable, and launcher

The function of the towed vehicle is to deploy the sonar arrays in a suitable attitude and position to transmit and receive the acoustic energy. The principal reasons for using a towed vehicle rather than hull-mounting the arrays are as follows:

1   It places the sensitive receivers as far as possible from the major source of noise, the survey ship.
2   It offers a more stable operating platform than a direct mounting to the ship.
3   It offers flexibility in the choice of survey ship.
4   The ship-fitting is faster and much cheaper.
5   It is relatively simple to gain access to the sonar arrays at sea for repair.

It would also be an advantage to be able to deploy the arrays deep enough to be below the break of the main thermocline, but this is not possible in tropical waters. Finally, the transducers actually handle quite a high power density, and a depth of a few tens of meters affords protection against cavitation.

A two-part dogleg tow configuration was considered for GLORIA II, consisting of a near vertical portion of towcable leading down from the ship to a depressor weight, followed by a neutrally buoyant umbilical leading to a neutrally buoyant vehicle. The operational disadvantages and the need to fair the near vertical part of the tow to avoid an excess of cable strumming led to this configuration being rejected in favor of a more simple one consisting of only a neutrally buoyant vehicle and a strong heavy towcable, as shown in Figure 1–6, which model trials had indicated would suffice. It gives good isolation of the important vehicle heading against ship yaw with periods shorter than about 120 seconds. Of course it gives no isolation at all against components of motion along the cable, including the fore and aft component of the orbital motion at the stern resulting from pitching of the ship. In practice the towing conditions most likely to produce noticeable vehicle instabilities and

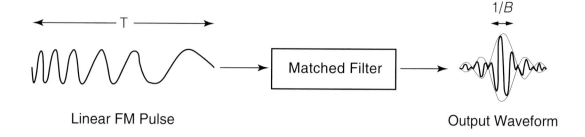

Linear FM Pulse   Output Waveform

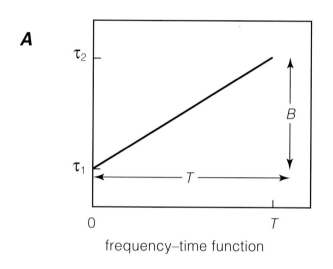

A

frequency–time function

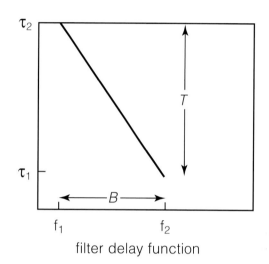

filter delay function

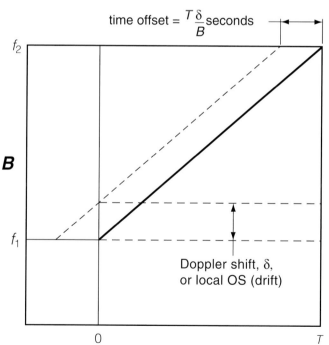

B

Figure 1–5. A, The basic idea of processing gain by pulse compression, using a linear FM (chirp) pulse. The output S/N ratio of a receiver matched to the pulse is proportional to the energy in the pulse. A simple way to visualize the action of a matched filter is to consider the transit time of the signal through the filter. The lower frequencies transit more slowly than the later arriving high frequencies. The filter is so proportioned that with the matched frequency function (chirp) arriving at the input, all the frequency components emerge together. It is, in fact, a standard procedure to design matched filters in this way (see O'Meara 1962), and for *BT* products less than 30 or so it is very effective. The *BT* product is the key parameter for coded pulse systems; it is the product of the duration and bandwidth, as the term implies. The sampling theorem indicates that at least *2BT* samples are needed to specify the signal completely. B, The equivalence of time delay and frequency offset in their effects on the linear FM signal. A Doppler shift caused by relative movement of the array and target along the line joining them means that the match between reference and echo occurs somewhat earlier or later, depending on the direction of movement. The match also occurs over a shorter time interval, resulting in a reduction of coherent energy in the output. The time offset is related only to the slope of the frequency-time curve, and in general the effects are small for sidescan sonar.

indeed to damage the cable are to be found when punching into a head sea. However, this simplest of cable configurations has proved to be practical and robust and to have served the system well.

The sonar arrays take up the main part of the vehicle ex-

tending 5.3 meters between the two main bulkheads. Fore and aft of these bulkheads are the nose and tail sections, providing both hydrodynamic fairing and space to mount the junction box and instrumentation. The transducer elements that make up the array (see Figure 1–7) are a close adapta-

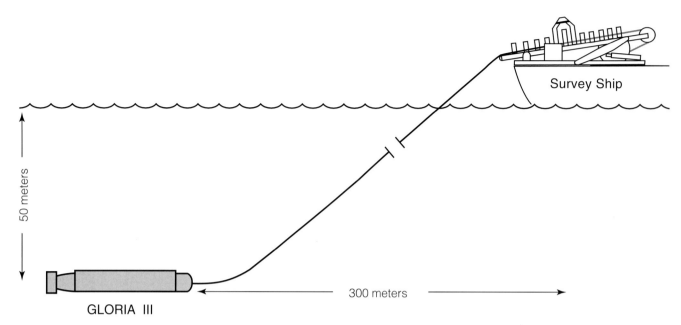

50 meters

GLORIA III

300 meters

Survey Ship

Figure 1–6. Diagram of the GLORIA III towing arrangement. The neutrally buoyant vehicle is towed by a single heavy armoured cable dominating the towing catenary. The most frequently used towing speed, weather permitting, is 8 kn and at this speed the cable angle is only a few degrees below the horizontal. The cable catenary levels out for the last 50–100 m. The depth of tow is not actively controlled and depends on the cable scope and speed.

tion of those used in GLORIA I. The original aluminum radiating head has been replaced by titanium with major gains in freedom from corrosion and its attendant problems, and the port and starboard elements are now tuned more closely to their proper frequency.

Inspection of the untuned admittance diagram (see Figure 1–8) will reveal that the original array cannot present an ideal nonreactive load to the driver amplifiers. The admittance diagram shows the electrical load that an array section presents at its terminals. At the lower (starboard) frequency it presents too high a reactance with a poor power factor, whereas at the higher frequency (port) the admittance, although it has a reasonable power factor, needs to have a larger real part. The desired effects can be achieved by rotating the admittance in the proper sense, and the modification to produce this effect is to alter the tail tuning mass. The desired effect is achieved by using different materials for the nut that applies the prestress to the stack as between the port and starboard transducers. Each element can handle 600 watts of acoustic power on a 20 percent duty cycle at an efficiency of 96 percent but at the usual operating depth of 50 m, the power rating is reduced to approximately 400 watts by the onset of cavitation, which leaves a 100 percent margin over the usual output power in operation. As a matter of interest, at peak power the radiating face has a peak acceleration of 1,000 g but with a peak-to-peak excursion of only 0.01 mm.

An array is made up of two rows, each of thirty elements. The backbone of the vehicle consists of thirty transducer housings bolted together. Figure 1–9 shows a cross section of the vehicle through the arrays. The housings are

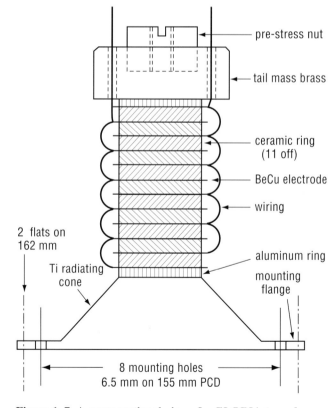

pre-stress nut

tail mass brass

ceramic ring (11 off)

BeCu electrode

wiring

2 flats on 162 mm

Ti radiating cone

aluminum ring

mounting flange

8 mounting holes 6.5 mm on 155 mm PCD

Figure 1–7. A cross-sectional view of a GLORIA transducer element. The watertight seal is placed on the flange away from the acoustic movement, contributing materially to the very high efficiency. The vehicle contains 120 such elements in two arrays of 60, port and starboard. Each array is a bank of two rows of thirty elements each. Figure 9 shows how they are mounted.

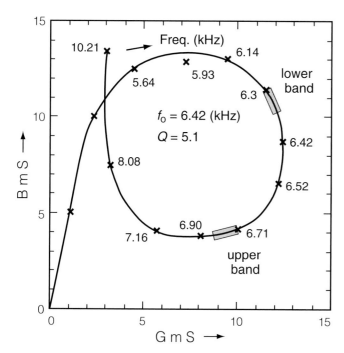

Figure 1–8. The electrical admittance diagram of a section of the GLORIA array. The ordinate (susceptance, in millisiemens) is the imaginary (reactive) part and the abscissa (conductance, in millisiemens) is the real (dissipative) part. Frequency increases clockwise along the curve. The circle is typical of the admittance of a series tuned electrical circuit. In fact, the mass of the cone and tail resonate with the stiffness of the stack of ceramic rings, and the overall analogy is of a spring-mass oscillator driven by an electrically activated spring. The electromechanical coupling makes this arrangement behave like a tuned electrical circuit.

approximately inverted triangular slabs made of a proprietary engineering nylon (Nylatron GSM), and each holds four transducer elements, a pair each of port and starboard back-to-back. Nylatron GSM has low weight, high specific strength, low water absorption, and complete freedom from corrosion or any galvanic reaction with metal. The blocks are stacked and bolted together at the corners and with a central prestressing bolt loaded to five metric tons. The upper and lower rows of elements are mounted at the minimum vertical spacing allowed by the element dimensions, which is 165 mm between centers.

The overall beam pattern in the vertical plane of this arrangement, allowing for the self-directivity of the elements, is 40° between half-power points. Some care was needed in deciding how much to cant the face of the array downward for the best distribution of sound energy in the quadrant within the minimum diameter envelope for the vehicle as a whole. The choice (20°) was complicated by the need to reduce the sound level in the vertical direction because in GLORIA I the images were frequently contaminated by second, third, and even fourth bottom echoes, which occurred whenever the seabed was flat enough to return a coherent echo, particularly if the sea state was mod-

erate. This effect is still present in GLORIA III but is less frequent and the higher multiples are rarer. The transducer faces are protected by bolt-on fiberglass covers shaped to conform to a cylinder 66 cm in diameter. The weight budget showed that there was not sufficient space within this diameter to provide all the buoyancy needed for neutrality, so the buoyancy blocks were extended upward. Note that the buoyancy, syntactic foam, has to be able to withstand the pressure at the full scope of the cable, nominally 400 m. One effect of this extended buoyancy is that the vehicle has great stiffness in roll, the roll period being only 1.85 seconds, and roll values seldom exceed 0.5° peak. The fiberglass transducer covers serve the additional function of separating the turbulent flow past the vehicle as far as practically possible from the sensitive transducer faces.

Electrically, the wiring harness divides the arrays horizontally into six sections per side, mainly to allow the acoustic beam to be stabilized against yaw (see next section) during the reception interval. It also allows some redundancy and flexibility in powering the arrays, and it allowed flexibility in laying out the cores in the main cable.

One of the main design objectives was, in the interests of reliability, to minimize the amount of complex instrumentation in the vehicle. Thus there are no transmit/receiver (T/R) switches or preamplifiers in the vehicle and a minimum set of instrumentation is provided – depth, water temperature, pitch, roll, and a compass. For its initial trials, the Mk II had a vertical reference gyro fitted, but it was never the intention to maintain such a power-hungry, short-lived, expensive, and noisy instrument in permanent operation once the adequacy of the configuration had been demonstrated. The minimum complexity decision carries the penalty of having to route the low-level sonar signals directly up the towing cable and through the ship to the receivers, which requires careful attention to layout, routing, and grounding. The vehicle tows a length of 50 m of thick braidline to afford additional short-period stability against yaw.

Vital to the success of the system was a reliable, robust, and simple method of deploying and recovering the vehicle in all but the most extreme weather conditions. The hydraulically operated launcher was specially designed for the system to achieve this. It consists of a bedframe with a moving carriage holding both the vehicle and (on its underside) the winch. The carriage tilts and slides under the action of a pair of hydraulically driven link arms. Starting at a tilt of about 12° it moves outboard and downward at the same time until it is 60° below the horizontal with the tail of the vehicle on or about the waterline. In this position, with the ship making about four to five knots through the water, launching is a simple matter of paying out the cable while the vehicle takes to the water. The cable is paid out to the desired scope and secured with the carriage retracted; then the connectors are plugged in and the system started. The whole process can take as little as twenty minutes before the first transmission. The only time of any danger to the vehicle is

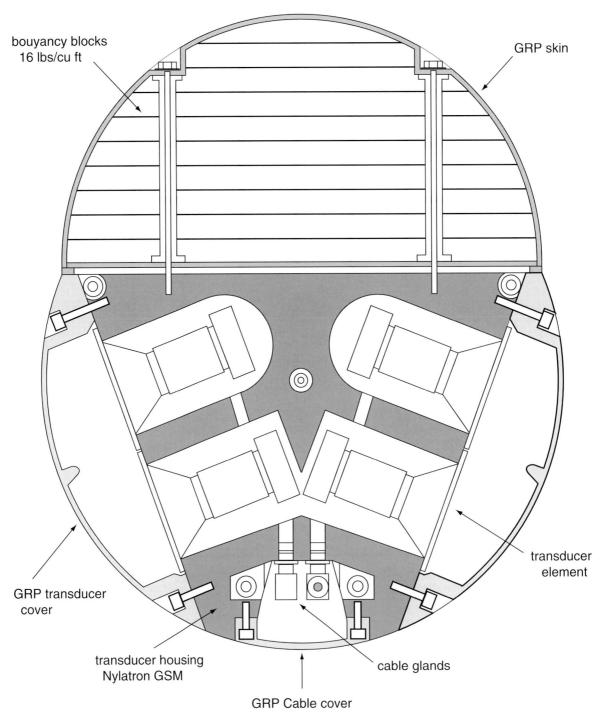

buoyancy blocks
16 lbs/cu ft

GRP skin

transducer
element

GRP transducer
cover

transducer housing
Nylatron GSM

cable glands

GRP Cable cover

Figure 1–9. Cross section of the vehicle through a transducer housing block, showing the construction of the main section of the vehicle. The transducer elements are mounted by their flanges with a stationary O-ring seal under the flange (not shown here). Nylatron GSM is a high strength engineering nylon with strength-to-weight ratio superior to mild steel and with very low water absorption. There are thirty such blocks in the vehicle. The buoyancy is closed-cell rigid polyester foam specified to withstand immersion to the full cable scope of 400 m.

while it is still on a very short scope and reluctant to drop back through the propellor wash. Recovery is much the same in reverse, except that with the winch having to pull the cable in, the danger of damage in the crucial region close to the ship during the last minute or so is potentially much

greater. To avoid this hazard, when the vehicle gets within about 50 m of the ship a drogue, or sea anchor, is added to the vehicle by looping the front end of its bridle around the cable and letting it stream aft. The drogue is a truncated cone of canvas on a 1-m diameter hard ring, dimensioned according to the *Admiralty Manual of Seamanship* (1967) recommendations for sea anchors. At 4.5 knots it provides an extra 250 kg of drag, which is sufficient to absorb the tendency of the vehicle to overrun the ship. The use of the drogue has made recovery a simple matter in all but a full-blown gale, and the launch is a more hazardous operation than the recovery. On the other hand, the launch can always be postponed until more favorable conditions can be found, whereas there is often no discretion over the timing of a recovery.

The winch is fitted with a watertight junction box for the cable connectors from the ship wiring. Connection and disconnection are both a matter of no more than two minutes' work, which is a small price to pay for the security and reliability of a fully mated set of connections, compared with slip rings, particularly for the very low level raw sonar returns.

The power amplifier system

Each section of the array is supplied by its own power amplifier (twelve in all) and has its own T/R switch, power factor correction, and time varied gain (TVG) amplifier. It will be noted from the admittance diagram of an array section (Figure 1–10) that the load has a significant capacitive component, for which the pulse power amplifier (PPA) has to provide reactive current. The tuning inductor provides correction for this. It also produces a slightly larger overall Q-factor, or narrower frequency band, but the reduction is insignificant in the context of GLORIA. As Figure 1–10 shows, the tuning inductor also forms one arm of a π-matching section on reception, to steer the maximum amount of received power into the receiver circuits. Note that the diodes, D1 and D2, act as an effective short circuit during transmission, but are open for reception, which protects the receiver circuit from the full transmitted power. The relay is provided to isolate the input of the TVG amplifier from any noise on the output of the PPA and does not play an essential part in the T/R action. This arrangement provides impedance protection for the TVG amplifier without the major power losses of using resistive protection, in which up to 10 percent of the transmitted power and 90 percent of the received power is lost.

The π-matching pad also serves as sea noise filter, retaining only a fairly broad band of signal and noise around the signal frequency, the effect of which is to enhance slightly the available dynamic range. The TVG function is achieved by a digitally controlled attenuator, which provides an 80 dB range in 1.5-dB steps, according to the following law:

$$30 \log_{10}R + 2\alpha R$$

with R being calculated from time at a velocity of 1,500 m/sec. The use of 1.5-dB steps instead of a smooth curve with time would at first sight seem to be open to the criticism that it will give rise to intensity banding on the image.

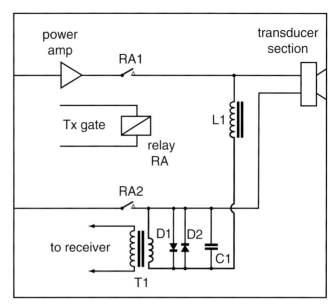

Figure 1–10. Schematic of the T/R switching arrangement. The tuning choke L1 balances the excess capacitance of the transducers and cable during transmission, when relay contacts RA1 and RA2 are closed. On reception the power amplifiers are isolated by the relay, and echo signals picked up by the transducers are converted into electrical form and fed via L1, C1, and the transformer T1 to the receiver. The diodes D1 and D2 protect the receiver during transmission by acting like a short circuit and placing L1 across the array section. As open circuits they play no role during reception.

This, however, is not the case because the pulse is still in its stretched FM form at the point where TVG is applied, so the gain steps appear as amplitude modulation on the FM pulse envelope at the input to the matched filter (correlator). The effect is that the gain variations are smoothed out over the two-second pulse length and are invisible in the correlator output. The filters, on the other hand, had to be designed with a little care, and it was prudent not to use the fastest cut-off function available, for example, a Butterworth response rather than elliptical. The reason is that there is a fundamental trade-off in filter design between the rate of amplitude cut-off and the phase distortion in the filter output; phase ripples in the filter response give rise to distortions in the correlator output that can blur the correlation peak, and increasing the filter complexity is expensive and only partially effective as a cure. The effect is not serious for the GLORIA parameters, and the impact, mainly a slightly wider noise band, is not noticeable. The effects of distortions on the correlator operation can be studied by the method of paired echoes (Anonymous 1981).

The TVG cards also incorporate the first stage of frequency shifting in the receiver chain, translating both port and starboard frequencies into a single intermediate frequency band (1887.5 – 1987.5 Hz). These frequencies and their place in the chain of signal frequencies are explained in the next section.

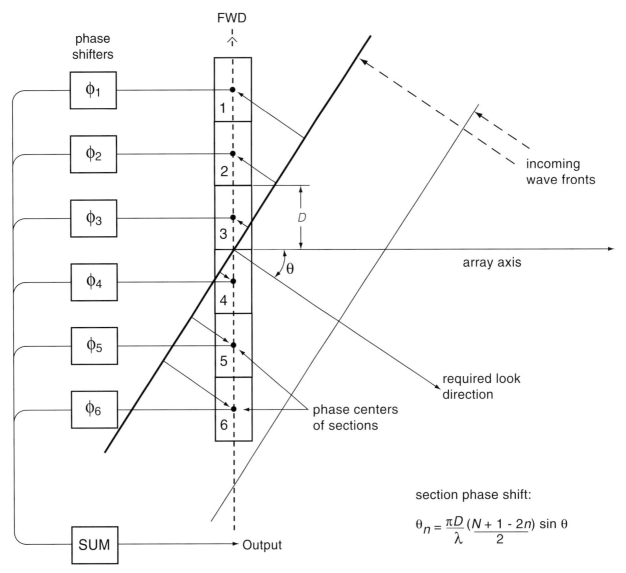

Figure 1–11. Illustration of the application of phase-shift beam steering to stabilize the received beam to the direction of transmission. The idea is to persuade the receiver that the array is looking not along its physical axis but in the required look direction. The angle θ is measured by a heading sensor. The phase shifts act to advance or retard a signal according to sign, and the phases are generated according to the formula shown to compensate for θ.

So far the signals from the individual sections of array have been kept separate. The next step combines them vectorially in the beam steering unit, whose job it is to maintain the sonar beam stabilized in the direction of transmission. With a horizontal beamwidth of 2.7° it is clear that any change over twenty-eight seconds of this magnitude in the heading of the vehicle will result in a loss of signal. The beam steering is to compensate for these losses of signal, or dropouts, rather than to maintain the sonar scan lines rigidly perpendicular to the ship's track (which would be a very demanding task).

A simple illustration of the way in which beam steering works is given in Figure 1–11. The sound arrives back perpendicular to the nominal direction of the array as shown.

If there is yaw it will arrive sooner at one extremity than it does at the center and later at the other extremity. Seen at its simplest, beam steering consists of inserting delays into the signals from separate sections so as to compensate for this variation in delay along the array. Of course it is impossible to realize a negative delay, so all sections are given an offsetting delay to keep the smallest positive, and the value of this offset clearly depends on the largest value of yaw to be compensated. While this scheme is conceptually simple, it is difficult actually to produce continuously variable delays for analog signals. Fortunately, in the case of GLORIA, there is no need to, because the bandwidth of the sonar signal at 100 Hz is only about 1.5% of the center frequency. From the engineering point of view this is a narrowband signal and has the property that delay and phase shift are interchangeable. The phase of a narrowband signal (ω = constant) is $\phi = \omega t$, so $dt = \delta\phi/\omega$. In this scheme a phase advance is equivalent to a negative delay. This paradox is resolved by consideration of the energy flow that is governed by the group delay, which is always positive, and not the phase delay. GLORIA exploits this property to effect the delay and sum beam steering, but the phase shifts are actually applied to the local carrier signal used in the mixers. The reason is that the local carrier is derived from a digitally divided frequency chain and it is technically a simple matter to shift the phase of such a signal with digital circuits. The mixer structure is of the suppressed carrier, multiplier type, and it is well known that phase is preserved in such a system.

Each array has six sections, so if we regard the phase center of the array as being at the junction of sections 3 and 4, numbering from the front, the phase shifts required are as follows:
for sections 3/4: $+/- \phi$
for sections 2/5: $+/- 3 \times \phi$
for sections 1/6: $+/- 5 \times \phi$
where ϕ is the quantity

$$\frac{\pi d \, \sin(\text{yaw})}{\lambda}$$

The sign in this schedule relates the port section signals to clockwise yaw seen from above, which is regarded as positive. The signs need to be reversed for the starboard sections.

The pulse power amplifiers (PPAs) are straightforward push-pull transformer-coupled amplifiers working in Class B. For the reader unfamiliar with electronic engineering, these terms simply refer to a circuit configuration in which two sets of power devices share the work, spreading the losses and increasing the efficiency. The supply voltage is a fairly high (for bipolar transistors) 125 volts so as to get the required output power without having to deal with very large currents. Even so, each half of the push-pull output stage has ten power transistors in parallel, with a small emitter degeneration resistor to force current sharing. Each transistor carries a peak current of 3.5 amperes and has to withstand a peak voltage of 250 volts. The large pulse of supply current ($>$160 amps) is delivered by a battery of nickel-cadmium cells that is float-charged from the ship's supplies during the interpulse period. This approach avoids heavy transient loads on the usually not very stiff ship's supply. With the possible threat of second breakdown in the output stage transistors and the very large reservoir of power in the battery, it is essential to have a good protection system for the amplifiers. There are circuits to check the cooling of the power stages, but the main line of defense is the short-circuit trip. This trip monitors the PPA input and output, and if, with suitable margins of time, there is an input and the output is too low for the normal

load, a short circuit is declared and a high-speed relay trip removes the drive from the PPA. This event is signaled both on the PPA rack in the container and in the laboratory, which helps to ensure that a breakdown in the cable does not destroy the PPAs. Of course a sensitive trip is more prone to false alarms, and a compromise is reached between margin of safety and false alarm rate. The battery and its charger are components of a commercially available unit sold by a supplier of emergency power systems, for example, for emergency lighting.

The laboratory systems

The laboratory systems include the following subsystems:
1 the signal generation chain including filters, mixers, gating, and timing
2 the sonar receivers, filters, amplifiers, and mixers
3 the signal correlator
4 signal detection, logging, and on-line display
5 onboard replay with basic image processing, hard copy output, and onboard mosaic
There is a close connection between the timing system and the correlator, as will become apparent.

THE SIGNAL GENERATION CHAIN

With a two-sided sonar one of the potential problems is cross talk between the two channels. Cross talk can arise in two ways. First, a sonar array cannot be completely insensitive in all unwanted directions and in particular, some sound escapes from the back of the array so that an extremely strong target to port will register on the starboard image. Because its signature will match the starboard receiver, it has to paint on the image. The only cure is to ensure that the array has enough front-to-back ratio to exceed the dynamic range of the reverberation. With a front-to-back ratio of more than 25 dB, which works in both directions, GLORIA is untroubled by this artifact, except for the occasional target close to the nadir where the front-to-back ratio is very much smaller. The other source of cross talk would occur when, for example, a port target insonified by the port transmission returns echoes that find their way into the starboard receiver, attenuated only by a single front-to-back ratio. This effect is avoided by separating the two operating frequencies far enough to eliminate the unwanted signal in the band definition filters; and this is the first line of defense. In a system such as GLORIA with a pulse of large *BT* product there is a further precaution to ensure that any residual energy does not register as a resolved target. This approach is to ensure that the port receiver is not matched to the starboard signal and vice versa, which is done by making one signal (the starboard) a rising frequency sweep and the other a falling sweep. The misplaced energy will then appear as noise and contribute marginally to the background.

Reference to the transducer admittance diagram will show that the available band of frequencies extends from 6.2 kHz

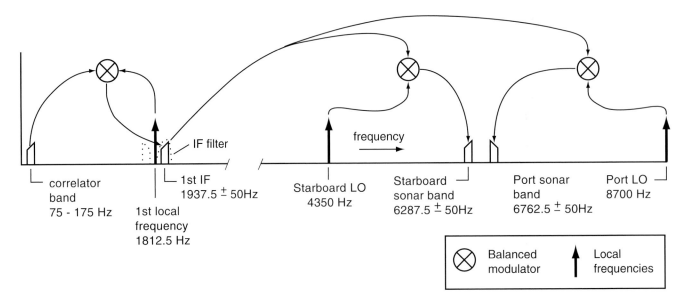

Figure 1–12. The GLORIA frequency generation and shifting scheme. The transmitter circuits work up the chain and the receiver circuits down the chain. Note how the selection of the upper and lower sidebands respectively for starboard and port signals in the second shifting stage generates up- and down-going chirp signals. All signal bands have selection filters as shown for the IF band, but most are omitted for clarity. Local oscillators are identified as LO.

to 6.9 kHz, and the frequency allocations are made to exploit all of this gap. However, the actual band definition filters operate on the signals after the frequency shifting stages, and the filtering at sonar frequency is aimed at removing noise and the more distant image frequencies. The reason is that the difficulty of designing a stable filter gets greater as its relative bandwidth is reduced. The fractional band, of course, increases after shifting to a lower frequency. Two stages of shifting are used to reduce the shift ratio, which would be an unwieldy 60:1 if we shifted directly between the correlator and the sonar bands. Thus the frequency-changing scheme is as shown in Figure 1–12. The correlator reference generator produces a sweep signal between 75 and 175 Hz, rising in frequency. This signal is mixed in the first shifter with a carrier of 1812.5 Hz, and the upper sideband from 1887.5 – 1987.5 Hz is selected by the intermediate-frequency (IF) filter, which is also an upsweep. The IF signal is mixed independently with two carriers at 4350 Hz and 8700 Hz, with the upper sideband being selected from the 4350 Hz mixer for the starboard signal and the lower sideband from the 8700 Hz mixer for the port signal. The starboard signal is an upsweep whereas the port signal is a downsweep.

THE RECEIVER CHAIN

The same local oscillator frequencies are used on reception to mix down to the correlator band. The sonar frequencies are as follows:

Port: 6712.5 – 6812.5 Hz (>)
Starboard: 6287.5 – 6387.5 Hz (<)
a minimum separation of 325 Hz nominally. Note that the arrows indicate the direction of frequency sweep.

The local frequencies share a divider chain from a crystal oscillator, which is predivided to 87 kHz, then divided by 10 and 20 to get the 8700 and 4350 Hz respectively. At the same time the 87 kHz is divided by 48 to get 1812.5 Hz.

THE CORRELATOR AND REFERENCE GENERATION

The wideband long FM pulse gives the GLORIA system some unique features in mapping sonars. Basically it multiplies the effective transmission power by a factor BT, which is 200 in the most common operating condition. The linear FM pulse has its own advantages in that it has a well-behaved correlation output under quite severe distortion conditions and it enjoys great immunity to impulsive noise spikes. It is very tolerant of Doppler shift, but this immunity has a side effect that needs to be watched for. If the ship is making ground across track as a result of a surface current, the returning signals will be Doppler shifted in frequency. This shift is carried as an absolute shift through the frequency changers and ends up as a much larger fractional shift at the correlator input. Clearly, as Figure 5–5B shows, this shift serves to delay or advance the time at which the signal frequency sweep comes into line with the reference. Thus the correlator produces its peak either earlier or later according to the sign of the radial velocity; that is, Doppler shift and time shift are ambiguous in a receiver matched to a linear FM sweep (to a first approximation). Similarly a shift in the frequency of one of the local oscillators is identical in effect to a Doppler shift. The delay is $T(df/B)$, which can be quite appreciable for a modest shift; for example, a 6.5-Hz shift, which is only 0.1% of the sonar frequency, produces a shift of 130 ms.

Mathematically the correlator performs the following action:

$$e(\tau) = \int_0^T s(t)r(t - \tau)dt \qquad (12)$$

in which $s(t)$ is the signal and $r(t - \tau)$ is the reference delayed τ seconds to match the echo return from a range $c\tau/2$. When the signal and the delayed reference match over the pulse length T, the product under the integral sign is always positive and the integral is at a maximum. At other times either side of this coincidence, the product under the integral will fluctuate in sign and the sum over T seconds will fall off in a time comparable with $1/B$ seconds. The correlation over the full T seconds (with T being 1, 2, or 4 seconds) has to be computed once for every sample of the matched filter output, which according to the sampling theorem is at least every $1/2B$ seconds. Also as a result of the sampling theorem, it requires at least $2BT$ samples to represent the signal without aliasing, and this is the number of multiply-accumulate operations needed to compute each correlator output sample. Thus the correlator has to perform $4B^2T$ multiply-accumulate operations per second. In fact, in the GLORIA system it is required to operate somewhat faster for two reasons. First, the matched filter output has to be reconstructed from the stream of output samples by an interpolation filter, and there is a complex relationship between the order of the output interpolation filter, the oversampling ratio, and the level of interpolation error. The oversampling ratio (OSR) is the ratio of actual sampling rate to the Nyquist rate. Similarly at the analog-to-digital converter there is a trade-off between the order of the antialias filter, the OSR, and the alias error. Second, there is a potential problem with the phase of the reference signal compared with the echo signal, which is of course quite random. The signal is shifted to a band B-Hz wide centered on f_0 Hz NB. Because it is a real signal it also occupies the same part of the negative frequency space. The matched filter output can be shown to be approximately

$$e(\tau) = \cos(2\pi f_0 \tau)\frac{\sin(\pi B\tau)}{\pi B\tau} \qquad (13)$$

The approximation is good over the main lobe of the sine-cardinal function for pulses with a BT product over about 10; the reason it is only an approximation is that the integration time is only the period of overlap between the two signals, which falls off linearly with mismatch time. The minimum value of f_0 for a real signal is $B/2$ when the cosine carrier will have only one cycle within the main lobe. Hence if ϕ is about $\pi/2$, the actual correlation peak will not attain its maximum value. To be sure the deficit is not too serious at no more than 3 dB (and no energy is lost), but by making $f_0/B > 1$, it is possible to achieve at least one carrier peak within 0.5 dB of the sine-cardinal maximum. However, if the resulting signal, somewhat above baseband, is viewed as a band-limited signal in its own right it is clear

the sampling frequency has to be higher, at 3 B for $f_0/B = 1$. The GLORIA signals are shifted down to the band 75 – 175 Hz, with $f_0/B = 1.25$, and the sampling rate is 500 Hz, which gives an OSR of at least 1.43 and a mean of 2.

The electronics of the correlator are fairly straightforward and use a special purpose multiplier-accumulator chip to perform the calculations. There are more efficient ways involving the fast Fourier transform but they do not have the conceptual simplicity of the direct method and the chip has some margin of speed in hand.

The reference pulse is stored in an EPROM. For the two-second pulse there are 1024 samples that are read out to the X-input of the correlator every 2 ms by an address counter called the stationary time series (STS) counter. A random access memory, also of 1024 samples for the two-second pulse, is used to store the signal samples. These are read out to the Y-input of the correlator, also in 2 ms, by a second address counter, the moving time series (MTS) counter. Starting with an STS count of zero, the counters run in synchronism until the last count in the STS sequence is reached. During this count the MTS counter receives an extra clock so that during the next cycle of correlation the MTS sequence will have moved by one sample relative to the STS sequence. Also during the last STS count, the signal sample at the MTS address, which is the oldest signal sample, is discarded and a newly converted sample is stored. This is the mechanism (see Figure 1–13) that slides the sonar sequence past the reference one to perform the real-time correlation. As well as storing a new sample during the last STS count, the MTS counter is used to access the STS EPROM. Because of the extra clock inserted into the MTS counter, the reference sample read out will advance by one for each cycle through the STS count; that is, a new reference sample is generated every 2 ms. Thus the reference signal is generated every two seconds (this is the connection between the correlator and the signal generation referred to earlier), and a gating arrangement selects exactly one pulse for transmission every TX cycle, which normally means once each thirty seconds. Other possible combinations of pulse length and repetition period can be chosen, but the USGS EEZ was surveyed almost entirely with the 2 s/30 s combination.

The reference pulse is stored at eight bits of significance and the signal at twelve bits. The relatively more coarse reference storage does not generate any noticeable quantization noise at the correlator output because of the smoothing effected by integration over two seconds. In fact, as few as four bits would suffice, though it is actually more convenient to use the industry standard eight. Each product can extend to twenty bits, which because of the sinewave signals is actually nineteen on average for a full-amplitude signal. For a full-scale signal in exact registration with the reference, the accumulation adds a further ten bits, so the absolute maximum correlator output is a twenty-nine-bit number (in 2 s – complement binary). In practice the reverberation is a noiselike signal and the correlator output al-

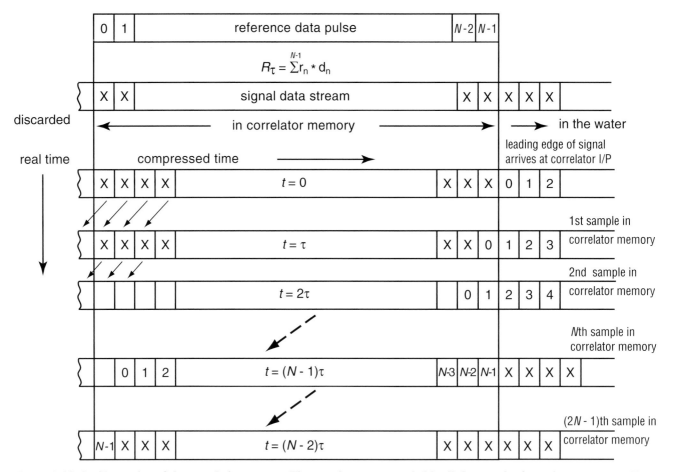

Figure 1–13. An illustration of the correlation process. The correlator memory holds all the samples from the most recent T seconds of seafloor echoes, as the moving time series (MTS). The reference is held in the same number of samples in a permanent memory as the stationary time series (STS). The signals are accessed simultaneously at high speed for presentation to the multiplier chip X- and Y-inputs by two counters, the MTS and the STS counters respectively. All samples are used in the time interval between the arrival of successive samples. At the end of the interval the accumulated product is the correlated (matched filter) output. During the last count of the STS counter (lasting about 2 microsecs), the MTS counter is incremented twice, so that it points at the oldest sample in the MTS memory. This sample is now discarded in favour of the most recent arrival, and the correlation cycle repeats. Because the counters are circular and return to zero when they overflow at a count of N, it can be seen that the STS sequence always starts at count zero whereas the MTS count in general does not, but starts instead at a count one higher than in the previous cycle. In this way the reverberation signals are effectively moved through the memory.

most never exceeds twenty-five bits. The most significant twelve of these are routed to the output, so the signal is at least twelve bits of significance higher than the reference quantization noise. To avoid the dire effects of overflow in the 2 s – complement accumulator on the output, a set of gates detects the approach of saturation and disables the multiplier for the rest of that cycle. This action is exactly the same as clipping an analog signal and the effects are identical. It is only rarely visible in the data when a strong sharply defined target occurs at close range, and it is greatly preferable to the effects of overflow.

THE SIGNAL FREQUENCY CHAIN, FILTERS, MIXERS, AND RELATED ELEMENTS

Earlier, brief reference was made to how the sonar signals are generated, using a two-stage signal mixing process (see

Figure 1–12). This scheme allows a single reference pulse to be stored and used in identical correlators for port and starboard. In addition the IF (intermediate frequency) band is the same for both sides. Sideband inversion takes place in the conversion to and from the sonar band (for *TX* and *RX* respectively).

All mixers are of the double-balanced suppressed carrier type, using a proprietary chip. The action is exactly the same as a four-quadrant multiplier, generating only the sum and difference signal frequencies with up to 60 dB of carrier (and signal) suppression. The unwanted sidebands, shown as dotted in Figure 1–12, are removed by the filters. This figure also indicates how difficult it would be to achieve proper unwanted sideband suppression with only one stage of frequency shift, because the sidebands would be only two to three percent of the sonar band apart.

The sonar band filters are four-pole-pairs narrowband designs modified for finite Q inductors, consisting of four parallel LC resonators top-coupled capacitively. The IF filter uses three stages of a proprietary universal active filter configured as a stagger-tuned triple. And finally the correlator band filter consists of a two-pole Butterworth active RC low-pass filter cascaded with a simple RC high-pass filter set at 50 Hz.

POSTCORRELATION TREATMENT OF THE SIGNALS

The first correlator used in the GLORIA system (in 1971) was an analog device based on magnetic recording principles. As a result the output was also analog and so were the following circuits – detectors, filters, and so forth used to produce the signal actually recorded on magnetic tape and displayed on electrosensitive paper recorders. When the digital correlator was introduced in 1981 it was as a plug-in replacement for the magnetic drum correlator, so the output was reconverted in a digital-to-analog converter (DAC) to analog form. Within a year the analog system of data recording was replaced by digital recording, which took as input the analog output of the postcorrelator detector and filter circuits. It was intended in due course to cut out the two conversion steps and to pass the digital data direct from correlator to digital recorder. However, the hybrid system worked well and at the time a purely digital system would have been unwieldy, so it has remained in place and still gives good results.

The digitizer takes 500 samples of the detected and filtered backscatter signal between zero and full range on each side, irrespective of the pulse repetition rate. Since the U.S. EEZ survey was conducted almost entirely at the 30-s rate, we will confine the following discussion to this setup. Thus nominally a sample is taken every 45 meters assuming a sound velocity of 1,500 m/s and ignoring any consideration of slant range effects (which lengthen the sample interval seriously at near-vertical incidence). Of course the 100-Hz pulse is inherently capable of nearly five times this rate of sampling, but before concluding that this ought to be done some thought should be given to the cross-range pixel dimensions. Over most of the swath this will be dominated by the horizontal beam pattern of 2.7^0, and even at half range (11.25 km) the pixel size in this direction is 500 m. This 11:1 average pixel aspect ratio is rather larger than one would wish to see and there is little point in extending it to 50:1 which would be the result of using the full rate sampling. A constant 500 samples per side irrespective of pulse repetition rate maintains a constant pixel aspect ratio.

A short digression on the topic of resolution might be in order at this point, as it is a concept widely misapplied to sidescan sonar and indeed not wholly appropriate to it in view of the existence of a much better performance measure, the modulation transfer function (MTF). Resolution in the classical Rayleigh sense refers to the ability of an optical system to separate the images of two closely spaced point sources (as shown in Figure 1–14A). In terms of a continuous two-dimensional illumination field, this concept is not too meaningful and has been largely discarded by optical engineers. What is required is some measure of how faithfully variations in object luminance are transferred to the image plane as a function of spatial frequency. The MTF fulfills this need (Melles Griot 1981). Figure 1–14B shows in simple terms how this concept applies to a lens. A test pattern ruled with intensity modulation of X percent at N lines per mm is imaged and the image is scanned with a microdensitometer. Generally the image modulation will be less than X percent; if, for instance, it is only $0.7X$ percent then the lens has an MTF of 70 percent at N lines per mm. Good-quality lenses are commonly sold with a set of MTF curves for on- and off-axis objects (see Figure 1–14C). The reader should have no difficulty in transferring these ideas to the realm of sidescan sonar. Sonar backscattering is not inherently limited in spatial frequency but the received signal is the convolution of the backscattering function and the probing pulse, so that spatial content is lost; Figure 1–15 illustrates this idea. The object of sidescan sonar is to elucidate the processes at work on the seafloor, so spatial frequency is clearly of prime importance. By contrast, for a mine-hunting sonar it is reasonable to consider resolution because the question of whether there is more than one mine in a contact could be of vital importance. Applying the concepts to GLORIA, the MTF is clearly different between the cross-range and downrange directions. Downrange it begins to fall off at 0.02 m^{-1} and cross-range it is 0.002 m^{-1} on average, degrading to 0.001 m^{-1} at full range. Finally, before returning to practical details, let us consider the topic of cross-range sampling at close range. The sampling is set, as mentioned earlier, by the distance traveled between transmissions – about 125 m at eight knots for the 30-s rate. With a 2.7^0 beam this means that the footprints of successive pulses on the seabed do not overlap until the range reaches about 2.5 km. At closer ranges triangular segments of the seafloor are not insonified. In terms of the foregoing discussion, this is an example of spatial aliasing. Actually 2.5 km is less than the optimum water depth for GLORIA, so it is not a really significant problem for GLORIA.

The master timing for the whole system is derived from the correlator clock, and the digital logging system simply accepts the various signals and formats and logs the data without exerting any control. The logging also drives the on-line display, thereby acting as a quick-look monitor for the correct operation of the whole system, because any degradation or failure of the timing, control, or sonar signals is immediately reflected in the on-line display, which is almost invariably the first indication of any trouble.

Another very important requirement of most surveys, the U.S. EEZ survey included, is the ability to display the images as soon as possible on board for a first cut at the scientific interpretation, so the logging system was designed with a view to producing a daily image mosaic on board.

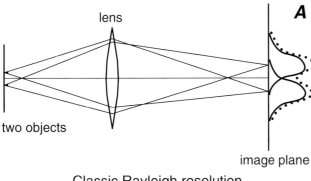

Classic Rayleigh resolution

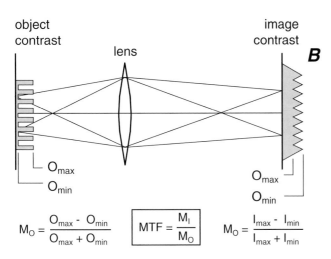

$$M_O = \frac{O_{max} - O_{min}}{O_{max} + O_{min}} \qquad \boxed{MTF = \frac{M_I}{M_O}} \qquad M_O = \frac{I_{max} - I_{min}}{I_{max} + I_{min}}$$

Modulation transfer function (MTF)

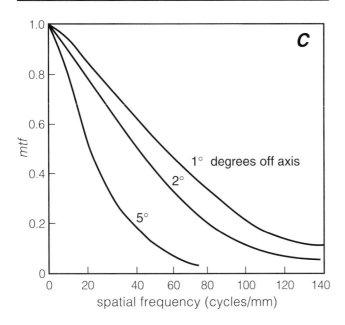

spatial frequency (cycles/mm)

Figure 1–14. A, The classical Rayleigh resolution criterion for two closely spaced point sources. The dotted line shows the combined illumination, and the criterion states that the sources are resolved when the valley is less than about 70% of the peaks. **B,** The concept of modulation transfer function applied to an optical lens. The object contrast has some spatial frequency spectrum, shown here as a square wave of contrast. Generally the image will show a similar pattern but with a smaller contrast ratio. The MTF for a given spatial frequency is the quotient of these two ratios at the frequency concerned. In the same way a sidescan system transfers backscatter contrast more or less efficiently to an image medium, and the contrast function is continuous and not a priori limited in spatial frequency. The image contrast attainable is governed by the spatial spread of the scanning mechanism along the two axes. **C,** Typical MTF curves supplied with a good optical lens.

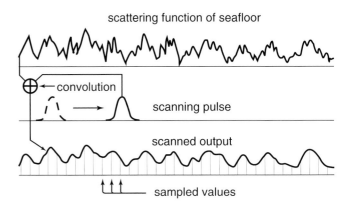

Figure 1–15. The sonar MTF. Note how convolution with the spatially smeared scanning pulse has removed the higher spatial frequencies from the scanned output. This figure should be examined in the light of the captions for Figures 2A and 14B.

This onboard mosaic is also a very important monitor of the quality control of the whole cruise. Thus the sonar data have to be available in convenient chunks soon after acquisition, preferably within twenty-four hours, which led at the time the system was put together to the idea of recording to data cartridges (DC300XL) at six hours of data per track, using the ECMA46 standard format. Two cartridges, each with two tracks filled, suffice to record a twenty-four-hour period. These are replayed daily, as described in the next section, to produce the day's strip of record. After the replay process the cartridges are downloaded to nine-track CCT (in multiple copies). The cartridges, which are a relatively expensive and bulky way of storing large volumes of data, are erased at the end of the cruise ready for recycling, but not until at least two tape copies are safely ashore at separate locations.

The sonar image waveform is sampled and digitized to twelve bits, allowing in theory some 4,096 different shades

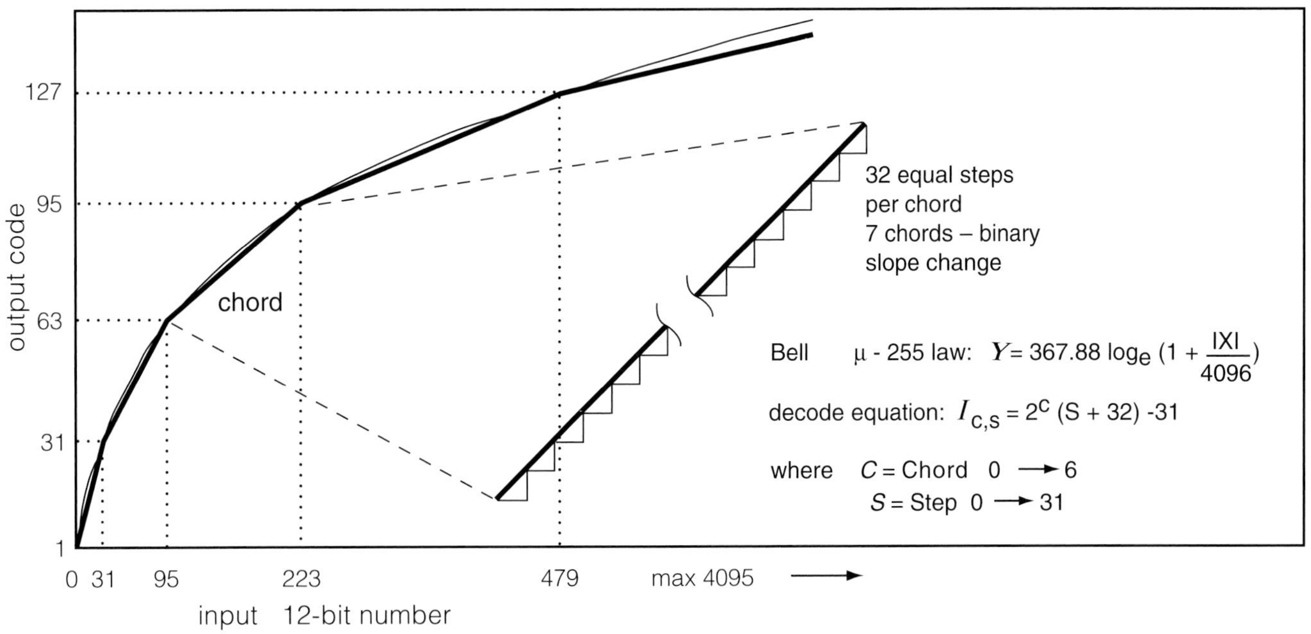
127
95
63
31
1
output code
chord
32 equal steps
per chord
7 chords – binary
slope change
Bell μ - 255 law: Y= 367.88 loge (1 + IXI/4096)
decode equation: I c,s = 2C (S + 32) -31
where C = Chord 0 → 6
S = Step 0 → 31
0 31 95 223 479 max 4095
input 12-bit number

Figure 1–16. The Bell μ-255 law and its approximate coding. A useful feature of the code for GLORIA is that, as well as only requiring one byte per sample instead of two, the step is approximately a constant proportion of the signal level (about 3.2%). This means that the quantization noise is a constant −36 dB rather than in the case of a linear code, negligible at the top of the scale but an unacceptable 0 dB at the lowest level.

of gray to be distinguished. This is well beyond the capacity of any display and also of the human eye, which can distinguish only sixteen to thirty-two shades (representable in at most five bits). If the data were simply coded linearly from twelve to five bits, there would be problems of thresholding and contrast. For example, the lowest 128 levels of the twelve-bit signal would be coded into the zero level of the five-bit one and for a fine sediment, which could be in this level, there would be no contrast at all. What is needed is a reasonably constant contrast for a given percentage modulation whether it occurs at the upper or lower end of the dynamic range. This problem has been faced by communications engineers, and their efforts have resulted in two companding laws. GLORIA uses the Bell μ-255 compression law, which is a piecewise approximation to taking the log (base 2) of the signal (Figure 1–16). It should be remembered, of course, that the samples are of the signal envelope only and are inherently positive. In this the quantization steps are fairly constant as a percentage of the actual value at about three percent of the sample. Thus the quantization errors are always about +/− 1.6 percent of the sample rather than varying from 0.025 percent at the top level to 100 percent at the zero level. This compression law reduces the number of bits per sample from twelve to eight, which is more convenient and efficient for tape recording. The law itself is illustrated in Figure 1–16. The 4,096 level input field is divided into 7 chords, each double the size of its predecessor, and each chord has thirty-two equal steps. The chord number is coded into the most significant three bits of the output byte, and the step within the chord is coded into the least significant five bits.

The core function of the logging system is to transfer the sonar samples and the specified auxiliary data to cartridge. Each record consists of 1,024 bytes arranged as a sixteen-byte header, 496 port samples starting at far range and ending at time zero (TZ), 496 starboard samples starting at TZ and ending at far range, and a trailer with full scale bytes framing the sonar at full range and at zero range. Note that only 496 samples are saved each side per pulse interval; they are acquired at the full rate of 500 per side per pulse, but the last four are discarded for various practical reasons. These samples are not worth anything because they correspond to targets well on into the eclipsing zone of the following transmission where the receiver has been muted for very nearly the whole of the echo period. The logger also assembles the data for the on-line recorder, which displays both a raw image complete with water column pixels and a slant-range corrected image using a depth keyed in by the operator. GLORIA does not rely on on-line bottom detection and does not throw away any water column pixels. The arguments against automatic bottom detection are that it is too unreliable and there are no useful algorithms yet to detect with certainty loss of bottom lock. Thus the user has no need to fear one unpleasant surprise when he looks at the data ashore, namely that bottom lock has been lost and there is no way to recapture the lost portion of the seabed image.

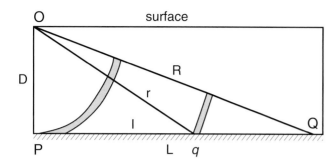

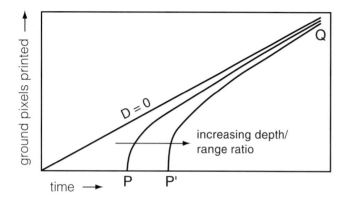

Figure 1–17. **The geometry of slant-range correction, assuming a flat seafloor and rectilinear propagation. By comparing this with Figure 1–1 the reader will see that this is the geometrical explanation of the features in that figure. The assumption of a flat seafloor is not intrinsic to slant-range correction, but if it is to be discarded in favor of full compensation then the bathymetry needs to be well known across the swath. When at sea and working on-line, it is generally convenient to assume a constant sound velocity, and the errors are quite modest. When working off-line with a powerful processor, this approach is less advantageous. Symbols in the figure are; O is position of sonar, D is water depth below sensor, P is nadir ground point, R is maximum slant range, Q is the point on seafloor illuminated by R, r is slant range of intermediate range, illuminating point q, L is maximum ground range (PQ), 1 is intermediate ground range (Pq), P' is the nadir point for a depth greater than D.**

The logger also carries out an approximate anamorphic correction using a nominal speed. Because of the geometry of the actual ground pixels and the proportioning of the recorder format and spot size, it always turns out that the recorder sonar lines are too narrow by three or four to one at eight knots. The operator has the option of keying in a value to override the default speed of eight knots. The correction is carried out by line repetition, the recorder buffer being read out to the recorder faster than real time so that on average the lines are repeated often enough to correct the distortion.

SHIPBOARD IMAGE CORRECTIONS, REPLAY, AND MOSAIC

As previously indicated, the data cartridges are downloaded every twenty-four hours to PC hard disk files from the DC300XL cartridges, a comparatively modest three megabytes (MB). The daily replay requires two further data input streams, navigation and along track depth. Normally these data sets are prepared on a shipboard format data tape and read by the replay PC, but if necessary hand entry can be used. The PC then processes the data at two-minute intervals and writes out image files.

The depth readings are used to compute a slant-range correction, which consists of mapping pixels incrementally from far range with pixel repetition at predetermined intervals according to the depth-to-range ratio. This approach assumes a flat seabed, which has to do in the absence of full

bathymetric knowledge, and rectilinear sound propagation as shown in Figure 1–17, and in fact it offers a very good approximation, certainly for shipboard use.

The anamorphic ratio is computed from the navigation record by smoothing the speed to eliminate large fluctuations and then computing the distance run between transmissions. Knowing the full-range distance, the scales are equalized by repeating the lines the correct number of times. Fractional repetitions are achieved by maintaining a running total of repetitions, which is incremented by the new ratio read from each image record and the line is then replayed repeatedly, decrementing the running total by one for each replay until the integer part is zero. This leaves a new fractional part to which is added the new ratio read from the next image record. On average each line is replayed a nonintegral number of times corresponding to the true anamorphic ratio.

The PC system plays the images out to a laser film recorder in which the beam from a HeNe laser is focused and scanned across a 35-mm film while its intensity is controlled by an acousto-optic modulator. The modulation is controlled by a feedback loop in which a beam splitter peels off half the optical power onto a photo cell whose output is used to control the modulator drive. Meanwhile the film is advanced by a precise step for each scan. The film used is Technical Pan which has good sensitivity to the 633 nm radiation of the laser and also has good definition and a very high ratio of density in the developed image. The actual download from image file to laser film writer takes one or two minutes for a six-hour pass of data. The processing time before replay depends on the power of the PC, but the time to download the cartridges has consistently been the limiting factor in processing. The waiting time for the depth and navigation tape is usually the longest delay of all.

Finally the film is developed, a process that requires meticulous care in the usage, strength, and temperature of the chemicals to achieve consistently high-quality results. Then prints are produced using a rapid processor, and the results after drying are set on the mosaic. The developed film is more permanent than the prints and it forms part of the cruise archive, whereas the prints do not. The laser filmwriter and the concomitant developer and printer are at the time of writing being phased out in favor of a thermal paper recorder output. This change is partly the result of increasing restrictions on the transport and storage of the chemicals and partly the result of great improvements in the gray-scale capabilities of the new generation of thermal paper recorders.

As described in a later chapter, once ashore the data are reprocessed with a wider range of image-processing tools, using much larger formats to include whole regions. However, the simple onboard processing can include options for radiometric corrections and other processing, and it is possible to achieve up to 95 percent of the processing at sea. In the final analysis the relevant processing depends on the intended purpose of the final product, which differs for onboard use and that ashore.

Conclusion

The GLORIA system is not, in its essential concept, complex, but there is considerable complexity in its details. It is also remarkable for the wide range of disciplines harnessed to achieve the objective of the system.

The last really major overhaul, which changed the system outright, was the development of Mk II in 1977. Since then every part except the arrays, vehicle structure, cable, and launcher has been changed at least once as improvements have been designed and incorporated. Significantly, the changes have been largely associated with the electronic systems, where the pace of change in the industry is possibly faster than in any other field. This underlines both the soundness of the original design and the strength of the system as an up-to-date tool. It is always in some respects a newly developed prototype taken to sea by the development engineers who remain responsible for its proper running. This arrangement ensures a constant exchange of ideas between the designers and scientists and also ensures that the scientists are fully aware of the lim-

its and powers of the GLORIA system, to the great benefit of both groups.

Acknowledgments

Over its history many people have made important contributions to the development of the GLORIA system and its successful deployment on the U.S. EEZ. In addition to Stuart Rusby who was responsible for the development of GLORIA I, I would like to mention the following with grateful thanks for their help: Jack Revie, Mark Carson, Roger Edge, Brian Barrow, Derek Bishop, Mac Harris, Jon Campbell, Andy Harris, Alan Gray, Steve Whittle, and many others for direct contributions to the technology; also to Sir Anthony Laughton, Arthur Stride, and the whole user community including the very significant USGS contingent for their constant support and encouragement. In the preparation of this manuscript I have received much help from Alan Gray and from Mike Conquer in the preparation of the diagrams, and great forbearance from the editors.

References

Admiralty Manual of Seamanship. 1967. Vol. 2: London: HMSO Books.

Chavez, P. S., Jr. 1986. Processing techniques for digital sonar images from GLORIA. *J. Photogram. Engrg. and Rem. Sens.* 52: 1133–45.

Chesterman, W. D., Clynick, P. R., and Stride, A. H. 1958. An acoustic aid to seabed survey. *Acustica.* 8: 285–90.

Cook, C. E., and Bernfeld, M. 1967. *Radar Signals.* New York: Academic Press.

O'Meara, T. P. 1962. Synthesis of 'band pass' all pass time delay networks with graphical approximation technology. *Hughes Res. Lab. Report.* 114.

Anonymous. *Optics Guide 2.* 1981. Irvine, Calif.: Melles Griot.

Rusby, J. S. M. 1970. A long range sidescan sonar for use in the deep sea. *Int. Hydrogr. Rev.* 47: 25–39.

Somers, M. L., Carson, R. M., Revie, J. A., Edge, R. H., Barrow, B. J., and Andrews, A. G. 1978. GLORIA II, an improved long range sidescan sonar. In *Oceanology International 1978.* Technical Sessions. London: J. BPS Publications Ltd.

Somers, M. L., and Stubbs, A. R. 1984. Sidescan sonar. *Proc. IEE,* Vol. 131(F), No. 3, June 1984, pp. 234–56.

Stein, S., and Jones, J. 1967. *Modern Communication Principles.* New York: McGraw-Hill; and many other good texts on communication engineering.

Stewart, J. L., and Westerfield, E. C. 1959. A theory of active sonar detection. *Proc. IEE.* 47:872–81.

Tucker, M. J., and Stubbs, A. R. 1961. Narrow beam echo-ranger for fishery and geological investigations. *Br. J. Appl. Physics.* 12: 103–10.

Urick, R. J. 1975. *Principles of Underwater Sound for Engineers,* Ed. 2. New York: McGraw-Hill.

2 Processing and manipulating GLORIA sidescan sonar images

Pat S. Chavez, Jr., Jeff A. Anderson, Stuart C. Sides, and JoAnn Bowell

U.S. Geological Survey, Flagstaff, Arizona

Abstract

Use of digital image data collected with remote sensing instruments has increased rapidly during the last two decades. Sidescan sonar imaging systems with digital capabilities are relatively new compared to airborne and satellite systems. The GLORIA (Geological LOng-Range Inclined Asdic) imaging system acquired digital capabilities in the early 1980s (see chapter by Somers and others in this volume for details on the GLORIA system). In this chapter we present several of the various digital image processing techniques developed by the U.S. Geological Survey (USGS) to correct, mosaic, enhance, and analyze digital sidescan sonar images. As with image data from other sensors, the processing is separated into the two different stages of preprocessing and information extraction. In the preprocessing stage, sensor-specific algorithms are applied to correct for both geometric and radiometric problems. This step is followed by digital mosaicking of the trackline strips into quadrangle format, which can be used as input to either visual or automatic digital interpretation. An automatic seam-removal procedure (ASRP) is presented as an alternative to our user-intensive digital feathering/stenciling procedure to help minimize tone- or seam-matching problems between images from adjacent tracklines.

Use of more than just local spatial variability information is important for all data sets, but especially for single-band data such as sidescan sonar. Therefore, the information extraction section discusses the generation and use of spatial variability at the local, intermediate, and regional scales for the automatic classification of GLORIA data. The sidescan sonar image – processing package is part of the USGS mini image processing system (MIPS) and is currently set up to handle and process data collected by any generic digital sonar imaging system.

Introduction

The use of digital data collected with remote sensing instruments has increased rapidly during the last two decades. One of the newer members of the remote sensing family of digital imaging systems is the sidescan sonar system; it is an active system using acoustical waves to produce images called sonographs (Sutton 1979). The sonographs are a measure of the reflectance properties of the seafloor's geomorphic features. This chapter describes several of the various digital image processing techniques developed by the U.S. Geological Survey (USGS) to correct, mosaic, enhance, and analyze digital sidescan sonar images. Development of a sidescan sonar processing package started in 1979 using some of the expertise developed to process and analyze side-looking airborne radar (SLAR); the sidescan processing package has been used by the USGS in an operational mode since the early 1980s (Teleki et al. 1981), with additions and improvements still being made. The package includes digital mosaicking and is now set up to handle data from any generic sonar imaging system. The sidescan sonar image – processing package used to correct for both radiometric and geometric distortions is part of the USGS mini image processing system (MIPS), which also includes both general and multispectral image-processing capabilities (Chavez 1984).

The characteristics and design of the GLORIA (Geological LOng-Range Inclined Asdic) sidescan sonar imaging system were presented by M. L. Somers in a previous chapter in this volume, to which we refer the reader for detailed information about the imaging hardware. The objective of this chapter is to present the core of the processing and manipulating capabilities developed over the years to work with GLORIA sidescan sonar images, including some information extraction capabilities. However, the current software package is set up to handle image data from any sidescan sonar imaging system and not just GLORIA.

The operational core processing capabilities used on the GLORIA sidescan sonar image data have been rather stable since about 1985, with some enhancements and improvements made to the sequence and algorithms used. Therefore much of the information presented in the first half of this chapter duplicates some of the information presented previously (Chavez 1986). The sidescan processing package within the USGS MIPS has been used in an operational mode to generate various products and several atlases of the Exclusive Economic Zone (EEZ Western Conterminous United States 1986; EEZ Gulf of Mexico and Eastern Caribbean Areas 1987; EEZ Bering Sea 1991 – these are published as *USGS Miscellaneous Investigations Series*). The information on digital mosaicking has been presented before but not formally published. The information on the automatic spatial variability and amplitude index (SVI/SAI) analysis was extracted from Chavez and Gardner (1994). In the following sections we will discuss the preprocessing requirements, digital mosaicking, and spatial analysis for information extraction and mapping using GLORIA sidescan images.

Data characteristics and test site

Sidescan sonar images have been used to extract information about the seafloor for over twenty years (Rusby 1970; Flemming 1976; Rusby and Somers 1977). The images used in this study were collected by the GLORIA system using a thirty-second pulse-repetition rate that insonifies a continuous 45-km-wide swath (Somers et al. 1978; Somers and Stubbs 1984). The images are from several different areas north of the Hawaiian Islands. The two quadrangles used are from 21° to 23° N latitude and 161° to 163° W longitude and 23° to 25° N latitude and 155° to 157° W longitude. The digital mosaics covering the quadrangles are in a Lambert conformal projection.

Digital image processing requirements

Digital data recorded by any imaging system requires two very different types of processing stages; first is the preprocessing stage followed by the information extraction stage (Chavez et al. 1977). The preprocessing stage is designed to correct an image for both radiometric and geometric problems or errors. The algorithms used are customized for the particular imaging system because each has its own set of unique data acquisition characteristics. In contrast, the information extraction stage uses algorithms that are more general and more a function of the application and independent of the imaging system (e.g., spatial filtering or automatic classification using spectral or backscatter information).

Preprocessing

As with most other digital remotely sensed image data, the preprocessing applied to the GLORIA sidescan sonar im-

ages is composed of three different types: (1) geometric corrections, (2) radiometric corrections, and (3) utility/miscellaneous other corrections. Keep in mind that the preprocessing requirements are very system dependent, even though every data set requires some preprocessing. The preprocessing steps presented here are for the GLORIA sidescan sonar images, but in general, are applicable to most sidescan sonar images.

Geometric corrections

Figure 2–1 shows an example of the original raw GLORIA sidescan sonar image without any corrections (a linear stretch has been applied for visual display purposes). Some of the geometric problems for which corrections are needed are (a) the water column offset, (b) slant-range to ground-range projection, (c) aspect or anamorphic ratio distortion, and (d) changes in the ship's velocity.

WATER COLUMN OFFSET

GLORIA, like all other sonar imaging systems, starts recording data as soon as it transmits an acoustical wave; therefore, a number of pixels (picture elements) on both sides of nadir do not contain information about the seafloor surface. Also, the actual nadir pixels are offset to the sides as a function of the water column or fish altitude. One of the first steps in the preprocessing sequence is to merge some of the information stored in a navigation file with the image file. This step is done by storing in the trailer section of each image line the date and time, latitude and longitude, and the bathymetric/fish altitude values at the nadir location. At times, especially with some of the deep tow and higher resolution sonars, the fish altitude values collected during the cruise have errors. Therefore, part of the generic sonar processing package within the USGS MIPS now includes an interactive water column digitizing program. The altitude values stored in the trailer of each line are used to compute the amount of offset in pixels and correct for the offset caused by the water column.

SLANT-RANGE DISTORTION

For efficiency the program that corrects for water column offset also simultaneously corrects for slant-range geometry distortions. The GLORIA system, as well as most other sidescan systems, collects image data using a near-range depression angle of approximately ninety degrees and a far-range depression angle of approximately five to ten degrees (Somers et al. 1978). Figure 2–2 shows the results of applying the water column and slant-range to ground-range corrections to the image shown in Figure 2–1. Note that the image shown in Figure 2–2 has also been through the shading corrections discussed later in the radiometric corrections section. The same linear stretch was applied to both images to make it easier to compare them with one another. Due to the large difference between the near- and far-range de-

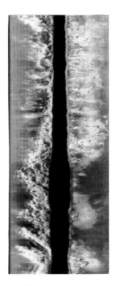

Figure 2–1. Example of the raw sidescan sonar image data collected by the GLORIA imaging system. The area shown is a portion from a trackline of data collected around the Hawaiian Islands (approximate latitude and longitude are 21.35° N and 168.04° W, respectively, with a heading of about 234°). Various geometric and radiometric problems can be seen in the image; the width of the image covers about 45 km.

Figure 2–2. This image shows the results of applying both the vertical and beam angle shading corrections along with the water column and slant-range to ground-range corrections. The shading corrections remove low frequencies in the across-track direction, including those related to power drop-off as a function of range and/or beam angle and artifacts that can be introduced due to the use of incremental TVGs. Note that, due to the extreme difference between near- and far-range depression angles, the pixels at near-range (nadir) are expanded much more by the slant-range to ground-range correction than pixels at far- or mid-range, which gives this area a more blocky appearance (the image width is 45 km).

pression angles, the pixels at near-nadir are duplicated more than those at mid- and far-range.

ANAMORPHIC/ASPECT RATIO AND SHIP
VELOCITY VARIATIONS

Another major geometric distortion present in most sidescan sonar images, including GLORIA images, is the aspect or anamorphic ratio between the along- and across-track directions. The sampling interval in the across-track direction for the raw GLORIA images generates pixels that have an approximate slant-range resolution of 45 m for the thirty-second pulse-repetition rate (the rate used to collect most of the GLORIA data). The program that corrects for both the water column and slant-range to ground-range distortions generated an across-track pixel resolution of 50 m (the output pixel size is a user-selected parameter; 50 m was used on most of the GLORIA data). The spatial resolution in the along-track direction is determined by the pulse-repetition rate used (thirty seconds) and the ship's velocity (7 to 10 kn). The average resolution in the along-track direction with the thirty-second pulse-repetition rate is approximately 125 m, which generates images with an aspect ratio distortion of about 2.5 (i.e., 125 vs. 50 m).

Of course, any change in the ship's velocity also changes the spatial resolution of the image in the along-track direction. The ship's velocity can be influenced by such variables as the direction and strength of the current and/or the wind pattern with respect to the ship's direction of travel, and whether the ship is going in a straight line or is in the pro-

cess of making a long turn. The velocity can vary from 7 to 10 kn, which will cause the pixel resolution in the along-track direction to vary from approximately 110 to 140 m for a thirty-second pulse-repetition rate, which in turn introduces an accordion effect into the geometry of the image in the along-track direction. The distortion is removed by using the latitude and longitude values stored in the trailers of the image lines to calculate the distance traveled over a thirty-minute time frame, ten minutes if a turn is detected. Given the distance traveled and the desired pixel size, it is easy to compute the number of pixels required for the particular thirty-minute segment. For most of the GLORIA data we have used a 50-m pixel size to keep from throwing out any pixels/information in the across-track direction. The same program corrects for distortions introduced by both the aspect ratio and changes in the velocity of the ship. Figure 2–3 shows the results of applying both these corrections to the image shown in Figure 2–2. Note that features within the image now have more familiar geometric shapes.

If maps with acceptable resolution and geometric characteristics are available for a given area, they can be used to select image-to-map control points. This approach would allow the sidescan sonar images to be registered to a map; however, most areas in the deep portions of the oceans do not have maps of sufficient resolution or detail for this type

Figure 2–3. This image shows the results of correcting the image data for both aspect ratio distortions and ship's velocity changes in the along-track direction. Features in the image, such as seamounts, will now have a more familiar geometric shape because the pixels are now square and the length-to-width (aspect) ratio is one. These corrections can be applied either as stand-alone corrections as shown here or included in the final geometric correction that maps the image into a user-specified projection and scale. The image covers the same 45 km shown in Figures 2–1 and 2–2.

of geometric correction. Therefore, only automatic corrections can be made using the navigation information that is available. Since the navigation information may be limited or inaccurate, the resulting corrected image may still have some geometric distortion. However, the corrections just discussed do remove the major sources of geometric errors and are relatively easy to apply in an operational mode. New systems that collect both sidescan sonar and swath bathymetry information simultaneously are being improved; once these systems are generating acceptable data they will enhance the geometric correction capabilities. The bathymetry data will be useful for correcting the sidescan sonar image data on a pixel-by-pixel basis for geometric displacements due to the topographic relief. This is the state of the art with current optical and radar imaging systems, except that the digital elevation model (DEM) is not collected simultaneously with the image data.

Radiometric corrections

The second major preprocessing phase involves radiometric corrections. This phase changes the digital number (DN) of a pixel rather than its spatial position, as is the case with geometric corrections. Most, if not all, algorithms dealing with radiometry change a pixel's DN value based on the pixel's current DN value, as a function of its spatial position, or as a function of both its DN value and spatial position. The radiometric corrections needed for the GLORIA images, as well as for images from most of the other sidescan sonar imaging systems, include (a) shading correction due to power drop-off from near- to far-range, as well as other lower frequency radiometric problems encountered in the across-track direction; (b) if needed, correction of speckle noise and blocky appearance due to extreme aspect ratio differences and type of resampling used (this includes extracting a narrow strip about the nadir, fixing the nadir's blocky appearance, and then putting it back into the image); and (c) striping noise removal.

(A) SHADING CORRECTION

Several different types of shading algorithms can be applied to sidescan sonar images to remove some of the across-track lower frequency shading problems. Within the USGS MIPS, we have implemented several algorithms and currently use the combination of two in an operational mode. Both must be applied before the data go through the slant-range to ground-range correction because they are designed to remove problems that are dependent on either the beam angle or vertical location of a pixel in the geometry of the original raw data. The first algorithm currently used in the USGS MIPS processing sequence is one in which the correction coefficients are computed as a function of beam angle. The algorithm is similar to the one discussed by Reed and Hussong (1989). The beam angle algorithm does a better job at removing the near-range radiometric problem, but does not do as good a job on problems that occur from mid- to far-range. The second algorithm is designed to remove problems that are exactly in the vertical direction, such as the striping seen when incremental time-varied gains (TVGs) are used. Also, the power drop-off problem is approximately vertical from mid- to far-range, and this algorithm can remove most of its undesirable effects. Figure 2–2 shows the results of applying these two shading corrections to the original GLORIA image shown in Figure 2–1 (recall that the image shown in this figure has also been through the water column and slant-range to ground-range corrections).

(B) SPECKLE NOISE REMOVAL

Speckle noise is a radiometric problem similar to that present in radar images (MacDonald and Lewis 1976; Rydstrom, La Prade, and Leonardo 1979). It is also similar to bit errors except that the contrast difference between the pixel with speckle and the surrounding local neighborhood is not as extreme. Therefore, pixels affected by speckle do not stand out in an image as much as pixels affected by bit errors, but their spatial frequencies are approximately the same. Note that the speckle noise seen in these data are caused by electronic problems and not seafloor specular response. Several methods can be used to suppress the effects of speckle on the visual quality of the image (especially when working with hard copy photographs that have been enlarged close to the limits supported by the resolution of the image). The method used on sidescan sonar images is one developed by the senior author for radar applications. It uses the combination of (1) a small three-by-three- or five-by-five-pixel high-pass filter (HPF) with an appropriate threshold to identify pixels with possible speckle noise and then (2) a small smoothing filter to remove their effects from the image (Chavez 1980). This procedure was compared with several other techniques and found to be at least as good, or better in most cases, than the other methods (Chavez and Berlin 1986).

The procedure is quite simple and straightforward and is as follows. The average of a small window subtracted from the pixel in the middle of the window provides information as to how much the DN differs from its very local neighborhood. In most images the DN value of a pixel varies by a small amount from pixels that are in close proximity (i.e., within a three-by-three- or five-by-five-pixel window).

Changes larger than a given limit are usually caused by something other than the natural scene characteristics in the image and can be considered noise. Using the histogram of a small high-pass spatial filtered image, an upper and lower threshold can be selected and used to suppress, or set to zero, all DN values in the original image that lie outside the threshold limits of the high-pass filtered image. Once this has been done a low-pass filter, or smoothing filter, can be used to fill in the "holes" in the original image created by setting those pixels with limits outside the high-pass filter threshold to zero. Two important aspects about the smoothing done at this stage are as follows: (1) only pixels with zero DN are smoothed, that is, replaced with the average of a very local neighborhood; and (2) only pixels in the very local neighborhood with nonzero DNs are used to compute the average. This approach results in the replacement of only the pixels that were identified as having speckle noise and leaves the rest of the original image data unchanged. Most of the time only a small percentage of the pixels are identified as having speckle noise; therefore, with this method only this small percentage of pixels is affected as compared with most other techniques that can affect most, if not all, the pixels in an image.

In cases where the speckle noise is considered extreme, two-by-two or three-by-three smoothing filter applied to the entire image may have to be considered. One example where this approach may be preferred is if, during the aspect ratio correction, the digital image is expanded by two to three times in the along-track direction to maintain maximum spatial resolution in the across-track direction. The automatic smoothing of all the pixels will help reduce the blocky appearance the image will have due to the digital enlargement in the along-track direction. One of the disadvantages of using this procedure is that all pixels are replaced by the average of their very local neighborhood. Also, if needed, a narrow strip about nadir can be extracted at this stage and processed to improve the very blocky appearance caused by the large digital enlargement required close to nadir, as compared to away from nadir, in the across-track direction when converting from slant-range to ground-range geometry. We have successfully used this nadir visual improvement procedure on GLORIA, EG&G, SeaMARC, and TOBI sidescan sonar images and generated results that had a better visual appearance than the noncorrected data.

(C) STRIPING

Most of the striping noise in sidescan sonar images occurs in the across-track or scanning direction. Figure 2–4A shows a portion of a GLORIA sidescan sonar image with some striping noise, similar to that seen in Landsat and other scanning imaging systems. To remove the striping from the image, a combination of a high- and low-pass spatial filtering procedure is used. The method is based on generating two separate images from the input data representing the high-frequency components, except for the noise frequency, and

A   **B**

Figure 2–4. A shows a GLORIA image with the striping problem that can be quite common in scanning imaging systems. B shows the results of applying the convolution filtering technique to remove the striping from the GLORIA image.

the low-frequency components without the noise. The results of the two spatial filters are then added to generate an image that is very similar to the original image, but without the noise (Chavez and Soderblom 1974). If the averages of the windows used for the high- and low-pass filters are represented by AVEH and AVEL, respectively, and the DN value of the pixel at the center of each of the two windows is represented by PTXH and PTXL, then as shown by Chavez (1986, p. 1137) the following can be used to generate the high- and low-frequency components:

$$HFC = PTXH - AVEH$$
$$LFC = AVEL \qquad (1)$$

The final results can then be generated as follows:

$$RESULTS = HFC + LFC \qquad (2)$$

The filter shapes used to remove the striping noise from the GLORIA sidescan sonar images were a one line by seventy-one sample high-pass filter and a nine line by seventy-one sample low-pass filter. For comparison, the shapes of the filters needed to remove the striping in Landsat images are a one line by 101 sample high-pass filter and a seven line by 101 sample low-pass filter; for TOBI sidescan sonar images the shapes currently being used are

1 line by 301 sample high-pass filters and 51 line by 301 sample low-pass filters. Figure 2–4B shows the results of applying this striping noise removal procedure to the portion of the GLORIA image shown in Figure 2–4A. As mentioned by Chavez (1986), this method is applicable to other noise patterns as well; the only difference is that the shape of the spatial filters must be appropriately selected for the characteristics of the noise pattern to be removed.

Digital mosaicking

Data from the GLORIA digital imaging system have become an important component within the marine program of the USGS for offshore mapping. The various digital processing techniques just described are routinely used to correct for geometric and radiometric distortions of sidescan sonar images. However, the resulting images are long strips in trackline format, which can be difficult to use for interpretation. Therefore, within the USGS MIPS software, part of the sidescan sonar processing package includes the capability to mosaic trackline images digitally into quadrangle format (i.e., two-by-two- or two-by-three-degree quadrangles for the GLORIA data). A few advantages of the digital quadrangle format over the long individual trackline strip format are as follows: (1) its shape and coverage is better from an interpretation point of view, (2) it is easier to merge digitally with other data sets, and (3) the user can generate several different products covering the quadrangle from the same input digital mosaic (Chavez, Schoonmaker, Jr., and Anderson 1987). Compared to making photo mosaics from the processed individual trackline image strips, where a new mosaic must be made for each enhancement, this is a better way to generate a digital database because the one mosaic can be used to generate many different products. A new mosaic does not have to be generated optically every time a different enhancement or analysis procedure is applied.

The procedure for digital mosaicking of any sidescan sonar image data, including the GLORIA data, is more difficult than that required for other types of remotely sensed image data, such as satellite images, because geometric control is usually available only at nadir and the navigation information often has both high- and low-frequency errors. The general processing sequence used in the USGS MIPS package to generate a digital mosaic from preprocessed trackline sidescan sonar image data is as follows:

1 Use the navigation information merged in the trailer of each image line during the preprocessing steps to project the image strip into the correct location within the quadrangle being generated. The current USGS MIPS package has a new geometric resampling program for applying the desired geometric transformation. Instead of using the general resampling geometric correction program used with satellite image data, which applies a warping transformation on a pixel-by-pixel basis, the current program applies the geometric correction on a line-by-line basis (i.e., projects the data a line at a time rather than a pixel at a time).

2 Once the individual image lines are projected based on the navigation information, which may include sensor pointing parameters such as yaw and pitch, the resulting image will more than likely have holes based on the difference between the yaw and pitch from line to line. The holes are made from the pixels with zero DN values between the image pixels; recall that all of the pixels within the quadrangle being generated are initialized to a DN value of zero. Within MIPS we use either a procedure that we call LPFZ (low-pass filter zeroes) or MFZ (median filter zeroes) to fill in the holes. Both the LPFZ and MFZ filters work identically to the one described previously to help remove speckle noise from the sidescan sonar image data. That is, the average or median value of a small window (three by three or five by five pixels) about a pixel is computed using only nonzero DN values, but only pixels with zero DN are replaced. Similar to radar speckle noise removal, several other methods can also be used to fill in the holes (Chavez and Berlin 1986). The procedure that gives the best results will be dependent on the quality of the input image data

and the amount of digital enlargement involved in correcting for the difference in resolution in the along-track versus across-track directions. Figure 2–5 shows a portion of an image strip geometrically projected before and after the MFZ has been applied. One note to make here is that we have given the new line-by-line geometric correction software the capability to resample an image into the desired projection, but rather than always having north at the top we can put it at any user-specified angle from the top. This approach makes the actual geometric correction process more efficient in both computer time and disk storage points of view. Having the general heading of the tracklines, rather than north, at the top of the mosaic means that the mosaic will be smaller because the tracklines, which usually do not run north to south, will not require a rotation and a larger digital array to project the data into during the geometric correction stage. The smaller array means less computer time and less disk storage space will be required.

3 In order to allow maximum flexibility for both geometric and radiometric matching of images on adjacent tracklines, we actually generate two digital mosaics. The preprocessed image data are used to generate an odd and even trackline mosaic. Making two mosaics within which the image data do not overlap each other gives us the flexibility to work on both the geometric and radiometric fit and matching all at once at a later stage. Once the odd and even mosaics have been made, the two are used as input to an interactive digital stenciling, or feathering, program within MIPS. This process allows the user interactively to select the digital cutline between adjacent strips for every pixel. This capability allows the user to generate what he or she feels is the best possible visual product; however, it is quite a user-intensive procedure.

Due to the user-intensive requirement of the interactive digital stenciling procedure, which can delay the generation of a final product to be used for interpretation and mapping, we have developed an automatic seam removal procedure (ASRP). This procedure is similar to one the senior author has been using on satellite image data for over fifteen years, but with the shape of the spatial filters adjusted for the GLORIA data characteristics. The ASRP can be very useful because it allows a good working product to be generated much more quickly and gets it into the hands of the users while the data are still relatively new. Similar to the digital stenciling procedure, the ASRP also uses the odd and even trackline mosaics as input, but instead of interactively selecting the cutlines to get a good tone match between images from adjacent tracklines, it uses low-pass and high-pass spatial filters. The general idea of automatic seam removal from digital mosaics is very similar to noise removal (e.g., striping removal discussed in the preprocessing section). The tone or brightness mismatch between adjacent strips of images is considered noise, and a low-pass and high-pass spatial filtering procedure can be used to remove the seam noise au-

Figure 2–5. A shows the results of applying the geometric correction to transform the image into the desired projection and scale. Note that, because the geometry correction program processes a line of data at a time, which allows it to take care of changes in pointing characteristics of the imaging system, "holes" are left in the resulting image. B shows the same image with the holes filled in using the LPFZ filtering option.

tomatically. Similar to the striping noise removal, the seam removal procedure used on the GLORIA images is identical to that used with either Landsat satellite images or any other sidescan sonar images; the only difference is the shape and size of the spatial filters used. Using the odd and even trackline digital mosaics as input, the sequence for the ASRP on the GLORIA sidescan sonar image data is as follows:

1   High-pass filter (HPF) both the odd and even digital mosaics with a 151-by-151-pixel window.

2   Digitally mosaic the HPF results of step (1) by dropping them together (i.e., automatically mosaic the two by sim-

ply having one image have priority over the other where they overlap).

3   Digitally mosaic the odd and even trackline mosaic used as input into the HPF by also dropping them together (i.e., normal nonfiltered mosaic of the odd and even tracklines).

4   Low-pass filter (LPF) the mosaic results of step (3) using the same size window (i.e., 151 by 151 pixels – it is important that zero DNs not be used when computing the average).

5   Merge the HPF (step 2 results) and LPF (step 4 results)

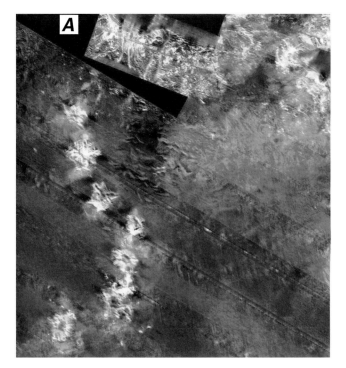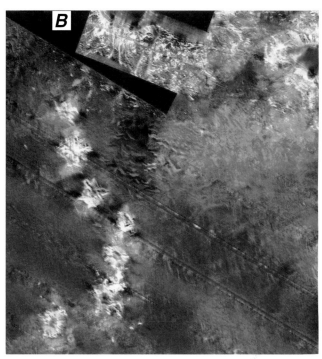

Figure 2–6. A shows the digital mosaic of the odd and even strips required for the quadrangle. Keeping the odd and even strips separated allows the user more flexibility in the final mosaicking stage and gives more options for taking care of radiometric and/or geometric matching problems. The odd and even mosaics can be combined to generate the mosaic of the entire quadrangle of interest as shown. The mosaic shown in B is the result of mosaicking the same odd and even files but using the automatic seam removal procedure (ASRP) to get a better tone match between adjacent strips. This procedure is an alternative to the more user-intensive digital stenciling that can be done to optimize both the tone and geometric match between strips.

mosaics by adding the two together (remember, if 127 was added to the HPF results to center the DNs about 127 to avoid negative values, it must be subtracted at this stage).

This procedure will usually generate a digital mosaic that is acceptable for analyses and interpretation in an efficient and timely manner. In many cases the results will be almost as good as the much more user-intensive interactive digital stenciling procedure; also, the overall time required to generate a finished product will be reduced. This method allows the user to get off the ship with a good working copy representative of what the final mosaic product will look like. Note that the results will have exactly the same geometric and radiometric characteristics as the mosaic generated with the user-intensive procedure in areas more than half the window size away from the overlap areas (i.e., both procedures will generate exactly the same results over most of the nonoverlapping areas, which is usually equal to about 70 to 80 percent of the mosaic).

The main difference between the two results will have to do with the geometric matching in the overlapping areas. Because we are not interactively selecting the cutlines on a pixel-by-pixel basis, the matching of spatial/geometric features will be different (i.e., the more time-consuming procedure allows the user to select the location of the cutlines so they minimize the effect of geometric mismatches). How-

ever, to help minimize geometric mismatches of this type, we have recently added a program to MIPS that allows us to identify, interactively, image-to-image control points on adjacent trackline images for use on improving the geometric match between tracklines in the mosaic. Only a few image-to-image control points are required; therefore, it is not a user-intensive procedure. This type of procedure is useful when the navigation information does not have the accuracy required for good geometric matching from strip-to-strip; this problem is especially seen with high-resolution sidescan image data. We use this program interactively to select common image tie points from strips adjacent to each other and adjust the latitude and longitude values for the lines within the image. The program corrects only for the low-frequency distortions in both the along- and across-track directions in the image; the high frequency information from line to line is preserved. This fact is very important; otherwise the resolution of the navigation data will be greatly reduced.

We will be working in the future on taking the corrected information stored in the trailers of each image line and using it to correct or adjust the actual navigation file. Shown in Figure 2–6A is the mosaic generated by just dropping the odd and even strips together; as you can see, in some areas there is a tone-matching problem between tracklines even with the extensive preprocessing that was done. Figure 2–6B shows the results of applying the ASRP to these data. The resulting mosaic has better tone matching between adjacent tracklines and was accomplished in an automatic mode with no user interaction required.

Information extraction

As mentioned, to work with and use remotely sensed image data, including sidescan sonar images, the user must separate the processing into two stages. The first stage includes the preprocessing requirements that are related specifically to the imaging system being used and are applicable regardless of the intended use of the data. The final result of the preprocessing stage is the cleaned-up digital mosaic considered the corrected data base; these are the data used as input to the second stage – the information extraction stage. In this stage the types of algorithms used to analyze and extract information from the image data are dependent on the application and/or interest of the user and often independent of the imaging system characteristics. For example, if the user is interested in structural information, he or she will use a high-pass spatial filter, regardless of whether the image data are from a sidescan sonar system, such as GLORIA, or from a satellite imaging system, such as Landsat. Many different algorithms and procedures can be used to extract information from remotely sensed image data. In this section we will discuss one method dealing with the extraction of spatial information from an image. Keep in mind that this is only one example of what can be done and there are other algorithms and techniques that can be applied to the data in the information extraction and analysis stage.

Most remotely sensed images contain both spectral and spatial information. The spectral information is related to various properties of the surficial materials and how they respond in the portion of the spectrum being used to record the data. The spatial information is related to the distribution, variations, and relief of the surficial materials from pixel to pixel. Therefore, the two main characteristics that determine a pixel's backscatter or brightness value, and consequently its DN in most remotely sensed images, are the physical properties of the surface and near-surface materials and topographic slope at the given pixel.

Since the mid-1970s, the large majority of remotely sensed image data have been collected in multispectral mode, such as those collected by the Landsat systems. Most of the analysis tools and methods make use of the spectral information contents of an image more than its spatial information. Because of this emphasis, the general trend has been to develop and use analysis procedures aimed at multiband data sets to extract information. However, a number of systems collect only a single-band image, including GLORIA and all the other sidescan sonar imaging systems, and it is especially important that as much spatial information as possible be available for the analysis of these data.

Figure 2–7. This view is the mosaic result of GLORIA side-scan images covering a quadrangle north of the Hawaiian Islands (23° to 25° N latitude and 155° to 157° W longitude). These are the data used as input into the spatial variability analysis and should be used to compare with the results shown in Figures 2–8 and 2–9.

Due to the importance of the spatial information in side-scan sonar images, we have included some of the material presented in a paper by Chavez and Gardner (1994). The objective of their paper was to present a method to extract spatial amplitude and variability information from remotely sensed digital image data at the local, intermediate, and regional scales. The paper discusses the use of the spatial information to extract geologic features automatically from GLORIA sidescan sonar images (Chavez and Gardner 1994). The method used high-pass filters (HPFs) to extract the spatial amplitude and variability information from the digital images. Both a spatial amplitude image/index (SAI) and a spatial variability image/index (SVI) at the local, intermediate, and regional scales were generated by using filters with three different window sizes. By computing the SAI and SVI at three different scales, a multiband data set representing spatial information was generated and analyzed using multispectral tools.

Previous studies have shown how the variance of an image is affected by scale variations; they were based on varying the spatial resolution of the pixels and not the size of the neighborhood about a pixel (Woodcock 1985; Woodcock and Strahler 1987; Townshend and Justice 1990). In the study by Chavez and Gardner the spatial resolution of the pixels was kept constant and the window size of the neighborhood about the pixel being processed was varied. The more heterogeneous an area of an image, the more spa-

SVI   **SAI**

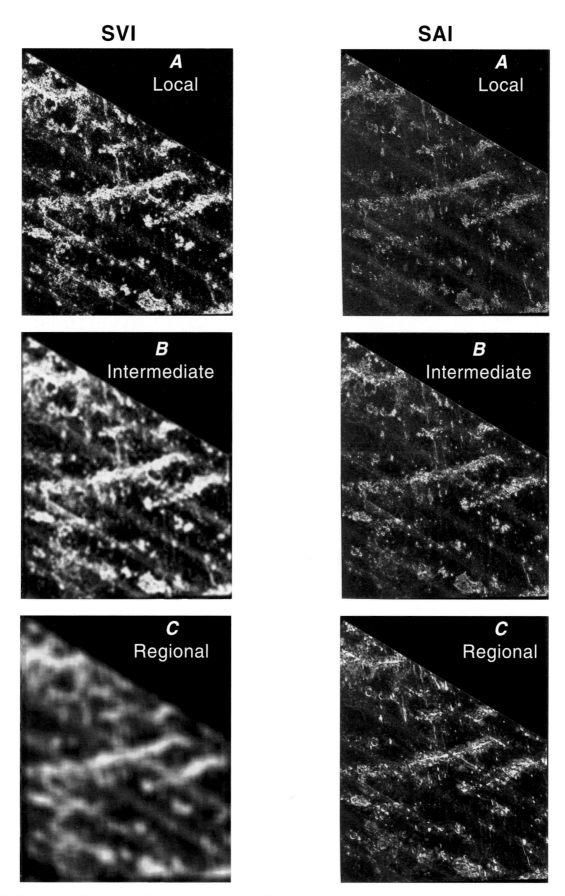

A
Local

A
Local

B
Intermediate

B
Intermediate

C
Regional

C
Regional

Figure 2–8. These views show the results of the spatial variability analysis that was applied to the mosaic shown in Figure 2–7. Spatial variability analysis was used to extract information dealing with spatial information at the (A) local, (B) intermediate and (C) regional scales. Shown are the SVIs on the left and SAIs on the right; they were generated from the mosaic shown in Figure 2–7.

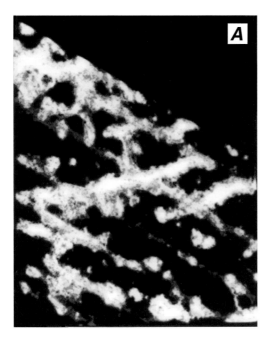
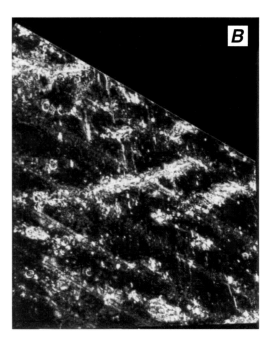

Figure 2–9. These views show the results of using an automatic clustering classification procedure on the three (A) SVI/left and (B) SAI/right images shown in Figure 2–8. By generating a multiband data set, some of the tools available for multispectral image analysis can be used on single-band sonar image data sets.

tially variable it will be; the more homogeneous an area, the less spatially variable it will be. Spatial filters, such as high or low pass, can be used to enhance this type of information. The HPF used in their study was a Laplace spatial filter; it computes the difference between a pixel and its surrounding neighborhood, with the neighborhood being determined by the window size of the filter.

The SAI is nothing more than the absolute value of the HPF results. The absolute value is used because only the amount of change or difference is desired and the direction of change is not needed. The SAI shows how much a pixel differs from the given neighborhood, but not how variable an image is within that same neighborhood; this information is given by the SVI. The SVI is basically the image generated by a filter that counts how many pixels within a selected neighborhood size exceed a specified threshold in the SAI image.

Several studies have been published that extracted textural information from Landsat, radar, or sidescan sonar images (Haralick and Shanmugam 1974; Shanmugan et al. 1981; Reed and Hussong 1989; Barber and LeDrew 1991). However, these studies incorporated only very local spatial information, which limits the representation of the spatial characteristics of a pixel and its neighborhood. Fractals can also be used to study the spatial variability of an image; however, fractals also look at only the very local information.

Generating the SVI and/or SAI at the local, intermediate, and regional scales allows the thematic units to be analyzed on their more complete spatial characteristics. Figure 2–7 shows the results of a mosaic generated from GLORIA

sidescan images covering a quadrangle north of the Hawaiian Islands. These data were used as input to the spatial variability analysis, and the results of the local, intermediate, and regional SAIs and SVIs are shown in Figure 2–8. The SVI results are highly correlated to the roughness of the image at the local, intermediate, and regional scales and can be used for mapping purposes. To show how the multiband/multispatial data set can be used with multispectral analysis techniques, the results of automatic clustering classification are shown in Figure 2–9. The image or mosaic being processed contains isolated seamounts, relatively young lava flows, Cretaceous ridges, and seafloor spreading fabric (Chavez and Gardner 1994). In general the features are volcanic in origin, but their relief has been modified by various submarine processes.

A comparison of the classified results in Figure 2–9 with the mosaic shown in Figure 2–7 shows that the classification based on the SVIs picks up all four main classes of features. One distraction in the classification is that it does pick out the nadir area. However, the classification results clearly delineate the Cretaceous ridges, the lava flows, and the seafloor spreading fabric where the fabric is well away from other features. The nadir problem can be solved either (1) by removing the nadir effect by using low-pass and high-pass filter noise removal techniques or (2) by combining its class with those of the surrounding areas. As stated in the paper by Chavez and Gardner (1994), "From a geological viewpoint, the SVI classification provides a rapid method of first-order automatic mapping of vast quantities of remotely sensed data into fundamental classes of terrains."

Summary

We have developed processing and analysis capabilities within the USGS MIPS for sidescan sonar images. The examples shown in this paper are GLORIA images, but the software package is set up to handle images from any generic sidescan sonar system. The processing sequence is separated into the two general stages of (1) preprocessing and (2) analysis and information extraction. The preprocessing algorithms applied are very sensor specific and aimed at correcting the image data for both geometric and radiometric distortions. The analysis and information extraction algorithms are very application specific and aimed at extracting and/or mapping various features within the image. The example presented in this chapter involving the use of spatial variability information to map several major geologic features automatically within the GLORIA image shows only one of the potential uses of the data and some of the analysis capabilities within the USGS MIPS.

The various products being generated show, often for the first time, detailed information of the ocean floor. The digital mosaics of GLORIA data give the user the large overview picture of the region, very similar to the regional views of land given by Landsat satellite images, and help identify areas of interest where further studies should be taken with higher-resolution data sets. The capabilities discussed in this chapter can help scientists map and understand the ocean floor and maximize the information that can be extracted from the digital sidescan sonar images. With the high cost involved in collecting these data, it is to the user's advantage to use capabilities such as those described here to extract the maximum amount of information from the data.

The software used to process these GLORIA data is part of the USGS MIPS developed in Flagstaff and has now been modified to handle any generic sidescan sonar image data set. As of this writing, other sidescan data sets that have been processed using the USGS MIPS package include Sea-MARC, EG&G, Datasonics SIS-1000, AMS-120, and TOBI. Work continues on refining and improving the capabilities with special emphasis in the near future on using image-to-image tie point information to improve the accuracy of the navigation data.

References

Barber, D. G., and LeDrew, E. F. 1991. SAR sea ice discrimination using texture statistics: A multivariate approach. *J. Photogram. Engrg. and Rem. Sens.* 57(4): 385–95.

Chavez, P. S., Jr., and Soderblom, L. A. 1974. Simple high-speed digital image processing to remove quasi-coherent noise patterns. *American Society of Photogrammetry Symposium,* Washington, D.C., pp. 595–600.

Chavez, P. S., Jr., Berlin, G. L., and Mitchell, W. B. 1977. Computer enhancement techniques of Landsat MSS digital images for land use/land cover assessments. *Sixth Remote Sensing of Earth Resources Symposium,* Tullahoma, Tenn., March 1977, pp. 259–75.

Chavez, P. S., Jr. 1980. Automatic shading correction and speckle noise mapping/removal techniques for radar image data. National Aeronautics and Space Administration, *JPL Publication 80-61,* pp. 251–62. Washington, D.C.: NASA.

———— 1984. U.S. Geological Survey mini image processing system (MIPS). U.S. Geological Survey *Open-File Report 84-880.* Reston, Va.: USGS.

———— 1986. Processing techniques for digital sonar images from GLORIA. *J. Photogram. Engrg. and Rem. Sens.* 52(8): 1133–45.

Chavez, P. S., Jr., and Berlin, G. L. 1986. Restoration techniques for SIR-B digital radar images. In proceedings *ERIM Fifth Thematic Conference: Remote Sensing for Exploration Geology,* Reno, Nevada, pp. 501–11.

Chavez, P. S., Jr., Schoonmaker, J. W., Jr., and Anderson, J. A. 1987. Digital mosaicking and merging of dissimilar data sets – GLORIA, magnetic, and bathymetric. *Proceedings Third McKelvey Conference,* Denver, Colorado, [abs].

Chavez, P. S., Jr., and Gardner, J. V. 1994. Extraction of spatial information from remotely sensed image data – An example: GLORIA sonar images. *Can. J. Rem. Sens.* v.20, p. 443–53.

EEZ-SCAN 84 Scientific Staff, 1986. Atlas of EEZ, Western conterminous United States. *USGS Miscellaneous Investigations Series I-1792.* Reston, Va.: USGS.

EEZ-SCAN 85 Scientific Staff, 1987. Atlas of EEZ, Gulf of Mexico and Eastern Caribbean Areas. *USGS Miscellaneous Investigations Series I-1864-A,B,* Reston, Va.: USGS.

EEZ Scientific Staff, 1991. Atlas of EEZ, Bering Sea. *USGS Miscellaneous Investigations Series I-2053.* Reston, Va.

Flemming, P. W. 1976. Side-scan sonar: A practical guide, *Int. Hydrogr. Rev.* 53: 65–92.

Haralick, R. M., and Shanmugam, K. S. 1974. Combined spectral and spatial processing of ERTS imagery data. *J. Rem. Sens. Environ.* 3: 3–13.

MacDonald, H. C., and Lewis, A. J. 1976. Operation and characteristics of imaging radar systems. *J. Rem. Sens. Electro Magnetic Spect.* July, 3(3): 23–45.

Reed, T. B., and Hussong, D. 1989. Digital image processing techniques for enhancement and classification of Sea-MARC II side scan sonar imagery. *J. Geophys. Res.* 94(B6): 7469–90.

Rusby, J. S. M. 1970. Along range side-scan sonar for use in the deep sea (GLORIA project). *Int. Hydrogr. Rev.* 47: 25–39.

Rusby, J. S. M., and Somers, M. L. 1977. The development of the GLORIA sonar system from 1970 to 1975. In *Voyage of Discovery,* ed. M. V. Angel, suppl. to *Deep Sea Res* 24: 611–25.

Rydstrom, H. O., La Prade, G. L., and Leonardo, E. S. 1979. *Radar Imagery Interpretation Adaptable to Planetary Investigations.* Goodyear, Ariz.: Goodyear Aerospace Corp., Arizona Division. 223 p.

Shanmugan, K. S., Narayanan, V., Frost, V. S, Stiles, J. A., and Holtzman, J. C. 1981. Textural features for radar image analysis. IEEE Transactions on Geoscience and Remote Sensing. GE-19 (3): 153–56.

Somers, M. L., Carson, R. M., Revie, J. A., Edge, R. H., Barrow, B. J., and Andrews, A. G. 1978. GLORIA II – an improved long range sidescan sonar. Proceedings of IEEE/IERE Subconf. on Ocean Instruments and Communication. *Oceanology International.* London: B.P.S. Pub. Ltd., pp. 16–24.

Somers, M. L., and Stubbs, A. R. 1984. Sidescan sonar. *Proc. IEEE* 131: 243–56.

Sutton, J. L. 1979. Underwater acoustic imaging. *Proc. IEEE,* April 1979, 67(4): 554–66.

Teleki, P. G., Roberts, D. G., Chavez, P. S., Jr., Somers, M. L., and Twichell, D. C. 1981. Sonar survey of the U.S. Atlantic continental slope; acoustic characteristics and image processing techniques. *Proceedings of 13th Annual Offshore Technology Conference,* May 1981, Houston, Tex., pp. 91–102.

Townshend, J. R. G., and Justice, C. O. 1990. The spatial varia-tion of vegetation changes at very coarse scales. *Int. J. Rem. Sens.* 11(1): 149–57.

Woodcock, C. E. 1985. Variograms and spatial variation in re-motely sensed images. *International Geoscience and Remote Sensing Symposium,* University of Massachusetts, Amherst, pp. 1078–83.

Woodcock, C. E., and Strahler, A. H. 1987. The factor of scale in remote sensing. *J. Rem. Sens. Environ.* 21(3): 311–32.

II U.S. East Coast EEZ

James M. Robb,[1] William P. Dillon,[1] Dennis W. O'Leary,[2] and Peter Popenoe[1]

[1]U.S. Geological Survey, Woods Hole, Massachusetts
[2]U.S. Geological Survey, Denver, Colorado

The USGS and the British Institute of Oceanographic Sciences (IOS) first surveyed parts of the U.S. East Coast continental slope between Georges Bank and the Blake Escarpment using GLORIA during October and November of 1979. That early survey gave Twichell and Roberts (1982) a new look at the overall geomorphic pattern of submarine canyons on the mid-Atlantic continental slope. They could see that many canyons extend only partially across the mid or lower slope. Therefore, some submarine canyons can be created on the continental slope, without connection to the shelf edge and independently of shelf or shallow water processes. That insight inspired other studies and led to major conceptual advances in the understanding of submarine canyon origins and growth.

The EEZ-SCAN survey of the U.S. East Coast areas took place from February to May 1987, using a much superior GLORIA system, which recorded the data in digital form rather than as analog photographic records. The 1979 images had been as tantalizing as they were effective; they provided a new, broad view, but commonly lacked enough contrast to show the details clearly. The digital processing techniques available eight years later produced much clearer images that could be readily manipulated and combined with other data, such as bathymetry.

In the spring of 1987, survey operations proceeded aboard R/V *Farnella* during five periods, each about twenty-five days in length, beginning off Georgia and working northeastward. Upon completion of the survey off Georges Bank, about 506,000 sq km of seafloor had been mapped with GLORIA. At the same time the GLORIA system was operating, the ship gathered other data to characterize the area, using airgun and 3.5-kHz seismic-reflection profiles, 10-kHz bathymetric profiles, and measurements of the total magnetic field. The spacing between the ship's tracklines was determined from line to line, based on the water depth and sound transmission characteristics, and varied between 11 and 28 km during the surveys.

Initially, stormy weather and rough seas challenged the watchstanders and the equipment that was under tow. Nonetheless, swaths of seafloor images appeared that were compiled day by day and line by line on big tables in the ship's laboratory. The slowly growing mosaic became a fascinating window to a submarine area that had been mapped using point and profile data but which had not been observable in panorama. The mosaic inspires a visceral appreciation of the sea bottom as a submarine landscape, one to be seen by human beings only with the aid of technological machinery. It is a landscape as varied and impressive as any on a continent, and the nature of the submarine geologic processes that work upon it can still, largely, only be inferred. The images show a marvelous variety of real features that ornament the East Coast continental rise, once described by Bruce Heezen (1968, p. 10) as "a broad, uniform, gently sloping, and smooth-surfaced wedge of sediments, 100 to 1000 km wide, and covered with . . . monotonously homogeneous gray lutites."

Probably the greatest impact of the East Coast continental rise survey is its display of the vast expanse and variety of submarine landslides. Many individual slide zones, each hundreds of km[2] in area, are distinctly different in appearance on the mosaic from intervening areas of seafloor. Most of the landslide zones had been identified by previous investigators, but their sum as seen in whole seems greater than their parts as described from individual studies. However, mass movements along this passive, progradational, "monotonously homogeneous" continental margin include even greater areas than the obviously anomalous patches casually seen in the mosaics would imply. Detailed image analysis by O'Leary (this volume) shows that most of the upper continental rise northeast of Hudson Canyon is an apron of slide debris, of complex history. The present planimetry of the debris sheets results from interplay with oceanic deep water transportation mechanisms that created large-area sediment drifts. Immense areas of mass movement on the southeastern U.S. margin are discussed by Popenoe and Dillon (this volume). These slides occurred in rapidly depositing sediment that was transported into the area by Gulf Stream currents; the cause of the slides may be linked to subsurface gas hydrates and the change in strength of hydrate-supported sediments in response to sea level changes. The potential connection between very large submarine landslides and subsurface gas hydrates along the continental margin would not have been recognized without the areal imagery and the well-processed subbottom profiles at the scale made possible during the EEZ-SCAN program.

A further revelation of the East Coast mosaics is the extent and complexity of continental rise deepwater channels and their relationships to canyons on the continental slope. On the rise south of Georges Bank and southern New England, channels commonly have uninterrupted individual paths across the rise with complexities that are related to the extensive mass movement deposits, as demonstrated by O'Leary (this volume). To the south and west of the Hudson Canyon and Channel, a system of east- to southeast-trending, parallel to dendritic channels appears to extend from the shelf break to more than 250 km offshore on the upper continental rise, then to feed into the southward-trending Wilmington Valley. On closer inspection, however, the development of a continuous dendritic system appears to have been stalled by massive sediment influxes that choked and modified each of the offshore channelways so they cannot be individually traced on available bathymetric representations or on the GLORIA images (Schlee and Robb 1991).

When compared with bathymetric contours from published contour maps, the East Coast EEZ mosaics identify some discrepancies where the routes of canyons or channels or locations of seamounts do not coincide, and thus they can focus attention on places that need inspection and correction in many other data sets. In most of those places, especially along the slope, modern multibeam bathymetric surveys are needed to resolve the submarine topography appropriately. Because the GLORIA returns are affected by sedimentary characteristics as well as topographic form, quantitative depth data are needed in many places to distinguish the topographic effects from sedimentary and shallow subsurface features, to resolve ambiguities, and to make greater use of the information in the GLORIA data.

Most important to us as citizens is that the GLORIA mosaics show us the floor of the deep ocean that lies near our most densely populated land areas, offshore of the eastern megalopolis. We are using (and contaminating?) these deepwater areas more and more. In these areas we need to know the centennial, decadal, and annual geologic occurrences that represent significance to human planning. We need to be aware of the size and frequency of potentially hazardous occurrences, and we also need to be able to evaluate the effects of our human activities on the little-known sea bottom. We need to know how many of the features that we can see on these pictures are due to ancient Pleistocene events and how many are due to present-day occurrences. At some existing dumpsites we will need to know where radioactive materials are possibly being moved and where they may be

moved to. The GLORIA images provide a basic framework upon which we can plan and carry out well-focused studies to answer these questions using swath-bathymetric measurements, high-resolution sidescan sonar imagery, high-resolution subbottom profiling, or specific sampling.

References

Cashman, K. V., and Popenoe, P. 1985. Slumping and shallow faulting related to the presence of salt on the continental slope and rise off North Carolina. *Mar. Petrol. Geol.* 2: 260–72.

EEZ-SCAN 87 Scientific Staff, 1991. Atlas of the U. S. Exclusive Economic Zone, Atlantic continental margin. *U.S. Geological Survey Miscellaneous Investigations Series I-2054*, 174 p., scale 1:500,000. Reston, Va.

Heezen, B. C. 1968. The Atlantic continental margin. *UMR Jour.* (University of Missouri at Rolla) 1: 5–25.

Hill, G. W., and McGregor, B. A. 1988. Small-scale mapping of the Exclusive Economic Zone using wide-swath side-scan sonar. *Mar. Geod.* 12: 41–53.

Hughes-Clark, J. E., O'Leary, D. W., and Piper, D. J. W. 1992. Western Nova Scotia continental rise: Relative importance of mass wasting and boundary-current activity. In *Geologic Evolution of Atlantic Continental Rise*, eds. C. W. Poag and P. C. de Graciansky, pp. 266–81. New York: Van Nostrand Reinhold.

Masson, D. G., Gardner, J. V., Parson, L. M., and Field, M. E. 1985. Morphology of upper Laurentian Fan using GLORIA long-range side-scan sonar. *Am. Assoc. Petrol. Geol. Bull.* 69: 950–9.

McGregor, B. A. 1987. Diversity of processes and morphology on the U.S. Atlantic continental slope and rise. In *U.S. Geological Survey Circular 961*, eds. D. W. Folger and J. C. Hathaway, pp. 21–35. Conference on continental margin mass wasting and Pleistocene sea-level changes, August 13–15, 1980. Reston, Va.: USGS.

O'Leary, D. W., Hughes-Clarke, J., Dobson, M., and Williams, S. R. J. 1988. GLORIA sonar survey of the New England-Nova Scotian continental margin: The rise revealed. *Geotimes* March, v. 33 pp. 22–4.

O'Leary, D. W., and Dobson, M. 1992. Southern New England continental rise: Origin and history of slide complexes. In *Geologic Evolution of Atlantic Continental Rise.* eds. C. W. Poag, and P. C. de Graciansky, pp. 214–65. New York: Van Nostrand Reinhold.

Popenoe, P., Coward, E. L., and Cashman, K. V. 1982. A regional assessment of potential environmental hazards to and limitations on petroleum development of the southeastern United States Atlantic continental shelf, slope and rise, offshore North Carolina. *U.S. Geological Survey Open-File Report 82-136*, 67 p., map scale 1:1,000,000. Reston, Va.: USGS.

Popenoe, P., and Manheim, F. T. 1991. Phosphorite deposits of the Blake Plateau as observed from the NR-1 submarine and DELTA submersible. *U.S. Geological Survey Circular 1062*, pp. 61–63. Reston, Va.: USGS.

Scanlon, K. M. 1982. Geomorphic features of the western North Atlantic continental slope between Northeast Channel and Alvin Canyon as interpreted from GLORIA II long-range sidescan-sonar data. *U.S. Geological Survey Open-File Report 82-728.* 8 p., 1 map. Reston, Va.: USGS.

——— 1984. The continental slope off New England: A long-range sidescan-sonar perspective. *Geo-Mar. Ltrs.* 4: 1–4.

Schlee, J. S., and Robb, J. M. 1991. Submarine processes of the middle Atlantic continental rise based on GLORIA imagery. *Geol. Soc. Am. Bull.* 103: 1090–1103.

Schlee, J. S., Dillon, W. P., Popenoe, P., Robb, J. M., and O'Leary, D. W. 1992. GLORIA mosaic of the deep sea floor off the Atlantic Coast of the United States. *U.S. Geological Survey Miscellaneous Field Studies Map MF-2211,* scale 1:2,000,000. Reston, Va.: USGS.

Schwab, W. C., Lee, H. J., and Twichell, D. C. eds. 1991. Submarine landslides: Selected studies in the U.S. Exclusive Economic Zone. *U.S. Geological Survey Bulletin 2002,* 204 p. Reston, Va.: USGS.

Teleki, P. G., Roberts, D. G., Chavez, P. S., Somers, M. L., and Twichell, D. C. 1981. Sonar survey of the U.S. Atlantic continental slope: Acoustic characteristics and image processing techniques. *Proc. Offshore Technol. Conf.* 4017: 91–102.

Twichell, D. C., and Roberts, D. G. 1982. Morphology, distribution, and development of submarine canyons on the United States Atlantic continental slope between Hudson and Baltimore Canyons. *Geology* 10: 408–12.

Twichell, D. C. 1987. Geomorphic map of the U.S. Atlantic continental margin and upper rise between Hudson and Baltimore Canyons. In *U.S. Geological Survey Circular 961.* eds. D. W. Folger and J. C. Hathaway, pp. 14–21. Conference on continental margin mass wasting and Pleistocene sea-level changes, August 13–15, 1980. Reston, Va.: USGS.

Twichell, D. C., and Polloni, C. F. 1993. CD-ROM Atlas of the deepwater parts of the U.S. Exclusive Economic Zone in the Atlantic Ocean, the Gulf of Mexico, and the Eastern Caribbean Sea. *U.S. Geological Survey Digital Data Series DDS-15* (Digital Data on CD-ROM). Reston, Va.: USGS.

3 The timing and spatial relations of submarine canyon erosion and mass movement on the New England continental slope and rise

Dennis W. O'Leary

U.S. Geological Survey, Denver, Colorado

Introduction

The origin of submarine canyons of the U.S. Atlantic continental margin remains a subject of controversy and speculation. Assumptions about their origin and function are based mainly on bathymetry and spatial relations observed on the middle to upper slope and on relations to stratigraphy of the rise, revealed chiefly by high-resolution seismic-reflection data. The largest canyons along the Atlantic margin (and elsewhere) are more or less aligned with the mouths of major rivers, except along Georges Bank. This association has led to the most generally accepted explanation for the origin of the canyons: that they originated by collapse of deltaic sediments deposited by rivers at the shelf edge during eustatic lowstands (Vail, Mitchell, and Thompson 1977; Reineck and Singh 1980; Coleman, Prior, and Lindsay 1983). The collapse, or slumping, is thought to have generated turbidity currents having sufficient erosive power to have carved canyons across the slope and, in many cases, out across the upper rise.

However, sidescan sonar surveys undertaken along the U.S. Atlantic slope during the 1970s (Twichell and Roberts 1982; Scanlon 1984) showed that extensive canyon erosion has occurred along the lower to middle slope and that a link to the shelf is not required for canyon development. The chutelike forms of many canyons and their lack of connection with the shelf edge suggested that they may have been initiated by mass movement low on the slope. Successive upslope failures, presumably generated in response to the initiating mass movement, could have created a headward-extending channel that, in some instances, breached the shelf break (Farre et al. 1983).

GLORIA sidescan sonar data acquired in 1987 from the New England continental margin (the slope and the continental rise to water depths of 4500 m; EEZ-SCAN 87 1991) call for a reevaluation of the putative relations between submarine canyon cutting and mass movement and of the sources and mode of transport effected by the canyons and

mass movement. These data show that there are complex relations between canyon erosion and mass movement. On a broad scale the processes are fundamentally distinct and one does not necessarily proceed directly from the other. The GLORIA data show that along the New England continental slope, the most extensive canyon cutting has occurred where mass movement is minor, whereas major mass movement has occurred apart from canyons. Although GLORIA data show that some canyons have functioned as debris conduits to the upper rise, much mass movement clearly postdates canyon erosion and does not seem to have utilized intercepted channels for downslope flow, transport, or generation of turbidity currents.

This report analyzes some of the major planimetric and geomorphic relations among canyon and mass movement features of the New England slope and rise as observed in GLORIA image data. The timing of canyon erosion and of mass movement, and the significance of morphology and spatial relations are discussed in light of GLORIA sonar image data.

Spatial relations of canyons and mass movement

GLORIA data (EEZ-SCAN 87 1991) show that the styles and the overall distribution of canyon erosion and mass movement along the New England continental slope are compartmentalized in two distinct domains controlled primarily by stratigraphy (O'Leary 1988, 1993). One domain comprises virtually the entire middle to upper slope along Georges Bank for a lateral distance of about 250 km (Figure 3–1). This domain is characterized by a succession of deep, narrow canyons having sharply etched branched or pinnate tributary networks. Most of the canyons head along the upper slope in water depths of about 300 m. The canyon heads are broad, scored bowls that open individually into narrow, sinuous axial canyons that extend more or less down to the foot of the slope. Most of the canyons that head along the continental slope converge or die out above the 3,000-m

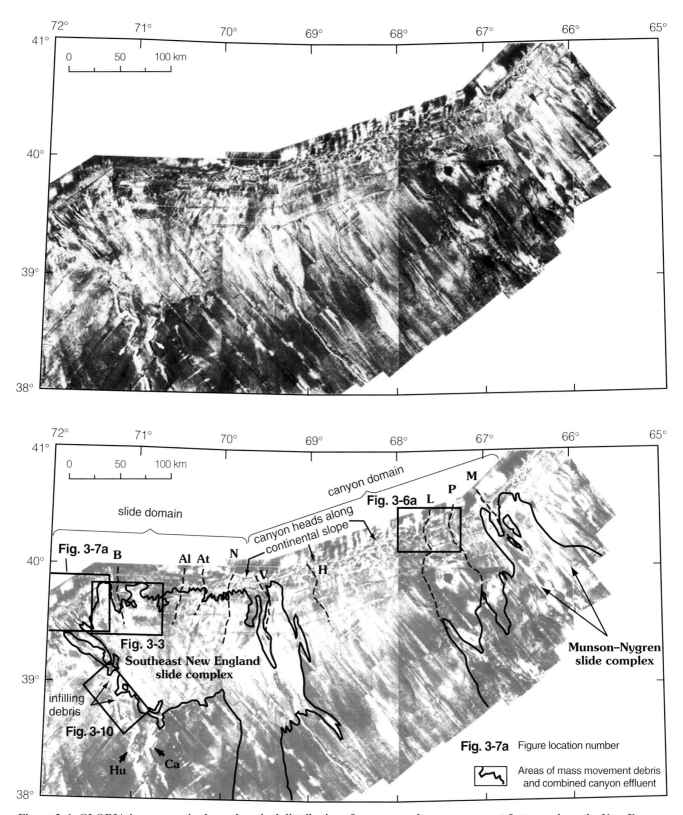

Figure 3–1. GLORIA image mosaic shows domainal distribution of canyons and mass movement features along the New England continental slope and rise. Major canyons include the following: Hudson (Hu), Carstens (Ca), Block (B), Alvin (Al), Atlantis (At), Nantucket (N), Veatch (V), Hydrographer (H), Lydonia (L), Powell (P), and Munson (M).

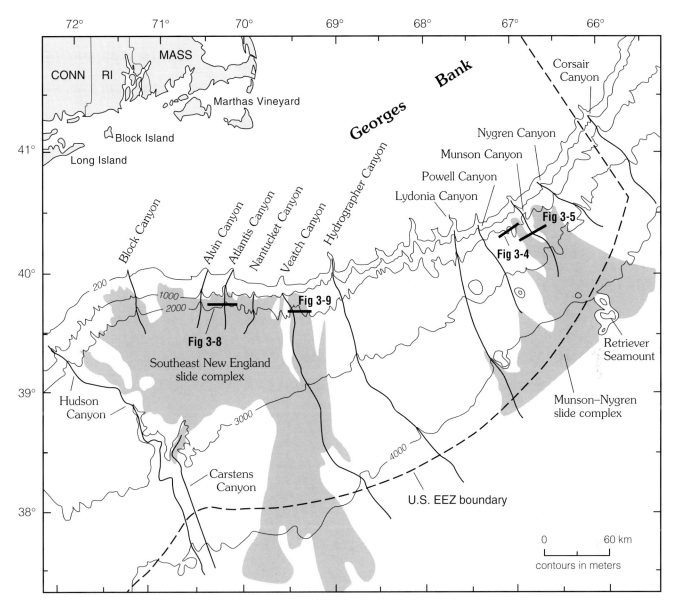

Figure 3–2. Map shows locations of major canyons and mass movements along the New England continental slope and rise.

contour (Scanlon 1984); either these canyons did not transport significant amounts of sediment or their deposits form
a smoothly coalesced apron that is graded to the rise and is
difficult to recognize in GLORIA images or in high-resolution seismic-reflection profiles.

The other major slope domain is located west of about
longitude 69°30′ W. Here, the pattern of close narrow
canyon incision that exists along Georges Bank is replaced
by areally extensive slab slides (the Southeast New England
slide complex; O'Leary and Dobson 1992) that occupy the
slope as far west as Hudson Canyon (Figures 3–1 and 3–2).
Within this domain along the southeast New England margin, large sections of the slope, as much as 20 km across
and 300 m thick, have slid away from scarps ranging in water depth from 750 to 1,000 m.

The debris field of the Southeast New England slide complex is a composite of at least four major slides. The debris
forms a more or less amalgamated sheet approximately 220
km wide and extends down the rise for a distance of about
220 km (Figures 3–1 and 3–2). Some tongues of debris 10
to 20 m thick may extend to even greater distances (EEZ-
SCAN 87 1991; O'Leary and Dobson 1992). GLORIA images also show that along the lower slope in this domain,
debris surrounds enclaves or promontories of intact strata
(Figure 3–3), a feature not seen along the lower slope or uppermost rise off Georges Bank.

The lower slope along Georges Bank is indented by broad
embayments, best shown in seismic-reflection profiles (Figure 3–4). Whether these embayments originated by mass
movement or are simply canyon segments later widened by

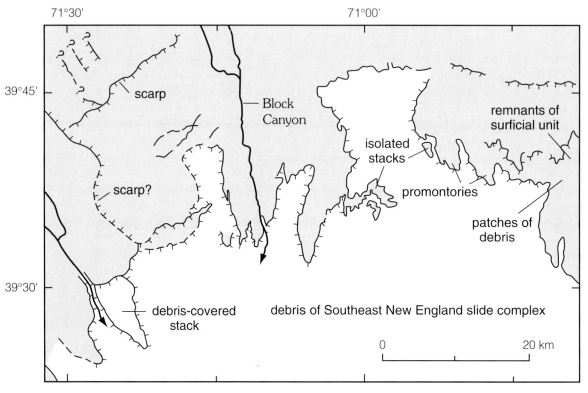

Figure 3–3. GLORIA image shows promontories and stacks of autochthonous strata within the Southeast New England slide complex along the base of the continental slope; see Figure 3–1 for location.

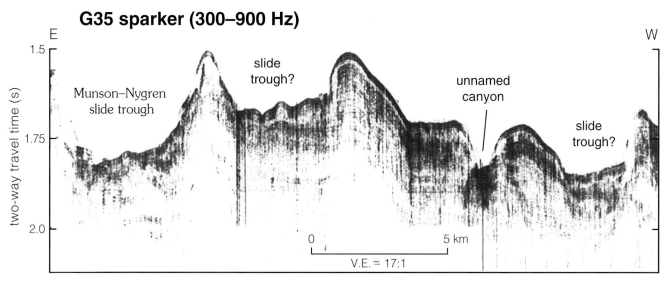

Figure 3–4. High-resolution seismic-reflection profile shows embayments along the lower continental slope of Georges Bank; see Figure 3–2 for location.

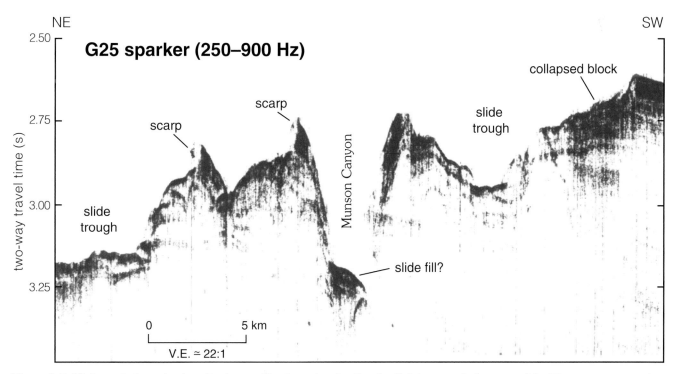

Figure 3–5. High-resolution seismic-reflection profile shows headwall and adjoining terrain features of the Munson-Nygren slide complex; see Figure 3–2 for location.

erosion and mass movement is not known. However, two of these embayments form the headward boundaries of the Munson-Nygren slide complex (O'Leary 1986a; O'Leary and Dobson 1992). High-resolution seismic-reflection profiles show details of the flanking scarps and the proximal debris of this complex (Figure 3–5). GLORIA images show that debris of the Munson-Nygren slide complex extends

from the base of the slope to beyond the EEZ boundary and beyond the limit of GLORIA coverage, a distance of at least 160 km (EEZ-SCAN 87 1991). The debris field flares out with distance from the slope, from a width of 19 km at the base of the slope to a width of at least 160 km near the EEZ border, where it merges with debris of Lydonia Canyon (Figure 3–1). The deep, relatively narrow embayed character of

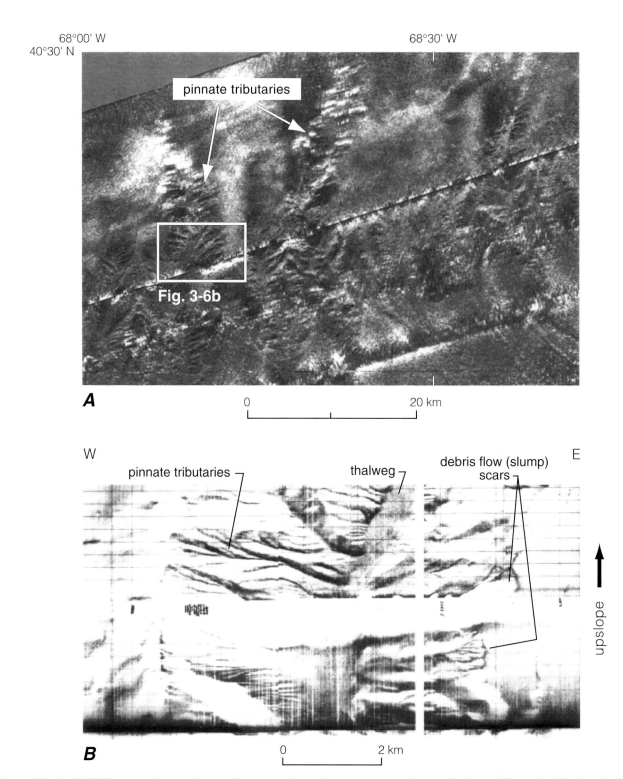

Figure 3–6. A, GLORIA image shows pinnate tributary pattern of canyons along Georges Bank; see Figure 3–1 for location. B, SeaMARC I image shows details of pinnate gullies, thalweg of canyon, and inferred debris flow scars near canyon rim; see Figure 3–6A for location.

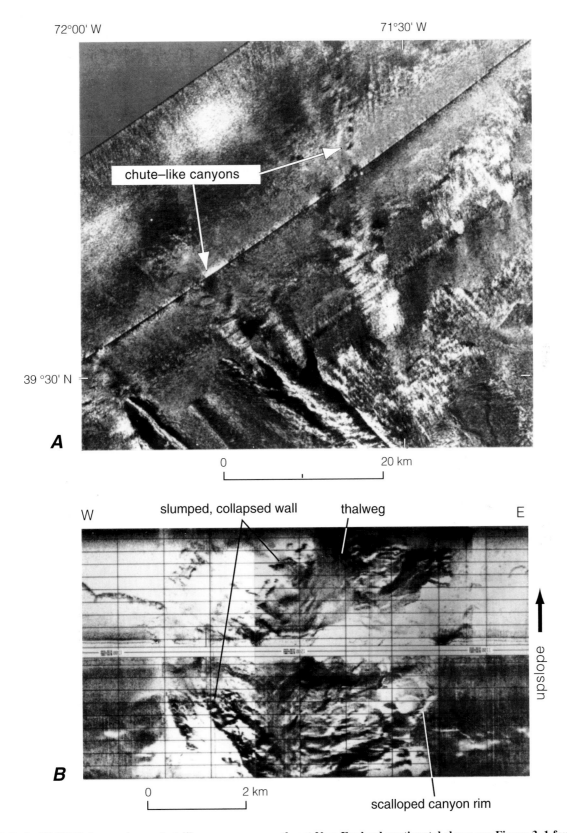

72°00' W   71°30' W

chute–like canyons

39 °30' N

A

0   20 km

W   slumped, collapsed wall   thalweg   E

upslope

B

0   2 km

scalloped canyon rim

Figure 3–7. A, GLORIA image shows chutelike canyons on southeast New England continental slope; see Figure 3–1 for location. B, SeaMARC I image shows slumped walls of chutelike canyon on southeast New England continental slope (upper part of Alvin Canyon shown).

the head zone of the Munson-Nygren slide, together with the flared, coalesced distal debris field, represents a geomorphic entity distinctly different in style from mass movement features that are located off the southeast New England slope west of 69°30′ W (Figure 3–1).

Although the two domains just described are fundamentally distinguished by the overall dominance of a particular style of mass movement versus the overall dominance of a particular style of canyon erosion, the distribution of canyons and mass movement is not spatially exclusive. Several major canyons cut the southeast New England slope in the area dominated by mass movement, and mass movement, as exemplified by the Munson-Nygren slide complex (O'Leary 1986a), may be significant along the canyon-dissected slope of Georges Bank. The differences in style, or the morphological characteristics, of the canyons and mass movements in each compartment or domain are distinctly exclusive.

Canyons cut in the slope along Georges Bank have widely extended pinnate tributaries (Figure 3–6); the canyons along the southeast New England slope do not have well-defined tributary nets but are defined by scalloped or cirquelike reentrants that suggest localized wholesale collapse of wall segments (Figure 3–7). The southeast New England slope canyons appear to be simple chutelike channels separated from each other by flat segments of slope that have typically undergone sliding. Atlantis Canyon, for example, is a scarp-flanked, nearly flat rubble-floored trough having an axial notch of nearly 15 m relief (Figure 3–8).

Although Atlantis and a few other canyons (Nantucket, Alvin, and Block; Figures 3–1 and 3–2) have clearly undergone extensive mass movement (their flanks are generally contiguous slump scars rather than rilled bowls), and although their debris has clearly fed the large debrites on the adjacent rise (O'Leary and Dobson 1992), the form and extent of the canyon-centered mass movement is markedly different from the open-slope mass movement that created the bulk of the debris. The open-slope mass movement is stratigraphically controlled slab sliding whereas the canyon-confined mass movement is localized slumping – possibly a slope adjustment to an initial deep cutting of the canyon axes by an erosive process.

Mass movement along Georges Bank consists primarily of small slumps and restricted debris flows that originated within the rilled headwalls and flanks of the canyons (Figure 3–6B). Only one instance of mass movement not associated with canyon erosion has been detected off Georges Bank, high on the slope (O'Leary 1986b).

The presence of large embayments having profiles similar to that of the Munson-Nygren slide excavation, located along the lower slope farther west (Figure 3–4) and downslope from pinnate canyon thalwegs (O'Leary 1986b; Figure 3–1), suggests that perhaps canyons such as Powell and Lydonia change from narrow downcut forms on the upper slope to broader, more open forms dominated by sidewall collapse on the lower slope. The conspicuous flared debris sheets debouched from Powell and Lydonia Canyons (Figure 3–1) suggest that along the lower slope erosion by channelized downcutting was later replaced by or largely dominated by mass movement and debris flow.

Timing

The Southeast New England slide complex clearly crosscuts pinnate canyons (e.g., Hudson and Veatch Canyons, Figure 3–1), and the Munson-Nygren slide complex crosscuts Munson Canyon (Figures 3–1 and 3–5). The crosscutting relations revealed by the planimetric GLORIA images indicate that the pinnate canyons predate the major mass movements of both the Munson-Nygren slide complex and the Southeast New England slide complex. Both Hudson and Veatch Canyons are partly filled with mass movement debris that has entered the canyons obliquely upon the upper rise (Figure 3–1; O'Leary and Dobson 1992). On the lower slope Veatch Canyon is also choked by debris derived from extensive failure and mass movement of strata adjacent to the canyon (Figure 3–9). This mass movement is not a consequence of canyon incision, however; it is the result of structural failure of a large area of the lower slope chiefly to the east of the canyon axis. Likewise, the incision of Munson Canyon on the upper rise is obliterated by the debris train of the Munson-Nygren slide (Figure 3–1); the source area of the slide complex dwarfs the canyon and is removed from the canyon.

GLORIA images and seismic-reflection profiles show that the mass movement debris lodged in the pinnate canyon channels remains essentially intact. Only in Hudson Canyon was there some reincision apparently by down-channel flow that probably cut through some of the infilling debris (Figure 3–10; O'Leary and Dobson 1992).

GLORIA data suggest that some of the larger pinnate canyons along Georges Bank also were funneling debris onto the rise during the period of major mass movement. For example, GLORIA images indicate that debris debouched from Powell Canyon merges with debris of the Munson-Nygren Slide Complex (Figure 3–1; EEZ-SCAN 87 1991), but resolution is too coarse to determine fine lapping relations that would clarify timing.

The relative age of the chutelike canyons located west of 69°30′ W is more problematic. These canyons have clearly fed the debris accumulation of the Southeast New England slide complex, so they were active at least during the time of latest debris emplacement. Seismic profile data show that the heads of Alvin and Atlantis Canyons are partly filled by Pleistocene strata – likely the same chronostratigraphic unit carried away lower on the slope by mass movement (O'Leary 1986b). Hence the chutelike canyons probably predate the mass movement of the Southeast New England slide complex.

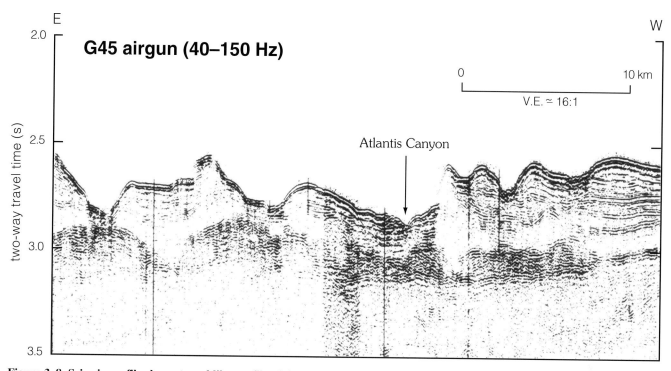

Figure 3–8. Seismic profile shows troughlike profile of Atlantis Canyon (compare with Figure 3–4); see Figure 3–2 for location.

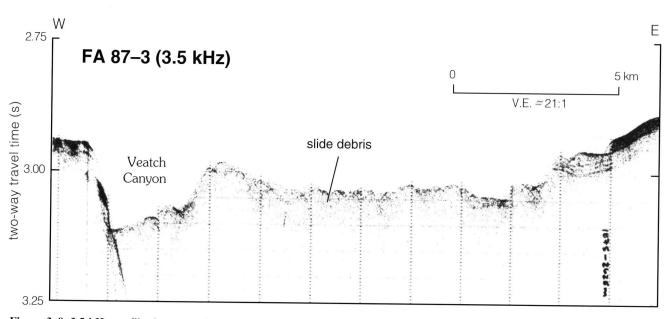

Figure 3–9. 3.5-kHz profile shows proximal debris filling Veatch Canyon on lower slope; see Figure 3–2 for location.

The actual ages of canyon erosion and of mass movement are still unknown. Ice-dropped clasts on the floor of Lydonia and other pinnate canyons suggest that erosion across the lower slope had ceased by approximately 18,000 years ago, when glacial ice was breaking up within or had left the Gulf of Maine.

Discussion

Two destructional processes have dominated in forming the present morphology of the New England continental slope: canyon cutting and mass movement. GLORIA data provide little indication of the probable causes of these processes;

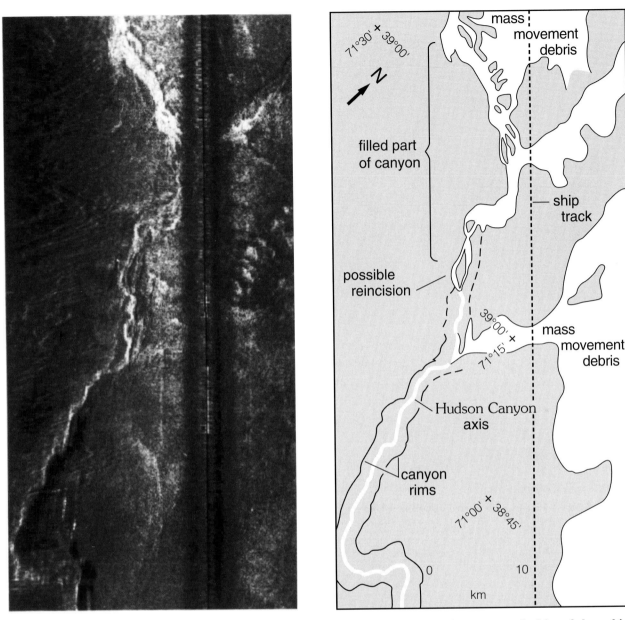

Figure 3–10. GLORIA image shows debris in Hudson Canyon on upper rise and features that suggest reincision of channel in the debris; see Figure 3–1 for location.

we do know, however, that the processes are segregated spatially but closely linked temporally.

The segregation or compartmentalization of process may reflect broad contrasts in form of the continental slope (Scanlon 1984), which in turn may reflect major contrasts in the Pleistocene depositional regime between Georges Bank and the southeastern New England shelf (O'Leary 1986b). The segregation of process on the slope is mirrored by segregation of process on the rise. For example, the amalgamated debrites of the Southeast New England slide complex have aggraded parts of the upper rise, whereas the rise-crossing channels that issue from the rilled slope along Georges Bank (Hudson Canyon not excepted) have eroded

the upper rise, increasing net relief. A feature common to all the rise-crossing channels is the uniformity of incision over distances of up to 100 km or more; none of the canyons of the New England margin have distributed sediment across the upper rise in the manner displayed by fan-building systems such as the Mississippi or Amazon Fans (Flood and Damuth 1987; Twichell et al. 1991). Turbidites, presumably fed through these rise-crossing channels, would issue on the lower rise or on the abyssal plain. The flared debris tongues that issued from Powell and Lydonia Canyons suggest that these canyons' debouched debris flows onto the rise rather than stacked fan lobes that are presumably characteristic of turbidity currents and submarine fan buildup. This is true also of the chutelike canyons to the west, although in their case, deposition has occurred at the foot of the slope rather than out on the rise within or at the mouth of a well-defined channel. The New England margin canyons were conduits for debris flows to the upper rise rather than turbidity currents in the commonly understood sense (i.e., producing overbank or levee deposits and spilling out to form shingled fanlike deposits).

Processes of canyon cutting and mass movement are somehow involved with late Pleistocene environmental conditions as indicated by the spatial association with both Pleistocene depositional form and stratigraphy. Among conditions that might have been significant are the following: (1) lowered sea level that led to clathrate destruction and consequent seafloor destabilization, (2) outwash deposition along the shelf edge and upper slope and resultant diagenetic processes such as compaction dewatering and/or creep, (3) formation of a foreland bulge fronting the last ice sheet that created a steeper upper continental slope than at present, (4) earthquake activity attending isostatic rebound during deglaciation.

The relatively narrow interval of time during which canyon incision and mass movement occurred, the temporal link to the late Pleistocene depositional environment, and the spatial association of process with Pleistocene stratigraphy and slope structure all indicate that a model of slope-rise morphogenesis must proceed from these conditions. Do presently accepted models of slope morphogenesis accord with these conditions?

Most models of submarine canyon formation employ the concept of runaway evolution: Once a process is initiated, a cascading sequence of events occurs until the complex of processes is played out or various controlling factors are modified by exogenous events. For example, canyons eroding headward by slumping are supposed to intercept longshore currents that thereupon amplify the process by providing sediment to the canyon heads, which generates erosive turbidity currents (Posamentier, Erskine, and Mitchum, Jr. 1991; Galloway, Dingus, and Paige 1991). This concept of canyon evolution leads to two seemingly incompatible processes: (1) erosion and deepening of the canyon axis and (2) filling and burial of the canyon. Along the New England slope, once the initiating process ceased and the late-glacial to early-postglacial energy dynamics dissipated, the canyon cutting and mass movement ended; canyon cutting did not proceed to an old age stage (in this case represented by infilling of the canyon from the rise up; Galloway et al. 1991). Accordingly, the submarine canyons are static forms – not necessarily relict – because they can be reactivated by an event, although it is likely to require particular environmental conditions to reinstitute such an event. There is thus no positive feedback, nor are the processes repetitive or cyclic, nor is channelized erosion by adventitious incorporation of extraneous processes a consequence of mass movement.

An appropriate model for the New England continental slope and rise might be the slope/apron system at or near the "mud-rich" apron end member of Nelson et al. (1991). The example presented by Nelson et al., (the Ebro apron off northeastern Spain) is located at the base of an "unusually steep" (4° to 7°), narrow (10 km) slope cut by relatively steep canyons described as "debris chutes," a setting very similar to the slope adjoining the Southeast New England slide complex. However, Nelson et al. (1991) infer the source of the Ebro apron as exclusively canyon-related (numerous mass movements marked by "multievent slide scars along an extensive (~100 km) canyon perimeter" (p. 184). Nelson et al. (1991) thereby distinguish aprons from large sheets of chaotic debris or "mass-transport beds" that result from single catastrophic failures on basin slopes. The distinction seems academic. In fact, Nelson et al. (1991) note that mass-transport beds and aprons are associated on passive margins. Although the Southeast New England slide complex certainly incorporates large "mass-transport beds," these are not single-event deposits; they represent multiple failures, and they are mixed with canyon-debouched flow deposits of the type that generated the Ebro apron. Therefore, I feel that the Southeast New England and the Munson-Nygren slide complexes (including the canyons that provided ancillary debris) created aprons, but perhaps aprons in an expanded sense that includes mass-transport beds. The mass-transport beds or debris sheets of the New England rise are probably very similar in gross composition and texture to the mega-debris flow deposits of the Oligo-Miocene Pindos foreland basin, Greece, described by Leigh and Hartley (1992).

Nelson et al. (1991) note that mud-rich aprons are most common on passive margins associated with prograding river deltas. They also note that in this environment slope failure is a fundamental process leading to sediment transport and canyon development. During late Pleistocene deglaciation, the New England margin was a depositional environment characterized by coalescing outwash deltas along the outer shelf and upper slope (O'Leary 1988). Failure of these deposits along the Southeast New England slope and dewatering and sapping of the deposits along Georges Bank led to the domainal distribution of landforms just described. Sequential open slope mass movement and local involvement of slope canyons led to formation of what is arguably a base-of-slope siliciclastic apron system. In this instance the character of the stratigraphy and the structure of the slope seem to have exerted a strong influence on the character and timing of landform development and on processes of sediment delivery to the rise.

Conclusions

The GLORIA data provide the information needed to work out the succession of events based on crosscutting relations among canyons and mass movement features of the U.S. Atlantic continental slope and rise. Although this information gives only relative timing, other data become more intelligible in light of the relations the GLORIA data reveal and clarify reasons for the observed spatial relations and timing of events. Possibly the most significant revelation of the New England GLORIA data is that large-scale, open slope mass movement is fundamentally unrelated to canyon formation, both spatially and temporally. However, both canyon erosion and mass movement along the New England margin share a common relationship with conditions that existed during a relatively short time interval in late Pleistocene time.

GLORIA data suggest that mass movement and canyon cutting along the New England margin are postdepositional and successive rather than cyclic and interdependent and that they are independent of sea level or depositional rates along the outer shelf. Debris of mass movement that intercepted canyons that had formed earlier did not form turbidity currents and did not utilize the intercepted channels as conduits to the lower rise. Canyon activity here did not result in significant aggradation or fan construction along the rise. Given these observations, it is clear that concepts of canyon formation and mass movement commonly applied to the U.S. Atlantic margin are in need of revision. Models that are in current vogue, including the stratigraphic sequence concept that depends heavily on conventional notions of canyon process (Vail et al. 1977), are based primarily on seismic-reflection profiles and well log data, supported by outcrops of presumably ancient analogs (e.g., Reineck and Singh 1980; Galloway et al. 1991; Posamentier et al. 1991). GLORIA provides the planimetric dimension missing in these models. A new model of canyon evolution and activity is required that takes the GLORIA data into account. At present, the mud-rich siliciclastic apron model of Nelson et al. (1991) seems best to explain the distribution of Pleistocene landforms and deposits along the New England slope and rise.

References

Coleman, J. M., Prior, D. B., and Lindsay, J. F. 1983. Deltaic influences on shelf-edge instability processes. In *The Shelfbreak: Critical Interface on Continental Margins.* eds. D. J. Stanley and G. T. Moore, pp. 121–37. Tulsa, OK: Society of Economic and Paleontologic Mineralogists Special Publication 33.

EEZ-SCAN 87 Scientific Staff 1991. Atlas of the U.S. Exclusive Economic Zone, Atlantic continental margin. *U.S. Geological Survey Miscellaneous Investigations Series I-2054,* 174 p., scale 1:500,000. Reston, Va.: USGS.

Farre, J. A., McGregor, B. A., Ryan, W. B. F., and Robb, J. M. 1983. Breaching the shelf-break: Passage from youthful to mature phase in submarine canyon evolution. In *The Shelf Break: Critical Interface on Continental Margins.* eds. D. J. Stanley and G. T. Moore, pp. 25–39. Tulsa, OK: Society of Economic Paleontologists and Mineralogists Special Publication 33.

Flood, R. D., and Damuth, J. E. 1987. Quantitative characteristics of sinuous distributary channels on the Amazon deep-sea fan. *Geol. Soc. Am. Bull.* 98: 728–38.

Galloway, W. E., Dingus, W. F., and Paige, R. E. 1991. Seismic and depositional facies of Paleocene-Eocene Wilcox Group submarine canyon fills, northwest Gulf coast, U.S.A. In *Seismic Facies and Sedimentary Processes of Submarine Fans and Turbidite Systems,* eds. P. Weimer and M. H. Link, pp. 247–70. New York: Springer-Verlag.

Leigh, S. and Hartley, A. J. 1992. Mega-debris flow deposits from the Oligo-Miocene Pindos foreland basin, western mainland Greece: Implications for transport mechanisms in ancient deep marine basins. *Sedimentology* 39: 1003–12.

Nelson, C. H., Maldonado, A., Barber, J. H., Jr., and Belen, A. 1991. Modern sand-rich and mud-rich siliciclastic aprons: Alternative base-of-slope turbidite systems to submarine fans. In *Seismic Facies and Sedimentary Processes of Submarine Fans and Turbidite Systems,* P. Weimer and M. H. Link, pp. 171–90. New York: Springer-Verlag.

O'Leary, D. W. 1986a. The Munson-Nygren slide, a major lower-slope slide off Georges Bank. *Mar. Geol.* 72: 101–14.

——— 1986b. Seismic structure and stratigraphy of the New England continental slope and the evidence for slope instability. *USGS Open-File Report 86-118,* 181 p. Reston, Va.: USGS.

——— 1988. Shallow stratigraphy of the New England continental margin. *U.S. Geological Survey Bulletin 1767,* 40 p. Reston, Va.: USGS.

O'Leary, D. W., and Dobson, M. R. 1992. Southeastern New England continental rise: Origin and history of slide complexes. In *Geologic Evolution of Atlantic Continental Rises.* eds. C. W. Poag and P. C. de Graciansky, pp. 214–65. New York: Van Nostrand Reinhold.

O'Leary, D. W. 1993. Submarine mass movement, a formative process of passive continental margins: The Munson-Nygren landslide complex and the Southeast New England landslide complex. In *Submarine Landslides: Selected Studies in the U.S. Exclusive Economic Zone,* eds. W. C. Schwab, H. J. Lee, and D. C. Twichell, pp. 23–39. U.S. Geological Survey Bulletin 2002. Reston, Va.: USGS.

Posamentier, H. W., Erskine, R. D., and Mitchum, R. M., Jr. 1991. Models for submarine-fan deposition within a sequence-stratigraphic framework. In *Seismic Facies and Sedimentary Processes of Submarine Fans and Turbidite Systems,* eds. P. Weimer and M. H. Link, pp. 127–36. New York: Springer-Verlag.

Reineck, H. E., and Singh, I. B. 1980. *Depositional Sedimentary Environments.* Heidelberg: Springer-Verlag. 549 p.

Scanlon, K. M. 1984. The continental slope of New England: A long-range sidescan-sonar perspective. *Geo-Mar. Ltrs.* 4: 1–4.

Twichell, D. C., and Roberts, D. G. 1982. Morphology, distribution, and development of submarine canyons on the United States Atlantic continental slope between Hudson and Baltimore Canyons. *Geology* 10: 408–12.

Twichell, D. C., Kenyon, N. H., Parson, L. M., and McGregor, B. A. 1991. Depositional patterns of the Mississippi Fan surface: Evidence from GLORIA II and high-resolution seismic profiles. In *Seismic Facies and Sedimentary Processes of Submarine Fans and Turbidite Systems.* eds. P. Weimer and M. H. Link, pp. 349–63. New York: Springer-Verlag.

Vail, P. R., Mitchum, R. M., and Thompson, S. 1977. Seismic stratigraphy and global changes of sea level. Part 4: Global cycles of relative changes of sea level. In *Seismic stratigraphy, application to hydrogen exploration,* ed. C. E. Payton, pp. 932–66. *Am. Assoc. Petrol. Geol. Bull.* 62.

4 Characteristics of the continental slope and rise off North Carolina from GLORIA and seismic-reflection data: The interaction of downslope and contour current processes

Peter Popenoe and William P. Dillon
U.S. Geological Survey, Woods Hole, Massachusetts

Introduction

In 1987, the U.S. Geological Survey (USGS) completed fieldwork on a Geologic LOng-Range Inclined Asdic (GLORIA) sidescan sonar survey of the U.S. Atlantic Exclusive Economic Zone (EEZ-SCAN 87 Scientific Staff 1991). The resulting GLORIA mosaic image from this survey covered the continental slope and rise off North Carolina, one of the best-studied and most intensely sampled margins of the world. Although a great number of studies have been conducted in this area, the GLORIA image presented the first truly integrated picture of the North Carolina margin that depicts the interplay of three different styles of slope and rise processes: (1) gravity-controlled processes including canyon, channel, and fan systems; (2) mass wasting processes including slumps, slides, and debris flows; and (3) contour current-controlled processes including the deposition of large sediment drifts and the scouring of large areas of seafloor. This chapter examines the GLORIA sidescan sonar image and the USGS seismic records to review the interplay of these systems in shaping the slope and continental rise of the segment of the margin off North Carolina. It reexamines the major rise features such as the canyons and debris slides and comments on previous studies that have been based on the interpretation of 3.5-kHz echo character.

Regional setting

Three major oceanic current systems affect the area of our study (Figure 4–1). First, north of Cape Hatteras the continental shelf is dominated by south-trending water drift that during winter is punctuated by intense southward pulses associated with storms (Bumpus 1973; Hunt, Swift, and Palmer 1977). The south-trending drift tends to turn northward at Cape Hatteras and be entrained in the north-flowing Gulf Stream (Bumpus 1973). The net effect of this drift on sedimentation is to sweep sediment southward from the

mid-Atlantic bight to the narrow shelf at Cape Hatteras (Hunt et al. 1977) and thence to be advected into the Gulf Stream or swept over the shelf edge.

Second, the northeast-flowing Gulf Stream sweeps across the Blake Plateau south of Cape Hatteras, effectively blocking terrestrial sedimentation to the outer continental shelf (Paull and Dillon 1985). This blocking of sedimentation to the outer shelf has created the sediment-starved, intermediate-depth Blake Plateau, which separates the continental slope into two components: the Florida-Hatteras Slope west of the current, and the lower continental slope and Blake Escarpment east of the current. At Cape Hatteras the Gulf Stream leaves the Blake Plateau and diverges from the continental slope northeastward across the continental rise. The Gulf Stream has been delivering sediment to the continental slope off the Cape Hatteras area nearly continuously since its initiation in the early Tertiary (Paull and Dillon 1985; Popenoe 1985; and references therein). Sediments are swept northward by the Gulf Stream to the oblique crossing of the current with the continental slope south of Cape Hatteras, where they are swept onto the continental slope and rise.

The third major current system is the deep Western Boundary Undercurrent (WBUC), which flows southwestward across the continental rise north of Cape Hatteras (McCave and Tucholke 1986). The WBUC is the dominant current on the lower slope and rise (McCave and Tucholke 1986), and much of the morphology of the continental slope and rise is due to this current. The WBUC impinges on the continental slope between Cape Lookout and Cape Hatteras, where it flows beneath the Gulf Stream before being diverted offshore along the northeast flank of the Blake Outer Ridge (Heezen, Hollister, and Ruddiman 1966). Measured near-bottom current velocities directly off Cape Hatteras (Betzer, Richardson, and Zimmerman 1974; Richardson 1977) near the base of the continental slope (2575 to 2800 m) show velocities at 100 m above the bottom range up to 47 cm/s ($\sim$ 1 kt). Evidence for the erosion and southward transport of sediments by this current is furnished by both the internal structure of the Blake Outer Ridge and by foraminifers of Cretaceous age found in the sediments of the ridge that can only

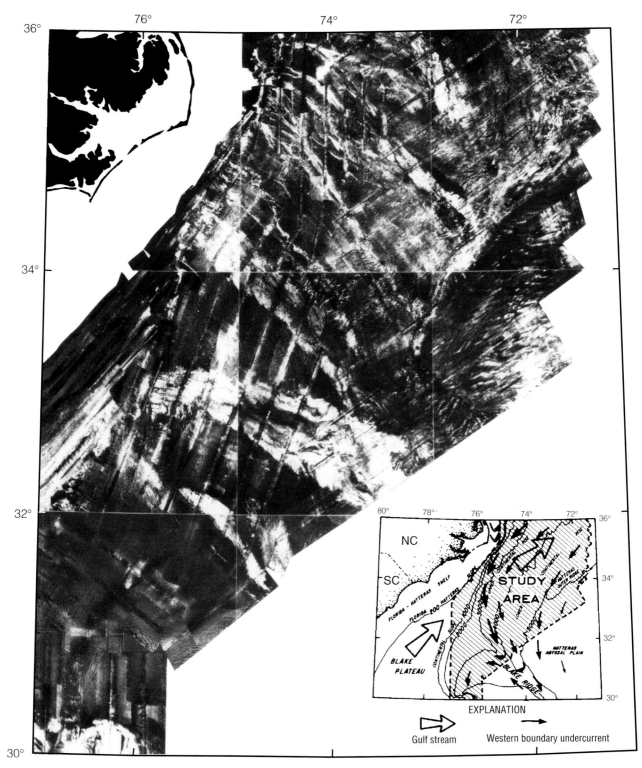

Figure 4–1. GLORIA mosaic of the continental slope and rise off North Carolina with index map.

have originally come from the continental slope at Cape Hatteras or from exposures on the Coastal Plain (Hollister and Heezen 1972).

Although the slope and rise off Cape Hatteras have received high sedimentation at the merging of the Gulf Stream and south-trending shelf currents throughout the Tertiary, the slope here is one of the steepest on the Atlantic margin, averaging 16° between Cape Hatteras and Cape Lookout but containing near-vertical cliffs. The steepness of the slope in this area and the presence of the Hatteras Canyon system has been attributed to a constant supply of sediment available for turbidity current flows (Newton and Pilkey 1969). In this area seismic profiles show sharp truncation of strata on the middle and lower slope and large debris piles of slumped material derived from the slope resting on the upper rise (Popenoe 1985). Camera transects show that the bottom at the base of the slope is composed largely of clay-gravel, a winnowed residual from upslope slumping. Seismic stratigraphic studies of the shelf and slope (Popenoe 1985) indicate that the continental slope at Cape Hatteras may have been eroded back as much as 20 km from its middle Miocene position.

The strength of the WBUC diminishes north of Cape Hatteras where the slope is less steep (averaging only about 8° north of 35°45′ N lat.). The slope here becomes progressively mantled northward by post-Miocene deposits that have been deeply incised by the canyons and mass wasting failures.

South of Cape Lookout, the slope becomes gentle (<1°) as it merges with the Blake Outer Ridge. Seismic records and sediment cores show (Menzies, George, and Rowe 1973; Fitchko 1976; Popenoe 1985; Snyder et al. 1993) the upper slope here is thickly mantled by post-Miocene strata and that the surface of the slope is draped with finely layered Holocene muds. Lower on the slope, between 3,000 and 4,000 m depth on the north flank of the Blake Outer Ridge, the seafloor is steeper and highly scoured by the WBUC, and older strata are exposed.

Methods

This study of the Carolina continental margin is based mainly on the interpretation of the GLORIA sidescan sonar mosaic of the U.S. Exclusive Economic Zone (EEZ-SCAN 87 Scientific Staff 1991) and on seismic-reflection records gathered during the collection of the GLORIA data. Fieldwork for this survey was carried out by the USGS in cooperation with the British Institute of Oceanographic Sciences in 1987 from the R/V *Farnella,* and the resulting GLORIA mosaic is shown in Figure 4–1. Figure 4–2 shows an interpretation of these data based on the correlation of features seen on the mosaic with seismic-reflection data gathered both during the GLORIA survey (3.5-kHz echo sounder and two-channel seismic reflection) and on previous USGS cruises (both single-channel and multichannel seismic reflection and mid-range sidescan sonar). Comparison of our interpretation with other published studies of features and processes of sedimentation on the North Carolina slope and rise based on 3.5-kHz echo character, piston cores, and other methods (for instance, Vassallo, Jacobi, and Shor 1984; Laine, Damuth, and Jacobi 1986; Embley and Jacobi 1986; Pratson and Laine 1989) was made after completion of our map. Figure 4–3 indicates the locations of figures and features discussed in the text and shows our geophysical trackline coverage within the study area.

The GLORIA sidescan sonar system creates an image of the seafloor that shows the strength of acoustic backscatter (echo strength) of sound pulses broadcast from a vehicle (fish) towed behind the survey ship. The backscatter intensity, or bottom return strength, is expressed on the GLORIA image as tonal variations; strong echos produce light tones on the image, and weak echos produce dark tones on the image. These variations in backscatter result from three factors: (1) the induration or hardness of the seafloor; (2) the topography of the seafloor, or angle of incidence of the seafloor to the transmitter/detector; and (3) the roughness or texture of the seafloor. Each pixel that makes up the GLORIA mosaic is the average return from an area about 125 m along track and 50 m across track, so features smaller than about the size of a football field are not discernable.

An important consideration in viewing the GLORIA mosaic is that the GLORIA signal penetrates the seafloor to varied subbottom depths depending on the type of subbottom material. As an example, piston cores taken within the Cape Fear slide track (Embley 1980) and from the area of the Hatteras Cone (Cleary, Pilkey, and Ayers 1977) show that these features are covered by up to 2.5 m of stratified clay (drape) that probably postdates the debris flow or fan complex, yet the GLORIA system clearly images details of the debris flow and fan. Gardner et al. (1991) found in modeling the GLORIA 6.5-kHz sidescan sonograph signal that the sonar energy is refracted into the sediment to depths of at least a few meters rather than scattered from the surface. Comparisons of backscatter intensity with piston core lithology showed that sand deposits with thin interbeds of silty clay produced lower backscatter than silty clay deposits with thin interbeds of sand. Thus, the cause of variations in backscatter intensity may correlate with lithostratigraphic variability that produces constructive or destructive interference from subsurface inhomogeneities.

In Figure 4–1, some of the strongest returns (white to light gray tones) are associated with submarine mass movement scoured bottom and associated debris flows, and the floors and walls of submarine canyons. These strong areas of backscatter empirically result from the predominance of sand and rubble in the debris flows that contrasts with the much weaker return (dark gray to black) from the more homogeneous soft, layered hemipelagic mud that dominates intercanyon areas or areas of active deposition. Strong returns are also produced by outcropping hard strata exposed by slumping or scour (as on the north-facing flank of the Blake Outer Ridge, the west-facing flank of the Hatteras Outer Ridge, and along the lower continental slope).

Canyon, channel, and fan systems

The GLORIA data were particularly effective in tracing the submarine valleys that cross the upper rise off North Carolina because of the light tones produced by valley floors

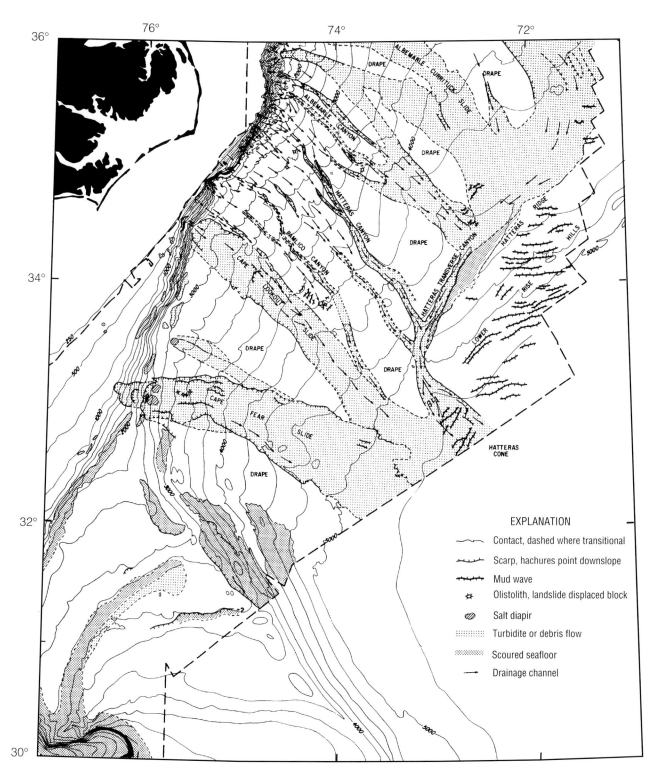

Figure 4–2. Interpretive map of the continental slope and rise off North Carolina based on the GLORIA mosaic.

and their deeply incised valleys. The canyons are less easily traced downslope across the lower rise because of lower channel relief relative to the divides and the mixing with turbidite and mass wasting debris. The channel system off North Carolina consists of the Hatteras Transverse Canyon on the lower rise and three major tributaries: Pamlico, Hatteras, and Albemarle Canyons (Rona, Schneider, and Heezen 1967; Newton and Pilkey 1969; Newton, Pilkey, and Blanton 1971), as well as a number of unnamed gullies. Downslope-trending lineations seen in the GLORIA image of the At-

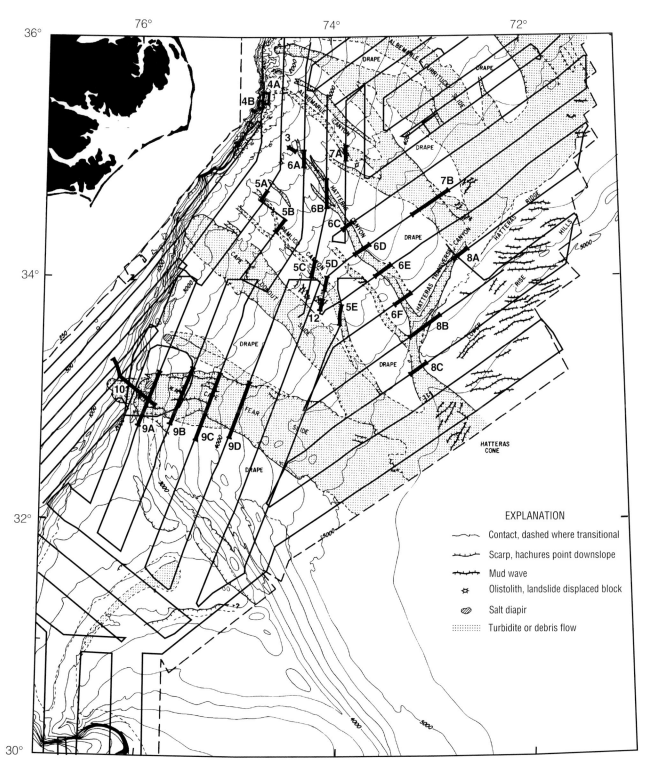

Figure 4–3. Generalized interpretive map showing locations of GLORIA tracklines, figures, and features discussed in text.

lantic margin (Schlee and Robb 1991; Schlee et al. 1992) and Pratson and Laine's map (1989) of seafloor echo character show that Norfolk, Washington, and possibly Baltimore Canyons off the middle Atlantic also feed sediment into upper Hatteras Transverse Canyon. Since most of these features are out of our study area and are very poorly defined on the lower continental rise, we do not include them in this discussion.

The canyons begin on the continental slope as steep pinnate to dendritic chutes with both rounded and V-shaped

Figure 4–4. Mid-range sidescan sonar images of the continental slope north of Cape Hatteras. Locations are shown on Figure 4–3. A shows a segment of the continental slope near lat. 35°30′ N, long. 74°45′ W that is incised with steep-walled pinnate chutes separated by sharp divides. The largest chute shown above the word *canyon* is over 500 m deep. Because the image is uncorrected for slant range, a profile of the bottom (marked *bottom* on photo) is shown on both sides of a centerline. The image is reverse-printed so that reflecting surfaces are white and shadows are dark similar to the GLORIA image.

bottoms (Figure 4–4). Off North Carolina, none of the canyons appear to indent the shelf edge deeply, as is common with the larger canyons to the north. While some of these chutes are separated by steep, sharp divides (Figure 4–4A), others are separated by large remnants of undissected slope that is draped by hemipelagic sediment. In Figure 4–2, forty of the larger chutes and gullies that feed into the Hatteras Canyon system were mapped us-

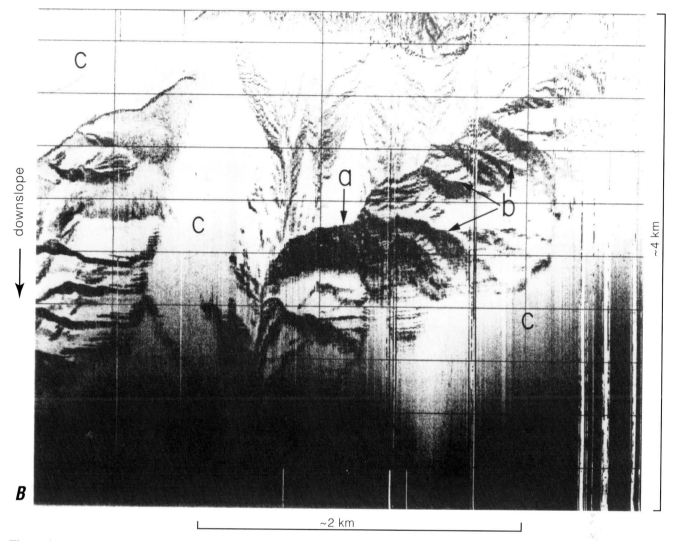

downslope

~4 km

~2 km

Figure 4–4. *Continued.* **B is a downslope view of a more dendritic feeder canyon. This image is printed so that reflecting surfaces are dark and shadows are light. The dark area at the bottom of the image is at the limit of the sidescan range. This canyon (*a*) is cut into an area of undissected slope that forms a smooth bottom between tributaries (*c*); (*b*) marks a feeder chute into the main canyon.**

ing the GLORIA image and our seismic-reflection records.

On the upper rise, which averages less than 1/2° in slope, the chutes and tributaries merge into valleys, the largest of which are Pamlico, Hatteras, and the Albemarle canyon drainage. The term *canyon* is misleading, since these valleys are deeply eroded only on the upper rise but broaden and shallow toward the lower rise. The valleys cross the rise along relatively straight southeast trends until they encounter the Hatteras Transverse Canyon, where they are deflected southwestward. The major contribution of the GLORIA survey (Figure 4–1) in defining the canyons is that we are able to trace the canyons and drainages across the rise and map them in their correct positions, which can differ by many kilometers from the positions of earlier surveys. For instance, the echo character map of Pratson and Laine (1989)

presented an excellent representation of the location and tributaries of Hatteras Canyon, but Pamlico Canyon and the upper tributaries of the Albemarle Canyon drainage were mapped incorrectly.

Pamlico Canyon

Of the three major canyons within our study area, the southernmost canyon, Pamlico Canyon, has the most clearly defined and deepest valley on the continental slope and the upper rise. Our data indicate that at 600 m water depth (near 34°56′ N lat, 75°13′ W long) the valley of Pamlico Canyon is incised 400 m into the slope and it is ~1 km across. This deep valley was not mapped on the Russell Bathymetric Map (National Ocean Survey 1981), but on the GLORIA image it is clearly present; it is so deep that it created a distortion, clearly visible on the image (football-shaped), that we were not able to remove with our geometric (slant-range) correction program (Chavez 1986).

On the upper rise at 3,050 m depth, Pamlico Canyon is about 5 km across and has cut a 225-m-deep, flat-floored valley into the seafloor. The canyon has an asymmetric profile with its southern bank steeper and crowned by a 40-m-high levee (Figure 4–5A). The northern bank is stepped where layered strata are truncated by erosion. As Pamlico Canyon crosses the rise, it broadens slightly and has less relief, so that at 3,400-m depth (Figure 4–5B) it has a more symmetrical 200-m-deep U-shaped valley, while at 3,900-m depth it is only about 100 m deep relative to the surrounding seafloor (Figures 4–5C and 4–5D) and forms a broad, shallow valley. Near 4000-m depth (Figure 4–5E), the canyon is difficult to trace either seismically or on the GLORIA image, because its previous depression has been filled with seismically transparent sediment, possibly debris from the Cape Lookout or an earlier sediment slide (discussed later). Still farther downslope the canyon funnels into the slide track of the Cape Lookout Slide before exiting the rise directly onto the Hatteras Cone (Figure 4–2).

Hatteras Canyon

Hatteras Canyon (Figure 4–6), the central canyon on Figure 4–2, has a very different profile than that of Pamlico Canyon. Hatteras Canyon drains a larger area of the continental slope causing several conspicuous and prominent tributaries to flow into the canyon on the upper rise. The canyon is also broader in cross section and shallower relative to the surrounding seafloor (Figures 4–6A and 4–8B), suggesting that the valley of the canyon both originated as a submarine canyon cut by turbidity flows and has acted as a debris chute for upslope mass wasting. This conclusion is supported by visible slide tracks on the upper rise, slide-scoured surfaces associated with the canyon (Figure 4–6A), and blocky rubble on the canyon floor (Figure 4–6B). On the lower rise the canyon has a rough, hummocky bottom suggesting blocky debris from upslope failure (Figures 4–6D and 4–6E). On the middle and lower rise, the general depression that forms Hatteras Canyon appears to be incised by a steep-walled, 100-m-deep, 1/2-km-wide inner canyon that leads into the Hatteras Transverse Canyon. This inner canyon is speculative, because it is too narrow to be well displayed by our data, but it may adjust the grade of the canyon floor (4,550 m depth on Figure 4–6F) to the 5,000-m depth of the floor of Hatteras Transverse Canyon (Figure 4–8B) where these two drainages intersect 35 km downslope.

Albemarle Canyon

What is called Albemarle Canyon, or Albemarle Transverse Canyon (Newton and Pilkey 1969), is actually three paral-

lel drainages, each separated by about 20 km (Figure 4–2), that occupy a broad depression on the upper rise. On the upper rise the valleys produce bright acoustic returns and are easily traced by both GLORIA data (see EEZ-SCAN 87 Scientific Staff 1991, p. 39) and by seismic-reflection data (Figure 4–7A). On the mid-rise these canyons lose their distinction when they are joined by the track of a 15-km-wide submarine sediment slide that originated at the base of the continental slope near 35°55′ N latitude. On the lower rise the combined former canyons and landslide track form a broad, shallow topographic depression that funnels debris into the Hatteras Transverse Canyon (Figure 4–7B). Within the bathymetric low, the three canyons continue only as small, shallow channels that are faintly visible on the GLORIA image (marked by arrows on Figure 4–2).

The lower reaches of both Albemarle and Hatteras Canyons display alternating light and dark stripes on the GLORIA image that are transverse to the canyon orientation (see EEZ-SCAN 87 Scientific Staff 1991, p. 41). This transverse pattern of stripes appears to be characteristic of depositional fans, because similar patterns are seen farther downslope on the Hatteras Cone. All of our transverse lines are oriented parallel to these features, so are poorly oriented to confirm their morphology seismically, but we believe that they represent low-relief, transverse sediment waves oriented perpendicular to the direction of down-canyon turbidity flow. On our original interpretation published in the GLORIA atlas (EEZ-SCAN 87 Scientific Staff 1991) these transverse wavelike features were interpreted as evidence of sediment waves constructed by turbidity currents on the surface of turbidite-debrite fans located where the canyons lose their relief on the lower rise. Jacobi and Hayes (1992) reached a similar conclusion for the source of bright and dark reflectivity stripes seen on GLORIA images off NW Africa, but the orientation of their crestlines was downslope.

Hatteras Transverse Canyon

Hatteras Transverse Canyon (Figure 4–2) is so named because the canyon trends nearly parallel to regional isobaths and nearly perpendicular to the downslope orientation of other canyons of the Atlantic margin (Rona et al. 1967). Echo character mapping and linearities seen on the GLORIA image off the mid-Atlantic (Pratson and Laine 1989; Schlee and Robb 1991; Schlee et al. 1992) indicate that the Hatteras Transverse Canyon is a conduit for mass movement debris and turbidite from as far north as Norfolk, Washington, and Baltimore Canyons, as well as Albemarle and Hatteras Canyons off North Carolina. Where debris flows and channelways originating on the upper rise off Virginia enter our study area near 36° N lat, 72° W long, the conduit is represented only by a broad depression whose edges are marked by minor scarps (Figure 4–2), but this depression must act as a shallow scupper to channel turbidity and mass movement debris from a large area of the mid-

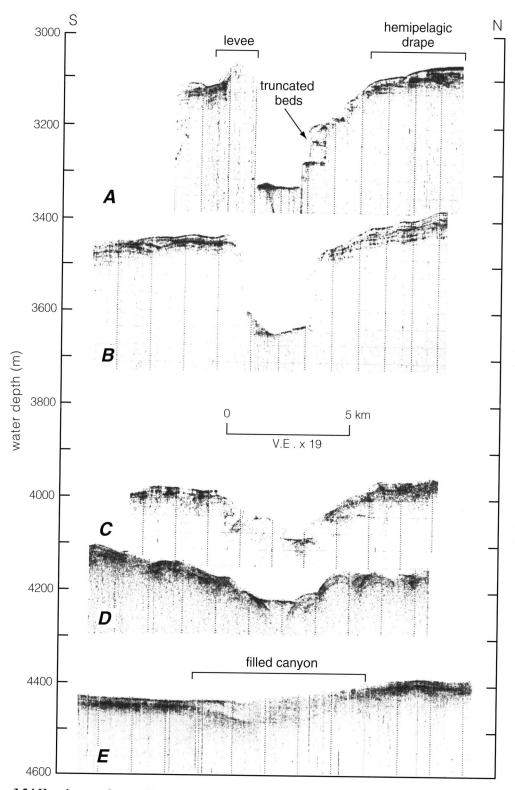

Figure 4–5. Five 3.5-kHz echo sounder profiles across Pamlico Canyon. Locations are shown on Figure 4–3. Profiles are arranged by depth with the shallowest crossings at the top and the deepest crossings at the bottom.

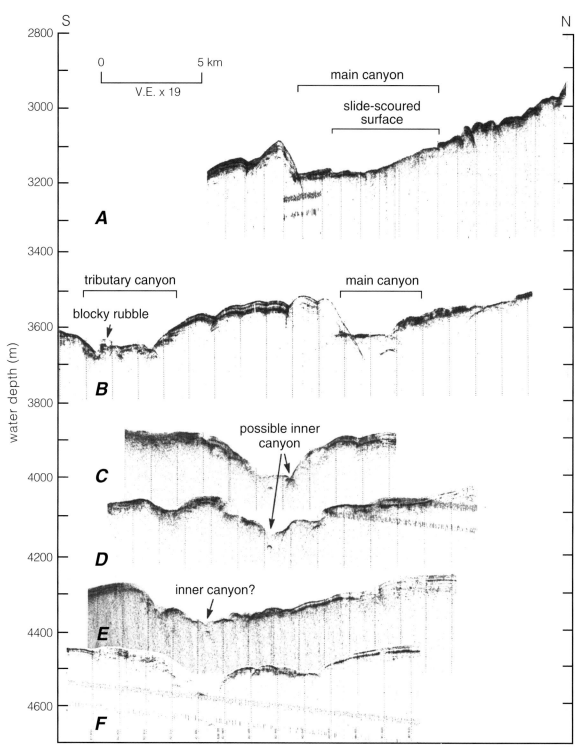

S   N

0   5 km

V.E. x 19

main canyon

slide-scoured surface

A

tributary canyon

blocky rubble

main canyon

B

possible inner canyon

C

D

inner canyon?

E

F

water depth (m)

2800 — 3000 — 3200 — 3400 — 3600 — 3800 — 4000 — 4200 — 4400 — 4600

Figure 4–6. Six 3.5-kHz echo sounder profiles across Hatteras Canyon. Locations are shown on Figure 4–3. Profiles are arranged by depth with the shallowest crossings at the top and the deepest crossings at the bottom.

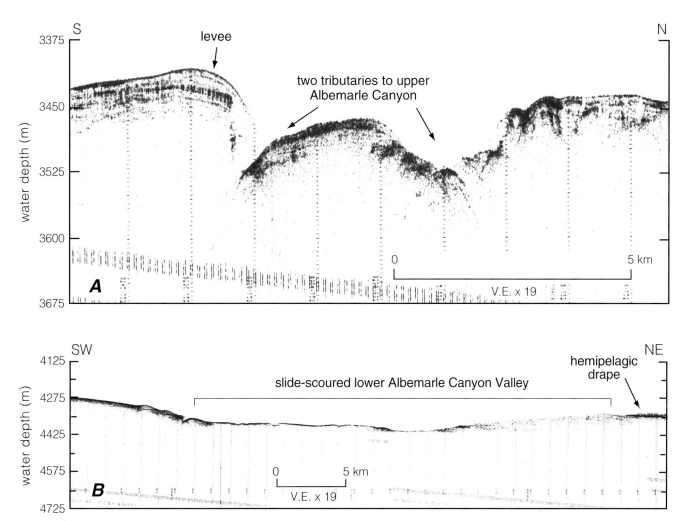

Figure 4–7. Two 3.5-kHz echo sounder profiles across upper and lower Albemarle Canyon. Locations are shown on Figure 4–3.

Atlantic region southward to the Hatteras Cone. The turbidites and debris flows that cross this area are evident on the GLORIA image (Figure 4–1) as a broad zone of downslope-trending streaks that lead into the upper part of Hatteras Transverse Canyon.

Although the Hatteras Transverse Canyon is poorly defined morphologically off Virginia, off North Carolina it has cut a deep channel where it is bordered by the Hatteras Outer Ridge. Here, over half of the relief on the west side of the Hatteras Outer Ridge is not depositional but due to the downcutting of the Hatteras Transverse Canyon. Northward, the west side of the Hatteras Outer Ridge is buried by ponded turbidites. Near its intersection with Albemarle Canyon (Figure 4–8A), the channel is narrow (~3 km) and has an irregular floor about 50 m deeper than the surrounding seafloor. Figure 4–8B shows a diagonal crossing of the channel near its intersection with Hatteras Canyon. Here, the channel is 350 m deep relative to its eastern edge and has a flat floor characterized by a strong echo, probably indicative of a hard sand bottom or debris flow deposits (Laine et

al. 1986), while the sides of the canyon show horizontal strata that have been truncated by the downcutting of the canyon. Farther downslope (Figures 4–2 and 4–8C), the channel constricts to about 5 km width, is 250 m below the adjacent seafloor, and bends southeastward toward the Hatteras Cone.

The Hatteras Cone

On the Hatteras Cone, the surface of the submarine fan is characterized on the GLORIA image by large stripes of bright and dark acoustic return that are oriented transverse to the fan. These stripes appear to represent transverse sediment waves, deposited by turbidity flow, because they look nothing like the distributary channels seen by GLORIA on the Mississippi Fan (Twichell et al. 1991). The stripes may represent low-relief wave crests, or alternatively, may reflect textural differences between troughs and crests.

Published studies of the Hatteras Cone (Cleary 1974; Cleary et al. 1977; Pilkey and Cleary 1986) have mapped

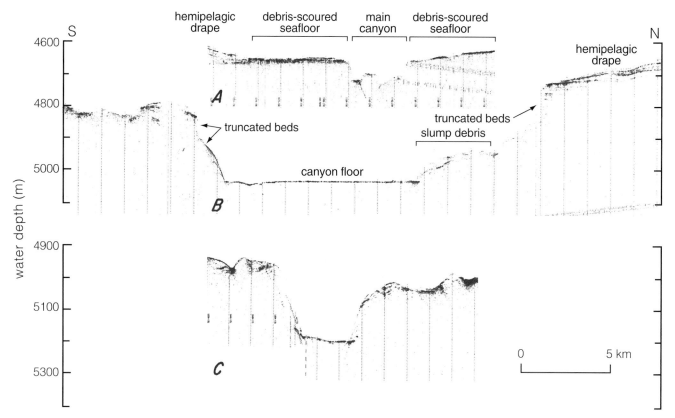

Figure 4–8. Three 3.5-kHz echo sounder profiles across Hatteras Transverse Canyon. Locations are shown on Figure 4–3.

the fan surface as an anastomosed system of braided channels separated by levees of varying relief. We have compared the GLORIA mosaic over the fan with the interpretation of Cleary et al. (1977) and, aside from the tracing of the main three channels, we find little correspondence of the system of mapped channels with the stripes seen on the GLORIA image. Cleary et al. (1977) state that as relief on the fan is low, the fan morphology was extremely difficult to interpret. We would interpret the GLORIA image to indicate that an anastomosing channel system is present but that it is dissecting sediment waves related to turbidity flow, rather than constructing levee systems.

Although Cleary et al. (1977) thought that the Hatteras Fan was extinct, based on the apparent absence of recent flows and the present importance of pelagic sedimentation, parts of the Hatteras Canyon system are apparently intermittently active at the present time. A turbidity flow may have been the cause of a submarine telephone cable break on the upper rise off Cape Hatteras in 1987 (Armand Silva, personal communication 1989); however, this turbidite surge may not have made it down-canyon to the Hatteras Cone.

Slumps, slides, and debris flows

Three major mass wasting complexes associated with large slope failures dominate the GLORIA sidescan sonar image

of our study area. From north to south the slope failures causing these major debris flows are the Albemarle-Currituck Slide (Bunn and McGregor 1980; Embley 1980; McGregor 1981; Popenoe, Coward, and Cashman 1982; Prior, Doyle, and Neurauter 1986; EEZ-SCAN 87 Scientific Staff 1991), the Cape Lookout Slide (Embley and Jacobi 1986; EEZ-SCAN 87 Scientific Staff 1991), and the Cape Fear Slide (Embley 1980; Carpenter 1981; Popenoe et al. 1982; Cashman and Popenoe 1985; EEZ-SCAN 87 Scientific Staff 1991; Popenoe, Schmuck, and Dillon 1993), named for the adjacent coastal features. In addition to these major slide complexes, several smaller unnamed slide complexes are mapped (Figure 4–2). If the "ill-defined bumps" in the intercanyon areas (discussed later) are related to failures, these failures are probably the most important process on the rise in delivering sediment to the deep sea.

The Albemarle-Currituck Slide

The head of the Albemarle-Currituck slide complex occurs just north of our study area as a series of scars that truncate bedding on the lower continental slope off Albemarle Sound (Bunn and McGregor 1980; McGregor 1981; Popenoe et al. 1982; Prior et al. 1986; EEZ-SCAN 1987 Scientific Staff 1991). The massive debris flow originating from this scar is over 190 km long (Embley and Jacobi 1977, 1986; EEZ SCAN 87 Scientific Staff 1991; Schlee and Robb 1991), entering our study area between 73° and 74° W long at 36° N lat. At this point the debris flow has a width of about 30 km. On the lower rise the debris flow loses its distinction as a separate flow as it merges with a turbidite apron (Vassallo et al. 1984; Pratson and Laine 1989) that flows into the Hatteras Transverse Canyon. It is apparent that Hatteras Transverse Canyon acts as a trap for both channelized and unchannelized flows.

The Cape Fear Slide

One of the largest and best-documented of the Atlantic margin landslides, the Cape Fear Slide (Embley and Jacobi 1977, 1986; Carpenter 1981; Popenoe et al. 1982, 1993; Cashman and Popenoe 1985; EEZ-SCAN 87 Scientific Staff 1991), occurs at 33° N lat, directly east of Cape Fear within our study area. The slide begins in an amphitheater-shaped headwall scarp that is over 50 km long and up to 120 m high near the intersection of the base of the slope with the upper rise in 2,600 m of water (Figure 4–2). This scarp encircles two large salt diapirs whose tops protrude above the surrounding seafloor, and a secondary complex of slumps and slide tracks extends more than 40 km upslope from the headwall scarp (Figures 4–2 and 4–11). Downslope, a broad trough over 150 m deep and more than 40 km across has been scoured into the seafloor (Figure 4–2). The mass wasting deposits associated with this trough extend for more than 250 km across the rise and onto the Hatteras Abyssal Plain (Embley and Jacobi 1977, 1986).

On the GLORIA sidescan sonar image, the path of the Cape Fear and Cape Lookout slides are represented by a light-toned (strong) acoustic return from the disturbed area that contrasts sharply with the darker (weak) return of the surrounding undisturbed sediments. Echo sounder profiles indicate that the slide track is cut into the bottom and bounded by abrupt scarps up to 150 m high where bedding is sharply truncated. The surface of the landslide debris within the trough is marked by longitudinal lines on the GLORIA image that represent both debris chutes between the diapirs and debris flow paths farther downslope.

Figure 4–9 shows four transverse seismic-reflection (3.5-kHz echo sounder) crossings of the Cape Fear Slide. Profile 4–9A crosses the amphitheater-shaped headwall area of the slide where two salt diapirs protrude above the surrounding seafloor. Removal of strata has scoured the bottom to a resistant layer between sidewalls, although near the north end of the profile a thin layer of mass movement debris has been covered by younger, well-layered sediment. Farther downslope (profile 4–9B) a trough over 25 km across and 100 m deep has been cut across the upper rise. Rough rubble and seismically transparent mass movement debris floor the trough, and this profile crosses a large displaced slide block over 20 m high. Disturbed bottom marks both sides of the trough, caused by both a spilling over of the turbidity flow and by downslope movement of the surficial layer similar to the rumpling of a carpet. Profiles 4–9C and 4–9D cross

the Cape Fear slide track 50 and 70 km downslope of the headwall scarp. At 50 km the trough is over 30 km wide and 150 m deep. Its northern sidewall shows truncated strata, while its southern boundary gradually slopes upward from the trough axis. At 70 km downslope the trough is 40 km across and has shoaled, partly because of a filling with mass movement debris. The fill within the trough is seismically transparent and produces a weak echo on the 3.5-kHz profile, making the seafloor difficult to trace. Still farther downslope where the Cape Fear Slide crosses the older Cape Lookout Slide, the trough is not present and mass movement debris forms an elevated sediment wedge that overlies the older deposits of the Cape Lookout Slide.

Although the Cape Fear Slide is typical of many of the Atlantic margin mass movement complexes, it is distinct because the head of the submarine landslide coincides with an area of salt diapir intrusion and because it occurs in an area where the shallow seafloor is frozen by a gas hydrate layer. Both the intruded salt diapirs (Dillon et al. 1983; Cashman and Popenoe 1985; Dillon and Popenoe 1988) and the gas hydrate layer (Carpenter 1981; Popenoe et al. 1993; Dillon et al. 1993) probably contributed to this massive slope failure. It is apparent from seismic profiles that the rising diapirs have fractured the slope, leading to a situation conducive to failure (Figure 4–10). The clathrate layer also probably contributed to the failure by helping to create a gas-charged lubricating layer along which the sediments failed at the uppermost part of the slide. Because clathrate fills the pore space within the shallow sediments, it is an effective barrier to the upward migration of gas from the deeper sediments. The clathrate may have formed a structural seal over the diapirs in which free gas was trapped. The effect of this pocket of free gas would be to reduce shear strength and create a zone of weakness immediately over and surrounding the diapirs. This condition may have been augmented by a eustatic drop in sea level, which would have caused a drop in pressure resulting in decomposition of the hydrate layer. The elevation of the salt diapirs above the seafloor within the failed area suggests that the diapirs uplifted the bottom before the failure and then were excavated by the slide.

Dillon et al. (1993) have shown that the presence of the Cape Fear and Cape Lookout Slides has affected the thickness of the gas hydrate layer (Figure 4–11). A distinct thinning of the gas hydrate zone has been mapped beneath the slide scars. The Cape Fear Slide, which is the younger of the two because it cuts across the Cape Lookout Slide at its distal end (Figure 4–2), shows the most thinning. The thinning represents not only the removal of clathrate within sediment removed by the slide, but a shallowing of the base of the hydrate layer, indicating that the removal of strata by mass movement causes a breakdown of hydrate, possibly as a result of pressure reduction, and that recovery from this breakdown is slow.

The age of the Cape Fear Slide is presently unknown, but piston cores indicate that the mass movement deposits on

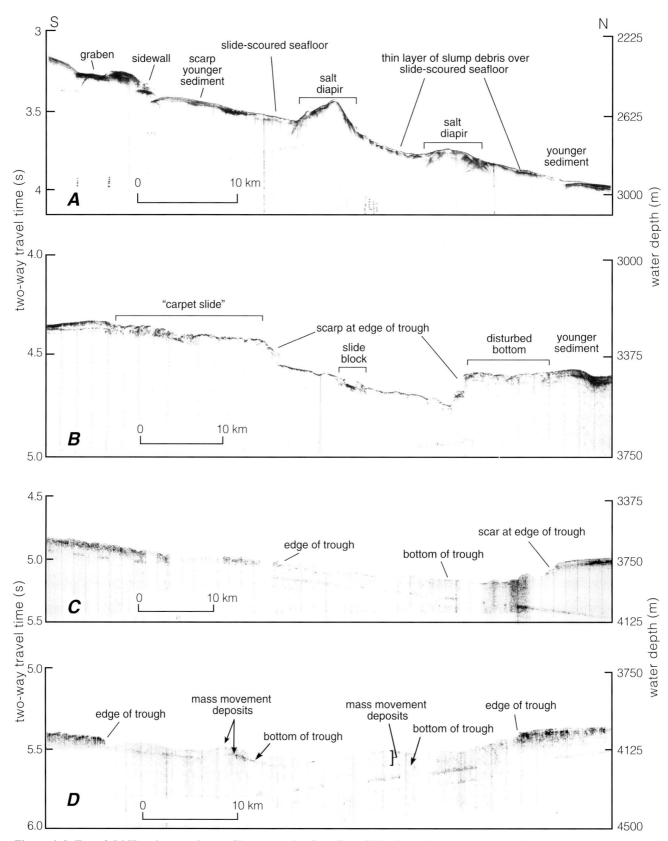

Figure 4–9. Four 3.5-kHz echo sounder profiles across the Cape Fear Slide. Locations are shown on Figure 4–3.

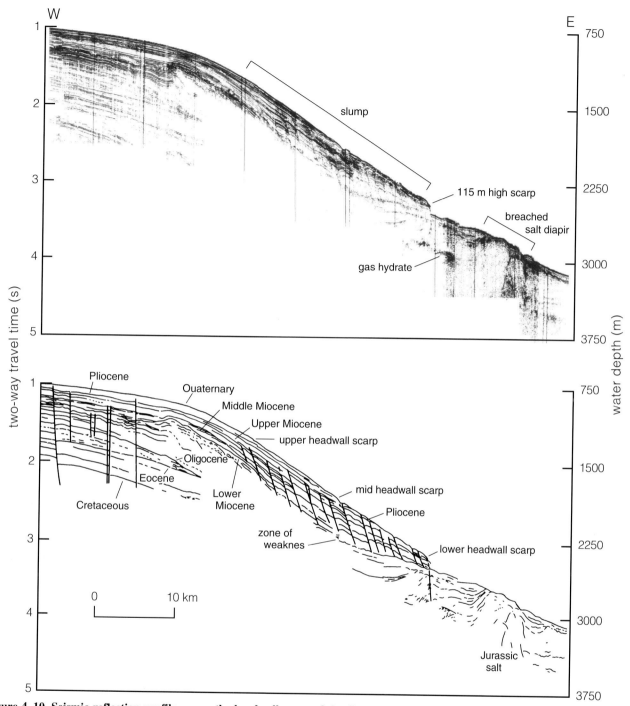

Figure 4–10. Seismic-reflection profile across the headwall scarp of the Cape Fear Slide. The slope above the 115-m-high headwall scarp is fractured by retrogressive slump faults.

the rise are overlain by a ubiquitous layer (>30 cm) of ooze (Embley 1980) that postdates the slide. Age dates of 12,124 and 20,830 years obtained from the base of this sediment layer suggest that the last major slope failure occurred during the Late Wisconsin low sea level stand (Embley 1980).

Deep penetrating multichannel seismic-reflection profiles indicate that the Cape Fear Slide trough on the rise is cut to

a surface that correlates with reflector "Blue," a surface that has been penetrated at DSDP* sites 102, 105, 106, 388, and 533 (Mountain and Tucholke 1985). Borehole correlations place this slide surface reflector roughly at the boundary of lower and upper Pliocene. It is apparent that the slide post-

*Deep Sea Drilling Project.

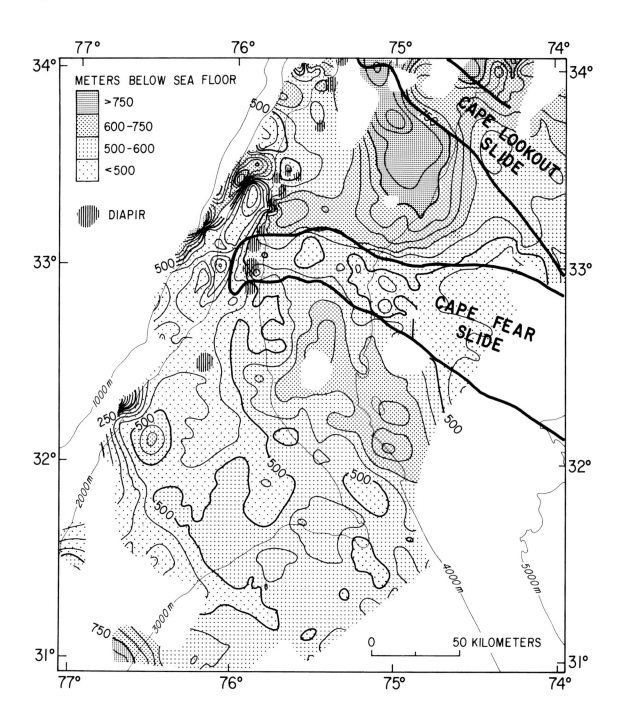

Figure 4–11. Location of landslide tracks plotted on a map of thickness of the gas hydrate layer (from Dillon et al., in press).

dates this surface by considerable time, because over 100 m of later sediment was removed by the slide. A similar late Pliocene age was obtained for the slip surface of the Albemarle-Currituck Slide by McGregor (1981).

The Cape Lookout Slide

The physiography of the Cape Lookout Slide is similar to that of the Cape Fear Slide. This slide originates at the base of the continental slope in an amphitheater-shaped scarp, and the slide has etched a 35-km-wide track across the rise. The greater age of the Cape Lookout Slide is demonstrated by the superposition of the Cape Fear Slide over the Cape Lookout Slide at its distal end and by gullies cut within the slide trough by later turbidity flow (Figure 4–2).

The infilling of the distal end of Pamlico Canyon on the lower rise occurs in an area of the rise that is characterized on seismic records by a hummocky bottom on both seismic

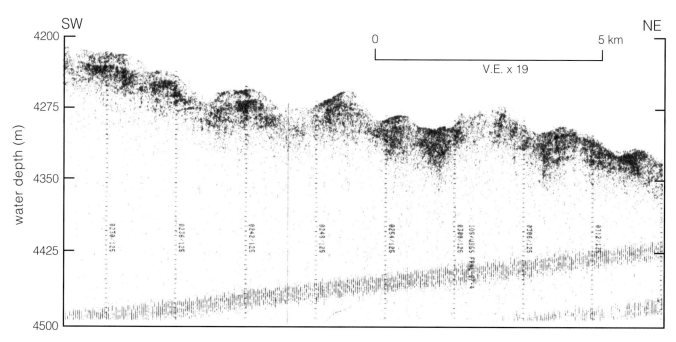

Figure 4–12. Hummocky, bumpy bottom characteristic of intercanyon areas on the mid-rise off Cape Hatteras. These "ill-defined bumps" probably represent sediment-draped blocky rubble derived from upslope failures. Location shown on Figure 4–3.

records (Figure 4–12) and on the GLORIA image (Figure 4–1). These hummocks occur throughout the area drained by the Hatteras canyon system, but are more characteristic of the middle to lower rise between Hatteras Canyon and the Cape Lookout Slide. They were noted by Newton and Pilkey (1969), who described them as "ill-defined humps" that are "probably slump structures." Embley and Jacobi (1986) interpreted these humps in the area of Figure 4–12 as "blocky slide debris," as did Vassallo et al. (1984) and Pratson and Laine (1989). On the GLORIA image the humps produce a mottled pattern that is variably linear, characteristic of mud waves, or circular, characteristic of blocky slide debris. Internal structure is ill-defined; some bumps appear to be accretionary, but most appear to be sediment draped over blocky rubble. We conclude that blocky rubble has been covered by a thick layer of ooze to form the humps.

The features shown in Figure 4–12 are out of the main track of the Cape Lookout Slide as defined by GLORIA, but are in a small drainage that lies between the Cape Lookout Slide and Pamlico Canyon, which can be traced on the GLORIA image. This drainage and Pamlico Canyon both originate on the continental slope below a segment of the slope that is shown on the Russell Bathymetric Map (National Ocean Survey 1981) as a huge amphitheater, although our seismic profiles show this amphitheater as an area cut by multiple deep canyons rather than an area of mass wasting. It is probable that the origin of the humps on the middle rise is an old landslide within this general area whose debris flows have been covered by hemipelagic ooze. The thick ooze layers that drape the humps imply that they are

relatively old. In other areas (Figure 4–13) there are a few very large blocks of displaced material on the upper and middle rise. There is little doubt that these large blocks are olistoliths, or exotic blocks transported by submarine gravity sliding from an origin on the continental slope.

Contour current-controlled processes

The North Carolina continental rise is swept by a complex current system that has piled up two large sediment drifts, the Hatteras Outer Ridge and the Blake Outer Ridge, and has scoured the seafloor in other places. The Hatteras Outer Ridge is mainly buried by ponded turbidites on its upslope side north of our study area and has been defined chiefly from subbottom data (Tucholke and Laine 1982; Mountain and Tucholke 1985; McMaster, Locker, and Laine 1989). The Blake Outer Ridge is actively accreting in some places (L. D. Keigwin, personal communication 1994), but subbottom seismic records show that its position has shifted southward through time so that the north flank of the ridge now exposes older sediments due to erosional scour by the Western Boundary Undercurrent.

The Hatteras Outer Ridge and lower rise hills

Seismic stratigraphic studies of the continental rise show that the Hatteras Outer Ridge (Rona et al. 1967) began developing as a significant sediment drift, deposited from the interaction of contour-following, southwestward-flowing

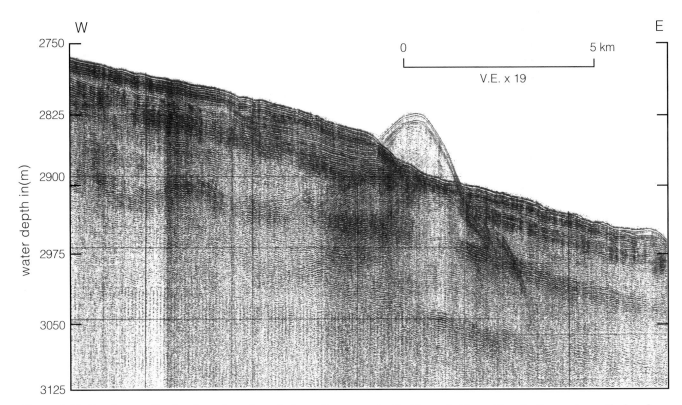

Figure 4–13. Sparker profile showing a large hyperbola on the upper rise that is probably a side echo from a large displaced block, or olistolith. The GLORIA image in this area shows a feature about 1 km across that lies just south of the trackline that is the probable source of this anomaly. The seismic profile shows that the feature projects at least 75 m above the bottom. Location shown on Figure 4–3.

abyssal currents (proto-WBUC) with Gulf Stream currents, in the early to middle Miocene (Tucholke and Laine 1982; Mountain and Tucholke 1985; McMaster et al. 1989). Accretion of the ridge was unusually rapid, ranging upward to 190 m/Ma, so that by middle Miocene the ridge formed a barrier nearly 1/2 km high and over 180 km long to the downslope transport of sediments (Tucholke and Laine 1982). The broad trough landward of the ridge became a basin into which turbidites from as far north as Hudson Canyon were funneled. These sediments completely filled the original trough in post-Miocene time, so that today the only morphological expression of the original ridge is east of the Hatteras Transverse Canyon. At Hatteras Transverse Canyon the relief of the western side of the ridge is over half erosional, formed by downcutting of the Hatteras Transverse Canyon into drift sediments, rather than original depositional relief (Figures 4–2 and 4–8).

The Hatteras Outer Ridge is poorly expressed on the GLORIA image because it is formed chiefly of fine-grained hemipelagic sediment that produces a uniform dark tone, similar to intercanyon areas of sediment drape on the rise. The scoured west-facing flank of the Hatteras Outer Ridge, where the Hatteras Transverse Canyon has cut into older sediments, shows as a light-toned area on the GLORIA image. Erosional truncations of sediments along the scoured

face of this flank are notable on seismic-reflection profiles that cross the ridge (Figure 4–8B). The east-facing flank of the ridge adjacent to the Hatteras Abyssal Plain is gentler in slope and represents the original depositional surface of the sediment drift. This flank is swept by the southwest-flowing WBUC and is mantled by the east-to-southeast-trending sediment waves known as the lower rise hills (Figure 4–2). These elongated ridges or hills are highly visible in the GLORIA image because they produce a striped pattern caused by (1) sediment wave flanks that face toward and away from the sidescan fish and (2) sediment textural differences on the wave crests and troughs.

The origin and orientation of the lower rise hills were hotly debated in the literature for a number of years (Ballard 1966; Heezen et al. 1966; Fox, Heezen, and Harian 1968; Rona 1969; Emery et al. 1970; Asquith 1979), but more recent detailed surveys (Tucholke and Laine 1982; Mountain and Tucholke 1985; McCave and Tucholke 1986; McMaster et al. 1989) and the GLORIA image (Figure 4–1) clearly show that these, and similar buried sediment waves noted in seismic records in this area, are abyssal current–controlled migrating sediment waves that are oriented dominantly east-west. Coring of the hills has shown that they are composed of primarily two types of sediment: hemipelagic mud (contourite) and turbidite, which flowed across the ridge in Pleistocene time eroding and filling channels and injecting fine-grained sand and silt into the WBUC (Asquith 1979). The mud waves range in height to over 100 m and typically have wavelengths from crest to crest of 8 to 10 km and ridge lengths in excess of 50 km.

The Blake Outer Ridge

Similar to the Hatteras Outer Ridge, the Blake Outer Ridge is a gigantic sediment drift that has formed since late Oligocene from deposition related to the interaction of the southward-flowing WBUC with the Gulf Stream (Markl, Bryan, and Ewing 1970; Markl and Bryan 1983; Mountain and Tucholke 1985). This enormous ridge dominates the southeastern quadrant of our study area and is the largest sediment drift known, constructed almost entirely of post-Miocene silty clays (Ewing and Hollister 1972). The ridge strikes southeast across the rise from the continental slope at the edge of the Blake Plateau, with its crest at about 2,000 m depth at 32°30′ N lat and its toe over 500 km away near 29° N lat in 6,000 m of water on the Hatteras Abyssal Plain. The northeastern flank of the ridge is relatively steep (>2.5° in places) and has been strongly eroded by the Western Boundary Undercurrent, exposing Oligocene and lower Miocene strata at the seafloor (Bryan and Markl 1966; Tucholke 1987). These more indurated sediments produce very bright, strong acoustic returns on the GLORIA image as well as alternating light and dark bands from erosion-exposed layered strata (Figures 4–1 and 4–2). The southwestern flank of the ridge is gentler and smoother, but current erosion here has also exposed older sediment ledges that form benches and knickpoints on the slope (shown as scour areas on Figure 4–2). The GLORIA image covers only the upper part of the Blake Outer Ridge, where no sediment waves are evident. Much of the ridge is smooth and featureless, producing a dark return from the smoothly layered mud that makes up the bottom.

Conclusions

The most important results of our study of GLORIA and seismic-reflection data on the Carolina margin are the visual depictions of many features of the margin that have been mapped or traced by other studies and the conformation of their nature and origin. Many excellent studies have been made of rise processes off North Carolina. Several, based primarily on the analyses of echo character as depicted by 3.5-kHz seismic records (Embley and Jacobi 1977, 1986; Vassallo et al. 1983; Pratson and Laine 1989), have produced regional maps of the surficial geology that are remarkably similar to our map, differing only in minor detail. Downslope-trending features dominate the GLORIA image, supporting the conclusion of Embley (1980) that mass flow deposits cover at least 40 percent of the continental rise.

Current-controlled processes are less conspicuous on the image. These are most important over two large areas, the Blake and Hatteras Outer Ridges, where currents drop due to the interaction of the Gulf Stream with the Western Boundary Undercurrent, or the currents increase due to topographic steering by the continental slope and Blake Outer Ridge. These intensified currents produce large areas of erosion and bottom scour in addition to sediment waves. Thus, the steep continental slope at Cape Hatteras appears to be due both to downslope and contour current – related processes.

Conclusions related to processes acting on the North Carolina rise are as follows: (1) Most turbidite flows originate on the continental slope and bypass the rise to be deposited on the Hatteras Abyssal Plain. Intercanyon sedimentation on the rise is chiefly finely layered hemipelagic ooze that drapes topography. On the upper rise there is little evidence of reworking or modification of mass wasting deposits by bottom currents. (2) Mass flow deposits appear to be a very important source of lower rise sediments. The large mass wasting failures originating at the base of the continental slope have cut deep troughs across the upper rise. The inertia of the passing debris flows on the upper rise has mobilized and stripped more material here than that stripped from the headwall areas where failure originated. Thus, failure must have been catastrophic, similar to an avalanche. Deposition of the mass wasting debris is mainly on the lower rise. (3) Contourites are important only in specific localities such as on the Blake Outer Ridge and on the Hatteras Outer Ridge. The major effect of contour currents at the present time appears to be erosion and scour of the seafloor.

The GLORIA survey of the Carolina margin took less than one month of total ship time. It has confirmed and refined the definition of features and provided information on their origin that augments over fifty years of previous oceanographic research involving thousands of man hours and hundreds of thousands of ship trackline miles. The GLORIA survey is definitive in tracing and correlating mappable features, such as canyon valleys and landslide tracks, from their origin on the slope to their terminus in the deep sea. It has settled questions concerning the orientation of mud waves on the Hatteras Outer Ridge, the correlation of drainage tracks across the rise, and the importance of slope failure as an agent in delivering sediment to the deep sea. The existence of many of the features and the processes that shaped them are now clearly displayed for us on the GLORIA mosaic.

Acknowledgments

All oceanographic research projects are team efforts, and this study is no exception. The area that we have studied contains data from three legs of the GLORIA cruises. We thank the captains and crews of the R/V *Farnella,* our coinvestigators from the IOS, and our USGS colleagues for many days of gallant dedicated effort in the winter storms of February and March 1987 off Cape Hatteras. Co-chief scientists included David C. Twichell, James M. Robb, John S. Schlee, Kathryn Scanlon, and Christopher F. Polloni of the USGS; and John B. Wilson, Quentin Huggett, and Stephen Williams of IOS. Our special thanks to Eric Darlington, Derek Bishop, and Andrew Harris, Electronic Engineers from IOS, who worked long hours without complaint to keep the GLORIA equipment running in the unpleasant environment of towering and confused 10-m-plus seas.

We thank John Schlee, David Twichell, Robert D. Jacobi, and Lincoln Pratson for their thoughtful reviews of the manuscript. Jeff Zwinakis drafted the illustrations, which were photographed by Dann Blackwood.

References

Asquith, S. M. 1979. Nature and origin of the lower continental rise hills off the East Coast of the United States. *Mar. Geol.* 32: 165–90.

Ballard, J. A. 1966. Structure of the lower continental rise hills of the western North Atlantic. *Geophysics* 31: 506–23.

Betzer, P. B., Richardson, P. L., and Zimmerman, H. B. 1974. Bottom currents, nepheloid layers and sedimentary features under the Gulf Stream near Cape Hatteras. *Mar. Geol.* 16: 21–9.

Bryan, G. M., and Markl, R. G. 1966. Microtopography of the Blake-Bahama region. *Lamont Doherty Observatory Technical Report no. 8, CU 8-66.* New York: Columbia University. 44 p.

Bumpus, D. F. 1973. A description of the circulation of the continental shelf of the east coast of the United States. *Prog. Oceanogr.* 6: 117–57.

Bunn, A. R., and McGregor, B. A. 1980. Morphology of the North Carolina continental slope, western North Atlantic, shaped by deltaic sedimentation and slumping. *Mar. Geol.* 37: 253–66.

Carpenter, G. B. 1981. Coincident slump-clathrate complexes on the U.S. continental slope. *Geo-Mar. Ltrs.* 1: 29–32.

Cashman, K. V., and Popenoe, P. 1985. Salt tectonic induced slumping on the continental slope and rise off North Carolina. *Mar. Petrol. Geol.* 2: 259–71.

Chavez, P. S. 1986. Processing techniques for digital sonar images from GLORIA: *Photogram. Rem. Sens.* 52: (8)1133–45.

Cleary, W. J. 1974. Hatteras deep-sea fan. *J. Sed. Petrol.* 44: 1140–54.

Cleary, W. J., Pilkey, O. H., and Ayers, M. 1977. Morphology and sediments of three basin entry points, Hatteras Abyssal Plain. *J. Sed. Petrol.* 47: 1157–70.

Dillon, W. P., Popenoe, P., Grow, J. A., Klitgord, K. D., Swift, B. A., Paull, C. K., and Cashman, K. V. 1983. Growth Faulting and salt diapirism: Their relationship and control in the Carolina Trough, eastern North America. In *Studies in Continental Margin Geology,* eds. J. S. Watkins and C. L. Drake, pp. 21–46. American Association of Petroleum Geologists, Memoir 34. Tulsa, OK: AAPG.

Dillon, W. P., and Popenoe, P. 1988. The Blake Plateau Basin and Carolina Trough. In *The Atlantic Margin: U.S.,* Vol. I-2: The Atlantic Continental Margin, eds. J. H. Grow and R. E.

Sheridan, pp. 291–328. Geological Society of America DNAG Series, Boulder, CO.

Dillon, W. P., Lee, M. W., Felhaber, K., and Coleman, D. F. 1993. Gas hydrates on the Atlantic margin of the United States – control on concentration. In *The Future of Energy Gasses,* ed. D. G. Howell, pp. 313–30. U.S. Geological Survey Professional Paper 1570. Reston, Va.: USGS.

EEZ-SCAN 87 Scientific Staff 1991. Atlas of the U.S. Exclusive Economic Zone, Atlantic continental margin. *U.S. Geological Survey Miscellaneous Investigations Series I-2054,* 174 p. Reston, Va.: USGS.

Embley, R. W., and Jacobi, R. D. 1977. Distribution and morphology of large submarine sediment slides and slumps on Atlantic continental margins: *Mar. Geotechnol.* 2: 205–28.

Embley, R. D. 1980. The role of mass transport in the distribution and character of deep ocean sediments with special reference to the North Atlantic. *Mar. Geol.* 38: 23–50.

Embley, R. W., and Jacobi, R. D. 1986. Mass Wasting in the western North Atlantic. In *The Geology of North America,* Vol. M: The Western North Atlantic Region, eds. P. R. Vogt and B. E. Tucholke, pp. 479–90. Boulder, CO: The Geological Society of America.

Emery, K. O., Uchupi, E., Phillips, J. D., Bowen, C. O., Bunce, E. T., and Knott, S. T. 1970. Continental rise off eastern North America. *Am. Assoc. Petrol. Geol. Bull.* 54: 44–108.

Ewing, J. I., and Hollister, C. D. 1972. Regional aspects of deep sea drilling in the western North Atlantic. In Initial Reports of the Deep Sea Drilling Project, Vol. 11, eds. C. D. Hollister and J. I. Ewing, pp. 951–73. Washington, D.C.: U.S. Govt. Printing Office.

Fitchko, R. M. 1976. Topography, shallow structure, and sedimentary processes of the Atlantic continental slope off the Carolina coast. Unpublished Master's Thesis, Old Dominion University, Norfolk, Va., 91 p.

Fox, P. J., Heezen, B. C., and Harian, A. M. 1968. Abyssal antidunes: *Nature.* 220: 470–2.

Gardner, J. V., Field, M. E., Lee, H., and Edwards, B. F. 1991. Ground-truthing 6.5 kHz side scan sonographs: What are we really imaging? *J. Geophys. Res.* 96(B4): 5995–74.

Heezen, B. E., Hollister, C. D., and Ruddiman, W. F. 1966. Shaping the continental rise by deep geostrophic contour currents. *Science* 152(3721): 502–8.

Hollister, C. D. and Heezen, B. C. 1972. Geologic effects of ocean bottom currents: western North Atlantic. In Gordon, A. L. (Ed.) Studies in Physical Oceanography – A Tribute to George Wüst on his 80th Birthday, 2. New York, N.Y.: Gordon and Breach, p. 37–66.

Hunt, R. E., Swift, D. J. P., and Palmer, H. 1977. Constructional shelf topography, Diamond Shoals, North Carolina. *Geol. Soc. Am. Bull.* 88: 299–311.

Jacobi, R. D., and Hayes, D. E. 1992. Northwest African continental rise: Effects of near-bottom processes inferred from high-resolution seismic data. In *Geologic Evolution of Atlantic Continental Rises,* ed. C. W. Poag, pp. 293–326. New York: Van Nostrand Reinhold.

Laine, E. P., Damuth, J. E., and Jacobi, R. 1986. Surficial sedimentary processes revealed by echo-character mapping in the western North Atlantic Ocean. In *The Geology of North America,* Vol. M: The Western North Atlantic Region, eds. P. R. Vogt and B. E. Tucholke, pp. 427–36. Boulder, Colo.: Geological Society of America.

Markl, R. G., Bryan, G. M., and Ewing, J. I. 1970. Structure of the Blake-Bahama Outer Ridge. *J. Geophys. Res.* 75: 4539–55.

Markl, R. G., and Bryan, G. M. 1983. Stratigraphic evolution of the Blake Outer Ridge. *Am. Assoc. Petrol. Geol. Bull.* 67: 666–83.

McCave, I. N., and Tucholke, B. E. 1986. Deep current controlled sedimentation in the western North Atlantic. In *The Geology of North America,* Vol. M: The Western North Atlantic Region, eds. P. R. Vogt and B. E Tucholke, pp. 451–68. Boulder, Colo.: Geological Society of America.

McGregor, B. A. 1981. Smooth seaward-dipping horizons – an important factor in seafloor stability. *Mar. Geol.* 39: M89–M98.

McMaster, R. L., Locker, S. D., and Laine, E. P. 1989. The early Neogene continental rise off the eastern United States. *Mar. Geol.* 87: 137–63.

Menzies, R. J., George, R. Y., and Rowe, G. T. 1973. Abyssal environment and ecology of the world oceans. New York: John Wiley & Sons, 488 p.

Mountain, G. S., and Tucholke, B. E. 1985. Mesozoic and Cenozoic geology of the U.S. Atlantic continental slope and rise. In *Geologic Evolution of the United States Atlantic Margin,* ed. C. W. Poag, pp. 293–342. New York: Van Nostrand Reinhold.

National Ocean Survey 1981. Bathymetric map of the Russell Quadrangle. *National Ocean Survey map NOS NI 18-5,* Scale 1:250,000. Washington D.C.: NOAA.

Newton, J. G., and Pilkey, O. H. 1969. Topography of the continental margin off the Carolinas. *Southeastern Geol.* 10: 87–92.

Newton, J. G., Pilkey, O. H., and Blanton, J. O. 1971. An oceanographic atlas of the Carolina continental margin. North Carolina Department of Conservation and Development, 57 p.

Paull, C. K., and Dillon, W. P. 1985. Structure, stratigraphy, and geologic history of the Florida-Hatteras shelf and inner Blake Plateau. *Am. Assoc. Petrol. Geol. Bull.* 64: 339–58.

Pilkey, O. H., and Cleary, W. J. 1986. Turbidite sedimentation in the northwestern Atlantic ocean basin. In *The Geology of North America,* Volume M: the Western North Atlantic Region, eds. P. R. Vogt and B. E. Tucholke, pp. 437–50. Boulder, Colo.: The Geological Society of America.

Popenoe, P., Coward, E. L., and Cashman, K. V. 1982. A regional assessment of potential hazards and limitations on petroleum development of the southeastern United States continental shelf, slope, and rise, offshore North Carolina. *U.S. Geological Survey Open-File Report 82-136,* 67 p., Reston, Va.: USGS.

Popenoe, P. 1985. Cenozoic depositional and structural history of the North Carolina margin from seismic stratigraphic analyses. In *Geologic Evolution of the United States Atlantic Margin,* ed. C. W. Poag, pp. 125–87. New York: Van Nostrand Reinhold.

Popenoe, P., Schmuck, E. A., and Dillon, W. P. 1993. The Cape Fear landslide: Slope failure associated with salt diapirism and gas hydrate decomposition. In *Submarine Landslides: Selected Studies in the U.S. Exclusive Economic Zone,* eds.

W. C. Schwab, H. J. Lee, and D. C. Twichell, pp. 40–53. U.S. Geological Survey Bulletin 2002. Reston, Va.: USGS.

Pratson, L. F., and Laine, E. P. 1989. The relative importance of gravity induced versus current controlled sedimentation during the Quaternary along the mideast U.S. outer continental margin revealed by 3.5 kHz echo character. *Mar. Geol.* 89: 87–126.

Prior, D. R., Doyle, E. H., and Neurauter, T. 1986. The Currituck slide, Mid-Atlantic continental slope – revisited. *Mar. Geol.* 73: 25–45.

Richardson, P. L. 1977. On the crossover between the Gulf Stream and the Western Boundary Undercurrent. *Deep Sea Res.* 24: 134–159.

Rona, P. A., Schneider, E. D., and Heezen, B. C. 1967. Bathymetry of the continental rise off Cape Hatteras. *Deep Sea Res.* 14: 625–33.

Rona, P. A. 1969. Linear "lower continental rise hills" off Cape Hatteras, *J. Sed. Petrol.* 39: 1132–42.

Schlee, J. S., and Robb, J. M. 1991. Submarine processes of the middle Atlantic continental rise based on GLORIA imagery. *Geol. Soc. Am. Bull.* 103: 1090–1103.

Schlee, J. S., Dillon, W. P., Popenoe, P., Robb, J. M., and O'Leary, D. W. 1992. GLORIA mosaic of the deep sea floor off the Atlantic coast of the United States. *U.S. Geological Survey Miscellaneous Field Studies Map MF-2211.* Reston, Va.: USGS.

Snyder, S. W., Kelchner, C. E., DeMaster, D. J., and Blair, N. E. 1993. Carbon sequestering on the Carolina slope: Evaluating source sites, sediment transport pathways, and flux magnitude. *Geol. Soc. Am. Abs. Prog.* 25(6): A-443.

Tucholke, B. E., and Laine, E. P. 1982. Neogene and quaternary development of the lower continental rise off the central U.S. east coast. In *Studies in Continental Margin Geology,* eds. J. S. Watkins and C. L. Drake, pp. 285–305. Tulsa, OK: American Association of Petroleum Geologists Memoir 34.

Tucholke, B. E. 1987. Submarine geology (Chapter 4); In *The marine environment of the U.S. Atlantic continental slope and rise,* eds. J. D. Milliman and W. R. Wright, pp. 56–113. Boston: Jones & Bartlett.

Twichell, D. C., Kenyon, N. H., Parson, L. M., and McGregor, B. A. 1991. Depositional patterns of the Mississippi Fan surface: Evidence from GLORIA II and high-resolution seismic profiles. In *Seismic Facies and Sedimentary Processes of Submarine Fans and Turbidite Systems,* eds. P. Weimer and M. H. Link, pp. 349–63. New York: Springer-Verlag.

Vassallo, K., Jacobi, R. D., and Shor, A. N. 1984. Echo character microphysiography, and geologic hazards. In *Ocean Margin Drilling Program Regional Atlas Series,* Atlas 5, eds. G. M. Bryan and J. R. Heirtzler, p. 40. Woods Hole, Mass.: Marine Science International.

III Gulf of Mexico and Caribbean EEZ

David C. Twichell, Kathryn M. Scanlon, and William P. Dillon

U.S. Geological Survey, Woods Hole, Massachusetts

The Gulf of Mexico Exclusive Economic Zone (EEZ) and parts of the northern Caribbean plate margin were surveyed using GLORIA (Geologic LOng-Range Inclined Asdic) during the U.S. Geological Survey EEZ-SCAN program. In the Gulf of Mexico, the first cruise was conducted in 1982 and three more were completed in the summer and fall of 1985. The survey of U.S. waters around Puerto Rico and the U.S. Virgin Islands was completed during a twenty-five-day cruise in the fall of 1985 as well. In addition, surveys were conducted in the Cayman Trough and north of Hispaniola during a transit in 1985 from the Gulf of Mexico to the Caribbean. In total, these surveys mapped approximately 600,000 km^2 of the U.S. Gulf of Mexico and Caribbean EEZ along survey tracklines that were spaced 10 to 30 km apart. The data collected included digital GLORIA sidescan sonar images, 40- to 160-in^3 airgun and 3.5-kHz seismic-reflection profiles, 10-kHz bathymetry profiles, and total magnetic field measurements.

The Gulf of Mexico is a small, geologically diverse ocean basin that can be divided into three distinct geologic provinces: a salt deformation province underlying the continental slope of the northern and western Gulf of Mexico, the Mississippi Canyon and Fan system in the central Gulf, and a carbonate province along its eastern and southern boundaries. The shallow structure and stratigraphy of the salt deformation province are the result of a complex interplay of varying sediment sources throughout the Tertiary and Pleistocene, tectonic deformation resulting from the sediment loading of the Jurassic-aged Luann salt, and autochthonous sedimentation associated with hydrocarbon seeps in this petroleum-rich province. The Mississippi Canyon and Fan system is the dominant feature in the central Gulf of Mexico where as much as 4.5 km of sediment has accumulated since the Pliocene. Although the fan has been studied extensively over the last three decades, the GLORIA survey provided the first complete acoustic image of this large passive-margin deep-sea fan. A complex set of depositional processes has contributed to recent sedimenta-

tion here. The Florida and Campeche escarpments, huge limestone cliffs in excess of 1,500 m high that mark the eastern and southern edges of the deep Gulf of Mexico, have undergone significant erosion since their initial formation.

Twichell and Delorey integrated the GLORIA imagery with multibeam bathymetry collected by National Oceanic and Atmospheric Association and 3.5-kHz and airgun subbottom seismic profiles to study sedimentary processes acting on a part of the salt deformation province in the northwestern Gulf of Mexico. They found that mass wasting is not uniformly distributed throughout the area and that there may be continuing tectonic activity. Additionally, these results show that integrated GLORIA and multibeam bathymetry can be used, in conjunction with sampling and high-resolution profiling, to identify and map the larger sites where hydrocarbons are escaping to the seafloor.

Twichell, Schwab, Kenyon, and Lee built upon the reconnaissance mapping of the Mississippi Fan done with GLORIA using 30-kHz sidescan sonar imagery, bathymetry, subbottom profiles, and cores to study the breaching of a channel levee. The high-resolution sidescan data and cores show that debris flows spill through the breach in the levee as well as choking the channel down-fan of the breach site. GLORIA imagery of the rest of this channel system shows other examples of channel choking. Mass wasting has occurred at a variety of scales on this fan (ranging from examples on seismic profiles to those recovered in piston cores) and appears to play a key role in redirecting channels and sites of sedimentation on this large passive-margin fan.

Twichell, Dillon, Paull, and Kenyon use the GLORIA imagery collected along the Blake, Florida, and Campeche escarpments to compare the morphologies of these three huge carbonate escarpments and infer the erosional processes that have shaped them. These limestone cliffs exceed 1,500 m in height and have average slopes greater than 40° along much of their lengths. Box canyons suggestive of groundwater sapping, terraces suggestive of differential erosion controlled by subtle lithologic differences in the strata exposed on the cliff face, and gullies suggestive of gravity-driven erosional processes are mapped. Tectonic, stratigraphic, and oceanographic processes acted to alter these spectacular cliffs.

The Caribbean plate margin is the only active plate margin in the North Atlantic. The part imaged during the EEZ-SCAN program is dominantly a strike-slip margin between the Caribbean and North American plates. The effect of tectonic forces is expressed both in the extreme bathymetry of the Puerto Rico Trench and in present seismic activity. Dillon, Edgar, and Scanlon address the three-dimensional aspects of the plate margin from the Cayman Trough to the Puerto Rico Trench. They studied three parts of the plate margin and found that variations in its trend and the presence of other features such as the Cayman spreading center and the Muertos Trough have resulted in a highly complex plate boundary zone. At the western end of the study area, the Cayman Trough is an active spreading center. A restraining bend in the plate boundary north of Hispaniola has created compression and an accretionary wedge. At the eastern end of the study area, Puerto Rico is bounded on the north by strike-slip motion along the Puerto Rico Trench and on the south by compression in the Muertos Trough. The GLORIA imagery from these three areas, seismic-reflection profiles, and seismicity data have been integrated to reveal the three-dimensional nature of this plate boundary and the relationship of the surficial geology to the deeper structures.

Scanlon and Masson discuss the effects of tectonic activity and lithology on the formation of submarine canyons on the north slope of Puerto Rico. They found that the sedimentary processes that have shaped the northern slope of Puerto Rico are strongly influenced by the tectonics of this active margin as well as by the lithology of the underlying strata. Canyons and sediment ponds indicate sediment pathways from the shelf edge to the trench floor. Large amphitheater-shaped scarps (up to 60 km across) on the slope suggest slumps are another sediment source, and seismic-reflection and drill-hole data indicate that these slumps are cut into the front of a former carbonate bank that tilted rapidly after the mid-Pliocene. The lithology of the carbonate strata appears to control both the locations of the amphitheater scarps and the unusual chutelike morphology of the canyons north of Puerto Rico. A complex, discontinuous sediment distribution along this active margin resulted from the combination of tectonic processes and the layered stratal nature of the slope.

Thus the EEZ-SCAN program in the Gulf of Mexico and the Caribbean addressed a variety of geological problems using diverse data types. These studies included plate interactions, salt tectonics, sedimentary processes of meandering channels in a passive-margin environment, and channel formation on an active margin. Selected references are listed below to provide a more complete overview of other studies that have stemmed from the EEZ-SCAN program.

References

Dillon, W. P., Edgar, N. T., Parson, L. M., Scanlon, K. M., Driscoll, G. R., and Jacobs, C. 1990. Magnetic anomaly map of the central Cayman Trough, northwestern Caribbean Sea. *U.S. Geological Survey Miscellaneous Field Studies Map, MF-2083 B*. Reston, Va.: USGS.

Dillon, W. P., Austin, J. A., Scanlon, K. M., Edgar, N. T., and Parson, L. M. 1992. Accretionary margin of north-western Hispaniola: Morphology, structure and development of part of the northern Caribbean plate boundary. *Mar. Petrol. Geol.* 9: 70–88.

Edgar, N. T., Dillon, W. P., Parson, L. M., Scanlon, K. M., Jacobs, C., and Holcombe, T. L. 1990. GLORIA sidescan-sonar image and interpretation of the central Cayman Trough, northwestern Caribbean Sea. *U.S. Geological Survey Miscellaneous Field Studies Map MF-2083 A*. Reston, Va.: USGS.

EEZ-SCAN 85 Scientific Staff 1987a. Atlas of the Exclusive Economic Zone, Gulf of Mexico. *U.S. Geological Survey Miscellaneous Investigations Series I-1864-A*, 104 p., scale 1:500,000. Reston, Va.: USGS.

EEZ-SCAN 85 Scientific Staff 1987b. Atlas of the Exclusive Economic Zone, Eastern Caribbean Areas. *U.S. Geological Survey Miscellaneous Investigations Series I-1864-B*, 58 p., scale 1:500,000. Reston, Va.: USGS.

Garrison, L. E., Kenyon, N. H., and Bouma, A. H. 1982. Channel systems and lobe construction in the Mississippi Fan. *Geo-Mar. Ltrs.* 2: 33–9.

Jany, I., Scanlon, K. M., and Mauffret, A. 1990. Geological interpretation of combined SeaBeam, GLORIA, and seismic data from Anegada Passage (Virgin Islands, North Caribbean). *Mar. Geophys. Res.* 12: 173–96.

Kenyon, N. H. 1992. Speculations on the geological causes of backscatter variation on GLORIA sonographs from the Mississippi and DeSoto Fans, Gulf of Mexico. *Geo-Mar. Ltrs.* 12: 24–32.

Masson, D. G., and Scanlon, K. M. 1991. The neotectonic setting of Puerto Rico. *Geol. Soc. Am. Bull.* 103: 144–54.

McGregor, B. A. 1987. A sonar view of the floor of the Gulf of Mexico. In *U.S. Geological Survey Yearbook, Fiscal Year 1986*, pp. 62–5. Reston, Va.: USGS.

McGregor, B. A., and Hill, G. W. 1989. Seafloor image maps of the U.S. Exclusive Economic Zone. *Hydrogr. J.* 53: 9–13.

Nelson, C. H., Twichell, D. C., Schwab, W. C., Lee, H. J., and Kenyon, N. H. 1992. Upper Pleistocene turbidite sand beds and chaotic silt beds in the channelized, distal, outer-fan lobes of the Mississippi fan: *Geology* 20: 693–6.

Paull, C. K., Spiess, F. N., Curray, J. R., and Twichell, D. C. 1990. Origin of Florida Canyon and the role of spring sapping on the formation of submarine box canyons. *Geol. Soc. Am. Bull.* 102: 502–15.

Paull, C. K., Twichell, D. C., Spiess, F. N., and Curray, J. R. 1991. Morphology and development of the Florida Escarpment: Observations on the generation of time transgressive unconformities in carbonate terrains. *Mar. Geol.* 101: 181–201.

Scanlon, K. M., and Masson, D. G. 1992. Fe-Mn module field indicated by GLORIA north of the Puerto Rico Trench. *Geo-Mar. Ltrs.* 12: 208–13.

Schwab, W. C., Danforth, W. W., Scanlon, K. M., and Masson, D. G. 1991. A giant slope failure on the northern insular slope of Puerto Rico. *Mar. Geol.* 96: 237–46.

Schwab, W. C., Lee, H. J., and Twichell, D. C. 1993. Submarine landslides: Selected studies in the U.S. Exclusive Economic

Zone. *U.S. Geological Survey Bulletin 2002,* 204 p. Reston, Va.: USGS.

Twichell, D. C., Parson, L. M., and Paull, C. K. 1990. Variations in the styles of erosion along the Florida Escarpment, eastern Gulf of Mexico. *Mar. Petrol. Geol.* 7: 253–66.

Twichell, D. C., Paull, C. K., and Parson, L. M. 1991a. Terraces on the Florida Escarpment: Implications for erosional processes. *Geology* 19: 897–900.

Twichell, D. C., Kenyon, N. H., Parson, L. M., and McGregor, B. A., 1991b. Depositional patterns of the Mississippi Fan surface: Evidence from GLORIA II and high-resolution seismic profiles. In: *Seismic Facies and Sedimentary Processes of Modern and Ancient Submarine Fans*, eds. P. Weimer and M. H. Link, pp. 349–63. New York: Springer-Verlag.

Twichell, D. C., Schwab, W. C., Nelson, C. H., Kenyon, N. H., and Lee, H. J. 1992. Characteristics of a sandy depositional lobe on the outer Mississippi fan from SeaMARC IA side-scan sonar images. *Geology* 20: 689–92.

5 Breaching the levee of a channel on the Mississippi Fan

David C. Twichell,[1] William C. Schwab,[1] Neil H. Kenyon[2] and Homa J. Lee[3]

[1]U.S. Geological Survey, Woods Hole, Massachusetts,
[2]Institute of Oceanographic Sciences, Southampton, United Kingdom
[3]U.S. Geological Survey, Menlo Park, California

Abstract

GLORIA images of the youngest channel on the Mississippi Fan indicate that it has not been a stable feature, but instead has shifted its course several times. A detailed study of a site of channel shifting found a complex stratigraphy that resulted from one episode of channel avulsion. The channel avulsion appears to have been initiated by a large mass flow that choked the channel below the point where the levee was breached and additionally spilled a large volume of material through the breach in the levee onto the adjacent fan. Subsequent flows were redirected through this breach in the levee and built a channel-levee complex over the mass movement deposits. A second phase of mass movement resulted from another large mass flow that came down the channel and triggered the collapse of part of the newly developed levee. In this case, locally derived levee sediment was mixed with allochthonous sediment from farther up the fan and was spread northward from the levee. This localized study suggests that fan stratigraphy is complex and variable at several scales, not just at the scale resolved in seismic stratigraphic studies (Weimer 1989) and that large mass flows capable of choking the channel system have been an important mechanism in redirecting sedimentation on the Mississippi Fan.

Introduction

Channels on deep-sea fans are conduits through which sediments are transported to the distal part of the fan (see Mutti and Normark 1987 and Shanmugam and Moiola 1991 for summaries). Early GLORIA sidescan sonar surveys of deep-sea fans such as the Amazon (Damuth et al. 1983a), Indus (Kenyon, Amir, and Cramp 1995), and Mississippi (Garrison, Kenyon, and Bouma 1982) showed that these leveed channels meander. These sidescan images often showed that many channels were present on the surface of fans, yet only

one appeared to have been active at a time (Damuth et al. 1983b). The overlapping relationships of unconformity-bounded sequences revealed on seismic-reflection profiles demonstrated the importance of channel shifting in the architecture of deep-sea fans (Feeley 1984; Damuth et al. 1988; Weimer and Buffler 1988; Weimer 1989). Although numerous studies recognize the importance of channel shifting (Weimer 1989; Flood et al. 1991; Droz and Bellaiche 1991; McHargue 1991), few have specifically investigated the point where the shifting occurs and the processes that lead to redirection of sediments to a different part of the fan (Manley and Flood 1988; Flood et al. 1991). On the Mississippi Fan, GLORIA images (EEZ-SCAN 85 Scientific Staff 1987) show that the most recently active channel has shifted its course several times (Twichell et al. 1991). We undertook a survey in one of these areas where the levee was breached to define the bathymetry, surficial geology, and shallow subsurface stratigraphy (Figure 5–1). The purpose of this chapter is to describe in detail the evolution of one of these channel-shifting episodes.

Methods

The base map used for this study is the GLORIA sidescan sonar imagery of the entire Mississippi Fan that was collected during the EEZ-SCAN project in 1985 (EEZ-SCAN 85 Scientific Staff 1987). This image reveals the channel system on the Mississippi Fan from near its source in the Mississippi Canyon to the distal reaches of the fan. GLORIA is a low-frequency (6.5 kHz) sidescan sonar system that insonifies the seafloor in swaths up to 45 km wide and is ideally suited for reconnaissance surveys of large areas (Somers et al. 1978). The GLORIA data were logged digitally and were processed and mosaicked using digital techniques described by Chavez (1986; this volume). In addition to the GLORIA image, 3.5-kHz subbottom profiles that provide a regional overview of the shallow subsurface stratigraphy were collected along the survey tracklines (EEZ-SCAN 85 Scientific Staff 1987).

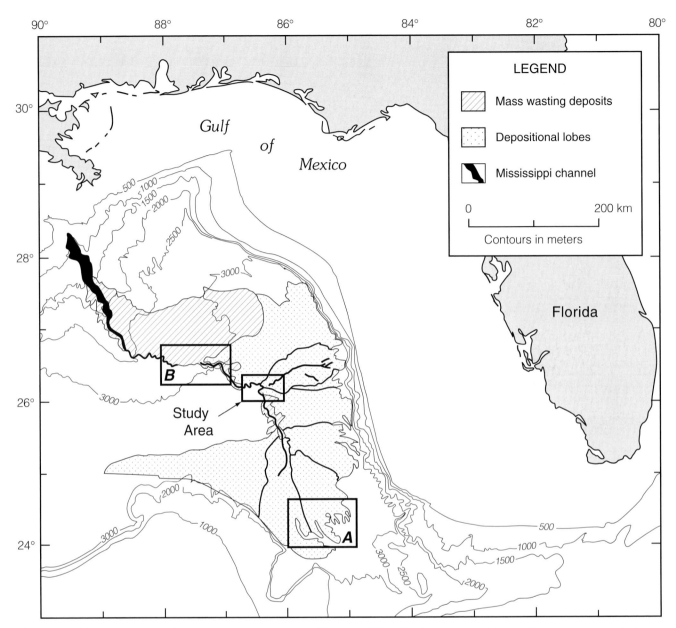

Figure 5–1. Map showing location of the study area on the Mississippi Fan. The boxes labeled A and B are other sites of channel choking, which are shown in Figure 5–10.

A site of channel avulsion was identified from study of the GLORIA image for survey with the higher-frequency SeaMARC IA sidescan sonar system, 3.5-kHz subbottom profiles, and piston cores (Figure 5–2). The SeaMARC IA system is a 27- to 30-kHz deep-towed instrument that insonifies the seafloor in swaths as much as 6 km wide (Kosalos and Chayes 1983). These data were also digitally logged (Danforth, O'Brien, and Schwab 1991) and were processed and mosaicked using techniques slightly modified from those described by Chavez (1986) to optimize the higher frequency data (Paskevich 1992). The 3.5-kHz subbottom profiles were collected along the SeaMARC IA survey lines as well as along other lines. Water depths digitized from these profiles were used to create a bathymetric map with the

sidescan sonar image serving as a guide to connect features between adjacent profiles. The subbottom information was used in concert with the sidescan sonar image to create an echo character map of the area as well as to establish the stratigraphic relationships of the different acoustic facies. Piston cores identified the lithologies of some of the acoustic facies that were defined.

Results

GLORIA overview

The GLORIA imagery of the study area (Figure 5–2) shows that part of the youngest channel that can be traced from

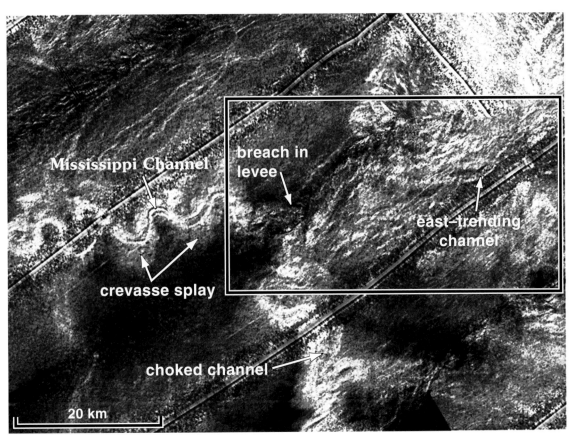

Figure 5–2. GLORIA sidescan sonar image of the study area showing a portion of the clearly defined Mississippi Channel with crevasse splays along the outsides of meander bends up-fan of the breach in the levee, a portion of the Mississippi Channel that has been choked by debris flow deposits down-fan of the breach, and the smaller east-trending channel. The box shows the location of Figures 5–3 and 5–6. Image location shown in Figure 5–1.

the Mississippi Canyon across the Mississippi Fan, here called the Mississippi Channel has crisp, clearly defined high-acoustic backscatter, continuous walls, and a low-backscatter floor. High-backscatter deposits extending down the levee flanks at the outsides of meander bends are interpreted to be crevasse splay deposits (Figure 5–2). Down-fan parts of this channel are poorly defined and have discontinuous walls. In these areas, a continuous zone of high-acoustic backscatter covers the channel floor, cuts the channel walls, and extends down the levee flanks. Near the transition from the up-fan clearly defined segment of the Mississippi Channel to the down-fan poorly defined segment, a region of mottled high and moderate backscatter extends eastward from the channel. This mottled region represents part of the surface of a younger deposit that extends 150 km, nearly to the base of the Florida Escarpment (Twichell et al. 1991). On the surface of this mottled deposit is a smaller east-trending channel (Figure 5–2) having clearly defined walls and a low-backscatter floor that can be traced from within about 15 km of the main channel to the edge of this deposit (Twichell et al. 1992).

Detailed study

A bathymetric map of the site of levee breaching and channel avulsion included part of the Mississippi Channel and the east-trending channel. For clarity, the discussion of the channels is divided into three parts: the up-fan portion of the Mississippi Channel, which represents the section of the channel west of the breach in the levee; the down-fan portion of the Mississippi Channel, which occurs south of the breach in the levee; and the east-trending channel, which extends from the site of channel avulsion (Figure 5–2).

UP-FAN PORTION OF MISSISSIPPI CHANNEL

A 10-km section of the sinuous Mississippi Channel up-fan of the breach in the levee has well-defined levees and is shown on the western side of Figure 5–3. The channel floor is 400 to 600 m wide, the levees rise 10 to 20 m above the thalweg, the thalweg gradient is less than 1 m/km (Figure 5–4), and the entire channel-levee complex rises about 80 m above the surrounding seafloor. The northeastern levee of the Mississippi Channel is breached where the channel makes an abrupt bend to the south, and here the northeast-

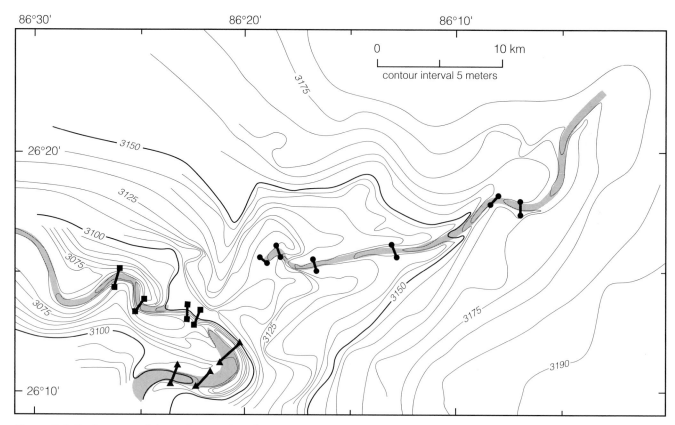

Figure 5–3. Bathymetry of the study area (location shown by the box in Figure 5–2). Channel floor shown by stipple pattern, and location of profiles used in Figure 5–4 are shown by solid lines.

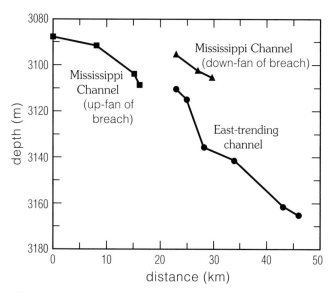

Figure 5–4. Plot of channel thalweg gradients based on the available echo-sounding profiles. Note that the channel thalweg south of the breach is 10 m shallower than the channels up-fan, presumably because of its being choked by a debris flow deposit.

ern levee as well as the channel are absent. The SeaMARC IA image of the up-fan portion of the Mississippi Channel shows that its floor is of a constant width and is represented by a mostly moderate-to-low-backscatter pattern with a few scattered high-backscatter patches. Gullies less than 200 m in length are extensive along the channel walls (Figure 5–5A). The levees to either side of the channel are characterized by discontinuous irregularly shaped patches of high backscatter within a zone of moderate backscatter (Figure 5–6). The subbottom profiles show the levees here tend to have a highly reflective surface with a prolonged subbottom return and some discontinuous highly reflective subbottom surfaces (Figure 5–7B, C). Such an acoustic facies is characteristic of much of the main channel of the Mississippi Fan (O'Connell and Normark 1986).

DOWN-FAN PORTION OF MISSISSIPPI CHANNEL

South, or down-fan, of the bend where the levee was breached, the Mississippi Channel does not have as much relief, and its thalweg is at least 10 m shallower than the channel up-fan of the breaching point (Figure 5–4). The sonographs show this portion of the Mississippi Channel is broader and more irregular in width (500 to 1,800 m wide), does not have walls that are as well defined, and has a higher-

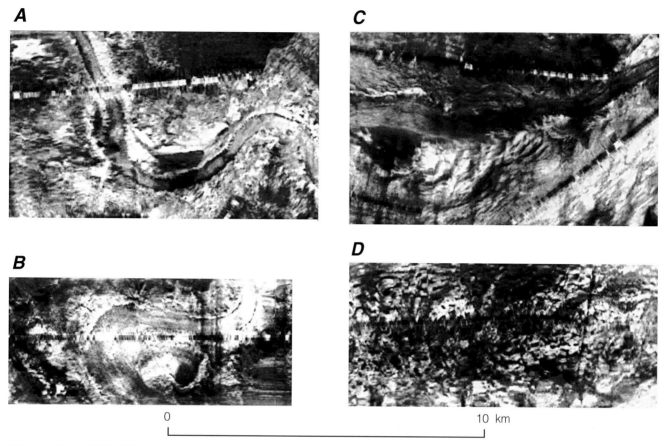

Figure 5–5. SeaMARC IA sidescan sonar images showing: A, main channel up-fan of the breach area with its high-backscatter floor, constant width, and gullied sidewalls; B, segment of choked channel down-fan of the breach area with its broader high-backscatter floor and poorly defined discontinuous sidewalls; C, eastward-extending channel with the possible sediment waves on its levee flanks; and D, the mottled area north of the channels. Figure locations shown in Figure 5–6.

backscatter channel floor (Figure 5–5B). The larger perspective provided by the GLORIA image shows that the high-backscatter material extends beyond the channel walls and covers much of the levees in this southern area (Figure 5–2). The subbottom profiles show that the Mississippi Channel down-fan of the breach in the levee is partially filled by sediment (Figure 5–7D). Part of the channel fill shows some subbottom reflections while in other areas the fill is acoustically transparent. This transparent unit is discontinuously distributed on the flanks of the levee as well and can reach a thickness in excess of 15 m.

EASTWARD-TRENDING CHANNEL

A smaller, less sinuous channel can be traced eastward from the breach in the Mississippi Channel nearly to the edge of the Mississippi Fan (Twichell et al. 1992). Within the study area, this channel is less than 400 m wide, has levees less than 10 m high, and its entire channel-levee complex rises about 50 m above the surrounding seafloor in the study area and decreases in relief to the east. The thalweg of the east-trending channel slopes continuously away from the Mis-

sissippi Channel and is at the same depth as the Mississippi Channel up-fan from the site of avulsion (Figure 5–4). This channel, for much of its length, has a moderate-to-low-backscatter floor and clearly defined channel walls (Figure 5–5C). Within the study area, the levees of the east-trending channel are characterized by a series of discontinuous sinuous lineations that converge in an easterly direction on the channel (Figures 5–5C and 5–6). Farther eastward, these wavelike features give way to the same discontinuous high-backscatter pattern that was seen on the levees of the Mississippi Channel (Figure 5–6). North and south of this east-trending channel and levee system are areas of mottled high and low backscatter (Figures 5–5D and 5–6).

The crest of the levee and the channel floor of the east-trending channel has a prolonged highly reflective signature on the subbottom profiles, while away from the channel on the flanks of this levee the echo character changes to one of multiple parallel subbottom reflections (Figure 5–7E). The mottled areas on the sidescan mosaic correspond with an acoustically transparent unit on the subbottom profiles (Figure 5–7A). This transparent unit has a diffuse rough basal

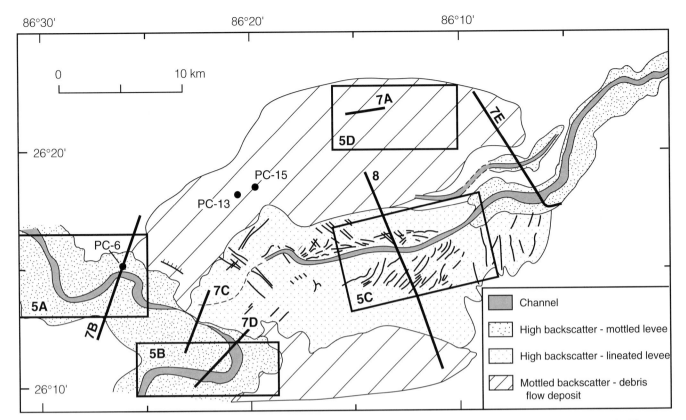

Figure 5–6. Interpretation of the SeaMARC IA image of the study area. Labeled boxes are locations of images shown in Figure 5–5, and labeled lines are locations of profiles shown in Figures 5–7 and 5–8. Solid lines in the high-backscatter lineated area represent lineations identified on the SeaMARC IA image that may represent sediment waves.

contact with the underlying sediments, and in some parts hyperbolae are found on the surface of this unit, which suggests that this mottled pattern has some fine-scale topography associated with it (Figure 5–7A). This unit reaches 20 m in thickness in the northern part of the study area, but over much of its extent the unit is 10 to 15 m thick. A second transparent deposit occurs south of the east-trending channel; it also shows a mottled surface on the sonographs, but its backscatter strength is not as strong as the deposit north of the levee.

STRATIGRAPHY

The stratigraphic relationships of the different echo character units identified on the 3.5-kHz profiles gives a relative chronology to the different events involved in this episode of channel avulsion. The Mississippi Channel and its associated levees, both up- and down-fan of the site where the levee was breached is the oldest unit identified in this area. The next oldest unit is the acoustically transparent unit that fills this channel down-fan of the breach and probably extends southeastward from the breach in the levee of the Mississippi Channel to form the transparent deposit south of the east-trending levee (Figure 5–6). The transparent layer in the southeastern part of the study area onlaps the flank of the levee of the Mis-

sissippi Channel, and it in turn is buried by the east-trending channel levee system (unit 1 in Figure 5–8). The east-trending channel-levee system is the next youngest in age (units 2 and 3 in Figure 5–8). Segments of abandoned channels that can be seen on the sidescan sonar mosaic show that this channel has not been stationary throughout its evolution (Figure 5–6), and Figure 5–8 shows an older leveed channel system (unit 2 in Figure 5–8) that is partially buried by the most recently active channel (unit 3 in Figure 5–8). The acoustically transparent layer along the north side of the east-trending channel, the youngest unit, onlaps the north flank of the east-trending levee system (unit 4 in Figure 5–8).

Cores recovered from the Mississippi Channel and its levees are composed primarily of clayey silts with some silt laminae approximately 1 cm thick. One core on the northern flank of the channel showed localized faulting. Consolidation testing shows that the sediments from core PC-6 (Figure 5–6) on the levees are overconsolidated (with overconsolidation ratios from 4.1 to 7.5), suggesting that the levees have been eroded. Cores from the levees of the east-trending channel recovered muddy sediment with some thin silt laminations. Sediments from the axis of the east-trending channel are homogeneous muds except for one 10-cm-thick graded silt bed.

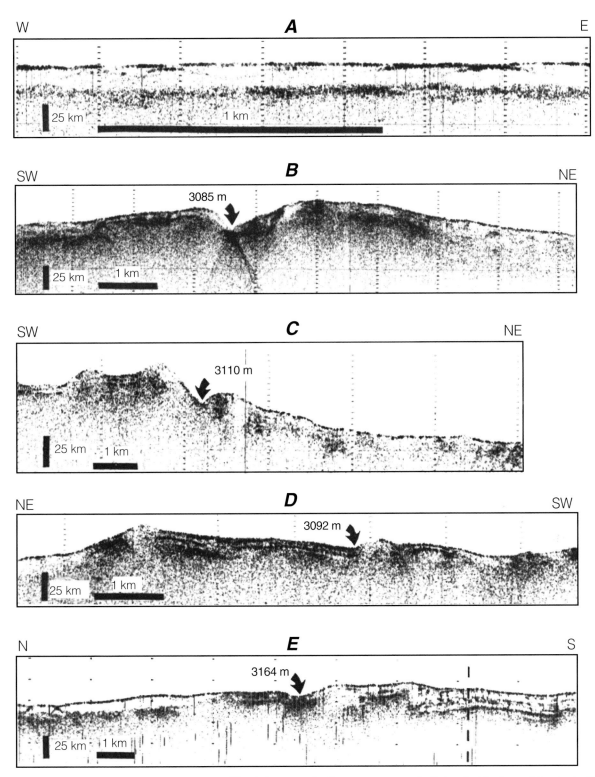

Figure 5–7. Five 3.5-kHz profiles showing: A, profile of the acoustically transparent unit that coincides in aerial extent with the mottled area shown on the sonographs, B, crossing of Mississippi Channel and levee up-fan of the breach, C, crossing of Mississippi Channel-levee system at the point where the levee was breached, D, choked channel down-fan of the breach showing the transparent fill in its floor, and E, crossing of the eastward-trending channel-levee system. Profile locations shown in Figure 5–6.

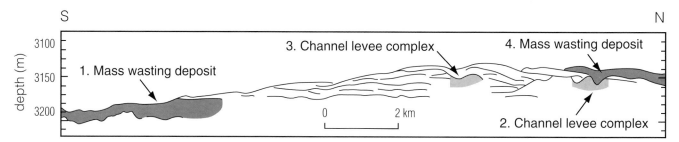

Figure 5–8. Line drawing interpretation of 3.5-kHz profile across the east-trending channel-levee system showing the stratigraphic relationship of this levee system with the two acoustically transparent debris flow deposits. Numbers refer to units discussed in the text. Profile location shown in Figure 5–6.

Discussion

Based on the geometry and acoustic character of the deposits determined from the sidescan sonar images, the subbottom profiles, and the sedimentology of the cored sediments, some inferences can be made as to the evolution of processes responsible for submarine fan channel avulsion and bifurcation. The levees of the Mississippi Channel were constructed by overbank deposition of fine-grained turbidites (Pickering et al. 1986), but the overconsolidated nature of the levee sediments that were cored in the study area suggests subsequent local erosion of the levees. The absence of sand beds in the channel floor indicates either a lack of sandy flows during the recent development of this channel or that sands bypassed this part of the channel. The faulting seen in one core suggests small-scale collapse of the levees, which is further supported by the presence of gullies on the channel walls and high-backscatter patches in the channel floor below some of the gullies that are shown on the sonographs (Figure 5–5A). Erosion of other parts of the levees of the Mississippi Channel has been identified by Prior, Adams, and Coleman (1983); Kastens and Shor (1985); and O'Connell, Ryan, and Normark (1991). The mechanism of erosion is not clear, but is probably due to episodic large mass flows overtopping the levee crests and scouring the levees.

The Mississippi channel-levee system down-fan of where the levee was breached presumably was constructed the same way as the channel farther up-fan, and subsequently this part of the channel was filled. The channel-filling unit is interpreted to be a debris flow deposit because the high-backscatter speckles seen in the channel floor on the sonographs are consistent with reflections from individual large blocks of debris in the channel (Figure 5–5B). Parts of the fill are acoustically transparent on the subbottom profiles as is the case with other debris flow deposits (Embley 1976). Unfortunately, no cores are available to confirm this interpretation.

The areas of mottled backscatter on the sonographs, which occur north and south of the east-trending channel-levee system, are interpreted to be debris flow deposits as well (Fig-

ure 5–6). The hummocky surface return and acoustically transparent nature of these deposits on the subbottom profiles is consistent with echo character studies of other debris flow deposits (Embley 1976). The laminated and bioturbated intervals in the cores seem inconsistent with a debris flow origin; however, these intervals may represent clasts and the homogeneous intervals may represent the matrix supporting these clasts. The volume of sediment in the deposit exceeds the volume of material from the levee, which suggests that the source for this debris flow was, in part, from farther up the fan and that levee material was incorporated as the flow overtopped and broke the levee. This deposit, then, is proposed to be a mix of exotic and locally derived material; however, a more detailed analysis of the cores is needed to confirm this interpretation.

The east-trending channel-levee system presumably was built by turbidites and other small mass flows. The graded silt from the channel floor is interpreted to be a turbidite; the laminated silt beds on the levees are consistent with turbidite deposition. The multiple reflections on the subbottom profiles are characteristic of levees as well (Normark, Piper, and Hess 1979), and the alternating dark and light lineations that converge toward the east on the sonographs may be sediment waves associated with flows that overtop the levees (Figures 5–5C and 5–6).

The stratigraphic evolution of this area suggests an active interplay of large mass movement events that choke the channels and smaller mass flows that either reoccupy existing channels or build new ones. Initially, the southern part of the Mississippi channel system was open and sediments transported through this channel ended up on the southern part of the fan (Figure 5–9A). The flows that were transported through this channel were small enough to be largely contained by the levees although some overtopping must have occurred for the levees to build and to create the crevasse splays, which are still preserved up-fan of the breach (Figure 5–2).

The next episode was the passage of a debris flow (or several flows) through the channel (Figure 5–9B) that was too large to be contained by the levees. The high backscatter associated with this down-fan part of the Mississippi

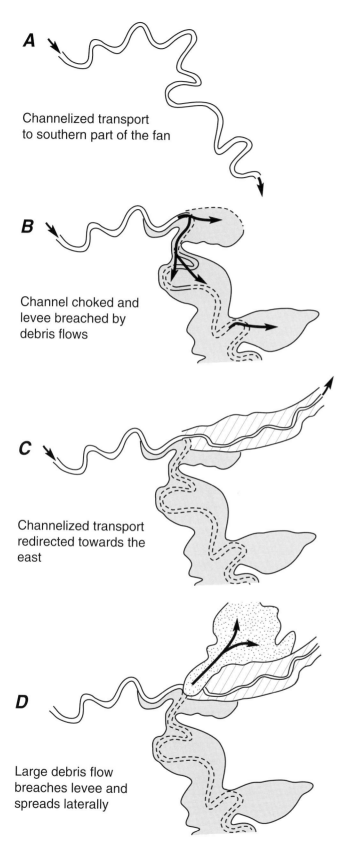

A

Channelized transport
to southern part of the fan

B

Channel choked and
levee breached by
debris flows

C

Channelized transport
redirected towards the
east

D

Large debris flow
breaches levee and
spreads laterally

Figure 5–9. Cartoon interpreting the evolution of the study area: A, sediment transport through the channel towards the south, B, choking of the southern part of the channel and breaching of the levee by a debris flow, C, building of the east-trending levee, D, collapse of the north side of the levee and abandonment of this channel system.

Channel on the GLORIA image (Figure 5–2) is inferred to show the extent of this debris flow deposit. Part of this debris flow, or a subsequent one, was unable to make the abrupt bend in the channel, breached the levee, and formed the deposit in the southeastern part of the study area, which onlaps the levee of the main channel system (Figure 5–6). The full extent of this deposit is unknown because its northern part is buried by the younger east-flowing channel-levee system (unit 1 in Figure 5–8).

Subsequent flows down the Mississippi Channel fed sediment eastward through the gap in the levee because the southern part of the channel was choked (Figure 5–9C). The axis of the southern extension of the channel was 10 m shallower than that of the channel farther up the fan. During this stage a new channel-levee system was built over the debris flow deposits that eventually extended eastward nearly to the Florida Escarpment (Twichell et al. 1992). During the development of this levee system, the channel shifted because abandoned sections of channel are apparent on the sidescan sonar image (Figure 5–6) and buried leveed channels are apparent on some of the subbottom profiles (unit 2 of Figure 5–8). However, its overall location remained fairly constant. Our data do not reveal the causes of these slight changes in the channel location. While this channel system was active, it must have been connected to the Mississippi channel-levee system, but this connection has since been obliterated.

Following the levee-building phase of the east-trending system, another large debris flow (or episode of several debris flows) removed part of the channel and its northern levee and deposited the acoustically transparent unit with the mottled surface along the north side of the east-trending channel-levee complex (Figure 5–9D). The engineering behavior of the cores suggests that the source area for this deposit was in part from farther up the fan and in part material from the breached levee (Figure 5–3). If a new channel-building phase were to start, it would presumably build to the northeast through this gap in the levee.

Since deposition of the debris flow deposit in the northern part of the study area, this section of the Mississippi Channel has been inactive. Cores show that a 0.5- to 1.5-m-thick veneer of Holocene-aged foraminifera-rich ooze covers the entire Mississippi Fan (Ewing, Ericson, and Heezen 1958; Nelson et al. 1992). The presence of some northwest-southeast-trending low-backscatter lineations along the north side of the levee systems (Figure 5–6) may indicate some current reworking of the bottom sediments; however,

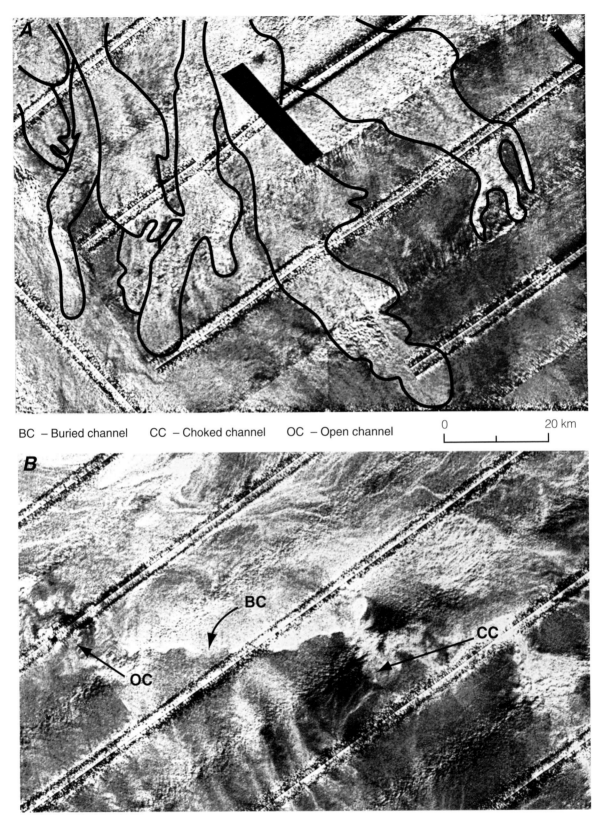

BC – Buried channel   CC – Choked channel   OC – Open channel

0   20 km

Figure 5–10. GLORIA images of two channel choking episodes: A, lower on the fan south of the study area, and B, up-fan of the study area where the channel was partially filled by a catastrophic mass movement (Walker and Massingill 1970). Image locations are shown in Figure 5–1.

94

the preservation of the channels suggests that reworking by bottom currents has not played a major role in modifying the seafloor.

This study of one channel-shifting episode on the Mississippi Fan suggests an active interplay of channel-levee growth, channel choking, and localized mass wasting in the evolution of this channel. The GLORIA imagery and 3.5-kHz subbottom profiles of the entire channel system provide an overview that suggests that other episodes of channel choking have affected the course of this channel as well. South of the study area on the lowermost part of the fan, the channels are too small to be resolved, but the overlapping and crosscutting of the different deposits suggests an active shifting of the channels that supplied these deposits (Figure 5–10A). The next choking episode was the topic of this chapter, and resulted in sedimentation being redirected eastward to a completely new section of the fan. The youngest choking episode within the Mississippi channel system was associated with a catastrophic mass flow deposit on the upper part of the fan. This deposit was initially mapped by Walker and Massingill (1970) and has been dated at 29,000 yr BP (Goodwin and Prior 1989). The GLORIA imagery and a detailed mapping of part of this area by Kastens and Shor (1985) show that part of this deposit buries 45 km of the channel and has partially filled another 75-km section downstream from this area (Figure 5–10B). If this mass movement deposit is associated with the excavation of the Mississippi Canyon, then the deposits lower on the fan predate this 29,000 yr BP time.

Studies of the Mississippi Fan based on seismic-reflection profiles have divided most seismic sequences on the fan into (1) a lower chaotic facies inferred to represent a period dominated by mass wasting and (2) an upper laminated facies inferred to represent a period dominated by turbidity current and hemipelagic sedimentation (Feeley 1984; Stelting et al. 1986; Weimer 1989). Our results, which address just the near-surface veneer of the youngest seismic sequence, suggest that there was an active interplay of mass wasting and turbidity current sedimentation throughout the evolution of this channel system. Deposition from small mass flows that are largely contained by the channel may comprise the bulk of the fan sediments, and the channel-choking episodes resulting from very large mass flows played a major role to redirect sedimentation to different parts of the fan.

Acknowledgments

We thank the officers and crew of the R/V *Farnella* for their capable assistance during the collection of both the GLORIA and SeaMARC sidescan sonar imagery. The success of the GLORIA field program was in large part due to the flawless operation of the GLORIA system by the technical and scientific staff of the Institute of Oceanographic Sciences. The effective operation of the SeaMARC IA system by the staff of Williamson and Associates resulted in an extremely successful cruise as well. J. Zwinakis and D. Blackwood are thanked for drafting and photography, and J. Robb, P. Valentine, P. Manley, and B. Dixon greatly improved this manuscript with their constructive reviews.

References

Chavez, P. S. 1986. Processing techniques for digital sonar images from GLORIA. *Photogram Engrg. Rem. Sens.* 52: 1133–45.

Damuth, J. E., Kolla, V., Flood, R. D., Kowsmann, R. O., Monteiro, M. C., Gorini, M. A., Palma, J. J. C., and Belderson, R. H. 1983a. Distributary channel meandering and bifurcation patterns on the Amazon deep-sea fan as revealed by long-range side-scan sonar (GLORIA). *Geology* 11: 94–8.

Damuth, J. E., Kowsmann, R. O., Flood, R. D., Belderson, R. H., and Gorini, M. A. 1983b. Age relationships of distributary channels on Amazon deep-sea fan: Implications for fan growth pattern. *Geology* 11: 470–3.

Damuth, J. E., Flood, R. D., Kowsmann, R. O., Belderson, R. H., and Gorini, M. A. 1988. Anatomy and growth pattern of Amazon deep-sea fan as revealed by long-range side-scan sonar (GLORIA) and high-resolution seismic studies. *Am. Assoc. Petrol. Geol. Bull.* 72: 885–911.

Danforth, W. W., O'Brien, T. F., and Schwab, W. C. 1991. USGS image processing system: Near real-time mosaicking of high-resolution side scan SONAR data. *Sea Technol.* 32: 54–9.

Droz, L., and Bellaiche, G. 1991. Seismic facies and geologic evolution of the central portion of the Indus Fan. In *Seismic Facies and Sedimentary Processes of Submarine Fans and Turbidite Systems,* eds. P. Weimer and M. H. Link, pp. 383–402. New York: Springer-Verlag.

EEZ-SCAN 85 Scientific Staff 1987. Atlas of the U.S. Exclusive Economic Zone, Gulf of Mexico and eastern Caribbean areas. *U.S. Geological Survey Miscellaneous Investigation Series I-1864-A,* 104 p. Reston, Va.: USGS.

Embley, R. W. 1976. New evidence for occurrence of debris flow deposits in the deep sea. *Geology* 4: 371–4.

Ewing, M., Ericson, D. B., and Heezen, B. C. 1958. Sediments and topography of the Gulf of Mexico. In *Habitat of Oil: Tulsa, Oklahoma,* ed. L. G. Weeks, pp. 995–1053. Tulsa, OK: American Association of Petroleum Geologists.

Feeley, M. H. 1984. Seismic stratigraphic analysis of the Mississippi Fan. Unpublished Ph.D. Dissertation, Texas A&M Univ., 208 p.

Flood, R. D., Manley, P. L., Kowsmann, R. O., Appi, C. J., and Pirmez, C. 1991. Seismic facies and Late Quaternary growth of Amazon Submarine Fan. In *Seismic Facies and Sedimentary Processes of Submarine Fans and Turbidite Systems,* eds. P. Weimer and M. H. Link, pp. 415–33. New York: Springer-Verlag.

Garrison, L. E., Kenyon, N. H., and Bouma, A. H. 1982. Channel systems and lobe construction in the Mississippi Fan. *Geo-Mar. Ltrs.* 2: 31–9.

Goodwin, R. H., and Prior, D. B. 1989. Geometry and depositional sequences of the Mississippi Canyon, Gulf of Mexico. *J. Sed. Petrol.* 59: 318–29.

Kastens, K. A., and Shor, A. N. 1985. Depositional processes of a meandering channel on Mississippi Fan. *Am. Assoc. Petrol. Geol. Bull.* 68: 190–202.

Kenyon, N. H., Amir, A. A., and Cramp, A. 1995. Geometry of the younger sediment bodies of the Indus Fan. In *Atlas of Deep Water Environments*, eds. K. T. Pickering, R. N. Hiscott, N. H. Kenyon, F. Ricci Lucchi, and R. D. A. Smith. London: Chapman and Hall. p. 89–93.

Kosalos, J. G., and Chayes, D. N. 1983. A portable system for ocean bottom imaging and charting. *Oceans 83 Proceedings,* p. 649–56.

Manley, P. L., and Flood, R. D. 1988. Cyclic sediment deposition within the Amazon deep-sea fan. *Am. Assoc. Petrol. Geol. Bull.* 72: 912–25.

McHargue, T. R. 1991. Seismic facies, processes and evolution of Miocene inner fan channels, Indus Submarine Fan. In *Seismic Facies and Sedimentary Processes of Submarine Fans and Turbidite Systems,* eds. P. Weimer and M. H. Link, pp. 403–14. New York: Springer-Verlag.

Mutti, E., and Normark, W. R. 1987. Comparing examples of modern and ancient turbidite systems: Problems and concepts. In *Marine Clastic Sedimentology: Concepts and Case Studies*, eds. J. K. Leggett, and G. G. Zuffa, pp. 1–38. London: Graham and Trotman.

Nelson, C. H., Twichell, D. C., Schwab, W. C., Lee, H. J., and Kenyon, N. H. 1992. Upper Pleistocene turbidite sand beds and chaotic silt beds in the channelized, distal, outer-fan lobes of the Mississippi fan. *Geology* 20: 693–6.

Normark, W. R., Piper, D. J. W., and Hess, G. R. 1979. Distributary channels, sand lobes, and mesotopography of Navy submarine fan, California Borderland, with applications to ancient fan sediments. *Sedimentology* 26: 749–74.

O'Connell, S., and Normark, W. R. 1986. Acoustic facies and sediment composition of the Mississippi Fan drill sites, Deep Sea Drilling Project Leg 96. In *Initial Reports of the Deep Sea Drilling Project, Leg 96*, eds. A. H. Bouma, J. M. Coleman, A. Meyer, et al., pp. 457–74. Washington, D.C.: U.S. Government Printing Office. p. 457–74.

O'Connell, S., Ryan, W. B. F., and Normark, W. R. 1991. Evolution of a fan channel on the outer Mississippi Fan: Evidence from side-looking sonar. In *Seismic Facies and Sedimentary Processes of Submarine Fans and Turbidite Systems,* eds. P. Weimer and M. H. Link, pp. 365–82. New York: Springer-Verlag.

Paskevich, V. 1992. Digital mapping of side-scan sonar data with the Woods Hole Image Processing System software. *U.S. Geological Survey Open-File Report 92-536,* 87 p, Reston, Va.: USGS.

Pickering, K. T., Coleman, J. M., Cremer, M., Droz, L., Kohl, B., Normark, W. R., O'Connell, S., Stow, D., and Meyer-Wright, A. 1986. A high sinuosity, laterally migrating submarine fan channel-levee-overbank: Results from DSDP Leg 96 on the Mississippi Fan, Gulf of Mexico. *Mar. Petrol. Geol.* 3: 3–18.

Prior, D. B., Adams, C. E., and Coleman, J. M. 1983. Characteristics of a deep-sea channel on the middle Mississippi Fan as revealed by a high-resolution survey. *Trans. Gulf Coast Assoc. Geol. Soc.* 33: 389–94.

Shanmugam, G., and Moiola, R. J. 1991. Types of submarine fan lobes: Models and implications. *Am. Assoc. Petrol. Geol. Bull.* 75: 156–79.

Somers, M. L., Carson, R. M., Revie, J. A., Edge, R. H., Barrow, B. J., and Andrews, A. G. 1978. GLORIA II – An improved long-range sidescan sonar. In *Proceedings, IEEE/IERE Subconference on Offshore Instrumentation and Communications,* Oceanology International, 1978, Technical Session J. London: BPS Publ. pp. 16–24.

Stelting, C. E., Droz, L., Bouma, A. H., Coleman, J. M., Cremer, M., Meyer, A. W., Normark, W. R., O'Connell, S., and Stow, D. A. V. 1986. Late Pleistocene seismic stratigraphy of the Mississippi Fan. In Initial Reports of the Deep Sea Drilling Project, eds. A. H. Bouma, J. M. Coleman, and P. W. Meyer, Vol. 96, pp. 437–56. Washington, D. C.: U.S. Government Printing Office.

Twichell, D. C., Kenyon, N. H., Parson, L. M., and McGregor, B. A. 1991. Depositional patterns of the Mississippi Fan surface: Evidence from GLORIA II and high-resolution seismic profiles. In *Seismic Facies and Sedimentary Processes of Submarine Fans and Turbidite Systems,* eds. P. Weimer and M. H. Link, pp. 349–63. New York: Springer-Verlag. p. 349–63.

Twichell, D. C., Schwab, W. C., Nelson, C. H., Kenyon, N. H., and Lee, H. J. 1992. Characteristics of a sandy depositional lobe on the outer Mississippi fan from SeaMARC IA sidescan sonar images. *Geology* 20: 689–92.

Walker, J. R., and Massingill, J. V. 1970. Slump features on the Mississippi Fan, northeastern Gulf of Mexico. *Geol. Soc. Am. Bull.* 81: 3101–8.

Weimer, P., and Buffler, R. T. 1988. Distribution and seismic facies of the Mississippi Fan channels. *Geology* 16: 900–3.

Weimer, P. 1989. Sequence stratigraphy of the Mississippi Fan (Plio-Pleistocene), Gulf of Mexico. *Geo-Mar. Ltrs.* 9: 185–272.

6 Morphology of carbonate escarpments as an indicator of erosional processes

David C. Twichell,[1] William P. Dillon,[1] Charles K. Paull,
and Neil H. Kenyon[3]

[1]U.S. Geological Survey, Woods Hole, Massachusetts
[2]University of North Carolina, Chapel Hill, North Carolina
[3]Institute of Oceanographic Sciences, Southampton, United Kingdom

Abstract

Large sections of the Blake, Florida, and Campeche Escarpments have been imaged by GLORIA, and this common data type allows a comparison of the morphology of these escarpments and inferences about the erosional processes that have shaped them. Four morphologic provinces have been identified as follows: (1) shallow valleys with tributary gullies, which coincide with areas of minimal erosion of the platform edge; (2) box canyons, which overlie areas of differential basement subsidence and fractured carbonate rocks; (3) straight terraced sections coinciding with areas of more uniform basement subsidence, but where varying lithologies exposed at the platform edge are being differentially eroded; and (4) straight unterraced sections where the lithologies of the carbonate rocks appear to be uniform. These different provinces are interpreted to be surficial expressions of processes that have shaped these escarpments through their histories.

Introduction

The escarpments at the edges of the Blake-Bahama, Florida, and Yucatan carbonate platforms are some of the largest cliffs on the surface of the earth. The Blake Escarpment extends along the eastern side of Florida and the northern Bahamas for about 450 km and has as much as 4,000 m relief. The Florida Escarpment extends along the western side of Florida for about 650 km and has about 2,000 m relief. The Campeche Escarpment rings the eastern and northern margins of the Yucatan Peninsula and has about 2,000 m relief (Figure 6–1). These escarpments formed initially by the growth of reefs and carbonate shelf-edge banks during the Mesozoic (Bryant et al. 1969), and subsequently were sculpted and steepened by a variety of erosional processes (Paull and Dillon 1980; Freeman-Lynde 1983; Freeman-Lynde and Ryan 1985; Dillon, Paull, and Gilbert 1985; Dil-

lon et al. 1988; Dillon and Popenoe 1988; Corso, Austin, and Buffler 1989; Paull et al. 1991a). Average gradients on these escarpments commonly exceed 20 degrees, and locally they are vertical (Jordan and Stewart 1959; Lindsay, Shipley, and Worzel 1975; Paull et al. 1990a; Dillon et al. 1993). One problem in studying these escarpments is the inability of conventional geophysical tools (particularly vertical-incidence seismic techniques) to image these cliff faces because of their steepness, which causes variable angles of refraction of seismic rays (Dillon et al. 1988). An understanding of the erosional processes affecting localized areas of these cliffs has been inferred from sampling and submersible observations (Freeman-Lynde 1983; Dillon, Valentine, and Paull 1987; Paull et al. 1990b), but, because of the length of these cliffs and because the base of the Blake Escarpment exceeds the depth of submersible dives, a full understanding of the processes that have shaped these cliffs has not been attainable by these methods.

GLORIA sidescan sonar images were collected along large sections of these three escarpments. This imagery provides a consistent and complete view of the three escarpments and permits comparing and contrasting the morphology and implied processes that have shaped them.

Geologic and oceanographic setting

The Blake, Florida, and Campeche Escarpments are part of a large early Cretaceous system of reefs that rings the Gulf of Mexico and extends northward under the Atlantic continental shelf and slope to Canada (McFarlan and Silvio Menes 1991). These escarpments initially probably developed by the vertical growth of reefs on subsiding, thinned continental crust (Antoine 1968), but in the Middle Cretaceous reef growth ended, probably due to a combination of sea level rise and subsidence of the platform edges (Locker and Buffler 1984; Dillon and Popenoe 1988; Paull et al. 1990b). The crust under these escarpments is not uniform and has influenced the subsequent evolution of the carbonate escarpments (Locker and Buffler 1984; Klitgord and Schouten 1986). The Jacksonville Fracture Zone (Figure

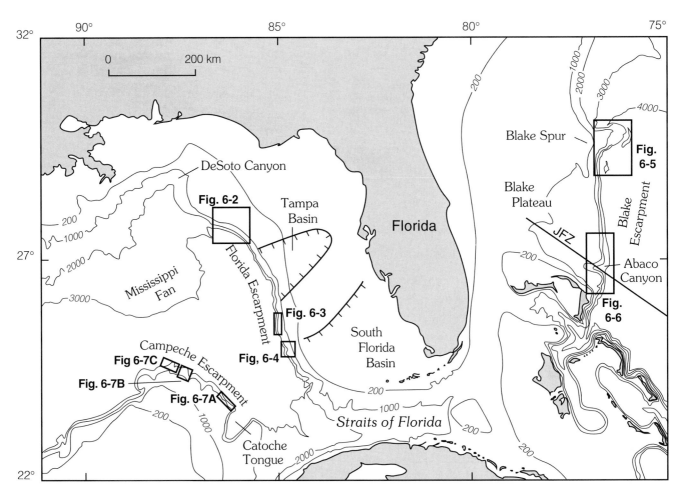

Figure 6–1. Location map showing the Blake, Florida, and Campeche Escarpments, some of the known tectonic features that underlie the escarpments, and figure locations. JFZ = Jacksonville Fracture Zone from Klitgord and Schouten (1986). Contours in meters.

6–1) is aligned with the southern edge of the Blake Plateau Basin, which underlies the Blake Plateau shoreward of the escarpment (Dillon and Popenoe 1988), and Abaco Canyon coincides with where this fracture zone crosses the escarpment (Mullins et al. 1982). Shoreward of the Florida Escarpment, the Tampa and South Florida Basins mark places where basement subsidence has been greater than under the intervening arches (Figure 6–1). The Campeche Escarpment and the platform shoreward of it lack adequate seismic coverage to map the presence and distribution of basins, but the Catoche Tongue at the southern end of the escarpment is thought to be fault controlled, suggesting that the crust under this platform is not uniform either (Shaub 1983).

Since the middle Cretaceous, and perhaps also simultaneously with their development, the escarpments have undergone erosion. The Blake Escarpment has undergone as much as 20 km of erosional retreat of its base (Paull and Dillon 1980), the Florida Escarpment as much as 8 km (Corso et al. 1989), and the Campeche Escarpment an unknown amount, but the presence of carbonate talus deposits in localized areas along its base (Halley and Schlager 1983)

suggests that at least parts of it have undergone some erosion as well.

The Florida and Campeche Escarpments were subjected to different geologic and oceanographic conditions than the Blake Escarpment during the late Cretaceous and Cenozoic. The Blake Escarpment was isolated from continent derived sediment by current flow through the Suwannee Strait in the late Cretaceous and since Paleocene/Eocene by the powerful Gulf Stream (Popenoe 1985; Dillon and Popenoe 1988). The eastern Gulf of Mexico, by contrast, has been a site of active terrigenous sedimentation throughout the Cenozoic, and the Florida and Campeche Escarpments both have thick accumulations of Pliocene and Pleistocene sediments associated with the Mississippi Fan onlapping their bases (Bouma, Stelting, and Coleman 1983/84). Sedimentation rates during the latest Pleistocene have been as much as 4 m/1,000 yr in the DSDP holes drilled nearest the escarpments (Kohl et al. 1986).

The physical oceanography of the Atlantic and Gulf of Mexico are different as well. Although strong surface currents flow over the platforms shoreward of all three escarpments (the Loop Current in the Gulf of Mexico and the Gulf

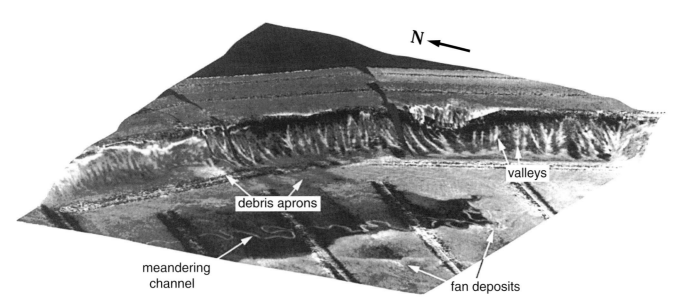

Figure 6–2. Oblique view of GLORIA image over bathymetry digitized from NOS bathymetric charts (NOS 1979, 1983) of part of the northern Florida Escarpment. Valleys with tributary gullies start on the upper part of the escarpment and end at its base where high-backscatter aprons extend away from the valley mouths onto the abyssal plain floor. Mass wasting is minimal on the slope above this part of the escarpment. Image covers an area 40 km wide by 64 km deep, and the vertical exaggeration is 5X. Figure location shown in Figure 6–1.

Stream and Antilles Current in the Atlantic), the deepwater currents and chemistry of the deep waters are different. A strong southward-flowing western boundary current has swept along the face of the Blake Escarpment and actively eroded the cliff face since the Oligocene (Paull and Dillon 1980). This bottom current also brings corrosive North Atlantic Deep Water in contact with the lower part of this escarpment, causing dissolution of its base (Paull and Dillon 1980). The Gulf of Mexico does not have strong bottom currents, and the bases of the Florida and Campeche Escarpments presently are above the lysocline; thus dissolution by bottom waters presently does not play a role in the erosion of these cliffs (Paull et al. 1991b).

Along the base of the Florida Escarpment, dense saline brines derived from evaporites under the Florida Platform are discharged along the base of this escarpment and are proposed to be actively dissolving and undercutting the base of this cliff (Paull and Neumann 1987). It is unknown whether similar seeps are present along the bases of the Blake and Campeche Escarpments.

Data

GLORIA data were collected along large sections of the three escarpments (Figure 6–1). The GLORIA survey of the Campeche Escarpment was conducted in 1982 and consists of only one line. This line imaged the escarpment mostly from its seaward side and also imaged a 5- to 15-km-wide swath of the slope above the escarpment. This survey line abuts the EEZ-SCAN survey of the Gulf of Mexico that was completed in 1985 when the abyssal plain floor below the escarpment was completely imaged (EEZ-SCAN 85 Scientific Staff 1987). The Florida Escarpment, from DeSoto Canyon to the Straits of Florida, was imaged in 1985 along with a 20- to 50-km-wide swath of the slope above the escarpment, and the entire Mississippi Fan along the base of the escarpment. The Blake Escarpment and Blake Spur were imaged with GLORIA in 1987 including a 10- to 25-km swath of the outer Blake Plateau above the escarpment and a similar swath of the abyssal plain floor below the escarpment (EEZ-SCAN 87 Scientific Staff 1991).

Besides the GLORIA imagery, the Blake and Florida Escarpments have complementary data sets whereas the Campeche Escarpment has not been as extensively studied. Bathymetry of the northern Blake Escarpment was compiled from a dense net of single-beam echo sounding profiles (Gilbert and Dillon 1981), and the stratigraphy of at least the upper half of the escarpment was derived from direct sampling of the escarpment face (Dillon et al. 1987). A multibeam bathymetric survey of the Florida Escarpment provides a detailed understanding of the morphology of a 220-km section of this escarpment between 25° and 27° N (Paull et al. 1991a; Twichell, Paull, and Parson 1991). The stratigraphy of this escarpment was developed using seismic stratigraphy (Antoine, Bryant, and Jones 1967; Mitchum 1978; Corso et al. 1989), dredge samples (Freeman-Lynde 1983), and sampling from the submersible ALVIN (Paull et

A

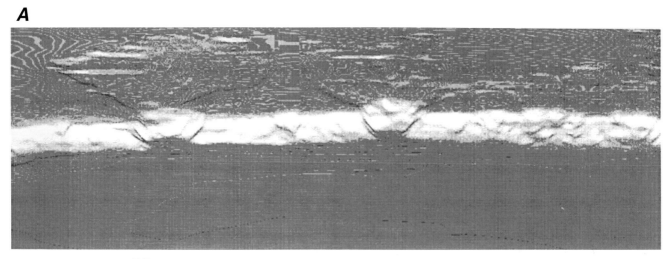

B   N ←

Figure 6–3. Oblique views of the Florida Escarpment using SeaBeam bathymetry and GLORIA imagery around 27° N showing A, the SeaBeam bathymetry alone and B, the GLORIA image superimposed on the SeaBeam bathymetry. This image shows two small box canyons and the surrounding straight terraced escarpment. Terraces are limited to the lower part of the escarpment. Scarps and deposits associated with mass wasting are present on the slope above the escarpment. Image covers an area 65 km wide by 56 km deep, the vertical exaggeration is 1X, and its location is shown in Figure 6–1.

al. 1991a). Because of the greater amount of data available from the Florida and Blake Escarpments, the discussion of these two escarpments is more complete, and that of the Campeche Escarpment is more speculative.

Results

Four distinctive morphological provinces have been identified in the data from these three escarpments. These provinces include the following: (1) sections cut by small valleys with tributary gullies, (2) sections incised by box canyons, (3) straight terraced sections, and (4) straight sections with no terraces. The distribution of these different provinces is discussed relative to each escarpment.

Florida Escarpment

Three of the morphologic provinces were identified on the Florida Escarpment (Twichell, Parson, and Paull 1990). The first province covers the area north of 27° N. Here the escarpment is relatively straight and has gradients less than 28° (Jordan and Stewart 1959). The GLORIA image shows that this section of the escarpment is dissected by numerous valleys spaced 1 to 5 km apart with tributary gullies feeding into them (Figure 6–2). Dredge and core samples from this part of the escarpment recovered Cenozoic sediments (Freeman-Lynde 1983), suggesting that this part of the escarpment has a thin discontinuous Cenozoic sediment cover. The presence of this sediment cover as well as possible reef structures in the underlying Cretaceous section (Antoine 1968) suggest that this part of the escarpment has not un-

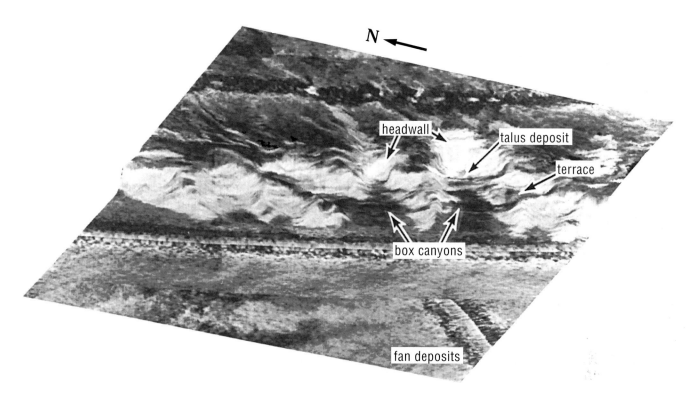

Figure 6–4. Oblique view of box canyons in the Florida Escarpment using SeaBeam bathymetry and GLORIA imagery showing the steep headwall scarps, the talus deposits with their high-backscatter surfaces below the headwall scarps, the terraced sidewalls, and the mass wasting of the slope above the escarpment. The image covers an area 26 by 26 km, the vertical exaggeration is 1X, and its location is shown in Figure 6–1.

dergone extensive erosion (Locker and Buffler 1984; Corso et al. 1989). The slope above this part of the escarpment is smooth and is interrupted by only a few mass wasting scarps (Twichell et al. 1990). Below the escarpment, high-backscatter aprons extend away from the valleys (Figure 6–2). These aprons are inferred to be carbonate debris shed off the escarpment that interfinger with late Pleistocene-aged fan deposits (Twichell et al. 1990).

The other two morphologic provinces occur south of 27° N. Here, part of the escarpment is relatively straight and terraced and parts are deeply incised by regularly spaced large canyons that are strikingly different from the valleys to the north. On the straight part of the escarpment, which occurs between 25.5° and 26.5° N, the lower part of the escarpment is terraced whereas the upper part is steeper and unterraced (Figure 6–3). Terraces have gradients of 10° to 20° while the terrace risers and the upper part of the escarpment have gradients of 30° to 42° as determined by the multibeam bathymetry (Twichell et al. 1991).

The third morphologic province is characterized by canyons that incise the edge of the platform as much as 15 km, are 1 to 3 km wide at their heads, and 3 to 7 km wide at their mouths (Figure 6–4). The floors of these canyons are flat and at the same depth as the adjacent abyssal plain floor except immediately below the canyon's headwalls where depressions occur that are as much as 80 m deeper

than the abyssal plain floor (Twichell et al. 1990). Below the headwalls, a high-backscatter return on the sidescan images is confirmed by bottom photographs to show the extent of talus deposits eroded from the headwalls (Paull et al. 1990a). Gradients of the headwalls exceed 40° and they do not have terraces, whereas the sidewalls are not as steep and have discontinuous terraces (Figure 6–4). These canyons, which are called box canyons to differentiate them from the valleys north of 27° N, are concentrated in two groups along this southern part of the escarpment, one group occurring between 26.5° and 27° N and the second between 24.3° and 25.6° N (Twichell et al. 1990). The slope above the escarpment south of 27° N is extensively scarred by mass wasting (Figures 6–3 and 6–4), commonly with the material being funnelled into the heads of the box canyons (Twichell, Valentine, and Parson 1993).

Blake Escarpment

The same three morphologic provinces that were identified on the Florida Escarpment are present on the Blake Escarpment. The first province of shallow valleys with tributary gullies occurs immediately south of the Blake Spur (Figure 6–5). These valleys do not extend above the top of the escarpment, and where they open onto the abyssal plain floor, high-backscatter aprons are present that are inferred

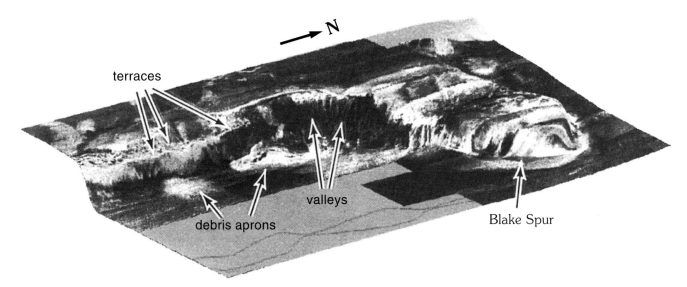

Figure 6–5. Oblique view of the Blake Spur and the Blake Escarpment immediately to the south based on bathymetry compiled by Dillon et al. (1993). Note the valleys with tributary gullies immediately south of the Blake Spur and the terraces on the upper part of the escarpment even farther to the south. Image covers an area 200 km wide by 120 km deep, the vertical exaggeration of this image is 5X, and its location is shown in Figure 6–1.

to be material derived from these valleys (Dillon et al. 1993). This morphologic province is not as extensive as it is on the Florida Escarpment, and its location immediately south of the Blake Spur suggests that its presence is caused by this part of the escarpment being sheltered from strong bottom currents (Dillon et al. 1993).

The second province, a straight escarpment with terraces, represents the section extending south from the area of shallow valleys nearly to Abaco Canyon (27.5° to 29° N). In contrast to the Florida Escarpment, however, the terraces are concentrated on the upper part of the Blake Escarpment, and the lower part is smooth. The bathymetry shows the lower part of the Blake Escarpment to be steeper than the upper part (Figure 6–5).

Between 27° and 27.5° N, three box canyons incise the Blake Escarpment. These canyons incise the escarpment 10 to 35 km, are 5 to 15 km wide, and have large cone-shaped deposits with high-backscatter surfaces extending out onto the abyssal plain from their mouths (Figure 6–6). These high-backscatter cones are interpreted to be debris aprons derived from the canyons (Dillon et al. 1993).

Campeche Escarpment

The GLORIA images of the Campeche Escarpment show two of the four morphologic provinces, but only one of the ones found on the other two escarpments. The southernmost 60 km of the escarpment that was imaged represents one morphological province. This southern section of the escarpment is relatively straight, and terraces were not imaged by GLORIA. The lack of terraces is assumed to be real, but in the absence of additional data their absence may be due to the instrument being saturated by the high amplitude of the returned signal from this steep, hard cliff masking the subtle backscatter differences associated with the terraces. The moderate backscatter stripes in the low-backscatter left side of the image that parallel the escarpment (Figure 6–7A) are artifacts that result from multiple travel paths of the sidescan signal. This section of the escarpment is interrupted by three small box canyons that indent the escarpment less than 2.5 km (Figure 6–7A). The 2-km offset in the face of the escarpment approximately in the center of the image may indicate the location of a fault (Figure 6–7A). Seismic data are not available to confirm this interpretation.

The remaining northern 140 km of the escarpment is incised by seven box canyons that appear similar to those on the Florida and Blake Escarpments. The southern four of these canyons are smaller and more closely spaced than the three to the north (Figure 6–7B). The southern canyons incise the escarpment about 4 km and are 3 to 5 km wide at their mouths. The northern three canyons incise the escarpment as much as 10 km and are 6 to 8 km wide at their mouths (Figure 6–7C). The narrow high-backscatter return from their headwalls suggests that these parts of the cliffs are steep, and the moderate backscatter areas below some of the headwalls may be talus deposits (Figure 6–7C). Discontinuous terraces are present along the sides of some of these canyons (Figure 6–7C). Scarps associated with mass wasting of the slope above the Campeche Escarpment are abundant along the northern part of the study area where the trackline crossed the escarpment and imaged the slope above the escarpment (Figure 6–7B). Sediments associated with

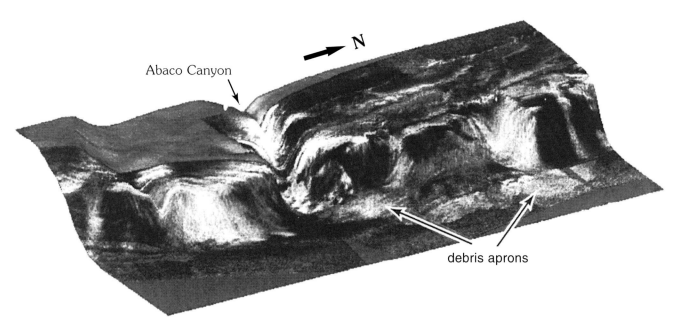

Figure 6–6. Oblique view of a section of the Blake Escarpment incised by box canyons. Bathymetry was compiled by Dillon and others (1993). Note the high-backscatter return from the canyon floors and the abyssal plain floor at the mouths of the canyons. The image covers an area 110 km wide by 50 km deep, the vertical exaggeration of this image is 5X and its location is shown in Figure 6–1.

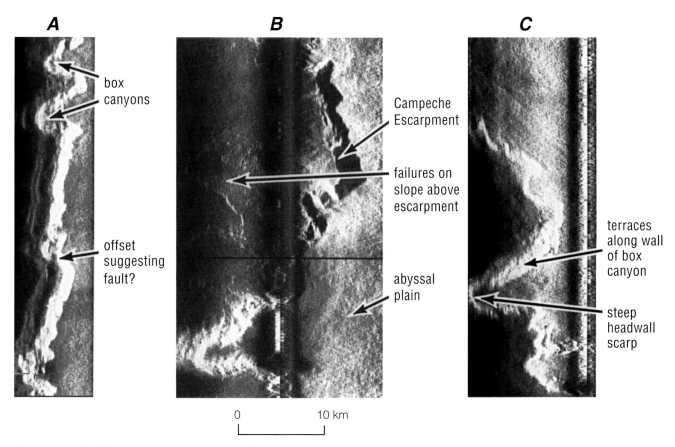

Figure 6–7. GLORIA images of the Campeche Escarpment showing A, straight section of the escarpment with small box canyons; B, larger box canyons and mass wasting scarps and deposits on the slope above the escarpment; and C, large well-developed box canyons with discontinuous terraces along their sides and talus deposits below their headwall scarps. The abyssal plain is to the right in each figure.

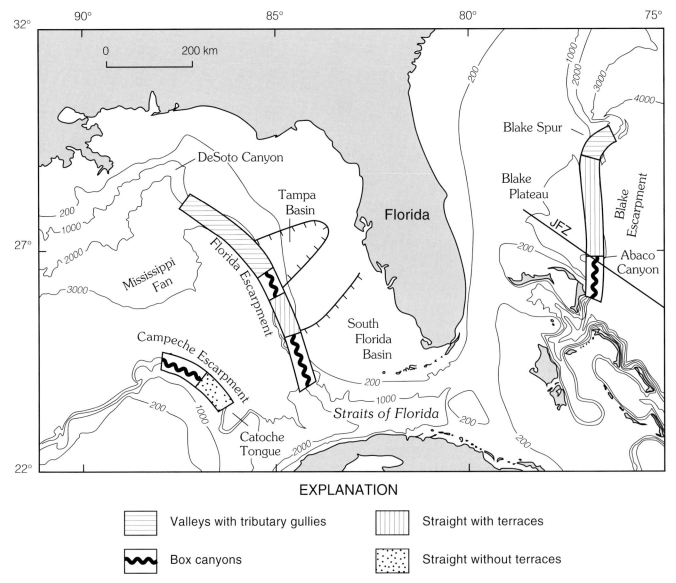

EXPLANATION

Valleys with tributary gullies	Straight with terraces
Box canyons	Straight without terraces

Figure 6–8. Summary map showing the distribution of the four morphologic provinces identified along the Blake, Florida, and Campeche Escarpments.

the Mississippi Fan lap up against the base of this escarpment as they do along the base of the Florida Escarpment.

Discussion

Four distinctive morphological provinces are found along these carbonate escarpments (Figure 6–8), and these differences suggest differences in the erosional processes that have sculpted these escarpments. The first of these is characterized by the small valleys with tributary gullies, which occur in areas that have undergone no, or only minor, erosion. The second is characterized by the box canyons, which indicate areas where fracturing of the carbonate strata due to differential subsidence of the underlying basement has

enhanced and focused other erosional processes at the sites of the canyons. The third is characterized by straight terraced sections, which suggest that varying lithologies have undergone differential erosion, but where the underlying basement has not been subjected to differential subsidence, so the rocks are not as fractured. The fourth province is characterized by straight unterraced sections and may indicate areas where a more uniform lithology is exposed than along the terraced sections.

The valleys with tributary gullies occur along the northern part of the Florida Escarpment and immediately south of the Blake Spur on the Blake Escarpment, but were not observed on the Campeche Escarpment (Figure 6–8). Their distribution along the Florida Escarpment coincides with the section of the escarpment where reeflike mounds are found along the platform edge on seismic profiles and rudists were cored from the cliff face (Bryant et al. 1969). This part of the escarpment is not as steep as the part south of 27° N, and core and dredge samples recovered Cenozoic sediments (Bryant et al. 1969; Freeman-Lynde 1983) rather than Cretaceous limestones. These observations suggest that this part of the escarpment has not been eroded during the Cenozoic. The valleys shown on the GLORIA image (Figure 6–2) do not reach the top of the Florida Escarpment, and the absence of mass wasting on the slope above this part of the escarpment suggests that the debris aprons along the base of the escarpment are derived from mass wasting of the Cenozoic veneer covering the escarpment rather than from the slope above the escarpment (Twichell et al. 1991). Along the Blake Escarpment, the section of the escarpment immediately south of the Blake Spur also is inferred not to have undergone erosion because of the presence of these gullied valleys (Figure 6–5). In this case, the lack of erosion may be caused by this section of the escarpment being sheltered by the Blake Spur from strong south-flowing bottom currents (Dillon et al. 1993).

The box canyons represent a distinctive morphology that is found on all three escarpments (Figure 6–8). Along the Blake and Florida Escarpments, their distribution is controlled by deeper structural features underlying the platform edges. Where the Jacksonville Fracture Zone crosses the Blake Escarpment, which marks the southern edge of the Blake Plateau Basin (Dillon and Popenoe 1988), three box canyons are present (Figures 6–6 and 6–8); their origin is attributed to the fracturing of the carbonate strata due to differential subsidence across the fracture zone (Mullins et al. 1982; Dillon et al. 1993). Along the Florida Escarpment, the extent of the box canyons coincides with the locations of basins under the West Florida shelf (Figure 6–8). Faulting of the Cretaceous and Jurassic strata is concentrated in the parts of the platform overlying the basins (Shaub 1984; Ball et al. 1988), and it has been attributed to differential subsidence of the basins (Twichell et al. 1990) although the collision of Cuba with the Florida-Bahama Platform in the Early Cenozoic may have contributed to the faulting as well (Klitgord and Schouten 1986). Differential subsidence of the basement may also explain the location of box canyons along the Campeche Escarpment, but other complementary data are not available to confirm this interpretation. The offset in the Campeche Escarpment shown in Figure 6–7A suggests that deeper structural control may be influencing the morphology of this escarpment. Fault control has been suggested for the origin of the Catoche Tongue (Shaub 1983).

The box canyons probably formed over a long period of time, and the presence of talus deposits at the base of their headwalls suggests that they continue to undergo erosion (Figures 6–4, 6–6, and 6–7C). Their headwalls are interpreted to be sites of ongoing erosion; otherwise, at least in the Gulf of Mexico, the deposits would have been rapidly buried during the Pleistocene by sediments associated with the Mississippi Fan. The presence of terraces along the walls of these canyons suggests that other erosional processes have continued to modify the walls of these canyons after they were initially cut (Figures 6–4 and 6–7C). Presumably differential erosion has enhanced subtle lithologic differences along these cliff faces (Twichell et al. 1990). Buried extensions of these canyons have been traced seaward of the Florida Escarpment under early Pleistocene sediments. The seismic coverage, however, is not adequate to map their full extent and define when they initially started being infilled (Lubinski and Twichell 1992).

The third morphologic province found on two of the three escarpments is the straight terraced sections (Figure 6–8). Along these sections of the escarpments, the Cretaceous limestones are exposed on the seafloor, but the erosion has not been as extensive as in the box canyons. The terraces may result from differential erosion of subtle lithologic differences in the strata exposed along these cliff faces, and may indicate sections of the platform that are not as heavily fractured as where the box canyons are found (Twichell et al. 1991). Along the Florida Escarpment, terraces occur on the lower part of the escarpment whereas the upper part is smooth, which suggests that the limestones exposed on the upper part of the escarpment are lithologically more uniform than the lower part (Figure 6–3). Alternatively, saline brines that cause corrosion of limestone by modifying its solubility have been observed seeping out along the base of the Florida Escarpment and along some of the terraces (Paull and Neumann 1987; Paull et al. 1991a). The distribution of terraces may indicate that brines are seeping out at several levels along the lower part of the escarpment rather than just at its base (Twichell et al. 1991).

Terraces also were observed along the Blake Escarpment, but here they occur on the upper rather than the lower part of the escarpment (Figure 6–5). If these terraces are attributed to differences in lithology alone, then the implication is that the lower part of this escarpment is uniform and the upper part is lithologically more varied. Additionally, the absence of terraces on the Blake Spur (Figure 6–5) implies that there are lateral as well as vertical differences in lithology on this escarpment. The absence of terraces on the lower part of the Blake Escarpment may indicate a uniform lithology or erosional processes that are intense enough to mask the lithologic differences. Scouring and dissolution of the lower part of this escarpment by bottom waters is an alternative explanation for the smooth, steep lower cliff face (Dillon and Paull 1980). The absence of clear evidence of terraces along the Campeche Escarpment (Figures 6–7A and 6–8) may be a function of the quality of the data or it may indicate a more uniform lithology or more uniform processes acting on this escarpment.

In summary, the comparison of these three escarpments allows us to assess the relative importance of the different processes that have been proposed for erosion of different localities along these three escarpments. Morphologically these escarpments can be divided into four provinces that suggest significant variability in the structural and depositional history of these platform edges. It appears that the box canyons owe their heritage, at least in part, to the tectonic activity of the basement that underlies these platform edges. Differential subsidence of the basement once the carbonate rocks were loaded on it resulted in the fracturing of carbonate rocks. Once fractured, other processes, either internal such as groundwater sapping (Paull and Neumann 1987; Paull et al. 1990a; Twichell et al. 1991) or external such as dissolution by bottom water have intensified erosion at these sites (Paull and Dillon 1980; Freeman-Lynde and Ryan 1985; Dillon et al. 1987, 1993). The terraced and unterraced sections of the different escarpment overlie sections where the basement has acted as a uniform, undisturbed block (Figure 6–8). The same erosional processes that acted in the box canyons probably were active along the straight sections; however, because the carbonate strata were not as fractured the morphological expression is different. Here, the terraces appear to reflect lithologic differences caused by depositional variations that have been accentuated by subsequent differential erosion of the cliff face. The sections of the escarpments with valleys and tributary gullies that were incised into Cenozoic sediment that blankets parts of the escarpments indicate areas where there has been minimal erosion. These valleys appear to cut just the Cenozoic veneer and presumably are the result of gravity-driven processes acting on the unconsolidated sediments on these steep slopes.

Acknowledgments

We would like to thank the captains and crews of the R/V *Farnella* and R/V *Atlantis II* who ensured the successful collection of the GLORIA imagery and SeaBeam bathymetry. We would also like to thank Tim LeBas of IOS for finding the GLORIA data of the Campeche Escarpment and John Risch for the work he did while an NAGT fellow at the USGS in Woods Hole, Massachusetts, processing the GLORIA imagery from the Blake Escarpment and compiling the digital bathymetry that was used to create Figures 6–5 and 6–6. Jeff Zwinakis and Dann Blackwood did the drafting and photography for this manuscript. Raymond Freeman-Lynde, Stanley Locker, Eric Schmuck, and Kathryn Scanlon provided constructive reviews of the manuscript.

References

Antoine, J. W., Bryant, W. R., and Jones, B. 1967. Structural features of continental shelf, slope, and scarp, northeastern Gulf of Mexico. *Am. Assoc. Petrol. Geol. Bull.* 52: 257–62.

Antoine, J. W. 1968. A study of the West Florida Escarpment. *Trans. Gulf Coast Assoc. Geol. Soc.* 18: 297–303.

Ball, M. M., Martin, R. G., Foote, R. Q., Leinback, J., Applegate, A. V., Nichols, D., O'Brien, T. F., Dodd, J. E., and Irwin, B. J. 1988. Seismic and subsurface structure and stratigraphy of the western Florida shelf. *U.S. Geological Survey Open-File Report 88-439,* 60 p. Reston, Va.: USGS.

Bouma, A. H., Stelting, C. E., and Coleman, J. M. 1983/84. Mississippi Fan: Internal structure and depositional processes. *Geo-Mar. Ltrs.* 3: 147–53.

Bryant, W. R., Meyerhoff, A. A., Brown, N. K., Furrer, M. A., Pyle, T. E., and Antoine, J. W. 1969. Escarpments, reef trends, and diapiric structures, eastern Gulf of Mexico. *Am. Assoc. Petrol. Geol. Bull.* 53: 2506–42.

Corso, W., Austin, J. A., and Buffler, R. T. 1989. The early Cretaceous platform off northwest Florida: Controls on morphologic development of carbonate margins. *Mar. Geol.* 86: 1–14.

Dillon, W. P., Paull, C. K., and Gilbert, L. E. 1985. History of the Atlantic continental margin off Florida: The Blake Plateau Basin. In *Geologic Evolution of the United States Atlantic Margin,* ed. C. W. Poag, pp. 189–215. New York: Van Nostrand Reinhold.

Dillon, W. P., Valentine, P. C., and Paull, C. K. 1987. The Blake Escarpment – A product of erosional processes in the deep ocean. In *Symposium Series for Undersea Research, NOAA's Undersea Research Program,* Vol. 2, No. 2, eds. R. A. Cooper and A. N. Shepard, pp. 177–90. Washington D.C.: National Oceanographic and Atmospheric Administration.

Dillon, W. P., Trehu, A. M., Valentine, P. C., and Ball, M. M. 1988. Eroded carbonate platform margin – The Blake Escarpment off southeastern United States. In *Atlas of Seismic Stratigraphy,* ed. A. W. Bally, pp. 40–7. Studies in Geology Series, No. 27, Vol. 2. Tulsa, OK: American Association of Petroleum Geologists.

Dillon, W. P., and Popenoe, P. 1988. The Blake Plateau Basin and Carolina Trough. In *The Geology of North America,* Vol. I-2, *The Atlantic Continental Margin, U.S.,* eds. R. E. Sheridan and J. A. Grow, pp. 291–328. Boulder, Colo.: Geological Society of America.

Dillon, W. P., Risch, J. S., Scanlon, K. M., Valentine, P. C., and Huggett, Q. J. 1993. Ancient crustal fractures control the location and size of collapsed blocks at the Blake Escarpment, east of Florida. *U.S. Geological Survey Bulletin 2002,* pp. 54–9, Reston, Va.: USGS.

EEZ-SCAN 85 Scientific Staff 1987. Atlas of the U.S. Exclusive Economic Zone, Gulf of Mexico. *U.S. Geological Survey Miscellaneous Investigation Series I-1864-A,* 104 p., Reston, Va.: USGS.

EEZ-SCAN 87 Scientific Staff 1991. Atlas of the U.S. Exclusive Economic Zone, Atlantic continental margin. *U.S. Geological Survey Miscellaneous Investigations Series I-2054,* 174 p., Reston, Va.: USGS.

Freeman-Lynde, R. P. 1983. Cretaceous and Tertiary samples dredged from the Florida Escarpment, eastern Gulf of Mexico. *Trans. Gulf Coast Assoc. Geol. Soc.* 33: 91–9.

Freeman-Lynde, R. P., and Ryan, W. B. F. 1985. Erosional modification of Bahama Escarpment. *Geol. Soc. Am. Bull.* 96: 481–94.

Gilbert, L. E., and Dillon, W. P. 1981. Bathymetric map of the Blake Escarpment. *U.S. Geological Survey Miscellaneous Field Investigations Map, MF-1362,* scale 1:250,000. Reston, Va.: USGS.

Halley, R. B., and Schlager, W. 1983. Alternative diagenetic models for Cretaceous talus deposits, Deep Sea Drilling Project site 536, Gulf of Mexico. In *Initial Reports of the Deep-Sea Drilling Project, Leg 77,* eds. R. T. Buffler, W. Schlager, et al., pp. 397–408. Washington, D.C.: U.S. Government Printing Office.

Jordan, G. F., and Stewart, H. B. 1959. Continental slope off southwest Florida. *Am. Assoc. Petrol. Geol. Bull.* 43: 974–91.

Klitgord, K. D., and Schouten, H. 1986. Plate kinematics of the central Atlantic. In *The Geology of North America,* Vol. M, *The Western North Atlantic Region,* eds. P. R. Vogt and B. E. Tucholke, pp. 351–78. Boulder, Colo.: Geological Society of America.

Kohl, B., Williams, D. F., Ledbetter, M. T., Constans, R. E., King, J. W., Heuser, L. E., Schroeder, C., and Morley, J. J. 1986. Summary of chronostratigraphic studies, Deep Sea Drilling Project, Leg 96. In *Initial Reports of the Deep Sea Drilling Project,* Vol. 96, eds. A. H. Bouma, J. M. Coleman, A. W. Meyer, et al. pp. 589–600. Washington, D.C.: U.S. Government Printing Office.

Lindsay, J. F., Shipley, T. H., and Worzel, J. L. 1975. Role of canyons in the growth of the Campeche Escarpment. *Geology* 3: 533–6.

Locker, S. D. and Buffler, R. T. 1984. Comparison of Lower Cretaceous carbonate shelf margins, northern Campeche Escarpment and northern Florida Escarpment, Gulf of Mexico. In *Seismic Expression of Structural Styles – A Picture and Work Atlas,* ed. P. W. Bally, pp. 123–8. Studies in Geology Series, No. 15, Vol. 2, American Association of Petroleum Geologists.

Lubinski, D. J., and Twichell, D. C. 1992. A preliminary Mid-Cretaceous to Late Pleistocene seismic stratigraphy for the deep eastern Gulf of Mexico adjacent to the Florida Escarpment. *U.S. Geological Survey Open-File Report 92-255,* 71 p. Reston, Va.: USGS.

McFarlan, E., and Silvio Menes, L. 1991. Lower Cretaceous, In *The Geology of North America,* Vol. J, *The Gulf of Mexico Basin,* ed. A. Salvador, pp. 181–204. Boulder, Colo.: Geological Society of America.

Mitchum, R. M. 1978. Seismic stratigraphic investigation of west Florida slope, Gulf of Mexico. In *Framework, Facies, and Oil Trapping Characteristics of the upper continental margin,* eds. A. H. Bouma, G. T. Moore, and J. M. Coleman, pp. 193–223. Studies in Geology 7. Tulsa, OK: American Association of Petroleum Geologists.

Mullins, H. T., Keller, G. H., Kofoed, J. W., Lambert, D. N., Stubblefield, W. L., and Warme, J. E. 1982. Geology of Great Abaco submarine canyon (Blake Plateau): Observations from the research submersible "ALVIN." *Mar. Geol.* 48: 239–57.

National Ocean Survey (NOS) 1979. *The Elbo: Bathymetric map NG-16-3,* Scale 1:250,000, Washington D.C.: National Oceanographic and Atmospheric Administration.

National Ocean Survey (NOS) 1983. *Lloyd: Bathymetric map NG-16-2,* Scale 1:250,000, Washington D.C.: National Oceanographic and Atmospheric Administration.

Paull, C. K., and Dillon, W. P. 1980. Erosional origin of the Blake Escarpment: An alternative hypothesis. *Geology* 8: 538–42.

Paull, C. K., and Neumann, A. C. 1987. Continental margin brine seeps: Their geological consequences. *Geology* 15: 545–8.

Paull, C. K., Speiss, F. N., Curray, J. R., and Twichell, D. C. 1990a. Origin of Florida Canyon and the role of spring sapping on the formation of submarine box canyons. *Geol. Soc. Am. Bull.* 102: 502–15.

Paull, C. K., Freeman-Lynde, R. P., Bralower, T. J., Gardemal, J. M., Neumann, A. C., D'Argenio, B., and Marsella, E. 1990b. Geology of the strata exposed on the Florida Escarpment. *Mar. Geol.* 91: 177–94.

Paull, C. K., Twichell, D. C., Spiess, F. N., and Curray, J. R. 1991a. Morphological development of the Florida Escarpment: Observations on the generation of time transgressive unconformities in carbonate terrains. *Mar. Geol.* 101: 181–201.

Paull, C. K., Commeau, R. F., Curray, J. R., and Neumann, A. C. 1991b. Seabed measurements of modern corrosion rates on the Florida Escarpment. *Geo-Mar. Ltrs.* 11: 16–22.

Popenoe, P. 1985. Cenozoic depositional and structural history of the North Carolina margin from seismic analyses. In *Geologic Evolution of the United States Atlantic Margin,* ed. C. W. Poag, pp. 125–87. New York: Van Nostrand Reinhold.

Shaub, F. J. 1983. Origin of the Catoche Tongue, southeastern Gulf of Mexico. In *Seismic Expression of Structural Styles; A Picture and Work Atlas,* ed. A. W. Bally, pp. 2.2.3–129–2.2.3–140. Studies in Geology 15, Vol. 2. Houston, TX: American Association of Petroleum Geologists.

Shaub, F. J. 1984. The internal framework of the southwestern Florida bank. *Trans. Gulf Coast Geol. Soc.* 34: 237–45.

Twichell, D. C., Parson, L. M., and Paull, C. K. 1990. Variations in the styles of erosion along the Florida Escarpment, eastern Gulf of Mexico. *Mar. Petrol. Geol.* 7: 253–66.

Twichell, D. C., Paull, C. K., and Parson, L. M. 1991. Terraces on the Florida Escarpment: Implications for erosional processes. *Geology* 19: 897–900.

Twichell, D. C., Valentine, P. C., and Parson, L. M. 1993. Slope failure of carbonate sediment on the West Florida Slope. In *Submarine Landslides: Selected Studies in the U.S. Exclusive Economic Zone,* eds. W. C. Schwab, H. J. Lee, and D. C. Twichell, pp. 69–78. U.S. Geological Survey Bulletin 2002. Reston, Va.: USGS.

7 Sedimentary processes in the salt deformation province of the Texas-Louisiana continental slope

David C. Twichell and Catherine Delorey

U.S. Geological Survey, Woods Hole, Massachusetts

Abstract

GLORIA sidescan sonar imagery, multibeam bathymetry, and seismic-reflection data collected from a 102-by-111-km area of the Texas-Louisiana continental slope were used to identify the surficial and deeper features that define sedimentary processes that have acted on this continental margin, which has been intensely deformed by salt tectonics. Distinctive high-backscatter patterns on the GLORIA images outline areas that appear to be associated with rough seafloor around hydrocarbon vents and areas of mass wasting. Integration of the images with the bathymetric and seismic-reflection data show that the vent-related features are associated with faults and diapirs suggesting that these are the conduits through which deeper hydrocarbons may be escaping to the seafloor. In addition, mass wasting deposits are found along the base of the Sigsbee Escarpment, which marks the edge of the salt province, but are absent (or below the resolution of the GLORIA system) below slopes within the salt province that are equal in steepness to the Sigsbee Escarpment. The distribution of mass wasting deposits suggests that, in the recent past, the Sigsbee Escarpment has been tectonically more active or has been subjected to different processes than the remainder of the salt province behind it.

Introduction

The continental slope south of Texas and Louisiana in the Gulf of Mexico is unique within the U.S. Exclusive Economic Zone because of the massive Jurassic-aged Louann salt which underlies it (Salvador 1991). The morphology of this continental slope is complex and dramatically different from other continental slopes due to the mobility of the salt. Rather than submarine canyons crossing the slope as seen off the east or west coasts of the United States, this area is characterized by numerous circular to irregularly shaped ba-

thymetric highs and depressions (Bryant, Simmons, and Grim 1991; Grim 1992). High-resolution seismic-reflection profiles, sidescan sonar images, submersible observations, and bottom sediment samples have been used to map the presence of hydrocarbon vents (Behrens 1988; Kennicutt et al. 1988; Anderson and Bryant 1990; Kennicutt and Brooks 1990; Roberts et al. 1990; Roberts, Cook, and Sheedlo 1992), areas of sediment eruption (Prior, Doyle, and Kaluza 1989; Neurauter and Bryant 1990), brine seeps (Aharon, Roberts, and Snelling 1992), carbonate buildups (Roberts, Sassen, and Aharon 1987; Roberts et al. 1989), sites of mass wasting (Bouma 1981; Coleman et al. 1989), and sediment pathways across the slope (Satterfield and Behrens 1990; Swenson and Bryant 1992). These studies have mapped the distribution of features associated with these processes in localized areas, usually related to lease block acquisition, but a regional overview of the distribution of these features, their correlation with the complex bathymetry, and their relation to the deeper subsurface geology has been lacking because of the localized acquisition of data. Vent rather than seep is used as a general term to describe sites where hydrocarbons are escaping the seafloor because the GLORIA images cannot differentiate between sites of seeping versus eruption.

The recent acquisition of multibeam bathymetry by the National Ocean Survey, NOAA (Grim 1992), over much of this continental slope provides a detailed understanding of the morphology of this slope for the first time. GLORIA sidescan sonar imagery collected by the USGS provides a nearly continuous acoustic image of this area (EEZ-SCAN 85 Scientific Staff 1987). The combination of these two data sets, in concert with available high- and low-frequency seismic profiles, has been used in a study of a 102-by-111-km area of the middle and lower slope (Figure 7–1).

We have relied on other detailed studies in the Gulf of Mexico that fall within the GLORIA coverage of the EEZ-SCAN project to calibrate the GLORIA image because ground-truth information is not yet available in the study area. Of particular note are a detailed high-resolution seismic study of a 25-by-33-km area of hydrocarbon vents on

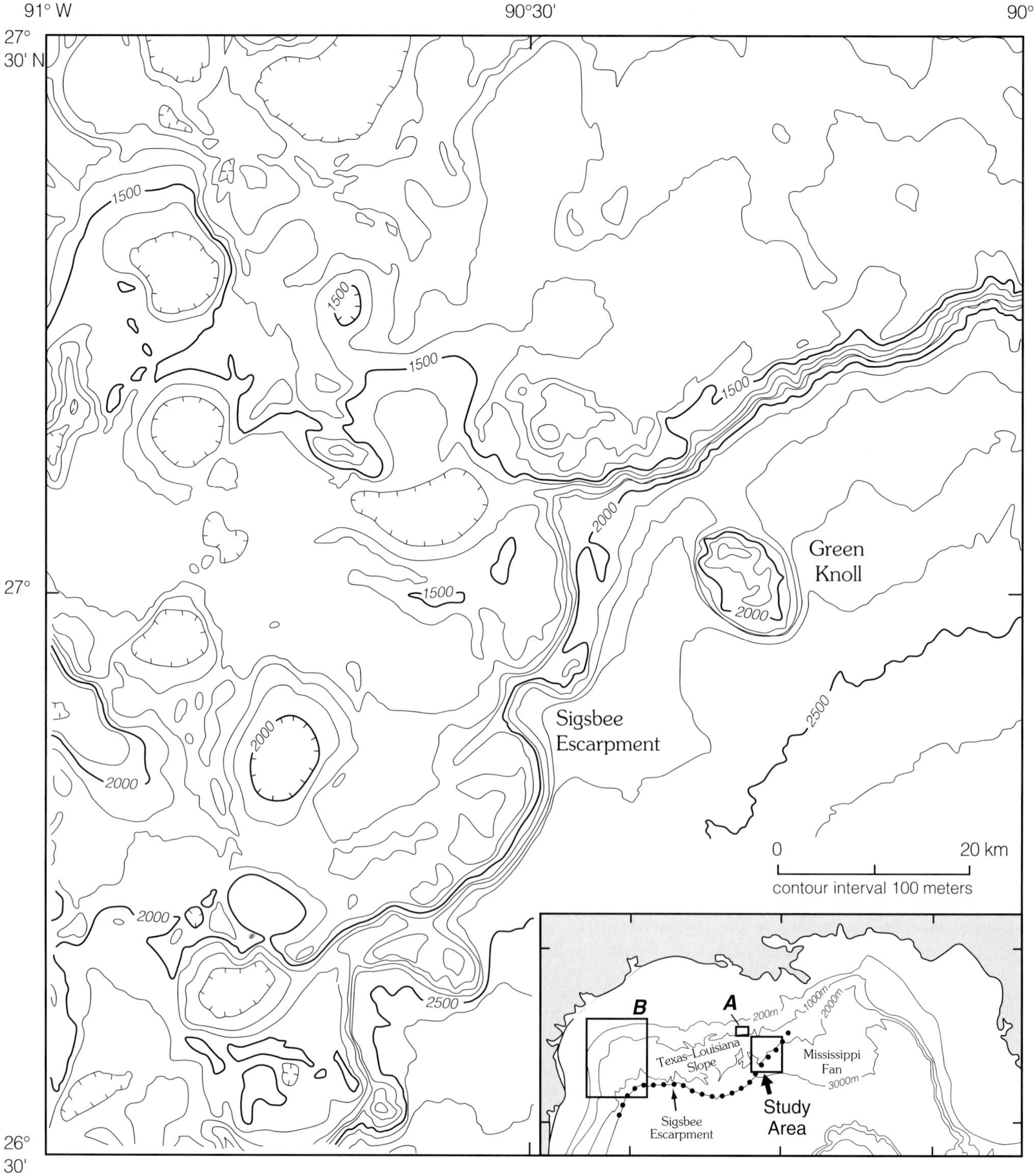

Figure 7–1. Map showing the bathymetry of the study area based on the NOAA multibeam bathymetry (contour interval 100 m). The index map shows the study area, the detailed seismic study by Behrens (1988) (box labeled A), the study by Rothwell et al. (1991) (box labeled B), and the extent of the Texas-Louisiana slope, Sigsbee Escarpment, and Mississippi Fan.

the upper slope (Behrens 1988) and a study of mass wasting deposits in the northwestern corner of the Gulf of Mexico (Rothwell, Kenyon, and McGregor 1991) (Figure 7–1). The detailed echo character study conducted by Behrens

(1988) used profiles spaced 1 to 5 km apart and identified an echo character type with a prolonged surface return and no subsurface reflections that coincided with the occurrence of hydrocarbon vents or carbonate buildups associated with

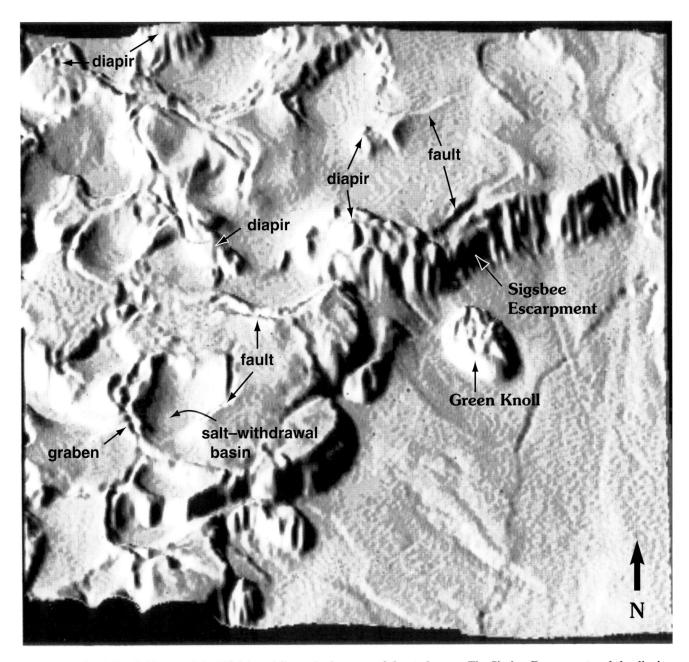

Figure 7–2. Shaded relief image of the NOAA multibeam bathymetry of the study area. The Sigsbee Escarpment and the diapirs mark places where salt is near the seafloor, and these areas have a rougher surface than the salt withdrawal basins and other areas where the salt is more deeply buried. This image covers the same area shown in Figure 7–1 and represents an area 111 km in the north-south direction and 102 km in the east-west direction.

these vents. In the western Gulf of Mexico, Rothwell et al. (1991) show a close correlation between the distribution of an acoustically transparent echo character type with areas of high-backscatter on the GLORIA image. Cores indicate that these areas represent debris flow deposits. These observations, in conjunction with available seismic data, are then used to infer the presence and distribution of sedimentary deposits derived from various processes in the study area.

Data

The data include multibeam bathymetry collected by NOAA (Grim 1992), GLORIA sidescan sonar images, 3.5-kHz high-resolution subbottom profiles, and airgun seismic-reflection profiles (EEZ-SCAN 85 Scientific Staff 1987). The multibeam bathymetry was collected digitally and processed by NOAA at a 250-m pixel size using a UTM projection, and was divided into areas representing half a degree of latitude by one degree of longitude. Two multibeam image files covered the study area and were combined, projected to an Albers equal area projection, and scaled to a 50-m pixel size by pixel duplication to match the scale and projection of the GLORIA image using the USGS mini image processing system (Chavez 1984). The GLORIA data were digitally collected during 1982 and 1985 (EEZ-SCAN 85 Scientific Staff 1987), then processed using techniques developed by Chavez (1986), and put in their proper geographic location using software developed by Paskevich (1992). Airgun seismic-reflection profiles and 3.5-kHz subbottom profiles were collected concurrently with the GLORIA imagery. The airgun used a 40, 80, or 160 in^3 chamber, the receiver was a two-channel hydrophone, and the data were printed on an analog paper recorder. Navigation during the GLORIA surveys was primarily by Loran C.

Results

The study area can be divided into two morphological provinces; the Texas-Louisiana Slope, which is underlain by salt, and the Mississippi Fan province, which lies seaward of the salt (Figure 7–1). These two provinces are separated by the Sigsbee Escarpment, which is 400 to 600 m high, has seafloor gradients that locally exceed 22° (Pratson 1993), and runs diagonally from the northeast to the southwest through the study area. The Mississippi Fan lies southeast of the Sigsbee Escarpment, and here the seafloor is deeper (2,100 to 2,950 m) and smoother than northwest of the escarpment. The Texas-Louisiana Slope is characterized by a more rugged bathymetry and depths that range from 950 to 2,150 m.

Notable morphologic features of the Texas-Louisiana Slope are the numerous circular to elliptical depressions. They are 2 to 15 km across, are mostly 60 to 180 m deeper than the surrounding seafloor, and have smooth floors (Figure 7–1). The bathymetrically high areas tend to be more irregular in shape than the depressions, and most of them are underlain by diapirs (Figure 7–2). Faults appear either as linear scarps and grabens or as arcuate terraces on a shaded relief image derived from the multibeam bathymetry (Figure 7–2). The arcuate terraces occur mostly on the southern or southeastern sides of the basins, and seismic profiles show these terraces to be associated with down-to-the-basin faults (Figure 7–3A). The other faults are not associated with the basins and instead appear as grabens or normal faults, and where seismic-reflection profiles are available, the grabens sit above diapiric intrusions (Figure 7–3B).

The seismic-reflection profiles show the relationship of the surface morphology to the deeper structure. The salt is part of a shallow salt sheet that is being pushed up and seaward by sediment loading (Seni 1991) and, with the exception of Green Knoll, is observed only on the profiles from the Texas-Louisiana Slope province. The salt has an irreg-

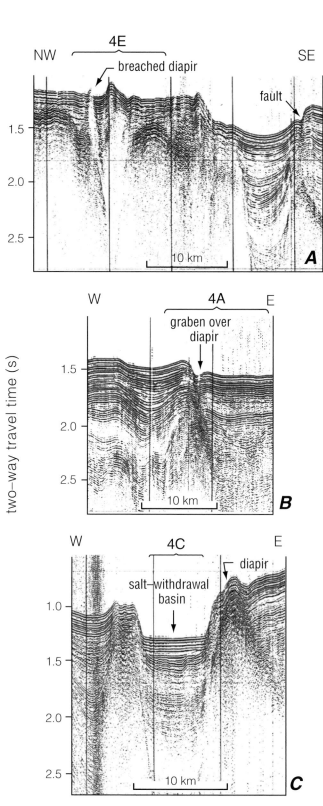

Figure 7–3. Single-channel airgun seismic-reflection profiles showing A, profile crossing a salt-withdrawal basin with a fault on its southeastern side and a breached diapir northwest of the basin (note the highly reflective surface of the buried salt); B, a narrow wall of salt separating two salt-withdrawal basins with a graben above it; and C, a salt-withdrawal basin with salt diapirs exposed or nearly exposed on either side. Profile locations shown in Figure 7–6.

ular, highly reflective surface on the seismic profiles (Figure 7–3A), which is shaped into salt diapirs and narrow salt walls that separate numerous areas of thick sediment fill (Figure 7–3). Most of the areas of thick sediment coincide with circular basins in the bathymetry called salt-withdrawal basins (Seni 1992). Our widely spaced profiles suggest that much of the recent sediment filling these basins is a drape that thins or pinches out along the sides of the basins and shows little internal deformation. Other studies of this area with more extensive seismic coverage have been able to map some of the sediment pathways deeper in the section through which sediments were transported across this margin (Swenson and Bryant 1992). Unconformities are present within the basin fill, but the reconnaissance nature of these data does not permit a detailed mapping of the structure or stratigraphy of the basins. Away from the basins, the limited seismic coverage shows the salt to be near the seafloor along the Sigsbee Escarpment as well as under some diapirs and along some faults (Figure 7–3). Over the diapirs and along the Sigsbee Escarpment the multibeam bathymetry showed a hummocky surface (Figure 7–2), and this morphology was used to map the spatial distribution of the salt subcrops. Most of the salt subcrop is along the Sigsbee Escarpment and upslope of the escarpment; however, one diapir, Green Knoll, is found seaward of the escarpment (Figure 7–2).

The 3.5-kHz profiles show five distinctive near-surface echo character types (Figure 7–4). Their distribution will be discussed in relation to the GLORIA image, and only the acoustic nature of each facies will be described here. The most common echo character type is the acoustically laminated facies (Figure 7–4A). This facies shows as a series of parallel or subparallel reflections that occur mostly in areas of smooth seafloor, but when relief is present, the laminations mimic the underlying morphology. The second echo character type is an acoustically transparent unit with a highly reflective, and usually rough, surface that occurs as lens-shaped deposits that can exceed 25 m in thickness (Figure 7–4B). The third unit also is acoustically transparent, but it has a smooth, moderately reflective surface (Figure 7–4C). It can reach 18 m in thickness. The fourth unit has a prolonged surface return with no subbottom reflections below it (Figures 7–4D and E). In most cases, this unit has a strong surface return, but in localized areas it has a weak

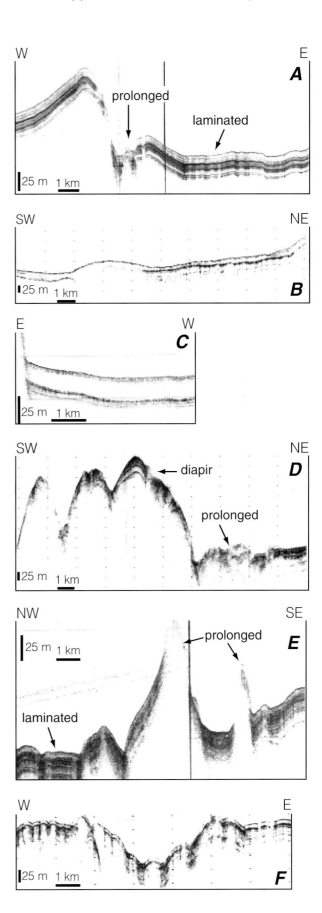

Figure 7–4. Echo sounding profiles showing the different echo character types. A shows the laminated facies characteristic of hemipelagic sedimentation, B shows the acoustically transparent facies with the highly reflective surface that is characteristic of debris flows, C shows the transparent facies with the weakly reflective surface that floors some salt-withdrawal basins, D shows the rough surface of a salt dome and the acoustically prolonged signature associated with vents, E shows the limited extent of the prolonged facies at the surface of salt diapirs, and F shows an area of hyperbolae that do not show on the GLORIA imagery and are of uncertain origin. Profile locations shown in Figure 7–6.

surface return. Moreover, this unit commonly has a rough surface, but not as rough as the hyperbolae that compose the fifth echo character type. Hyperbolae are associated mostly with areas of high relief such as the Sigsbee Escarpment or the flanks of salt diapirs, but locally they occur in areas of seemingly flat seafloor as well (Figure 7–4F).

The GLORIA imagery and 3.5-kHz profiles show a good correlation of backscatter patterns with some, but not all, of the echo-character types. In the area on the upper slope studied by Behrens (1988) (Figure 7–1), for example, there is a close correlation between the extent of the high-backscatter areas on the GLORIA image and the echo character type showing the prolonged surface return with no subsurface reflections (Figure 7–5). This same relationship holds for the Texas-Louisiana Slope part of the study area (Figure 7–6). High-backscatter fanlike patches that extend from the base of the Sigsbee Escarpment onto the Mississippi Fan (Figure 7–7) coincide with the extent of the acoustically transparent unit with the highly reflective surface (Figure 7–4B) and are similar to the debris flow deposits mapped by Rothwell et al. (1991) in the western Gulf of Mexico. The granular low-to-moderate-backscatter pattern that covers the Mississippi Fan province and much of the remainder of the Texas-Louisiana Slope province coincides with the distribution of the laminated facies on the subbottom profiles and presumably represents hemipelagic deposition (Figure 7–6).

Some of these echo character types, however, are not differentiated by the GLORIA system. The acoustically transparent unit with the weak surface return (Figure 7–4C) occurs on the floors of three salt-withdrawal basins, yet these three are not distinguished from the others, which are all floored by the laminated facies (Figure 7–4A). Where hyperbolae coincide with areas of relief such as along the Sigsbee Escarpment or over salt domes, the GLORIA image shows a strong acoustic return, often with discontinuous bands trending either parallel or perpendicular to the slope. The bands that parallel the slope are terraces and those perpendicular to the slope are gullies; the hyperbolae on profiles crossing these areas are indicating this rough surface. Where hyperbolae occur in areas of flat seafloor in the Texas-Louisiana Slope province (Figure 7–4E), GLORIA does not show a signature that can be distinguished from that associated with the laminated facies. In most cases, this hyperbolic echo character type occurs along the upper flanks of salt-withdrawal basins, but its origin is not understood.

The integration of the GLORIA imagery, bathymetry, subsurface geology, and echo character is summarized in Figure 7–6, and the relationship of the GLORIA image to the bathymetry is shown in Figures 7–7, 7–8, and 7–9. The Sigsbee Escarpment shows on the GLORIA image either as an area of high backscatter or low backscatter depending on whether the system is looking onshore at the escarpment or is looking offshore down over the top of it (Figure 7–7). Along the base of the escarpment, the high-backscatter

patches that fan out from valleys in the escarpment coincide with the distribution of lenses of acoustically transparent material on the echo sounding profiles (Figure 7–4B), and they are interpreted to be debris flow deposits. These debris flow deposits are common along the base of the Sigsbee Escarpment and around Green Knoll (Figures 7–6 and 7–7).

Faults that are identified on the seismic profiles and multibeam bathymetry usually show as discontinuous high-backscatter patches on the GLORIA images (Figures 7–6 through 7–9). Where the faults are linear, such as the one shown in Figure 7–7, small circular to elliptical high-backscatter patches are found closely associated with the fault. The terraces associated with the arcuate normal faults on the southern sides of some salt-withdrawal basins (Figure 7–9) tend to have larger high-backscatter patches that are more irregular in shape. The discontinuous distribution of these patches argues against their being solely a consequence of seafloor steepness. In Figure 7–8, for example, the part of the basin wall that abuts a salt-related feature (Figure 7–3C) shows discontinuous high-backscatter patches whereas the southern flank of this basin, which is removed from underlying salt but is of equal steepness, has a moderate-backscatter signature. The high backscatter along part of the basin wall may indicate areas where salt is exposed on the seafloor. Along the southern rim of this basin, the high-backscatter patches are present on the terrace at the top of the basin wall that is associated with a fault. Where the high-backscatter patches cross the tracklines, the 3.5-kHz profiles show a prolonged acoustic return (Figures 7–4A, D, and E). In Figure 7–9, discontinuous linear high-backscatter patches radiating from a salt dome may indicate the presence of radial faults that cannot be resolved by the sparse seismic coverage.

The surfaces of diapirs that breach, or nearly breach, the seafloor have a moderate backscatter signature with some high-backscatter patches (Figures 7–7 through 7–9). On Green Knoll (Figure 7–7) one of the high-backscatter patches coincides with where brine seeps were identified during submersible dives (Aharon et al. 1992). In this area, numerous cirquelike features were observed during the dives; these features were inferred to be associated with dissolution of the salt and are assumed to be the cause of the high-backscatter signature.

The salt-withdrawal basins all have featureless, moderate-backscatter floors and discontinuous patches of high acoustic backscatter around their margins (Figures 7–6, 7–8, and 7–9). All but three of the basins are floored by the laminated echo character facies (Figure 7–4A), and the other three have an acoustically transparent bed (Figure 7–4C) at the surface that cannot be differentiated by GLORIA. The high-backscatter patches surrounding the basins have already been discussed and consequently will not be discussed here. The laminated echo character facies that covers the rest of the salt deformation province and much of the Mississippi Fan surface within the study area shows on the GLORIA image as a granular acoustic signature of moderate intensity.

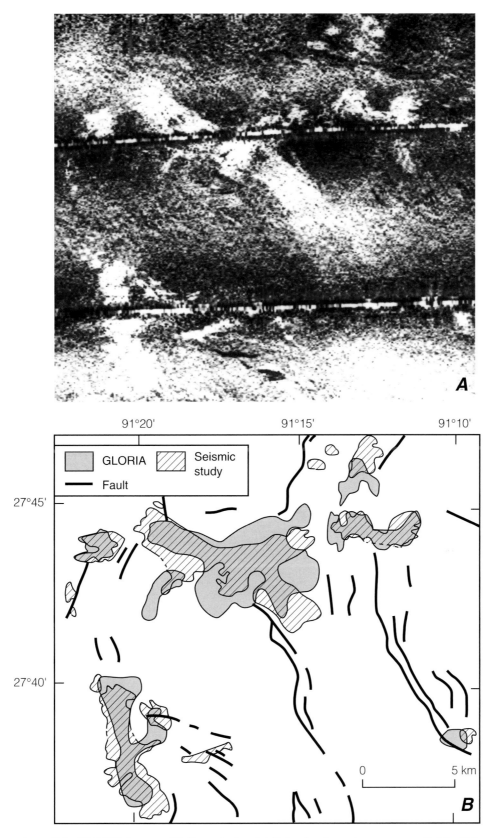

Figure 7–5. Comparison of A, GLORIA image, with B, part of a detailed echo character map compiled by Behrens (1988). In B, GLORIA pattern refers to high-backscatter patches; seismic study pattern refers to acoustic wipeout areas mapped by Behrens (1988) on 3.5-kHz subbottom profiles. Note that most of the faults are not apparent on the GLORIA image, but that the distribution of the acoustic wipeout areas corresponds closely with the high-backscatter patches on the GLORIA image.

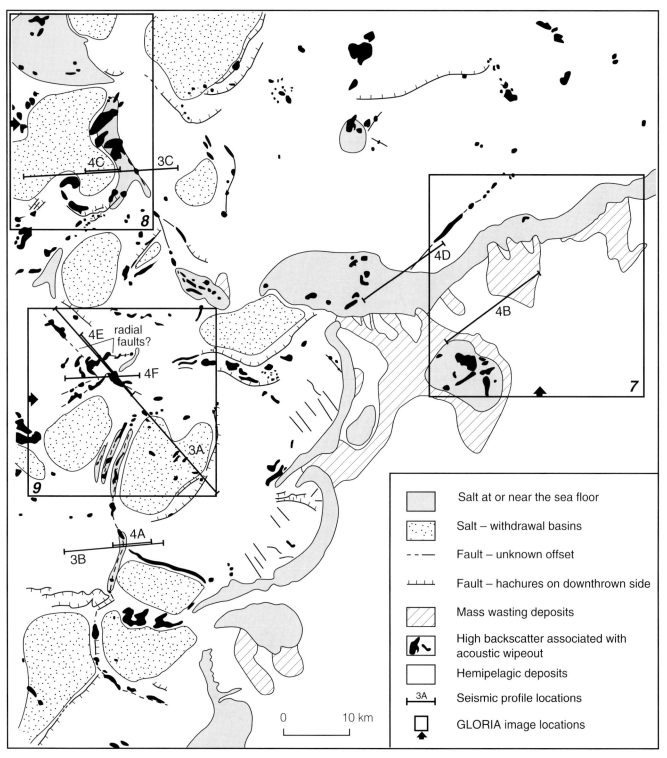

Figure 7–6. Interpretive map showing the distribution of salt-withdrawal basins, locations where salt is predicted to be near the seafloor, faults, high-backscatter patches associated with acoustic wipeout (probably indicating hydrocarbon vents around salt diapirs and faults), debris flow deposits, and areas of hemipelagic sedimentation. Profile locations shown in Figures 7–3 and 7–4 are shown by labeled lines, and images shown in Figures 7–7, 7–8, and 7–9 are shown by labeled boxes.

Legend:

Salt at or near the sea floor

Salt – withdrawal basins

Fault – unknown offset

Fault – hachures on downthrown side

Mass wasting deposits

High backscatter associated with acoustic wipeout

Hemipelagic deposits

3A — Seismic profile locations

GLORIA image locations

0   10 km

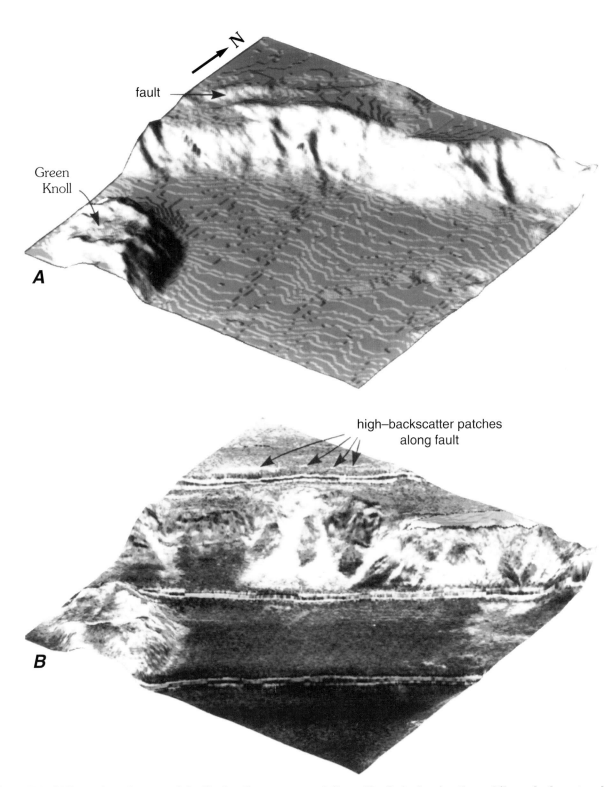

Figure 7–7. Oblique view of a part of the Sigsbee Escarpment and Green Knoll: A, showing the multibeam bathymetry alone illuminated from the lower left of the image and B, the GLORIA image draped over the bathymetry showing the high-backscatter areas on top of Green Knoll and along a linear fault above the Sigsbee Escarpment that presumably reflect carbonate buildups; and high-backscatter deposits along the base of the Sigsbee Escarpment and in an apron around Green Knoll that represent debris flow deposits. A 33-by-33-km area is shown in this image, and the vertical exaggeration is 9X. Location shown in Figure 7–6.

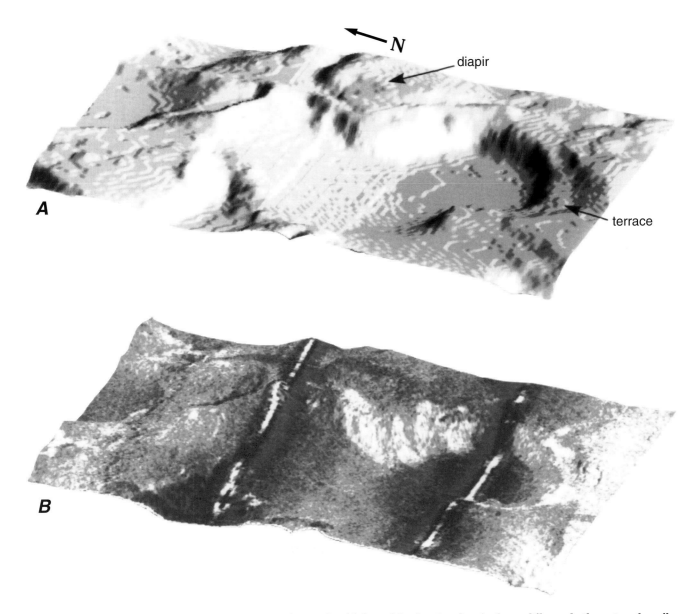

Figure 7–8. Oblique view of a salt dome and parts of two salt-withdrawal basins showing A, the multibeam bathymetry alone illuminated from the left side of the image, and B, the GLORIA image draped over the bathymetry showing the patchy distribution of the high-backscatter features on top of the salt dome, along parts of the wall of the salt-withdrawal basin, and on terraces surrounding the basins. Seismic profile shown in Figure 7–3C, and 3.5-kHz profile shown in Figure 7–4C cross the salt-withdrawal basin in the forefront of the scene. A 32-by-21-km area is shown in this image, and the vertical exaggeration is 9X. Location shown in Figure 7–6.

Discussion

The use of multiple data sets was needed to show the relationship of the surficial geology and morphology to the subsurface geology. This approach was necessary because of the known relationship in this area of surficial processes with deeper structures. For example, Bouma, Stelting, and Leg 96 sedimentologists (1986) have demonstrated the association of mass wasting with rising salt diapirs. Behrens (1988), Roberts et al. (1990, 1992), and Prior et al. (1989), to mention a few, have shown an association of chemosynthetic communities, carbonate buildups, gas hydrates, and seabed eruptions with the escape of hydrocarbons from the seafloor, and Bouma et al. (1986) and Aharon et al. (1992) have shown the effects of salt dissolution on the seafloor.

The GLORIA image, in concert with the 3.5-kHz sub-bottom profiles, appears to show effectively the spatial distribution of many surficial geologic features and when integrated with the bathymetry, shows their relation to the sea floor morphology. GLORIA, however, does not identify all

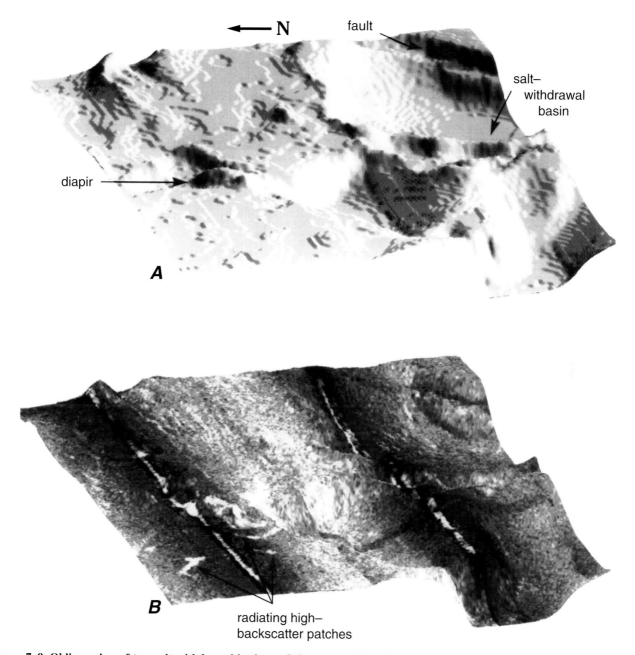

Figure 7–9. Oblique view of two salt-withdrawal basins and the surrounding high areas showing A, the multibeam bathymetry alone illuminated from the left side of the image and B, the GLORIA image draped over the bathymetry. Note that the basin floors are featureless on the GLORIA image and that the high-backscatter patches are found along parts of the basin walls and irregularly distributed on the plateaus surrounding the basins. Seismic profile shown in Figure 7–3A runs diagonally across this scene from the upper right to the lower left. The normal fault on the seismic profile shows on the surface as the terrace along the wall of the basin in the upper right corner of the image. A 28-by-28-km area is shown in this image, and the vertical exaggeration is 9X. Location shown in Figure 7–6.

the geologic features reported in other studies. The features that GLORIA images are unable to identify fall into three categories: (1) those that are below the resolution of the system, (2) linear features that are oriented perpendicular to the survey tracks, and (3) echo character patterns that appear acoustically similar on the sidescan image. Because of the

reconnaissance nature of the GLORIA system, the minimum size of a detectable feature is about 100 m in the across-track direction and 300 m in the along-track direction (Kleinrock 1992). For example, the sediment eruption mapped by Prior et al. (1989) was only about 300 m in diameter and thus is unlikely to show on GLORIA images. Small brine-filled pockmarks (MacDonald et al. 1990) also fall below the resolution of the GLORIA system. Faults like the ones shown in Figure 7–5 that trend perpendicular to the survey tracks do not show while those oriented parallel to the tracks are well imaged (Figures 7–7 and 7–9). Where the high-backscatter patches are associated with the faults, they show up regardless of their orientation relative to the tracklines as long as the patches are large enough (Figures 7–7 and 7–9). Some of the facies identified on the 3.5-kHz profiles, particularly the transparent facies with the weakly reflective surface and the areas of hyperbolae on a flat seafloor, are indistinguishable on the GLORIA images from the laminated facies. A more comprehensive study of this problem is needed, but apparently the acoustic character of these facies appears the same as the laminated facies to the GLORIA system.

The close correlation of the high-backscatter patches on the GLORIA image with the locations of hydrocarbon vents on the upper slope (Figure 7–5) and brine seeps on Green Knoll (Aharon et al. 1992) suggests that GLORIA effectively identifies the surficial extent of the areas affected by the larger vents and seeps. The cause of the high-backscatter signature remains to be confirmed by ground-truth studies, but some speculations can be made from the existing data. Many of the vents that have been mapped have authigenic carbonates or gas hydrates associated with them (Brooks et al. 1984; Kennicutt et al. 1988; Roberts et al. 1989, 1990, 1992). These carbonates and hydrates are lithified and presumably would have a higher acoustic impedance than the surrounding hemipelagic sediments. The 6.5-kHz frequency of the GLORIA system does penetrate the seafloor (Gardner et al. 1991); thus some of the high backscatter areas may be buried, or at least their surficial extent may not be as large as shown by GLORIA. In other cases, such as the top of Green Knoll, the high backscatter may be a consequence of fine-scale bathymetric differences that have been observed from a submersible but are below the resolution of GLORIA and just show as high backscatter (Aharon et al. 1992).

If the high-backscatter patches identified on the GLORIA image are in fact associated with vents and their lithified features, then the locations of these patches identify the sites of conduits through which deeper-seated fluids are escaping to the seafloor. The high-backscatter patches can be divided into two categories: those associated with salt structures that breach or nearly breach the seafloor and those associated with faults. The association of vents with shallow salt structures has been observed in several places on the Texas-Louisiana Slope (Roberts, Singh, and Coleman 1986; Behrens 1988; Kennicutt et al. 1988). On top of Green Knoll the dense saline brines and associated dolomite and ferrous hydroxide deposits (Aharon et al. 1992) coincide with the location of high-backscatter patches on the GLORIA image (Figure 7–7). Perhaps the randomly placed high-backscatter

patches on the tops of salt domes, especially when not associated with faults, are indicating fine-scale bathymetric features associated with salt dissolution and cap rock formation whereas those patches around the flanks of the diapirs and those associated with faults may indicate sites where more deeply derived fluids are escaping.

The association of the high-backscatter patches on the GLORIA images with faults has not been recognized in the Gulf of Mexico before, although the association of authigenic carbonates with faults in the accretionary prism off the coast of Oregon has been inferred (Carson et al. 1991). In the Gulf of Mexico, though, where the seismic and multibeam data show the surficial expression of faults, the GLORIA images consistently show discontinuous high-backscatter patches (Figures 7–6 and 7–7). Some of the patches are distributed in linear bands that radiate from diapirs. These patches may indicate radial faults, but their absence on the multibeam bathymetry suggests that the offsets at the seafloor are small (Figures 7–6 and 7–9). If in fact this association of surficial features with deeper structures holds, then the combined use of the GLORIA imagery with the bathymetry would provide a means of mapping the slope on a regional scale and outline areas of high probability where given seafloor types may occur. So far, this type of systematic mapping has been limited to studies of lease blocks and thus has been of restricted regional extent (e.g., Behrens 1988; Anderson and Bryant 1990; Roberts et al. 1990, 1992).

The surficial geology, in addition to showing the relationship of vents to deeper structures, also appears to show differences in tectonic activity within the study area. Evidence of mass wasting, at a scale that can be resolved by GLORIA, seems to be confined to the base of the Sigsbee Escarpment and around Green Knoll and is absent from the flanks of the salt-withdrawal basins (Figure 7–6). Mass wasting has been documented in seismic data (Bouma 1981) and in DSDP cores from Orca and Pigmy Basins (Bouma et al. 1986), yet within the study area the 3.5-kHz profiles and GLORIA image show no convincing evidence of mass wasting off the walls of the salt-withdrawal basins (Figures 7–3, 7–6, 7–8, and 7–9). Seafloor gradients of the flanks of some of the salt-withdrawal basins are equal to those found on the Sigsbee Escarpment (Pratson 1993), yet large-scale mass wasting deposits are not found at the base of these slopes (Figures 7–8 and 7–9). With the seafloor gradients being similar, the absence of large-scale mass wasting deposits in the salt-withdrawal basins may reflect a difference in the tectonic activity between the Sigsbee Escarpment and the salt-withdrawal basins. Perhaps the salt along the Sigsbee Escarpment and in Green Knoll is still mobile and the failures are triggered by the mobility of the salt in these areas whereas the salt-withdrawal basins are no longer subsiding and thus their flanks are tectonically more stable and subject to only small failures that are not resolved by GLORIA (Bouma 1982; Doyle, Kaluza, and Roberts 1992). Mass wasting deposits may have been prevalent early on in the evolution of the salt-withdrawal basins when salt movement was more extensive, but presumably now the tectonic activity of these basins has decreased and their fill has changed primarily to hemipelagic sedimentation. The DSDP hole in Pigmy Basin indicates that the uppermost 140 m of basin fill consists of silts with a few thin fine-sand beds (Bouma et al. 1986), and the fine-grained and undisturbed nature of these sediments supports the inference that these basins recently have been tectonically stable.

Conclusions

1 GLORIA images are effective at identifying the large areas of inferred hydrocarbon vents, but small vents and areas of sediment eruption are below the resolution of the system. Faults were preferentially imaged when oriented roughly parallel to the survey track and often were characterized by linear belts of high-backscatter patches.

2 GLORIA appears to be effective for mapping the larger-scale features associated with hydrocarbon vents. The cause of the acoustic signature, however, remains to be defined by future ground-truth work.

3 Deposits associated with hydrocarbon vents have high acoustic backscatter on the GLORIA imagery and occur as irregular patches over areas where the salt is near the seafloor, as discontinuous linear bands presumably associated with faults, and in discontinuous arcuate bands on terraces associated with normal faults along the downslope walls of some of the salt-withdrawal basins. Thus the surficial geology seems to record some of the conduits through which the escaping hydrocarbons and formation fluids are focused.

4 Mass wasting is extensive along the Sigsbee Escarpment, but is uncommon around the salt-withdrawal basins even though seafloor gradients in both areas are similar. Perhaps the absence of mass wasting deposits indicates that these basins are no longer tectonically active or at least are below the scale of GLORIA resolution.

5 Understanding the surficial geologic processes acting on the Texas-Louisiana continental slope requires integration of the surface and subsurface geology.

Acknowledgments

We thank the captain, crews, and scientific staffs of the R/V *Farnella* for the successful collection of the GLORIA and seismic-reflection data. Paul Grim of the National Ocean Survey of NOAA made the gridded multibeam bathymetric data available to us. Deborah Hutchinson, Eric Schmuck, Lincoln Pratson, and Harry Roberts provided helpful, constructive reviews of this manuscript.

References

Aharon, P., Roberts, H. H., and Snelling, R. 1992. Submarine venting of brines in the deep Gulf of Mexico: Observations and geochemistry. *Geology* 20: 483–6.

Anderson, A. L., and Bryant, W. R. 1990. Gassy sediment occurrence and properties: Northern Gulf of Mexico. *Geo-Mar. Ltrs.* 10: 209–20.

Behrens, W. E. 1988. Geology of a continental slope oil seep, northern Gulf of Mexico. *Am. Assoc. Petrol. Geol. Bull.* 72: 105–14.

Bouma, A. H. 1981. Depositional sequences in clastic continental slope deposits, Gulf of Mexico. *Geo-Mar. Ltrs.* 1: 115–21.

Bouma, A. H. 1982. Intraslope basins in northwest Gulf of Mexico: Key to ancient submarine canyons and fans. *Geologic Evolution of Continental Margins,* eds. J. S. Watkins and C. L. Drake, pp. 567–81. American Association of Petroleum Geologists Memoir 43, Tulsa, OK.

Bouma, A. H., Stelting, C. E., and Leg 96 Sedimentologists 1986. Seismic stratigraphy and sedimentary processes in Orca and Pigmy Basins. In *Initial Reports of the Deep-Sea Drilling Project, Leg 96,* eds. A. H. Bouma, J. M. Coleman, A. H. Meyer, et al., pp. 563–76. Washington, D.C.: U.S. Government Printing Office.

Brooks, J. M., Kennicutt, M. C., Fay, R. R., McDonald, T. J., and Sassen, R. 1984. Thermogenic gas hydrates in the Gulf of Mexico. *Science* 225: 409–11.

Bryant, W. R., Simmons, G. R., and Grim, P. 1991. The morphology and evolution of basins on the continental slope Northwest Gulf of Mexico: *Trans. Gulf Coast Assoc. Geol. Soc.* 41: 73–82, Houston, Tx.

Carson, B., Holmes, M. L., Umstattd, K., Strasser, J. C., and Johnson, H. P. 1991. Fluid expulsion from the Cascadia accretionary prism: Evidence from porosity distribution, direct measurements, and GLORIA imagery. *Phil. Trans. Royal Soc. London A.* 335: 331–40.

Chavez, P. S. 1984. U.S. Geological Survey Mini Image Processing System (MIPS). *U.S. Geological Survey Open-File Report 84-880,* 12 p. Reston, Va.: USGS.

Chavez, P. S. 1986. Processing techniques for digital sonar images from GLORIA. *Photogram. Engrg. Rem. Sens.* 52: 1133–45.

Coleman, J. M., Bouma, A. H., Prior, D. B., and Roberts, H. H. 1989. Nearsurface geology of the Gulf of Mexico continental slope. *Offshore Technol. Conf.* 21: 641–8.

Doyle, E. H., Kaluza, M. J., and Roberts, H. H. 1992. Use of manned submersibles to investigate slumps in deep water Gulf of Mexico. *Civil Engrg. Oceans.* 5: 770–82.

EEZ-SCAN 85 Scientific Staff 1987. Atlas of the U.S. Exclusive Economic Zone, Gulf of Mexico and eastern Caribbean areas. *U.S. Geological Survey Miscellaneous Investigations Series 1-1864a,* 104 p. Reston, Va.: USGS.

Gardner, J. V., Field, M. E., Lee, H. J., Edwards, B. E., Masson, D. G., Kenyon, N. H., and Kidd, R. B. 1991. Groundtruthing 6.5-kHz side scan sonographs: What are we really imaging?: *J. Geophys. Res.* 96: 5955–74.

Grim, P. 1992. Atlas of NOAA's multibeam sounding data in the Gulf of Mexico Exclusive Economic Zone, Vol. 59 p. Rockville, Mass.: NOAA/NOS/Coast and Geodetic Survey.

Kennicutt, M. C., Brooks, J. M., Bidigare, R. R., and Denoux, G. J. 1988. Gulf of Mexico hydrocarbon seep communities. I. Regional distribution of hydrocarbon seepage and associated fauna. *Deep-Sea Res.* 35: 1639–51.

Kennicutt, M. C., and Brooks, J. M. 1990. Recognition of areas effected by petroleum seepage: Northern Gulf of Mexico Continental Slope. *Geo-Mar. Ltrs.* 10: 221–4.

Kleinrock, M. C. 1992. Capabilities of some systems used to survey the deep-sea floor. In *CRC Handbook of Geophysical Exploration at Sea*, ed. R. A. Geyer, pp. 35–86. Boca Raton, Fla.: CRS Press.

MacDonald, I. R., Reilly, J. F., Guinasso, N. L., Brooks, J. M., Carney, R. S., Bryant, W. A., and Bright, T. J. 1990. Chemosynthetic mussels at a brine-filled pockmark in the northern Gulf of Mexico. *Science* 248: 1096–9.

Neurauter, T. W., and Bryant, W. R. 1990. Seismic expression of sedimentary volcanism on the continental slope, Northern Gulf of Mexico: *Geo-Mar. Ltrs.* 10: 225–31.

Paskevich, V. 1992. Digital mapping of side-scan sonar data with the Woods Hole Image Processing system software. *U.S. Geological Survey Open-File Report 92-536*, 87 p. Reston, Va.: USGS.

Pratson, L. F. 1993. Morphologic studies of submarine sediment drainage. Unpublished Ph.D. Dissertation, Columbia University, New York, 103 p.

Prior, D. G., Doyle, E. H., and Kaluza, M. J. 1989. Evidence for sediment eruption on deep sea floor, Gulf of Mexico. *Science* 243: 517–19.

Roberts, H. H., Singh, I. B., and Coleman, J. M. 1986. Distal shelf and upper slope sediments deposited during rising sea level, north-central Gulf of Mexico. *Trans. Gulf Coast Assoc. Geol. Soc.* 36: 541–51.

Roberts, H. H., Sassen, R., and Aharon, P. 1987. Carbonates of the Louisiana Continental Slope. *Offshore Technol. Conf.* 19: 373–82.

Roberts, H. H., Sassen, R., Carney, R., and Aharon, P. 1989. Carbonate buildups on the Continental Slope off central Louisiana. *Offshore Technol. Conf.* 21: 655–62.

Roberts, H. H., Aharon, P., Carney, R., Larkin, J., and Sassen, R. 1990. Sea floor responses to hydrocarbon seeps, Louisiana Continental Slope. *Geo-Mar. Ltrs.* 10: 232–43.

Roberts, H. H., Cook, D. J., and Sheedlo, M. K. 1992. Hydrocarbon seeps of the Louisiana continental slope: Seismic amplitude signature and seafloor response. *Trans. Gulf Coast Assoc. Geol. Soc.* 42: 349–61.

Rothwell, R. G., Kenyon, N. H., and McGregor, B. A. 1991. Sedimentary features of the South Texas Continental Slope as revealed by side-scan sonar and high-resolution seismic data. *Am. Assoc. Petrol. Geol. Bull.* 75: 298–312.

Salvador, A. 1991. Triassic-Jurassic. In *The Geology of North America.* Vol. J, *The Gulf of Mexico Basin*, ed. A. Salvador, pp. 131–80. Boulder, Colo.: Geological Society of America.

Satterfield, W. M., and Behrens, E. W. 1990. A late Quaternary canyon/channel system, northwest Gulf of Mexico continental slope. *Mar. Geol.* 92: 51–67.

Seni, S. J. 1992. Evolution of salt structures during burial of salt sheets on the slope, northern Gulf of Mexico. *Mar. Petrol. Geol.* 9: 452–68.

Swenson, J. P., and Bryant, W. R. 1992. Sediment pathways of the northwest Green Canyon area, offshore Louisiana. *Trans. Gulf Coast Assoc. Geol. Soc.* 4: 707–15.

8 Sedimentary processes in a tectonically active region: Puerto Rico North Insular Slope

Kathryn M. Scanlon[1] and Douglas G. Masson[2]

[1]U.S. Geological Survey, Woods Hole, Massachusetts
[2]Institute of Oceanographic Sciences, Southampton, United Kingdom

Abstract

GLORIA long-range sidescan sonar data extending from the shelf edge north of Puerto Rico across the insular slope to the floor of the Puerto Rico Trench reveal numerous canyons, amphitheater-shaped scarps, and sediment ponds. Interpretation of these data, in light of previously collected geophysical and sample data, indicates that tectonic forces have played a major role in the depositional and erosional history of the study area. Regional northward tilting of the flat, shallow-water Tertiary carbonate bank north of Puerto Rico created the insular slope and set the stage for the development of submarine canyons. The distinctive morphology of these canyons can be attributed to the character of the strata into which they have been cut. The front of the former carbonate bank has been eroded by large-scale mass wasting, which has left large amphitheater-shaped scarps on the lower slope. Shelf-derived sediment, along with material eroded from the canyon walls, is transported through the canyons to the basin and ridge province at the base of the slope where most sediment is trapped in basins south of the trench floor. The discovery of a depositional lobe on the trench floor confirms the presence of a turbidite entry point and suggests that some sediment reaches the trench floor as well.

Introduction

The deepest part of the Puerto Rico Trench, which has depths in excess of 8,000 meters (m), lies 150 km north of the island of Puerto Rico (Figure 8–1). The area between the island and the trench is made up of a steep insular slope, extending to a depth of about 7,000 m, and an 80-km-wide basin and ridge province that separates the trench floor from the insular slope. Although sediment of shallow-water origin, derived from the vicinity of Puerto Rico, has been identified in cores taken in the trench, little attempt has been made to understand the processes that have operated to trans-

port the sediment across the intervening morphologically complex area.

The major aims of this study are to describe the morphology of the insular slope north of Puerto Rico and the patterns of sedimentation and erosion seen there and to infer the processes that have created them. It is also necessary to discuss the tectonic setting of the area, because tectonic activity is primarily responsible for the overall morphology and sedimentation pattern. Regional northward tilting of the northern insular slope of Puerto Rico since early Pliocene and seismic activity which continues to the present day are probably the two most important tectonic influences (Masson and Scanlon 1991).

The study area: background

Morphology

The study area includes the seafloor between the island of Puerto Rico and the Puerto Rico Trench floor and extends west to Mona Canyon and east to the U.S. Virgin Islands (Figure 8–1). Bathymetric maps and seismic-reflection profiles (e.g., Maley, Sieber, and Johnson 1974; Tucholke and Ewing 1974; National Ocean Survey 1980; Trumbull 1981), give the general form of the seafloor in the study area. The insular shelf north of Puerto Rico is narrow, generally less than 5 km wide, whereas farther east the Virgin Islands Shelf widens to about 25 km. The shelf edge occurs at a water depth of 100 m or less. The insular slope is steep and convex in cross section. The upper slope extends from the shelf edge to about 3,000 m depth and has an average gradient of about 4°. The lower slope extends from 3,000 m to 7,000 m and has an average gradient of 15° or more. Locally the slope can be vertical (Gardner, Glover, and Hollister 1980). Numerous canyons, up to 400 m deep (Trumbull 1981), have been recognized on the upper slope. Mona Canyon is a wide, flat-floored canyon near the western edge of our study area (Figure 8–1), which differs from the other canyons north of Puerto Rico in that it is larger and probably fault-controlled

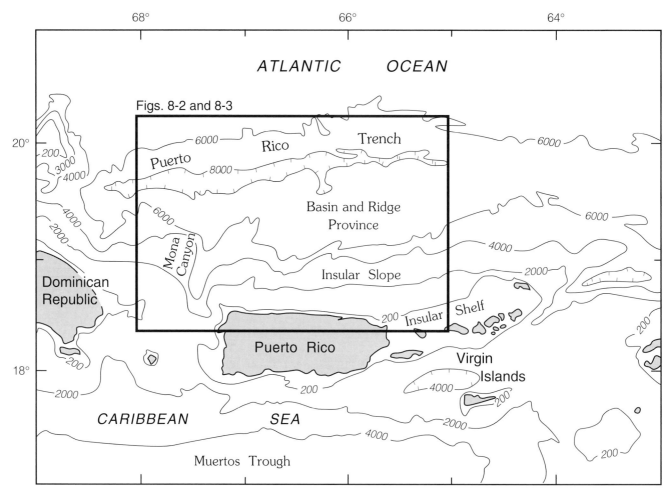

Figure 8–1. Location map showing major bathymetric features of the Puerto Rico region. The outline of Figures 8–2 and 8–3 is also shown.

(Gardner et al. 1980). Between the base of the insular slope (at about 7,000 m) and the trench floor is a region referred to as the basin and ridge province (Masson and Scanlon 1991). Within this province east-southeast-trending ridges exhibit several hundred meters of local relief, and basins containing flat-lying sediments are distributed irregularly in the interridge areas (Masson and Scanlon 1991). The Puerto Rico Trench lies north of the basin and ridge province. It is flat-floored, sediment-starved, and, at 8,000 m deep, the deepest part of the Atlantic Ocean.

Tectonic setting

Puerto Rico and the Virgin Islands lie within the North American – Caribbean plate boundary zone, which extends from the Puerto Rico Trench in the north to the Muertos Trough in the south (Figure 8–1) (Burke et al. 1984; Dillon, Edgar, Scanlon, and Coleman, this volume). Until Eocene time the region experienced north-south convergence; since the end of the Eocene the relative plate motion has been

mainly left-lateral slip (Burke et al. 1984; Stein et al. 1988). Deformation due to plate interactions is complex and extends across the entire plate boundary zone. Recent explanations for the observed complexity include Heubeck and Mann's proposal (1991) that the Caribbean Plate may actually consist of three or more subplates, and Masson and Scanlon's suggestion (1991) that one or more microplates may exist and move semiindependently within the plate boundary zone. Seismic activity extending to depths of at least 150 km confirms that deformation is active to the present day (Frankel, McCann, and Murphy 1980; Sykes, McCann, and Kafka 1982; Mann and Burke 1984).

An important aspect of the regional tectonics is the large amount of late Tertiary tilting, which has affected the northern insular slope of Puerto Rico. Birch (1986) has suggested that this tilting could have caused as much as 8 km of subsidence of the lower slope since the middle Eocene. Certainly, the occurrence of in-place shallow-water Miocene to early Pliocene limestones at depths in excess of 5 km demonstrates at least 5 km of subsidence in less than 4 million years (my) (Moussa et al. 1987). Further evidence for large-scale subsidence comes from the recognition of karst topography in Tertiary limestones at depths to 2,100 m (Heezen et al. 1985).

Rocks

The mountainous core of Puerto Rico consists mainly of Cretaceous and early Tertiary volcanic rocks that have been folded into an anticlinorium and intensely faulted. This core is flanked by Tertiary sedimentary rocks, predominantly shallow-water limestone (Monroe 1980). Following the nomenclature of Moussa et al. (1987) and the lithologic descriptions of Monroe (1980), the lowermost of the Tertiary sedimentary units is the San Sebastian Formation (0 to 300 m thick), a sandy carbonaceous clay, locally containing conglomerates and lignites. It is overlain by the Lares Limestone (0 to 500 m thick), a hard, very pure, fossiliferous limestone in beds 10 to 30 cm thick, followed by the Cibao Formation (170 to 300 m thick), a calcareous clay and earthy limestone. The Los Puertos Limestone (90 to 150 m thick) is indurated in its upper part and chalky in the lower parts and forms a transition between the underlying earthy Cibao Formation and the overlying Aymamon Limestone, a very pure, thick-bedded, massive, fossiliferous limestone, about 200 m thick. An unconformity separates the Aymamon Limestone from the uppermost of the Tertiary sedimentary units, the Quebradillas Limestone, a thin-bedded, sandy limestone and chalk, about 170 m thick.

On the northern flank of Puerto Rico the Tertiary sedimentary units mapped on land can be traced offshore for 50 km using seismic-reflection profiles (Moussa et al. 1987). They are parallel bedded and have a regional northward dip of 4°, approximately the same angle as the seafloor of the upper slope. The shallow-water depositional environment and parallel-bedded character of the limestones require that the 5 km of subsidence recorded on the lower slope must be a postdepositional event; that is, it must have occurred after early Pliocene.

Several publications have described rocks dredged from our study area (Schneidermann, Beckmann, and Heezen 1972; Shido, Miyashiro, and Ewing 1974; Fox and Heezen 1975; Perfit et al. 1980). The majority of the samples are from the steep lower slope. East of Mona Canyon (Figure 8–1), most of the lithologies recovered from the lower slope are middle Tertiary limestones or pre-Eocene rocks of similar aspect to those known in Puerto Rico. West of Mona Canyon, a more diverse suite of rocks has been recovered, including calcarenites, calcilutites, marbles, and low-grade metavolcanics and schists. The affinities of these rocks appear to be with northern Hispaniola and the carbonate banks of the southern Bahamas rather than Puerto Rico (Perfit et al. 1980).

Only two dredge hauls have been recovered from the basin and ridge province (Shido et al. 1974; Perfit et al. 1980). One, located at 19°20′ N, 65°32′ W, contained an assemblage of rocks typical of the lower slope, even though it was collected from a ridge separated from the lower slope by a 15-km-wide sedimentary basin. The second, located just to the east of the GLORIA coverage at 19°36′ N, 65°04′ W, recovered a different assemblage consisting of tholeiitic volcanic rocks that represent oceanic material derived from the North American Plate.

Sedimentary processes

Sediment deposition and movement on the insular shelf north of Puerto Rico have been the subjects of several studies (e.g., Schneidermann, Pilkey, and Saunders 1976; Pilkey, Trumbull and Bush 1978; Grove, Pilkey, and Trumbull 1982; Pilkey, Bush, and Rodriguez 1987). The dominant shelf-sediment type is calcareous skeletal sand, but a significant amount of terrigenous material is deposited near the numerous river mouths during storms. These sediments are believed to move both laterally along the shelf and seaward into canyon heads, but the relative importance of these two processes is not known. The mouths of the rivers that contribute these sediments are spaced about 5 to 10 km apart along the coast and some are known to have shifted in location by several kilometers (Bush, Pilkey, and Rodriguez 1986; Rodriguez et al. 1992) during the Holocene.

Canyons were recognized on the insular slope north of Puerto Rico as early as the mid-1950s (Northrop 1954; Ewing and Heezen 1955). The upper parts of three canyons off northwestern Puerto Rico were studied in the 1970s by Gardner et al. (1980) using research submersibles. They concluded that the canyons are not active during the present high sea level stand; indeed the upper parts of the canyons appeared to be filling with terrigenous mud. However, past episodes of erosion, which incised the canyons into the consolidated Tertiary limestones, were inferred from the occurrence of scarps, ledges, and perched limestone boulders on the canyon walls.

The distribution of the most recent turbidites across the basin and ridge province and onto the main abyssal floor of the Puerto Rico Trench was described by Doull (1983) using a suite of forty-four piston cores. She was able to demonstrate that turbidite sands consisted primarily of shallow-water carbonate material derived from the Puerto Rico Shelf. Rock fragments, including some of lithified limestone, are present in the turbidites but are not an important constituent. By mapping the thicknesses of basal sand layers, Doull was able to identify five distinct entry points for turbidity currents into the Puerto Rico Trench, distributed along the 360 km length of the trench studied. One of the turbidites entered the trench from all five entry points, suggesting simultaneous failure of a considerable length of the Puerto Rico Slope. Such a widespread failure is likely to have been triggered by a large seismic event.

67°   66°

20°

19°

Figure 8–2. Digital mosaic of GLORIA data from the Puerto Rico Trench and the insular slope north of Puerto Rico. Location shown in Figure 8–1. North of 19° N the ship's tracks are oriented NW-SE; south of 19° N they are E-W. Interpretation of this mosaic is shown in Figure 8–3.

Data

Data for the present study were collected during a thirty-day cruise on the research vessel *Farnella* in late 1985 and consist of GLORIA long-range sidescan sonar imagery (Figure 8–2), 80-in^2 airgun seismic reflection profiles, and 3.5-kHz and 10-kHz echo sounder profiles (EEZ-SCAN 85 Scientific Staff 1987; Scanlon, Masson, and Rodriguez 1988). All of these data were collected simultaneously along the *Farnella* tracklines shown in Figure 8–3. In addition, data from the archives of the Woods Hole Oceanographic Institution and Lamont-Doherty Geological Observatory were used to aid in the interpretation of the GLORIA mosaic. These include echo sounder profiles, core descriptions, and bottom photographs. Technical details about the GLORIA system and information about processing and mosaicking the images can be found in the introductory chapters of this book.

Interpretation

Light and dark areas in the GLORIA images represent areas of the seafloor that produce high and low backscatter, respectively. The intensity of backscatter is affected by the angle of the seafloor relative to the incidence angle of the sonar signal (i.e., bathymetry) and changes in the character (e.g., composition or degree of lithification) of the sea bottom. Small-scale roughness or inhomogeneities on the surface of the seafloor, of similar magnitude to the GLORIA wavelength of 23 cm, have usually been considered to be the principal contributor to backscatter intensity (see Huggett and Somers 1988). However, it is now recognized that, in sedimented areas, significant subseabed penetration of energy may occur at the GLORIA frequency. Penetration to at least 5 m is well documented (Gardner et al. 1991; Huggett et al. 1992). Although taking penetration into account makes interpretation more complicated, this phenom-

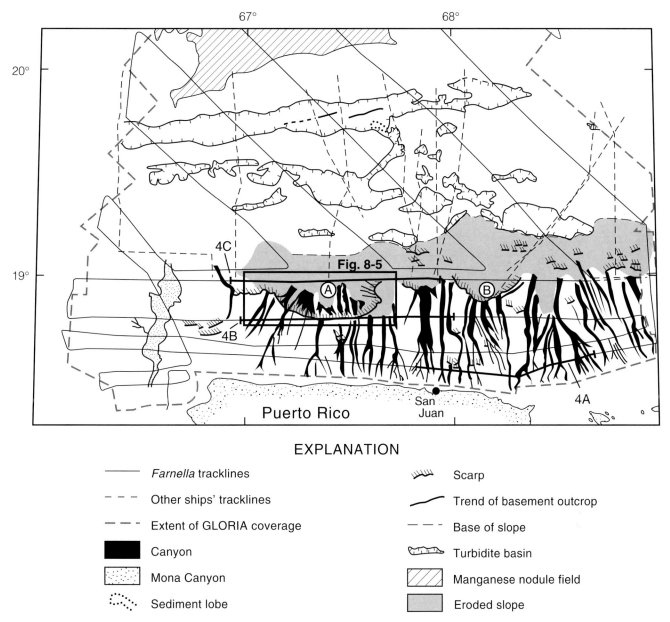

Figure 8–3. Interpretation of the GLORIA mosaic shown in Figure 8–2. Seismic profiles, collected simultaneously with the side-scan data, and previously collected echo-sounder profiles were used to aid interpretation. Box outlines location of Figure 8–5. Amphitheaters A and B are indicated by letters A and B, respectively. Labels show locations of seismic-reflection profiles shown in Figures 8–4A, B, and C.

enon can be exploited. For example, GLORIA has been used to map debris flows buried by as much as 3 m of turbidites off west Africa (Masson et al. 1992). Clearly, in interpretations of backscatter patterns, the assumption that GLORIA shows only seabed character and presently active sedimentary processes is incorrect. The GLORIA signal may be an integrated response from complex subseabed sediments representing several depositional and erosional episodes. At least some elements of the following interpretation must be viewed in these terms.

The upper slope

The upper slope extends from the shelf edge to a water depth of about 3,000 to 4,000 m. The GLORIA mosaic of the upper slope reveals dozens of downslope-trending canyons (Figure 8–2), which can be traced for 30 to 50 km, across the entire width of the upper slope. In cross section, the canyons range from V- to U-shaped, with the latter being more common. Canyons appear on the GLORIA image as areas of high backscatter that correspond to the canyon

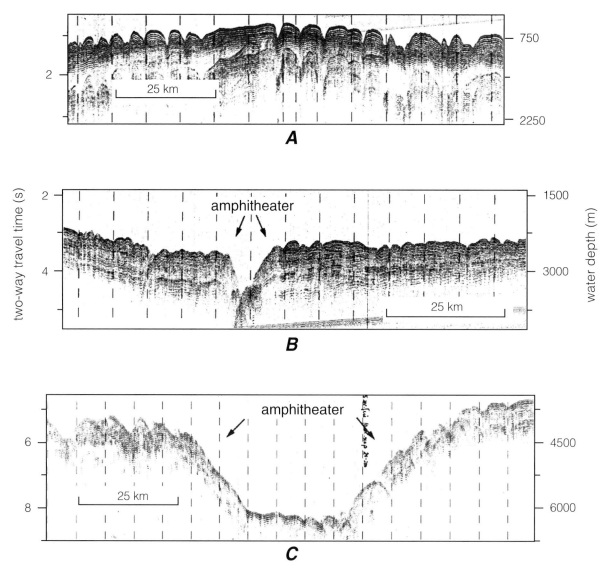

Figure 8–4. Three airgun profiles collected approximately parallel to the regional bathymetric contours on the insular slope north of Puerto Rico. Locations of the profiles are shown in Figure 8–3. Canyons are generally U-shaped and are more deeply incised on the shallow upper slope (A) than they are on the deep upper slope (B). C shows the full breadth and depth of the larger of the two amphitheaters (amphitheater A) shown in Figure 8–5.

floors and walls from rim to rim, and not just to the canyon floors. We believe the high backscatter results from outcrops of the Tertiary limestone beds into which the canyons are cut and from talus and coarse sediments within the canyons. Limestone outcrops, talus, and coarse sediments have been observed during submarine dives in the upper parts of canyons in the western part of our study area (Gardner et al. 1980; Heezen et al. 1985). Accumulations of shelf-derived sediments have been reported in some canyons (Gardner et al. 1980), but these are generally thin, and the GLORIA signal probably penetrates them.

Near the top of the slope (Figure 8–4A), the canyons are narrow (100 to 500 m wide) and closely spaced (generally

less than 4 km apart). At a regional water depth of between 500 m and 800 m, the canyons are incised as much as 400 m into the dipping limestone strata. The depth of canyon incision gradually decreases downslope, with canyons at 2,250 m typically being incised to only about 100 m (Figure 8–4B). At this depth, many of the canyons have merged into the broader, straighter canyons that typify the deep part of the upper slope. Some "canyons" on the deep upper slope have little or no bathymetric relief in our seismic profiles, but show clearly on the GLORIA mosaic as stripes of high backscatter and are continuous with deeply incised canyons on the shallow upper slope. In these cases, the high backscatter may indicate areas where a thin layer of unconsolidated

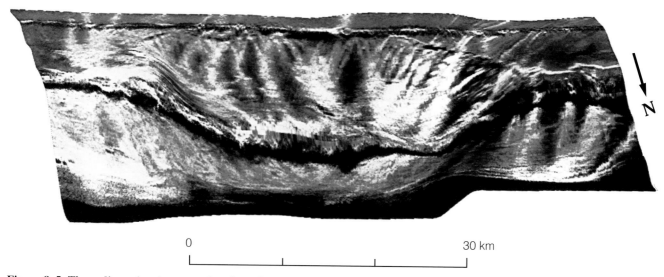

0   30 km

Figure 8–5. Three-dimensional perspective view of the amphitheater scarp, viewed from the north. A digital mosaic of GLORIA data has been merged with digital bathymetry (Schwab et al. 1991) to produce this figure. Area of figure shown in Figure 8–3.

sediment has been stripped away, exposing the underlying consolidated Tertiary units; alternatively, it may arise from trails of debris or coarse-grained sediment derived from the canyons above.

The lower slope

Below about 3,000 to 4,000 m the character of the slope changes: it becomes steeper (up to 15°) and exhibits high backscatter in the GLORIA image (Figure 8–2). We interpret this to mean that the unconsolidated sediment and at least part (and in some places probably all) of the Tertiary carbonate section have been removed by mass wasting processes. Rocks dredged from this region are Eocene to Miocene carbonates and pre-Eocene rocks of Puerto Rican affinity, which supports this interpretation (Perfit et al. 1980).

Two large amphitheater-shaped scarps (A and B in Figure 8–3) are seen in the GLORIA images on the lower slope north of Puerto Rico. Amphitheater A is about 60 km across and up to 2,250 m deeper than the surrounding seafloor (Figure 8–4C). Amphitheater B is smaller; it is about 30 km across and 1500 m deep. Based on seismic-reflection profiles, an estimated 1,500 km^3 of sedimentary section have been removed from the larger amphitheater. A system of canyons has developed within amphitheater A (Figure 8–5), whereas the interior of amphitheater B appears to have an irregular, high-backscatter surface with no canyons. More recent slumping may have occurred in the smaller amphitheater B, and sufficient time has not yet passed to allow the creation (or re-creation) of canyons within that feature. This implies that modification of the amphitheaters may be an ongoing, presently active process. In addition to the two large amphitheater-shaped scarps, several small (one to five

km across) downslope-facing scarps have been identified (Figure 8–3).

The basin and ridge province and the Puerto Rico Trench

GLORIA data from the basin and ridge province show generally subtle backscattering contrasts, partly because the ship's tracks were oriented at a high angle to the bathymetric ridges, which minimized the effect of those ridges on backscatter. In addition, 3.5-kHz echo sounder profiles show that the basin and ridge province is covered by a veneer of sediment, thus reducing expected contrasts between the ridges and the sedimentary basins. However, by comparing GLORIA and profile data, the basins are seen to correspond to areas of lower backscatter, the majority of which can be outlined with reasonable confidence (turbidite basins, Figure 8–3). Note that only those basins crossed by at least one profile that verifies the cause of the lower backscatter are shown in Figure 8–3. The ridges are oriented 95° to 105° subparallel to the trench, which is oriented 86° (Masson and Scanlon 1991). The basins are bounded by, and their shape is controlled by, the ridges. The basin floors are shallower than the trench floor by up to 1,000 m, with each basin having its own characteristic depth. The sediment surface in some basins is tilted very slightly toward the south.

No obvious sediment transport pathways or lineated sediment fabrics are seen in the GLORIA data from the basin and ridge province. As noted previously all of the basins including the trench floor exhibit uniform, low backscatter. The single exception is a small lobate area of slightly higher backscattering seen on the southern side of the trench floor at 66°20′ W, which corresponds to the location of a major turbidite entry point indicated by sedimentological studies (Doull 1983).

The Puerto Rico Trench floor is clearly recognizable in the GLORIA mosaic as a band of low backscatter, trending 86° at about 19°45′ N (Figure 8–2). The flat floor is as much as 18 km wide but narrows to nothing at the eastern edge of the study area. Variations in the width of the trench floor may be related to the intersection of the trench with the slightly oblique structures of the basin and ridge province. The basement ridges seen protruding through the sediment in the floor of the trench near 66°30′ (Figures 8–2 and 8–3) are probably partially subducted blocks of oceanic crust from the North American Plate.

Discussion

Morphology of canyons

The canyons on the insular slope north of Puerto Rico differ from most other canyons that have been mapped using GLORIA data. Most canyons imaged by GLORIA are defined by subtle variations in backscatter intensity and their walls are dissected by numerous gullies oriented at a high angle to the canyon axes. Canyons that show such features are found in a variety of tectonic and sedimentologic settings, for example, off the east and west coasts of the United States (Twichell and Roberts 1982; Scanlon 1984; EEZ-SCAN 84 Scientific Staff 1988), the Bering Sea (Bering Sea EEZ-SCAN Scientific Staff 1991; Carlson, Karl, and Edwards 1991), and the Bay of Biscay (Belderson and Kenyon 1976). In contrast, the Puerto Rico canyons are steep-sided features with unusually high backscatter and a complete lack of side gullies. Geological conditions that might be unique to the Puerto Rico area and provide an explanation for the unique nature of the Puerto Rico canyons include structural control, the tectonic tilting experienced by the area, and the nature of the strata being eroded.

Structural control is unlikely, because north-trending lineaments (such as faults or joints) are lacking both onshore and offshore. Monroe (1980) suggested jointing as a possible cause of NNE-, WNW-, and west-trending landforms in the Tertiary carbonate province of northern Puerto Rico. However, he found no north-trending structures. Side-looking radar imagery of Puerto Rico shows east-striking bedding planes and irregular karstic weathering within this province, but likewise, no northerly-trending structures (Scanlon and Southworth 1989; Scanlon in press).

Tectonic tilting, as has occurred on the Puerto Rico slope, is not typical of most other continental or insular margins. Although we do not know the age of the Puerto Rico canyons, their formation certainly postdates the initiation of tilting, because the area was previously a flat, shallow-water carbonate platform (Monroe 1980). Although tilting by itself cannot explain the unusual morphology of the canyons, it did create an environment conducive to the formation of canyons. Earthquakes associated with the tectonic

forces that caused the tilting (Dillon et al. this volume) probably triggered mass wasting of the unconsolidated sediments on the insular slope and caused fracturing of the underlying carbonate rocks. Also, the steeper slopes that resulted from tilting accentuated canyon excavation by mass wasting and gravity flow processes.

As noted by Monroe (1980), the Tertiary carbonates of Puerto Rico erode either by dissolution or by spalling of large blocks. Observations made from submersibles in the canyons similarly indicate block-by-block failure as the primary process controlling canyon enlargement (Gardner et al. 1980; Heezen et al. 1985). This type of failure does not produce gullies. In contrast, canyons in most other locales imaged by GLORIA have formed in relatively unconsolidated sediment that erodes grain by grain, a process that is conducive to gully formation. Moreover, McHugh, Ryan, and Schreiber (1993) found that some submarine canyons off New Jersey change from V-shaped with side gullies where they cut into siliciclastic sediments to U-shaped without gullies where they incise an Eocene chalk unit. GLORIA imagery of the short segments of the canyons in the chalk (EEZ-SCAN 87 Scientific Staff 1991) shows high backscatter features with straight sides, similar to the Puerto Rico canyons.

We believe that tectonic tilting and associated earthquakes set the stage and provided a mechanism for the initiation of submarine canyons on the Puerto Rico Slope, but it is the character of the strata that controls the morphology of the canyons. We suggest that the canyons began to form soon after the deposition of the Quebradillas Limestone (at about 3.3 to 3.6 Ma, according to Moussa et al. 1987) when tectonic tilting began. Unconsolidated sediments on the tilting platform became unstable, and earthquakes triggered submarine slides that removed any unconsolidated sediment in their paths. The stress release caused by unroofing resulted in exfoliation and spalling of the Tertiary carbonates along the slide scars. Earthquakes helped to move the loose blocks downslope and further to fracture the carbonate beds in place. A similar cycle involving unroofing, exfoliation, and fracturing by diagenesis has been suggested by McHugh et al. (1993) for canyons cut into chalk off New Jersey.

Relationship of canyons to rivers

It is tempting to try to correlate the locations of present-day rivers on Puerto Rico with the locations of canyons on the insular slope as has been suggested by Gardner et al. (1980). We have identified more than fifty canyons within the study area on the insular slope, but less than a dozen rivers of significant size flow into the sea north of Puerto Rico. Canyons are uniformly distributed across the study area. Rivers, in contrast, are absent in the eastern part of the study area where no large landmass lies above the insular slope. Nevertheless, this eastern segment of the slope is incised by canyons in a manner that appears to be identical to the rest of the slope, demonstrating that rivers are not essential for canyon formation. Furthermore, not all of the canyons reach the uppermost slope; some canyons head 20 km or more seaward of the shelf edge. Previously suggested links between rivers and canyon heads are most likely based on coincidence; the high density of canyons makes it inevitable that each river will debouch near a canyon head. Moreover, it is known that at least some of the river mouths have migrated several kilometers laterally (Bush et al. 1986; Rodriguez et al. 1992) during the Holocene, making it unlikely that a river would be associated exclusively with any one canyon through time.

Mass wasting on the lower slope

The lower slope has a markedly different appearance from the upper slope in the GLORIA mosaic (Figure 8–2). Canyons are either not present or are poorly defined on the lower slope, which shows higher backscatter and appears rougher than the intercanyon areas on the upper slope. In some sections of the study area large amphitheater-shaped scarps mark the transition from upper to lower slope. It is evident from profiles crossing the slope that the lower slope is much steeper than the upper slope (15° and 4°, respectively).

The change in steepness between the upper and lower slope may be a relict of the morphology of the presubsidence carbonate bank (Figure 8–6). We can reasonably assume a slope angle of 11° for the front of a carbonate bank (Read 1985). We can also assume that the present upper slope strata were deposited nearly horizontally (Monroe 1980; Moussa et al. 1987). A tilt of 4° would then yield the present configuration of a 4° upper slope angle, a 15° lower slope angle, and shallow-water carbonates at about 5,000 m water depth about 60 km north of Puerto Rico. The tilted bank front has been modified by mass wasting processes, including stripping of the unconsolidated sediment cover and slumping of parts of the Tertiary strata (as in the amphitheater areas). The high backscattering properties and rough appearance of the lower slope in the GLORIA mosaic can be attributed to exposed carbonate and basement rock outcrops.

An important question regarding mass wasting of the insular slope is whether the sediment removal was by large catastrophic events or by slower headward erosion. We suggest that, in our study area, headward erosion produces canyons such as are common on the upper slope. Small canyons are also present within the large amphitheater (A in Figure 8–3) on the lower slope, suggesting that headward erosion is possible here, also. However, the amphitheaters themselves are an order of magnitude larger than the upper slope canyons, and yet they formed within the same length of time (<4 my). The process that formed the amphitheaters must be very different from that which created the canyons. Perhaps the amphitheaters result from large-scale mass wasting events (Schwab et al. 1991) whereas the canyons on the

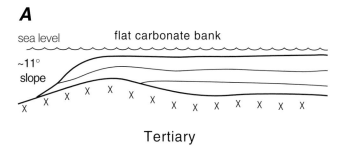

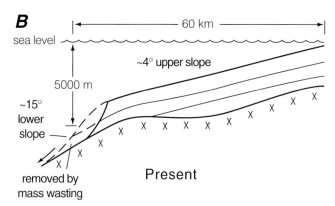

Figure 8–6. Schematic diagram of A, the carbonate bank north of Puerto Rico before late Tertiary tilting, and B, the insular slope today, after 5 km of tilting. The present configuration of a 4° upper slope angle, a 15° lower slope angle, and shallow-water carbonates at about 5000 m water depth about 60 km north of Puerto Rico can be explained by this simple model.

upper slope and within the amphitheaters have been created by headward erosion. Each amphitheater (Figure 8–3) may have been formed in a single event or several somewhat smaller events. Amphitheater B (Figure 8–3), which has no canyons within it, may be more recently formed, and insufficient time has passed for the slow headward erosion processes to create canyons.

An alternative scenario might be that the amphitheater scarps represent an advanced stage of the scalloped margins typical of some mature carbonate banks (Mullins and Hine 1989). They suggest that large (tens of kilometers across) convex embayments are common in carbonate banks throughout the Caribbean and may represent the beginning of the disintegration of those banks. SeaMARC II mosaics from the nearby southeast Bahamas reveal scalloped edges along the banks nearest Hispaniola. Mullins et al. (1991, 1992) attribute the formation of these scallops to tectonic activity associated with the interaction here of the North American and Caribbean Plates. The Tertiary bank north of Puerto Rico would have existed in a similar tectonic environment and may have formed similar scalloped edges. The amphitheaters could then be large preexisting scallops that have been lowered to their present positions by tilting. However, if the amphitheaters were merely relicts of the shape of the former bank edge, then we would expect to see a continuous scarp, similar to the headwall of the amphitheaters, defining the entire bank edge. Since we do not (Figure 8–3), we believe it is most likely that the amphitheaters formed during or after tilting. Earthquakes associated with the tectonic tilting probably triggered the mass wasting events that created the amphitheater scarps. Since the region is still a seismically active plate boundary, earthquakes may trigger such events in the future.

Sediment transport from shelf to trench floor

Sediments on the shelf north of Puerto Rico move both offshore and laterally and eventually find their way into the heads of canyons (Pilkey et al. 1978). The canyons then act as conduits to the lower slope. Shallow marine Tertiary carbonates (Perfit et al. 1980) typical of the lower slope have been found on a ridge that is separated from the lower slope by a 15-km-wide basin. Shelf sediments are found in turbidites on the floor of the Puerto Rico Trench (Doull 1983), 150 km from the shelf and north of the basin and ridge province. Clearly, material has been transported from the shelf and slope across the basin and ridge province to the Puerto Rico Trench, but no simple transport paths or channels across the basin and ridge province are seen. A turbidity current crossing the basin and ridge province would have to surmount the sills between adjacent basins, which can rise hundreds of meters above the basin floors. This scenario suggests either large, thick turbidity flows, which simply swamp the entire basin and ridge system, or flows with momentum sufficient to run up and over the sills. The steepness of the lower slope and the wide area covered by some of the turbidites in the trench (Doull 1983) suggest that elements of both the above scenarios are possible.

Conclusions

We believe that the morphology of the canyons on the insular slope north of Puerto Rico is controlled by the nature of the strata (i.e., parallel-bedded carbonate rocks) into which they have been cut. Earthquakes resulting from the tectonic tilting of the former shallow carbonate bank provide a mechanism for the initiating of canyon formation. They also contribute to a cycle of canyon cutting involving stress release by unroofing, fracturing by earthquakes, and mass wasting. Differences in erosional style between the upper slope (where canyons are numerous) and the lower slope (where evidence of large-scale collapse is seen) can be attributed to the steepness of the slope, which may be a relict of the morphology of the Tertiary carbonate platform.

Although the canyons do not show a one-to-one relationship to major rivers on Puerto Rico, they do serve as conduits across the insular slope for river-derived sediment.

Much of this sediment is trapped in basins in the basin and ridge province, but some reaches the floor of the Puerto Rico Trench in the form of turbidites. Our data suggest that at least some canyons have been recently active and confirm the presence of one turbidite entry point onto the trench floor.

The morphologic features of the north Puerto Rico Insular Slope are relatively young (less than 4 my) and have formed rapidly in response to abrupt oversteepening of the former carbonate bank. Because the tectonic setting of Puerto Rico is the same today as it was during the creation of the canyons, they are probably undergoing modification by mass wasting at the present time. Earthquakes have most likely been triggers in the past for large-scale slope failures (e.g., at the amphitheater scarp sites) and could trigger similar events in the future.

Acknowledgments

We thank the officers and crew of the R/V *Farnella* for a successful cruise. W. Dunkle provided data from the Woods Hole Oceanographic Institution archives that aided the interpretation of the GLORIA mosaic. Reviews by and discussions with D. C. Twichell and W. P. Dillon have greatly improved this manuscript. We also thank C. M. McHugh and C. C. Almy for helpful reviews.

References

Belderson, R. H., and Kenyon, N. H. 1976. Long-range sonar views of submarine canyons. *Mar. Geol.* 22: M69–M74.

Bering Sea EEZ-SCAN Scientific Staff 1991. Atlas of the U.S. Exclusive Economic Zone, Bering Sea. *U.S. Geological Survey Miscellaneous Investigations Series I-2053*, 145 p., scale 1:500,000. Reston, Va.: USGS.

Birch, F. S. 1986. Isostatic, thermal, and flexural models of the subsidence of the north coast of Puerto Rico. *Geology* 14: 427–9.

Burke, K., Cooper, C., Dewey, J. F., Mann, P., and Pindell, J. L. 1984. Caribbean tectonics and relative plate motions. In *GSA Memoir 162: The Caribbean South American Plate Boundary and Regional Tectonics*, Eds. W. E. Bonini, R. B. Hargraves, and R. Shagam, pp. 31–63. Boulder, Colo.: Geological Society of America.

Bush, D. M., Pilkey, O. H., and Rodriguez, R. W. 1986. Impact of river mouth migration on continental margin sedimentation (abs.) *AAPG Ann. Mtg., Prog. Abs.* 70(5): 570.

Carlson, P. R., Karl, H. A., and Edwards, B. D. 1991. Mass sediment failure and transport features revealed by acoustic techniques, Beringian Margin, Bering Sea, Alaska. *Mar. Geotechnol.* 10(1/2): 33–51.

Doull, M. E. 1983. Turbidite sedimentation in the Puerto Rico Trench abyssal plain. Master's thesis (unpublished), Duke University, Durham, North Carolina, 124 p.

EEZ-SCAN 84 Scientific Staff 1988. Physiography of the western United States Exclusive Economic Zone. *Geology* 16: 131–4.

EEZ-SCAN 85 Scientific Staff 1987. Atlas of the U.S. Exclusive

Economic Zone, Eastern Caribbean. *U.S. Geological Survey Miscellaneous Investigations Series I-1864 B*, 58 p., scale 1:500,000. Reston, Va.: USGS.

EEZ-SCAN 87 Scientific Staff 1991. Atlas of the U.S. Exclusive Economic Zone, Atlantic Continental Margin. *U.S. Geological Survey Miscellaneous Investigations Series I-2054*, 174 p., scale 1:500,000. Reston, Va.: USGS.

Ewing, M., and Heezen, B. C. 1955. Puerto Rico Trench topographic and geophysical data. *GSA Special Paper 62*, pp. 255–68. Boulder, Colo.: Geological Society of America.

Fox, P. J., and Heezen, B. C. 1975. Geology of the Caribbean Crust. In *The Caribbean and Gulf of Mexico*, eds. A. E. M. Nairn and F. G. Stehli, pp. 412–66. New York: Plenum.

Frankel, A., McCann, W. R., and Murphy, A. J. 1980. Observations from a seismic network in the Virgin Islands Region: Tectonic structures and earthquake swarms. *J. Geophys. Res.* 85(B5): 2669–78.

Gardner, W. D., Glover, L. K., and Hollister, C. D. 1980. Canyons off northwest Puerto Rico: Studies of their origin and maintenance with Nuclear Research Submarine NR-1. *Mar. Geol.* 37: 41–70.

Gardner, J. V., Field, M. E., Lee, H., Edwards, B. E., Masson, D. G., Kenyon, N. H., and Kidd, R. B. 1991. Ground-truthing 6.5-kHz side scan sonographs: What are we really imaging? *J. Geophys. Res.* 96(B4): 5955–74.

Grove, K. A., Pilkey, O. H., and Trumbull, J. V. A. 1982. Mud transportation on a steep shelf, Rio de la Plata Shelf, Puerto Rico. *Geo-Mar. Ltrs.* 2(1-2): 71–5.

Heezen, B. C., Nesteroff, W. D., Rawson, M., and Freeman-Lynde, R. P. 1985. Visual evidence for subduction in the western Puerto Rico Trench. *Géodynamique de Caraïbes Symposium*, Paris: Feb. 5-8 1985, Editions Technip, p. 287–304.

Heubeck, C., and Mann, P. 1991. Geologic evaluation of plate kinematic models for the North American – Caribbean plate boundary zone. *Tectonophysics* 191: 1–26.

Huggett, Q. J., and Somers, M. L. 1988. Possibilities of using the GLORIA system for manganese nodule assessment. *Mar. Geophys. Res.* 9: 255–64.

Huggett, Q. J., Cooper, A. K., Somers, M. L., and Stubbs, A. R. 1992. Interference fringes on GLORIA side-scan sonar images from the Bering Sea and their implications. *Mar. Geophys. Res.* 14: 47–63.

Maley, T. S., Sieber, F. D., and Johnson, G. L. 1974. Topography and structure of the western Puerto Rico Trench. *Geol. Soc. Am. Bull.* 85: 513–18.

Mann, P., and Burke, K. 1984. Neotectonics of the Caribbean. *Rev. Geophys. Space Phys.* 22(4): 309–62.

Masson, D. G., and Scanlon, K. M. 1991. Neotectonic setting of Puerto Rico. *Geol. Soc. Am. Bull.* 103(1): 144–54.

Masson, D. G., Kidd, R. B., Gardner, J. V., Huggett, Q. J., and Weaver, P. E. 1992. Saharan continental rise: Facies distribution and sediment slides. In *Geologic Evolution of Atlantic Continental Rises*, eds. C. W. Poag and P. C. de Graciansky, pp. 327–343. New York: Van Nostrand Reinhold.

McHugh, C. M., Ryan, W. B. F., and Schreiber, B. C. 1993. The role of diagenesis in exfoliation of submarine canyons. *Am. Assoc. Petrol. Geol. Bull.* 77(2): 145–72.

Monroe, W. H. 1980. Geology of the middle Tertiary formations of Puerto Rico. *U.S. Geological Survey Professional Paper 953*, Washington, D.C.: U.S. Government Printing Office.

Moussa, M. T., Seiglie, G. A., Meyerhoff, A. A., and Taner, I. 1987. The Quebradillas Limestone (Miocene-Pliocene), northern Puerto Rico, and tectonics of the northeastern Caribbean margin. *Geol. Soc. Am. Bull.* 99: 427–39.

Mullins, H. T., and Hine, A. C. 1989. Scalloped bank margins: Beginning of the end for carbonate platforms? *Geology* 17: 30–3.

Mullins, H. T., Dolan, J., Breen, N., Anderson, B., Gaylord, M., Petruccione, J. L., Wellner, R. W., Melillo, A. J., and Jurgens, A. D. 1991. Retreat of carbonate platforms: Response to tectonic processes. *Geology* 19: 1089–92.

Mullins, H. T., Breen, N., Dolan, J., Wellner, R. W., Petruccione, J. L., Gaylord, M., Anderson, B., Melillo, A. J., Jurgens, A. D., and Orange, D. 1992. Carbonate platforms along the southeast Bahamas – Hispaniola collision zone. *Mar. Geol.* 105: 169–209.

National Ocean Survey 1980. *Bathymetric Map of the Caribbean Region*. Washington, D.C.: U.S. Department of Commerce, NOAA. Mercator 1:2,500,000 scale.

Northrop, J. 1954. Bathymetry of the Puerto Rico Trench. *Trans. Am. Geophys. Union.* 35: 221–5.

Perfit, M. R., Heezen, B. C., Rawson, M., and Donnelly, T. W. 1980. Chemistry, origin and tectonic significance of metamorphic rocks from the Puerto Rico Trench. *Mar. Geol.* 34: 125–56.

Pilkey, O. H., Trumbull, J. V. A., and Bush, D. M. 1978. Equilibrium shelf sedimentation, Rio de la Plata shelf, Puerto Rico. *J. Sed. Petrol.* 48(2): 389–400.

Pilkey, O. H., Bush, D. M., and Rodriguez, R. W. 1987. Bottom sediment types of the northern insular shelf of Puerto Rico: Punta Peñon to Punta Salinas. *Miscellaneous Investigations Map I-1861*, Reston, Va.: U.S. Geological Survey.

Read, J. F. 1985. Carbonate platform facies models. *Am. Assoc. Petrol. Geol. Bull.* 69: 1–21.

Rodriguez, R. W., Webb, R. M., Bush, D. M., and Scanlon, K. M. 1992. Marine geologic map of the north insular shelf of Puerto Rico: Rio de Bayamon to Rio Grande de Loiza. *U.S. Geological Survey Miscellaneous Investigations Map I-2207*, scale 1:40,000. Reston, Va.: USGS.

Scanlon, K. M. 1984. The continental slope off New England: A long-range sidescan-sonar perspective. *Geo-Mar. Ltrs.* 4(1): 1–4.

Scanlon, K. M., Masson, D. G., and Rodriguez, R. W. 1988. GLORIA side-scan sonar survey of the EEZ of Puerto Rico and the U.S. Virgin Islands. *Transactions of the 11th Caribbean Geological Conference*, Barbados, 1986, pp. 32: 1–32:9.

Scanlon, K. M., and Southworth, C. S. 1989. Combined radar and GLORIA mosaics of Puerto Rico, the U.S. Virgin Islands, and surrounding deep ocean areas: Tectonic interpretations (abs.). *Abstracts, 12th Caribbean Geological Conference*, Christiansted, St. Croix, p. 151 University of Puerto Rico, Mayaguez, P.R.

Scanlon, K. M. (in press). Side-looking airborne radar mosaic of Puerto Rico. In *Mineral Resource Assessment of Puerto Rico*. Reston, Va.: U.S. Geological Survey.

Schneidermann, N., Beckmann, J. P., and Heezen, B. C. 1972. Shallow water carbonates from the Puerto Rico Trench Region. *Sixth Caribbean Geological Conference Volume*, Margarita, Venezuela, pp. 423–5. Universidad Central de Venezuela, Caracas, Venezuela.

Schneidermann, N., Pilkey, O. H., and Saunders, C. 1976. Sedimentation on the Puerto Rico insular shelf. *J. Sed. Petrol.* 46(1): 167–73.

Schwab, W. C., Danforth, W. W., Scanlon, K. M., and Masson, D. G. 1991. A giant slope failure on the northern insular slope of Puerto Rico. *Mar. Geol.* 96: 237–46.

Shido, F., Miyashiro, A., and Ewing, M. 1974. Basalts and serpentinite from the Puerto Rico Trench: 1. Petrology. *Mar. Geol.* 16: 191–203.

Stein, S., DeMets, C., Gordon, R. G., Brodholt, J., Argus, D., Engeln, J. F., Lundgren, P., Stein, C., Wiens, D. A., and Woods, D. F. 1988. A test of alternative Caribbean Plate relative motion models. *J. Geophys. Res.* 93(B4): 3041–50.

Sykes, L. R., McCann, W. R., and Kafka, A. L. 1982. Motion of the Caribbean plate during last 7 million years and implications for earlier Cenozoic movements. *J. Geophys. Res.* 87: 10656–76.

Trumbull, J. V. A. 1981. Oceanographic data off Puerto Rico and the Virgin Islands. *Lawrence Berkeley Laboratory Publication 360*, Berkeley, CA. Mercator projection, 1:781,786 scale, 1 sheet.

Tucholke, B. E., and Ewing, J. I. 1974. Bathymetry and sediment geometry of the Greater Antilles Outer Ridge and vicinity. *Geol. Soc. Am. Bull.* 85: 1789–802.

Twichell, D. C., and Roberts, D. G. 1982. Morphology, distribution, and development of submarine canyons on the United States Atlantic continental slope between Hudson and Baltimore Canyons. *Geology* 10: 408–12.

9 A review of the tectonic problems of the strike-slip northern boundary of the Caribbean Plate and examination by GLORIA

William P. Dillon,[1] N. Terence Edgar,[2] Kathryn M. Scanlon,[1] and Dwight F. Coleman[1]

[1]U.S. Geological Survey, Woods Hole, Massachusetts

[2]U.S. Geological Survey, Reston, Virginia

Introduction

The Caribbean region, south of Cuba (Figure 9–1A), forms one of the distinct lithospheric plates of the Earth's surface (Case and Holcombe 1980). Targets of a scale appropriate for GLORIA imaging are provided by tectonic disruptions of the seafloor along the plate's northern edge. We selected three areas to survey using GLORIA, which allows us to examine the variety of structures produced along this active plate boundary (Figure 9–1B). In the central Cayman Trough, plate motion and geometry cause extension, which creates a short spreading axis that is not connected to the world rift system; GLORIA is used to analyze the crustal structures that are created. Off northwestern Hispaniola, an irregularity in the plate boundary results in compressional motion, and GLORIA is used to analyze the accretionary wedge that is formed by sediments that are scraped off the North American Plate as it is forced against the Caribbean Plate. North of Puerto Rico, the plates appear to slide past each other with neither compression nor extension, yet, surprisingly, a major oceanic trench exists, which exhibits the world's greatest negative free-air gravity anomaly. Structural trends displayed by GLORIA and earthquake distribution are used to hypothesize the plate interactions that form the trench and analyze the response at a corner of a plate (the North American Plate) that is being overrun by another plate (the Caribbean Plate).

Tectonic setting of the Caribbean Plate

The Caribbean Plate is marked by clearly defined subduction zones to its east and west (Figure 9–1B). On its west, the Cocos Plate is being subducted beneath the Caribbean, and on its east the North and South American Plates are being subducted. These regions are marked by volcanic chains, accretionary wedges of sediments that are scraped off the down-going plate and dipping zones of earthquakes, all im-

plying subduction of oceanic lithosphere beneath the Caribbean in a direction normal to the plate boundary. In contrast, along the northern and southern boundaries of the Caribbean Plate, the North American and South American Plates slide westward relative to the Caribbean, with an overall motion that is essentially strike-slip. However, because of significant irregularities in the shape of the northern and southern boundaries, those regions actually display extremely complex structures.

Along the northern boundary of the Caribbean Plate, where we made our surveys, a roughly left-lateral strike-slip motion is confirmed by most large-scale studies (Molnar and Sykes 1969; Pindell and Dewey 1982; Sykes, McCann, and Kafka 1982; Burke et al. 1984; Duncan and Hargraves 1984; Mann and Burke 1984; Mattson 1984; Burke 1988; Stein et al. 1988; Ladd et al. 1990; Pindell and Barrett 1990). The present northern plate boundary was initiated during Eocene time. Before that time, the motion of the Caribbean Plate relative to the North American Plate had been northeasterly with subduction, whereas, subsequently, the Caribbean Plate moved easterly with respect to the North American Plate, with primarily transcurrent motion (Malfait and Dinkelman 1972; Bralower et al. 1993; Caceres et al. 1993). At the Eocene reorganization, Cuba and the Yucatan Basin were sheared off from the Caribbean Plate and added to the North American Plate to form the present, straighter plate boundary that extends from Central America to the Virgin Islands along the Cayman Trough and north of Hispaniola and Puerto Rico (Figure 9–1B). We divide the northern plate boundary into a western region from Central America to central Hispaniola in which the pattern of plate motion is reasonably well agreed on, and an eastern region from eastern Hispaniola to the Virgin Islands in which varying tectonic models exist in the literature.

In the western region, the seismicity (Figure 9–2) generally is shallow-focused and marks a fairly clearly defined plate boundary as indicated in Figure 9–1B (although a set of earthquakes off the boundary occurs at Jamaica). The most common view is that the seismicity pattern is best explained by a short, north-trending spreading axis at the middle of the Cayman Trough connecting two strike-slip faults,

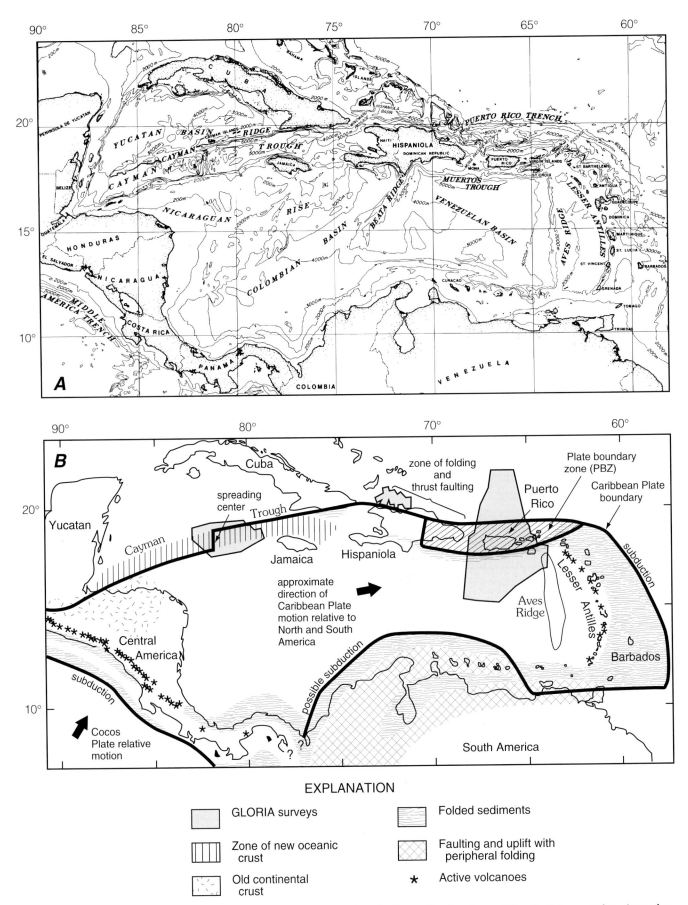

Figure 9–1. A, Physiography and geography of the Caribbean region. B, Generalized geology of the Caribbean region. Areas imaged by GLORIA are shown stippled. The entire central Cayman Trough image is shown in Figure 9–4 and the entire northwestern Hispaniola image in Figure 9–8. Several portions of the Puerto Rico image are shown (Figures 9–13, 9–14, and 9–15).

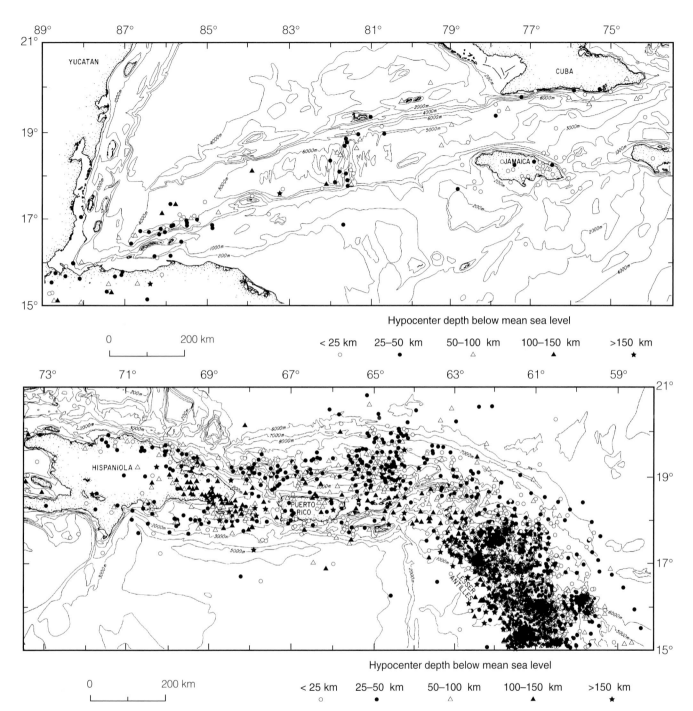

Figure 9–2. Seismicity of the northern boundary of the Caribbean Plate. Data were provided by the National Geophysical Data Center, Boulder, Colorado. The plot shows only earthquakes having locations recorded by six or more stations. Because arbitrary depths of 5, 10, and 33 km are sometimes assigned to earthquakes for which hypocenter depths are not known, all earthquakes indicating such depths were deleted. Locations having recorded depths less than 0 also were deleted.

one on the south side of the Cayman Trough to the west of the spreading center and the other on the north side of the trough to the east, forming the pattern of the plate boundary shown in Figure 9–1B. As the North American Plate moves westward relative to the Caribbean, a continual opening takes place at the offset (spreading center, Figure 9–1B)

and this opening is filled with upwelling mafic asthenospheric material, which hardens to form new oceanic crust (Holcombe et al. 1973; Perfit 1977; MacDonald and Holcombe 1978; CAYTROUGH 1979; Holcombe and Sharman 1983; Mann et al. 1983; Rosencrantz and Sclater 1986; Rosencrantz et al. 1988).

In the eastern region of the northern Caribbean Plate boundary, the seismicity occurs across a much broader zone and many deeper hypocenters appear, in comparison to the western region (Figure 9–2). The shallower earthquakes (less than about 50 km) are considered to define lithosphere of a broad plate boundary zone (Figure 9–1B, diagonally hatched region). The plate boundary zone (PBZ) appears to be detached from the adjacent major plates to the north and south. The northern detachment occurs to the north of Hispaniola, Puerto Rico, and the Virgin Islands, forming the southern boundary of the North American Plate in the area; it is associated with the Puerto Rico Trench east of Hispaniola (~ 68.5° W). The southern detachment bounds the PBZ to the east of the Beata Ridge (east of ~ 71° W, Figure 9–1A), where the PBZ seems to be detached from the Caribbean Plate along a fault system associated with the Muertos Trough (Figure 9–1A). The deeper earthquakes of the eastern region seem to be associated with interactions of plates at depths beneath the lithosphere of the PBZ. These interactions have been explained by a variety of plate models. The models seem to fall into three categories that require either (1) a component of north-south compressional motion and subduction (Figure 9–3A); (2) transcurrent overriding of a downbent flap (Figure 9–3B); or (3) long-distance westward movement of a slab that is subducted from the east (Figure 9–3C).

GLORIA surveys, description, and interpretation

Cayman Trough region

The central part of the Cayman Trough was acoustically imaged using GLORIA (Figure 9–1; Edgar et al. 1991). Single-channel seismic and magnetic intensity data also were collected during the sidescan survey. The image of the surveyed area (Figure 9–4), interpretations of the GLORIA imagery, and bathymetry have been published by Edgar et al. (1991), the magnetic anomaly field was mapped by Dillon et al. (1993), and detailed bathymetry was published by Jacobs et al. (1989).

THE FLOOR OF THE TROUGH

Refraction measurements (Ewing, Antoine, and Ewing 1960) and gravity models (Dillon, Vedder, and Graf 1972) show that the crust is relatively thin beneath the floor of the Cayman Trough. GLORIA shows that the morphology of the floor of the trough containing its spreading axis and most recently formed crust is dominated by north-trending ridges in the central part, with much smoother seafloor to the east and west (Figure 9–4). Our seismic profiles and those of Rosencrantz and Sclater (1986) show that the smoother seafloor is underlain by a sediment-covered basement having more modest relief than in the central region. The

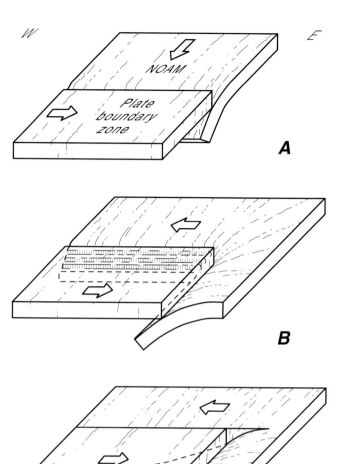

Figure 9–3. Sketches of three styles of models that have been proposed for plate interactions in the northeastern Caribbean.

GLORIA imagery demonstrates that the trend of the subdued morphology of the seafloor is the same as that of the central region. Because we are observing an active spreading center (at about longitude 81°40′ W), the central region represents younger rocks, with age increasing progressively to the east and west (assuming no ridge jumps).

The rift valley at the spreading axis in the Cayman Trough is more than twice the average water depth of the rift valleys on the midocean ridge system. The rift valley appears on the sonograph as a 5- to 15-km-wide, north-trending, bright band of reflection at about longitude 81°40′ W (Figures 9–4 and 9–5) and is clearly distinguishable on the swath bathymetric map as a depressed linear zone (Figure 9–6). The walls of the rift valley are subparallel; the eastern wall trends 353° and the western wall trends 358°, resulting in a widening of the valley southward. A crustal block, called Mt. Dent by Edgar et al. (1991), interrupts the west wall at

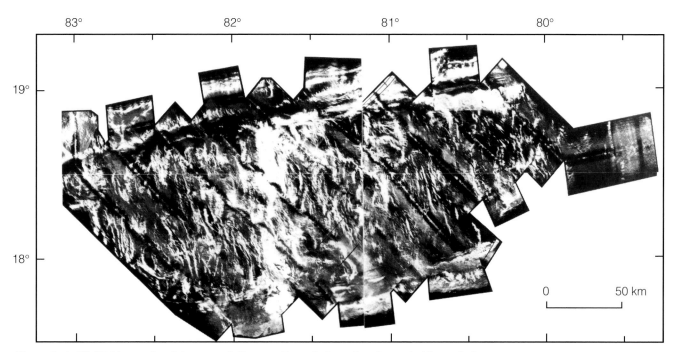

Figure 9–4. GLORIA mosaic of the central Cayman Trough. Location shown in Figure 9–1.

latitude 18°22′ N and diminishes the width of the valley to 5 km (Figures 9–4, 9–5, and 9–6). Topographic trends within the rift valley north of Mt. Dent approximate the trend of the rift valley, whereas south of Mt. Dent (latitude 18°20′ N) a ridge trending 337° dominates the topography of the valley. This ridge extends from the intersection of the west wall of the valley and Mt. Dent to where it intersects uplifted gabbroic crustal blocks (CAYTROUGH 1979) of the southeastern rift valley. A dormant spreading axis may lie along these blocks and possibly may be traced to the southeastern extremity of the rift valley on the basis of topography. The uplifted gabbroic blocks confine the deep rift-valley floor to a 2- to 4-km-wide strip adjacent to the western wall, south of Mt. Dent.

Nodal basins (for summary of the subject, see Fox and Gallo 1986) lie at the northern and southern intersections of the rift valley with the principal transform faults. These flat-lying basins approximate right triangles in plan view and lie on the side of the spreading axis where active transform motion occurs. They are evident on the swath bathymetry (Figure 9–6) and on the sidescan imagery (Figure 9–4), where they appear as very strong roughly triangular acoustic reflections (bright areas). They are the deepest part of the rift valley, lying at depths greater than 6,000 m. At most locations in the world, the surface of nodal basins is sediment that covers basalt flows (Fox and Gallo 1986), but the strong acoustic reflections returned from the basins in the Cayman Trough suggests that there is little sediment covering the basalt there.

Topographic lineaments of ridges and valleys within older oceanic crust east and west of the rift valley have trends that vary between about 335° to 358° (range of 23°, mean 350°). A few lineaments are widely disparate from this trend, such as those that trend 015° evident in the southwestern part of the mosaic. Offsets in topographic lineaments, marking possible faults, can be observed in some areas (Figures 9–4 and 9–5), but displacements are generally less than 10 km.

The principal transform fault forming the North American – Caribbean Plate boundary to the east of the northern rift-transform intersection, the Oriente transform fault, is clearly imaged along the northern trough margin (Figures 9–4 and 9–5). The comparable transform fault, the Swan Island transform fault, is also evident, but not as well displayed, in the short segment of the mosaic west of the southern rift-transform intersection. Similar lineations are not imaged by GLORIA on the northwestern or southeastern boundaries of the trough.

Sediment cover over crustal rock is indicated by the lower level of acoustic reflection away from the rift valley (Figure 9–4). The pelagic sediment in the survey area is about 250 m thick at maximum; however, ponded sediment can be about 400 m thick.

TROUGH MARGINS

One of the most striking features of the northern margin of the Cayman Trough is a broad shelf of sediment that fills a crustal depression midway up the slope of the north wall (Figure 9–5). It divides the north wall into an upper and lower escarpment. The sidescan sonar mosaic displays the escarpments as bright returns and the sediment shelf as dark

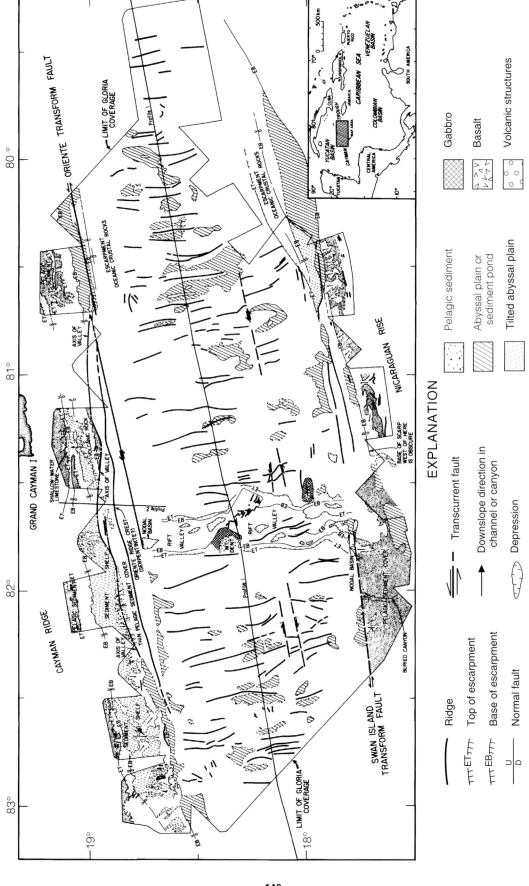

Figure 9–5. Interpretation of the Cayman Trough GLORIA mosaic. Seismic data and samples were used in the interpretation.

140

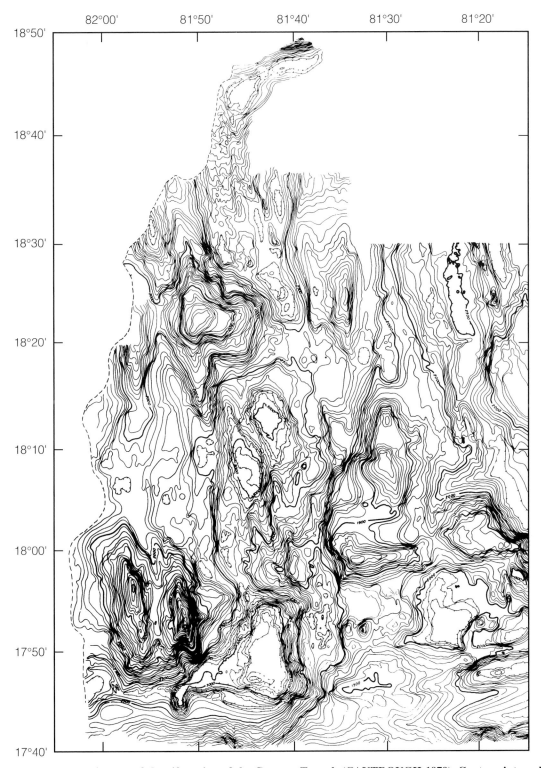

Figure 9–6. Swath bathymetric map of the rift region of the Cayman Trough (CAYTROUGH 1979). Contour interval is fifty fathoms.

(Figure 9–4). The sediment shelf extends over 100 km along the northern wall of the Cayman Trough and is best developed in the region west of the rift-transform intersection, where it is about 8 km wide. A southerly tilt is consistent throughout most of the extent of the shelf and increases to both the east and west. Fine parallel bright reflections on the face of the upper escarpment separated by dark bands suggest rock outcrops covered with sediment. Samples of

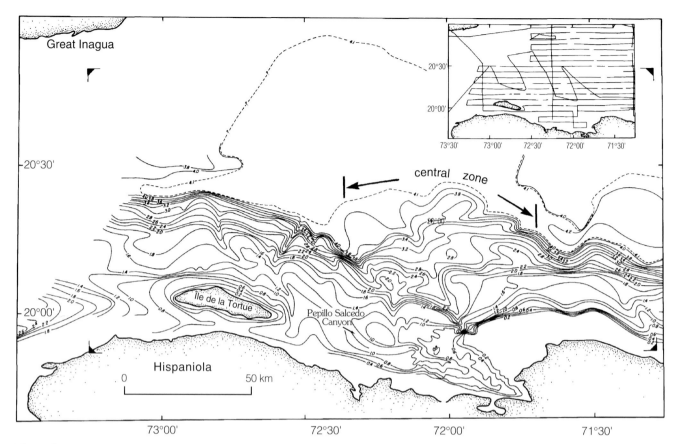

Figure 9–7. Bathymetry of the region off northwestern Hispaniola. Contour interval is 0.2 km. Data lines for control are shown in inset map. Contouring also was guided by the GLORIA mosaic.

Miocene shallow-water limestone were collected from the scarp by the submersible ALVIN at a water depth of about 3,000 m (Emery and Milliman 1980; Figure 9–5), suggesting substantial subsidence of the Cayman Ridge since that time.

Seismic-reflection profiles indicate the presence of a major ridge extending along the Cayman Trough northern margin a distance of about 100 km on either side of the rift-transform intersection. The ridge, called Oriente Ridge (Figure 9–5; Edgar et al. 1991) has been interpreted as a serpentinite extrusion on the basis of its smoothly rounded topography and the evidence of a single dredge sample. Serpentinite may have formed along the margin by conversion of peridotite as a result of sea water circulating through fractures. Only a weak acoustic return is imaged from Oriente Ridge, perhaps because its sides slope gently, and a thin layer of pelagic sediment may cover the surface. No sediment-filled trough exists on the wall of the southern margin of the Cayman Trough. West of the southern ridge-transform intersection, the south wall of the trough is characterized by uniformly low acoustic returns, possibly caused by sediment cover thick enough to reduce reflectance but insufficiently thick to be detected on the seismic data. East of the rift-transform intersection, the south wall is marked by a 15-km-wide band of ridges trend-

ing parallel and subparallel to the wall of the Nicaraguan Rise (Figures 9–4 and 9–5). These ridges extend eastward to about longitude 80°45′ W, where a major northeasterly-trending ridge extends to about longitude 79°30′ W. This structure has been sampled by dredging and a typical suite of oceanic rocks was recovered, mostly from the top and south-facing steep escarpment (Perfit and Heezen 1978).

Western Hispaniola region

The GLORIA survey north of western Hispaniola (Figure 9–1B) occurs at a place where the trend of the plate boundary is changing from a restraining bend situation to the east to what appears to be nearly pure transcurrent motion to the west. The boundary here seems to be on land or just offshore (Edgar 1991), so that the GLORIA images primarily show deformation of sediments on the North American Plate where that plate is being crushed in the restraining bend.

MORPHOLOGY

The insular slope off northwestern Hispaniola is divided into three sharply demarcated morphologic zones as indicated by our bathymetric map (Figure 9–7), GLORIA mosaic (Figure 9–8), and GLORIA interpretation (Figure 9–9). In the

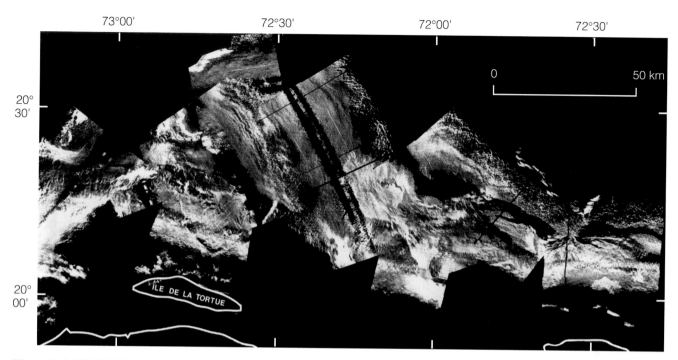

73°00' 72°30' 72°00' 72°30'

0   50 km

20°
30'

20°
00'

ÎLE DE LA TORTUE

Figure 9–8. GLORIA image of the region off northwestern Hispaniola (location shown in Figure 9–1).

central zone (between longitudes 71°43′ W and 72°22′ W), the insular slope forms a gently dipping irregular surface with approximately margin-parallel hills and valleys; the average gradient is about 4°. In contrast, to the east and west, the overall surface morphology is smoother but much steeper and the slope is convex in cross section. The lowest parts of the slope present a declivity of about 9° to 12° west of the central zone and about 13° to 16° to the east of the zone.

STRUCTURE AND TECTONICS

The GLORIA mosaic (Figure 9–8) has been interpreted with the use of a variety of seismic-reflection data, including four deep-penetration multichannel seismic profiles (Figure 9–9; Austin 1983; Dillon et al. 1992). A seismic profile that displays the typical structural style across the central zone shows a series of upstepping anticlinal ridges (Figure 9–10 – note high vertical exaggeration; location on Figure 9–9). These are the margin-parallel hills observed in the bathymetry and GLORIA image (Figures 9–7 and 9–8). This is the characteristic structure of a tectonic accretionary wedge. As the plates move together, the anticlines form sequentially by folding of basin sediments against the backstop formed by Hispaniola. In such a tectonic setting, a fold is formed at the foot of the insular slope, it grows to some limit, then it is sheared off at its base and added to the bottom of the thrust stack of anticlines that forms the middle and upper insular slope, and a new fold begins to grow to seaward. The series of older folds, separated by faults, is elevated by wedging of sediments beneath it along the new thrust. The thrust faults may merge downward into a main décollement.

The accretionary wedge areas to the east and west are very different in morphology and structure from the central zone (Figure 9–7). In addition to being much steeper and convex in cross section, no broad fold ridges appear and the internal structure is very poorly imaged by seismic-reflection profiles, suggesting complex deformation and fracturing of the strata. Because the central zone and flanking zones of the insular margin differ extremely in structure, we infer the presence of a tear fault that forms a boundary between them. This inference is supported, at least to the east of the central zone, by the presence of a boundary ridge separating the zones (Figure 9–9). Obviously, the entire region imaged by GLORIA is involved in the same overall plate interactions and no evidence of independent motions of microplates appears in the seismicity. The reasons for the differences between zones in structure and slope morphology are unclear but may relate to changes in the dip of the main décollement, changes in the angle at the front of the tectonic accretionary wedge between the seafloor and décollement (known as the critical taper), or differences in the materials that have been swept into the wedge on the North American Plate (Dillon et al. 1992).

Despite the dominantly strike-slip nature of the plate motion along the northern Caribbean Plate boundary, the structures observed in GLORIA and seismic-reflection profiles cannot be related to strike-slip movement. In a zone of strike-slip motion, we would expect linear or anastomosing fault scarps of near-vertical faults, perhaps with flower structures. These are not the characteristic structures of this region. Where imaged, this insular margin is characterized by curving folds (in plan view), which apparently are separated

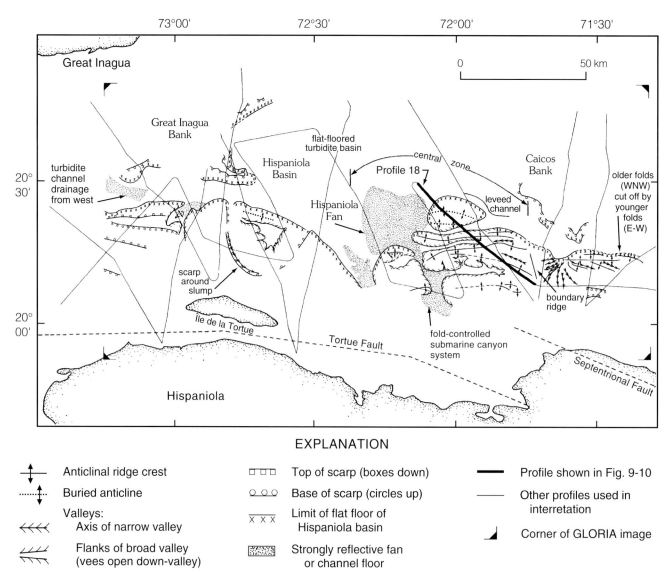

EXPLANATION

Anticlinal ridge crest	Top of scarp (boxes down)
Buried anticline	Base of scarp (circles up)
Valleys: Axis of narrow valley	Limit of flat floor of Hispaniola basin
Flanks of broad valley (vees open down-valley)	Strongly reflective fan or channel floor
	Profile shown in Fig. 9-10
	Other profiles used in interretation
	Corner of GLORIA image

Figure 9–9. Interpretation of the GLORIA image shown in Figure 9–8. Seismic profiles used in interpretation are shown as light solid lines. The location for profile 18, shown in Figure 9–10, is indicated by a heavy line.

by shallowly dipping thrust faults. Consequently, the structure of the insular margin must be explained almost completely by reference to compressional motion. The compressional component of motion along this part of the plate boundary is due primarily to an asperity or restraining bend at the boundary, although a minor general convergence between the North American and Caribbean Plates or eastern Caribbean subplate in this region is possible (Sykes et al. 1982; Mann, Burke, and Matumoto 1984; Mann and Burke 1984; Mattson 1984; Heubeck and Mann 1991).

Faulting of the strata in the accretionary wedge apparently occurs nearly aseismically, because the historical seismicity is concentrated south of the insular slope, near the present coastline (Figure 9–2). Apparently the main transcurrent motion occurs near the coast; the inferred positions of the plate boundary faults from Edgar (1991) are shown in Figure 9–9

(Tortue and Septentrional faults). The main shortening of sediments atop the basement occurs at the insular slope, indicated by the GLORIA and seismic data, whereas the main shortening of basement apparently occurs farther south within the Caribbean Plate across most of western Hispaniola (Biju-Duval et al. 1983; Mann et al. 1984; McCann and Sykes 1984).

SEDIMENTATION

Sedimentation in the deep Hispaniola Basin is dominated by turbidity currents whose flow patterns are controlled by the major tectonic features disclosed by GLORIA. Turbidity current deposits make up 75% to 80% of the accumulation (Ditty et al. 1977) and generate the laminated appearance of reflections of the basin fill (Figure 9–10). Turbidity currents enter the basin from several directions, including from the

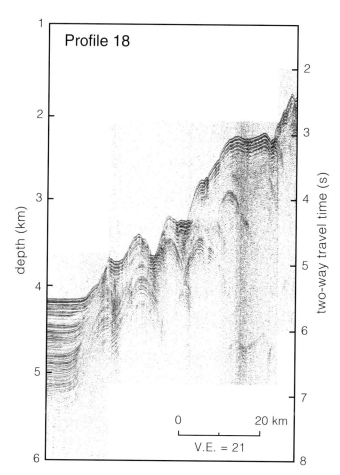

Profile 18

0   20 km

V.E. = 21

Figure 9–10. Seismic-reflection profile 18 showing thrust ridges in the tectonic accretionary wedge off northwestern Hispaniola. Location shown in Figure 9–9.

carbonate banks to the north, but two source directions seem dominant on the basis of GLORIA results and studies of cores (Bennetts and Pilkey 1976; Ditty et al. 1977). The primary direction is from the south, from the Hispaniola Fan, and the other is from the west, through the deep-sea channel between Hispaniola and Great Inagua Bank. Both of these areas are marked by bright seafloor reflections (Figure 9–8; indicated by dot patterns in Figure 9–9). Analyses of the distribution of sand layers in cores indicate that the Hispaniola Fan is the dominant source of sand to the basin, because sediments near the fan contain the greatest proportion of sand layers – more than 70% (Bennetts and Pilkey 1976, their Figure 5). The dominance of the Hispaniola Fan as a sand source also is shown by the distribution of the average mean grain size of samples from the bases of all turbidites, which decreases away from the mouth of the fan channel (Figure 9–11; Ditty et al. 1977). Cuba probably is the source of turbidites that enter the Hispaniola Basin from the west. According to Bennetts and Pilkey (1976), a mappable turbidite unit becomes finer eastward (dashed contours in Figure 9–11) – clearly indicating a western source.

The canyon that supplies sediment to the Hispaniola Fan is marked by complex morphology that is dominated by the tectonics of the insular margin. Channel patterns are controlled by folding, but influenced by erosion. One turbidite derived from the canyon/fan source contained significant terrigenous components (27% quartz plus feldspar), but also contained foraminifera and pteropods (43%), indicating a probable slope origin (Bennetts and Pilkey 1976). Thus, some sediments are derived from land and probably passed directly through channels like the Pepillo Salcedo Canyon (Figure 9–7), whereas other sediment grains are slope-derived, deposited temporarily on the slope, and then released by mass movements. Small gullies indenting the flanks of the anticlinal hills, and observable in the GLORIA mosaic, probably result from minor slumping on the flanks of the growing anticlines facilitated by tectonic oversteepening.

A meandering leveed channel or channels on the deep basin floor off the eastern part of the central zone (Figure 9–9) may indicate flow of turbidity currents westward toward a region of bright return from the basin floor (sand deposits?). Alternatively, the contours show a general slope of the basin floor to the east (Figure 9–7), which might control the direction of flow.

Tectonic oversteepening, which may cause slumping from the growing anticlinal hills, also probably accounts for larger-scale mass movements on the steep slopes of the eastern and western zone of the survey. One very large mass movement feature, about 35 km across, is interpreted north of the Ile de la Tortue. A fold at the base of this feature may result from crumpling as the block slid downslope and/or from folding due to the overall compressional motion related to plate interactions. A reentrant just east of the central zone (just east of the boundary ridge, Figure 9–9) is marked by a radial drainage pattern of gullies and may have been caused by mass movement.

Puerto Rico region

The GLORIA and single-channel seismic-reflection surveys of the Puerto Rico region (Figure 9–1B; EEZ-SCAN 85 Scientific Staff 1987; Scanlon, Masson, and Rodriguez 1988) extends from oceanic crust of the North American Plate, across the Puerto Rico Trench and Muertos Trough (including the insular slopes of Puerto Rico and the U.S. Virgin Islands), into the Venezuela Basin on the Caribbean Plate (Figure 9–12). At the Puerto Rico Trench the survey extends from about longitude 67.5° W to 65.5° W, and at the Muertos Trough it extends from about 68.5° W to 64° W. This extensive coverage provides an excellent overview of morphologic and structural features of the plate boundary zone in the Puerto Rico region.

NORTH AMERICAN PLATE

North of the Puerto Rico Trench, GLORIA shows numerous basement ridges with little or no sediment cover, trend-

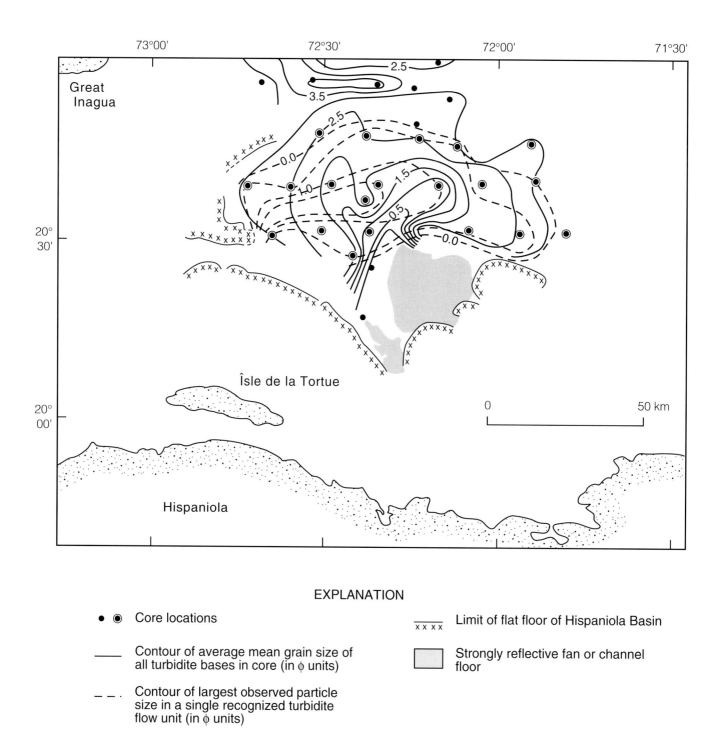

EXPLANATION

● ◉ Core locations

—— Contour of average mean grain size of all turbidite bases in core (in φ units)

– –. Contour of largest observed particle size in a single recognized turbidite flow unit (in φ units)

x x x x Limit of flat floor of Hispaniola Basin

▨ Strongly reflective fan or channel floor

Figure 9–11. Relation of sediment textural distribution to the boundary of the tectonic accretionary wedge (line with x's) and fans/channels off the northeastern Hispaniola insular margin. Sediment textural data were adapted from Bennetts and Pilkey (1976) and Ditty et al. (1977).

ing about 025° (Figure 9–13). They appear to be ridges formed by the seafloor spreading process that creates ocean-floor crust, and thus they are normal to the local opening direction, which we would infer to be 115°. Regional mapping based on broadly spaced tracks of magnetic data has suggested approximate ocean opening directions of 130°

(Klitgord and Schouten 1986a, b; Vogt 1986) or 125° (Hall and Westbrook 1990). We are convinced that the 115° value is correct locally. Near the western edge of the survey area, near 20° N, a series of en echelon basement ridges, individually trending 150° but together defining a feature at 135°, are distinctly oblique to the spreading fabric. This fea-

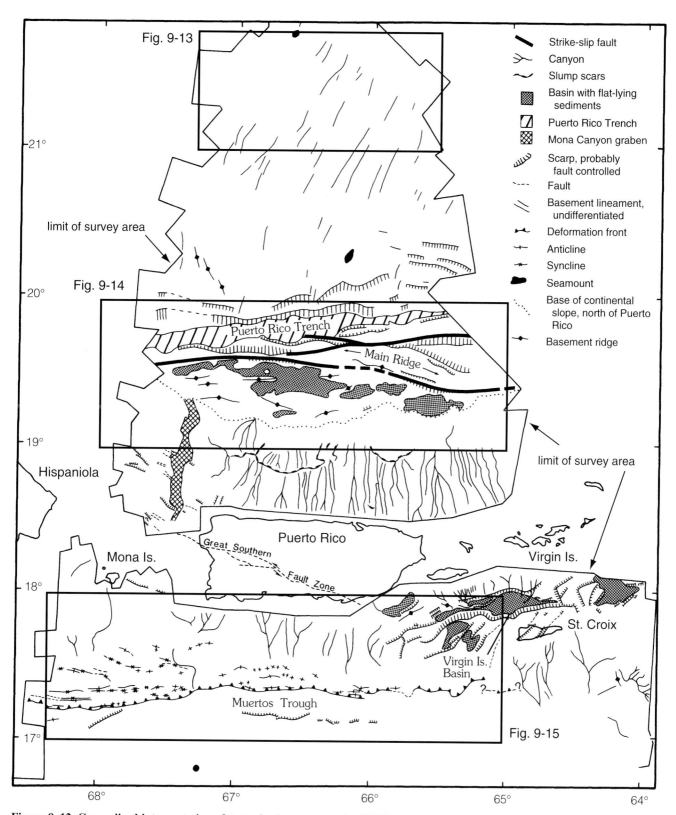

Figure 9–12. Generalized interpretation of tectonic elements seen in GLORIA mosaic.

147

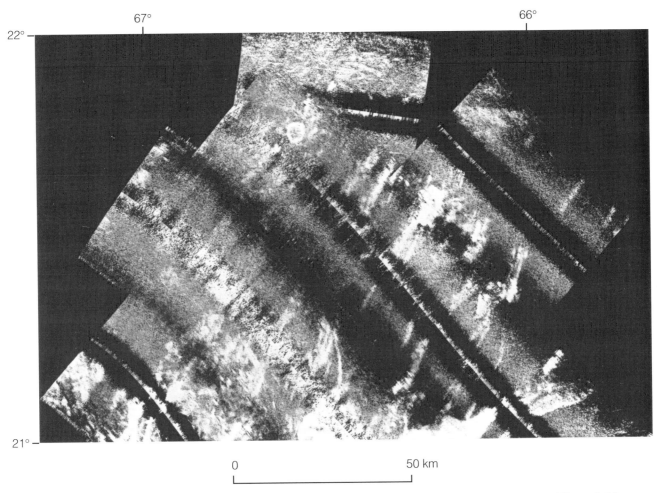

Figure 9–13. GLORIA mosaic of basement ridges on the North American Plate. Location of image shown on Figure 9–12.

ture, which is also marked by a bathymetric high, has been identified as the northwesterly extension of a North Atlantic fracture zone, Barracuda Ridge, by McCann and Sykes (1984).

The southern margin of the North American Plate oceanic crust is strongly downfaulted along the northern side of the Puerto Rico Trench with the formation of some horst and graben structures as well as the predominant down-to-the-south normal faults. Such structures are commonly found where an oceanic plate bends into a trench and result from tension in the upper part of the plate due to bending (Hilde 1983). The trend of the normal faults is parallel or subparallel to the trench. Their trend changes abruptly (from 085° to 105°) at 65°45′ W (Figures 9–12 and 9–14), mirroring a change in trend of the trench that occurs just outside the eastern edge of our study area. The character of the faulting also changes at this point. To the west, it consists of a narrow zone made up of two major fault scarps, but to the east, it is much broader and is made up of a complex horst and graben province.

PUERTO RICO TRENCH AND BASIN AND
RIDGE PROVINCE

The flat, sediment-covered floor of the Puerto Rico Trench (indicated by diagonal hatching labeled "Puerto Rico Trench" in Figure 9–12) is easily recognized in the GLORIA mosaic as a band of low backscatter (dark band) that trends 086° crossing the surveyed area at 19°45′ N (Figure 9–14). Between the trench and the steep insular slope north of Puerto Rico (19° N to 19°30′ N) lies a 50-km-wide zone of ridges and sediment-filled basins. All of the basins, including the trench, contain largely flat-lying undeformed sediments. The ridges and sedimentary basins have been mapped in considerable detail by careful integration of seismic-reflection profile and sidescan data (Scanlon and Masson this volume). An interesting feature of these ridges and basins is that they have a dominant 095° to 105° trend, distinctly oblique to the axis of the Puerto Rico Trench, which trends 086°.

Within the basin and ridge province, a set of straight to slightly sinuous lineaments, trending 085° to 105°, are interpreted to be the traces of strike-slip faults (Masson and

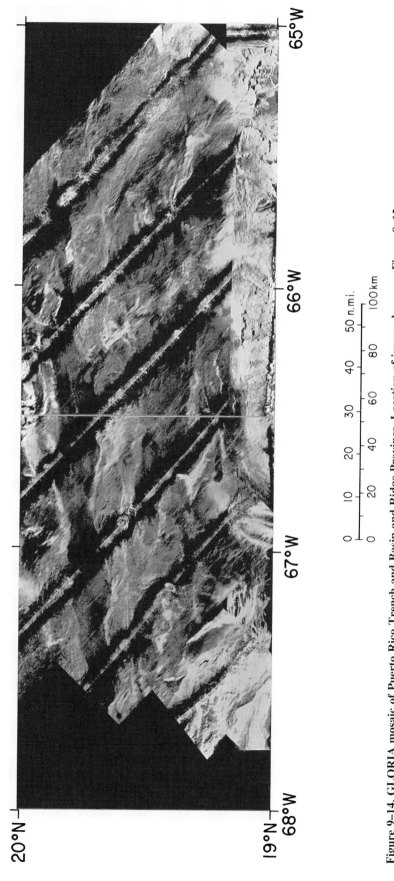

Figure 9–14. GLORIA mosaic of Puerto Rico Trench and Basin and Ridge Province. Location of image shown on Figure 9–12.

149

Scanlon 1991). They can be traced, more or less continuously, for about 250 km across the entire width of the GLORIA mosaic (Figure 9–14).

INSULAR SLOPE

An important aspect of the regional tectonics is the large amount of late Tertiary tilting that has affected the northern insular slope of Puerto Rico. Until early Pliocene the region north of Puerto Rico was a flat, shallow carbonate bank. Today, the region tilts northward at about 4°, lowering the distal parts of the former carbonate bank to water depths of more than 5,000 meters. The tilted sedimentary rock units are exposed on land on Puerto Rico and are predominantly shallow-water Tertiary carbonate units (Monroe 1980). They are parallel-bedded and have a regional northward dip of 4°, the same as the seafloor of the upper slope. Moreover, seismic-reflection data and dredged rock samples confirm the presence of the 3.3- to 3.6-million-year-old Quebradillas Limestone 50 km north of the island, in water depths greater than 5,000 m (Schneidermann, Beckmann, and Heezen 1972; Moussa et al. 1987). The shallow-water origin and parallel (initially horizontal) bedding of the Quebradillas require tilting after its deposition in mid-Pliocene. Further evidence for large-scale subsidence comes from the recognition of karst topography in Tertiary limestones at depths to 2,100 m (Heezen et al. 1985).

Two large (several 10s of kilometers across) amphitheater-shaped slump scars are seen in the GLORIA mosaic of the insular slope north of Puerto Rico. Headwall scarps of these features are marked with lines having square teeth, labeled "slump scars," in Figure 9–12. Failure of the Tertiary units probably resulted from the oversteepening of the slope by tilting and was perhaps triggered by earthquakes (Scanlon et al. 1988; Scanlon and Masson this volume). Preexisting faults, identified in a multichannel seismic-reflection profile (Moussa et al. 1987) may have significantly reduced the strength of the strata (Schwab et al. 1991) and controlled the locations of the scarps.

Scarps, trending northwest-southeast and probably fault-controlled, are seen on the shelf and upper slope in the vicinity of Mona Canyon (Figure 9–12). Their trend is similar to that of the Great Southern Puerto Rico Fault Zone, an active fault zone that cuts across the southwestern part of Puerto Rico (Figure 9–12; Garrison et al. 1972), and it is likely that they are related to it. Mona Canyon (diagonal cross-hatching, Figure 9–12) is much wider than the other canyons of the Puerto Rico insular slope and has a rectilinear plan and vertical sides. The canyon appears to be controlled by north-south-trending normal faults, locally offset by the northwest-southeast-trending faults.

MUERTOS TROUGH

The Muertos Trough is an approximately east-west-trending trough that lies at the base of the insular slope south of Puerto Rico and eastern Hispaniola. In the area south of Puerto Rico,

the GLORIA data (Figure 9–15) reveal that the trough is bordered on its south side by north-facing scarps, presumed to be normal faults associated with bending of the Caribbean Plate beneath the Greater Antilles. The north side of the trough is defined by a deformation front characterized by folds trending subparallel to the trough (Masson and Scanlon 1991). The zone of folding is about 40 km wide south of Mona Island and narrows eastward until it disappears south of the Virgin Islands Basin (Figure 9–12), indicating a decreasing amount of compression from west to east.

ANEGADA PASSAGE

The Virgin Islands Basin (Figure 9–12) is over 4,000 m deep and separates St. Croix on the south from the rest of the Virgin Islands to the north. The basin is about 90 km long east to west and about 20 kilometers wide. Several smaller, more shallow basins occur nearby. Together they constitute the ENE-trending Anegada Passage (Figure 9–1A), a bathymetric/topographic feature that follows the southern edge of the Virgin Islands platform, connecting the eastern Muertos Trough with the Puerto Rico Trench. Normal faults recorded in seismic-reflection profiles and mapped on the GLORIA mosaic clearly show the extensional nature of the basins (Jany, Scanlon, and Mauffret 1990; Masson and Scanlon 1991). The western limit of this extensional regime occurs at about longitude 65°30′, approximately the eastern limit of deformation north of the Muertos Trough (Figure 9–12).

Tectonics of the northern Caribbean Plate boundary

Patterns of seismicity

WESTERN REGION – CENTRAL AMERICA TO WESTERN HISPANIOLA

The seismicity of the western region of the northern Caribbean Plate boundary (Figure 9–2) appears to be related to a reasonably clear and well-understood set of plate arrangements and motions. Shallow earthquakes signify a strike-slip motion between plates with a left offset that results in extension in the Cayman Trough. This scenario has been reviewed in the section on tectonic setting and will not be repeated. This view of a simple offset at the plate boundary that results in opening has been challenged with the proposal that the eastern part of the Cayman Trough floor might move as a semiindependent microplate by a partial detachment along the southern side of the eastern part of the trough, extending through Jamaica (Rosencrantz and Mann 1991), but this concept has been questioned on the basis of the seismicity distribution and pattern of nodal basins (Edgar and Dillon 1992; Rosencrantz and Mann 1992). Although some minor complications exist, such as transverse structures formed by deformation of the floor of the Cayman Trough off eastern Cuba (Calais et al. 1989; Calais and Mercier de

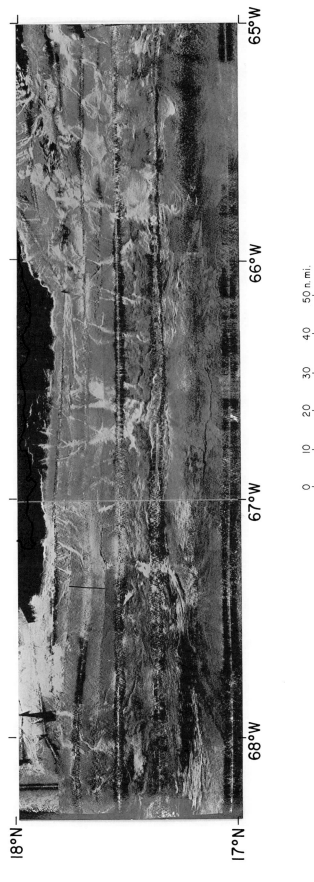

18°N

17°N

65°W

66°W

67°W

68°W

0   10   20   30   40   50 n.mi.

0   20   40   60   80   100 km

Figure 9–15. GLORIA mosaic of Muertos Trough deformation front. Location of image shown on Figure 9–12.

Lépinay 1990a, 1991), the simple concept of transcurrent motion along a fairly narrow and well-defined strike-slip fault zone, offset at the Cayman Trough spreading center, appears to explain the plate boundary situation adequately as far east as northwestern Hispaniola.

EASTERN REGION – EASTERN HISPANIOLA TO THE NORTHERN LIMIT OF THE LESSER ANTILLES

The earthquake pattern along the eastern part of the northern Caribbean Plate boundary (Figure 9–2) appears to disclose the following: (1) a major zone of earthquakes that dips southward from the Puerto Rico Trench beneath the PBZ – the zone has a western termination near 70° W; (2) a minor and shallower set of earthquakes that dips northward from the Muertos Trough on the southern side of the PBZ; and (3) some shallow earthquakes scattered through the PBZ, especially east of Hispaniola. These represent three distinct sets of interactions, respectively: (1) the movement of the North American Plate beneath the PBZ, (2) the movement of the Caribbean Plate beneath the PBZ, and (3) fracturing of rocks and interaction of microplate(s) within the PBZ.

Previous tectonic explanations of the three sets of plate interactions for the eastern region

MOVEMENT OF THE NORTH AMERICAN PLATE BENEATH THE PLATE BOUNDARY ZONE

The first of the interactions, that of the movement of the North American Plate beneath the PBZ, has long puzzled researchers, and subtly different models have been proposed. Some researchers (e.g., Murphy and McCann 1979; Frankel, McCann, and Murphy 1980; Sykes et al. 1982; McCann and Sykes 1984; McCann and Pennington 1990) have emphasized a southward subduction model (Figure 9–3A) to create the southward-dipping zone of seismicity (a true Wadati-Benioff zone model). Although these authors certainly recognized that the dominant motion is transcurrent (Caribbean Plate plus PBZ moving relatively eastward compared to North America), a very significant component of compressive motion would be required. This subduction model does not seem to explain the abrupt westward termination of deep earthquakes beneath central Hispaniola. A somewhat comparable model (Schell and Tarr 1978) proposes a more truly strike-slip motion along the plate boundary with subduction mainly from the east and preservation of a downbent portion of the North American Plate that is fractured into horizontal steplike blocks under the northern margin of the PBZ (Figure 9–3B). This part of North American lithosphere is postulated to survive because it is "above the critical melting depth" according to Schell and Tarr (1978). A third recent model (Calais, Bethoux, and Mercier de Lépinay 1992) proposes that the Caribbean has simply overridden the North American Plate, and that part of North

America forms a shallowly dipping detached slab beneath the PBZ (Figure 9–3C). This model does preserve the basic strike-slip motion of the plate boundary and accounts for the westward termination of the zone of deep seismicity. However, it is not clear why such a slab should not have entered the asthenosphere and caused melting, creating volcanoes, as similar slabs have undoubtedly done to the south as they dived beneath the Lesser Antilles Island Arc. The authors recognize this difficulty and suggest that the volcanism might be inhibited by dehydration of the slab. Furthermore, a major problem with this model is that it would not generate the observed southward-dipping zone of seismicity along the plate boundary.

Each of the above models seems to solve some, but not all, aspects of the seismicity and plate interaction problem of the North American Plate – PBZ boundary. A solution will be suggested here that combines aspects of all of them.

MOVEMENT OF THE CARIBBEAN PLATE BENEATH THE PLATE BOUNDARY ZONE

South of the PBZ, the seafloor is depressed into the Muertos Trough, and, as noted previously, a minor earthquake zone seems to dip northward from this trough. Analysis of seismic profiles (Ladd, Worzel, and Watkins 1977; Ladd and Watkins 1978, 1979; Ladd, Shih, and Tsai 1981) and long-range sidescan sonar (EEZ-SCAN 85 1987; Masson and Scanlon 1991) suggest minor compression and folding of sediments in the Muertos Trough south of eastern Hispaniola and Puerto Rico. An analysis of four published plate models (Heubeck and Mann 1991) also predicts recent (since 5 Ma) compression in this region. Ladd and Watkins (1978) concluded that the Caribbean Plate is being subducted beneath the PBZ south of Puerto Rico and eastern Hispaniola, east of the Beata Ridge (Figure 9–1A). Heubeck and Mann (1991) have proposed that any subduction ends near the ridge because that feature forms a boundary between eastern and central subplates of the Caribbean that move semi-independently. These authors also suggested that the western Caribbean (west of the longitude of the Cayman trough spreading center; Figure 9–1B) probably forms a third subplate.

PLATE BOUNDARY ZONE

Within the PBZ of the eastern Caribbean northern plate margin are two distinct regions, one in Hispaniola and the other to the east. The island of Hispaniola consists of an agglomeration of terranes that accumulated before the onset of strike-slip motion along the northern Caribbean Plate boundary in Eocene time (Mann, Draper, and Lewis 1991a). Much of the geology of the island is best explained by approximately north-south crustal shortening (Bowin 1975; Biju-Duval et al. 1983; Edgar et al. 1991; Mann et al. 1991a; Mann, McLaughlin, and Cooper 1991b). The northern insular margin shows clear evidence of thrusting (Figure 9–4; Austin 1983; Dillon et al. 1991, 1992). The tectonic stresses along this part of the northern Caribbean Plate boundary have been quite variable over time as shown by variable uplift in Cuba (Calais and Mercier de Lépinay 1990b) and Hispaniola (Mann et al. 1991a), by variations in rates of tilting of deep-sea strata north of Hispaniola (Dillon et al. 1992), and by development of folds in the eastern Cayman Trough (Calais and Mercier de Lépinay 1990a). These tectonic variations might be caused by slight changes in directions of plate motion (including slight rotations of all or parts of the Caribbean) or by interaction of irregularities on the plate boundaries as one plate slides past the other causing relative movement of releasing and restraining bends. One major restraining bend along the plate boundary has been long known; this is formed by the major right offset bend of the North American – PBZ boundary north of eastern Hispaniola (Figure 9–1; Bracey and Vogt 1970; Mann et al. 1984). This restraining bend clearly seems to be the cause of some of the compressional structures within Hispaniola. A recent detailed study of the compressional/extensional structures of the northern Caribbean Plate boundary (Calais and Mercier de Lépinay 1993) draws the conclusion that, despite the complexity, most structures can be explained by movements related to rigid plate tectonics about the appropriate pole of rotation.

East of Hispaniola, in the region including Puerto Rico and the Virgin Islands platform, a counterclockwise microplate rotation has been proposed. This motion has been interpreted on the basis of inferred structural patterns of shortening in the Muertos Trough region and extension in the Virgin Islands area (Byrne, Suarez, and McCann 1985; Mauffret et al. 1986; Jany et al. 1987; Mauffret and Jany 1990; Masson and Scanlon 1991). Paleomagnetic studies in Puerto Rico appear to confirm this counterclockwise sense of rotation (Fink and Harrison 1972; Vincenz and Dasgupta 1978; Reid and Plumley 1991). A careful analysis of several plate models by Heubeck and Mann (1991) suggested that the best analysis of motion in the Puerto Rico and Virgin Islands (PR/VI) part of the PBZ can be accomplished by assuming the presence of a PR/VI microplate. An unresolved issue appears to be the location of the western limit of this microplate rotation. Subsidence of the northern Puerto Rico insular margin (rapid since Pliocene) has been attributed to transtension (Jany et al. 1990) or deep tectonic erosion of the PBZ by the North American Plate (NOAM) (Birch 1986).

Although microplate rotation seems to explain successfully the inferred extensions in the Virgin Islands region, another recent model (Speed and Larue 1991; Larue, Joyce, and Ryan 1991) explains the extension in the Virgin Islands area by a general transtension across the plate boundary region. This model of overall regional motion seems less convincing than a transcurrent one because the continuous nature of the subducted North American Plate that dips beneath the PBZ, which is indicated by the distribution of earthquakes, does not seem to suggest extension. Furthermore,

an extensional model would require an upward bulge of dense mantle that would generate a gravity high, whereas the Puerto Rico Trench is characterized by the largest negative gravity anomaly in the world (Bowin 1976; Molnar 1977). For these reasons a model requiring general extension seems unsuccessful.

Analysis of tectonics on the basis of GLORIA and seismicity for the eastern region

APPROACH TO ANALYSIS

Obviously, models of the plate interactions in the eastern part of the northern Caribbean Plate boundary are contradictory. Therefore, we will analyze the distribution of earthquakes in light of structural trends deduced from GLORIA, many of which clearly are related to major plate interactions. Earthquake distribution along the northern Caribbean Plate boundary (Figure 9–2) indicates a broad range of earthquake depths. Shallow seismicity (less than about 50 km) occurs where surface plates are stressed and fractured, in part by being bent as the plates are being subducted. Deeper earthquakes are considered to indicate the locations of portions of the Earth that are brittle; presumably these are cold, subducted lithospheric plates. The deeper earthquakes may be due to (1) overthrusting, in which the subducting plate slides beneath the overlying plate in contact with it; (2) intraplate compression, in which the slab fails due to being forced through the asthenosphere (downward or horizontally); (3) intraplate extension due to gravity effects on the colder, denser downgoing slab, which probably has undergone phase transformations; or (4) stresses caused by volume changes that result from such phase transformations. Stein, Wiens, and Engeln (1986) contend that, in the Lesser Antilles at least, most recent seismicity is related to intraplate normal faulting. However, to the west (beneath eastern Hispaniola), Calais et al. (1992) report that deeper events show compressive focal mechanisms. In any case, because an earthquake marks the location of brittle lithosphere, the earthquake distribution outlines the regions occupied by lithosphere, either at the surface or included in the asthenosphere in the stressed plate boundary region. The overall pattern of seismicity is shown in Figure 9–2, which represents a plot that has been rigorously edited to remove, insofar as possible, spurious hypocenters (see caption, Figure 9–2).

EVIDENCE FROM GLORIA AND SEISMICITY

As noted previously, GLORIA shows lineations in seafloor basement north of the Puerto Rico trench that trend NNE (025°; Figure 9–13). These are considered to represent a pattern of seafloor ridges that disclose the ocean-opening structure and may represent a grain of weakness in the crust. Examination of the seismicity patterns also seems to indicate the presence of offsets of seismic activity along the same 025° trend. We have used boundaries following this trend

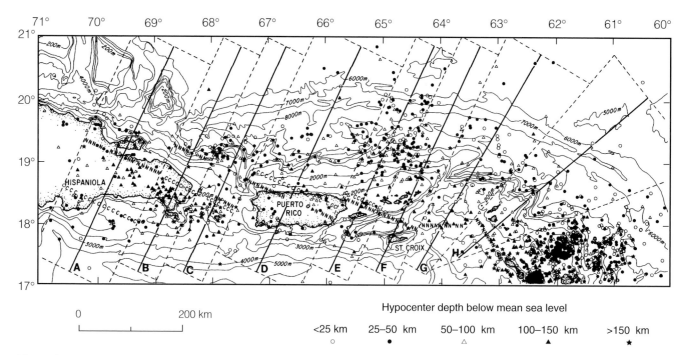

Figure 9–16. Seismicity in the plate boundary zone region. Figure also shows compartments in which hypocenters were projected for plotting Figure 9–18 and the southern limits of overridden NOAM (lines of Ns) and northern limit of CARIB (lines of Cs).

of offset to divide the earthquakes into compartments labeled A through H (Figure 9–16). For example, note that seismicity of compartment F does not extend northeast as far as that of compartments E and G on each side (Figure 9–16) and that the boundaries of earthquake activity seem to trend 025°. A series of left offsets of the earthquake pattern seem to mark the boundaries of compartments E, D, C, and B. We inferred that the compartments of seismicity might represent the locations of slabs of North American Plate (NOAM) that have been overridden by the plate boundary zone (PBZ) and Caribbean Plate (CARIB) as a result of the relatively westward motion of NOAM. If the slabs dip parallel to their trend (NNE or 025°), their dip direction would be SSW (25° + 180° = 205°). However, the faults that indicate the direction of strike-slip motion in the Puerto Rico north slope strike 085° to 105°, so a dip of 205° is not normal to the NOAM/CARIB boundary.

In order best to estimate the direction of dip for NOAM slabs, we selected a series of points located along 18° N latitude at the center of each of the eight inferred compartments. For compartments A through F, a series of vertical planes were passed through these points on bearings ranging from 000° to 045° at 7.5° intervals. Earthquake hypocenters within a swath equal to the width of the compartment at 18° N were projected onto these planes, and a series of plots was made. True direction of dip was indicated by the plot direction that showed a dipping group of hypocenters having the shallowest apparent dip angle and the minimum group thickness. This test showed that the dip directions of

slabs A through F actually were approximately 205°, parallel to the trends of the ocean floor grain. Therefore, in Figure 9–17 we display plots of hypocenters projected to planes oriented NNE-SSW (025° to 205°). For compartments G and H, proceeding around the corner of the plate, we performed the test for planes bearing 22.5° to 67.5° for G and 45° to 90° for H. True bearings of these slabs appear to be 030° (dip direction = 210°) for G and 050° (dip direction = 230°) for H, as used to plot Figure 9–17. The same logic was applied to analysis of the dip direction of the CARIB beneath the PBZ, and, within the compartments that seem to have dipping slabs of CARIB (A through E), a bearing of 025° also appears to apply (dip direction here is reverse to that of NOAM – NNE or 025°). The agreement in trend between NOAM and CARIB slabs is surprising because CARIB would not necessarily be expected to have the same grain of weakness as NOAM (no grain is clearly identifiable in CARIB either on GLORIA or magnetics).

The patterns of seismicity are simplest to the east, so we will begin our discussion with compartment H and proceed westward. Compartment H is located just north of the clearly defined Wadati-Benioff zone of the Lesser Antilles; these islands form a typical volcanic island arc above the subducting plate. As seen in Figure 9–2, the southern boundary of compartment H was chosen at a marked change in earthquake density, and generally, the region of the CARIB corner (in compartments G and H) has much lower earthquake density than the region to the south and also somewhat lower earthquake density than the region immediately to the west (com-

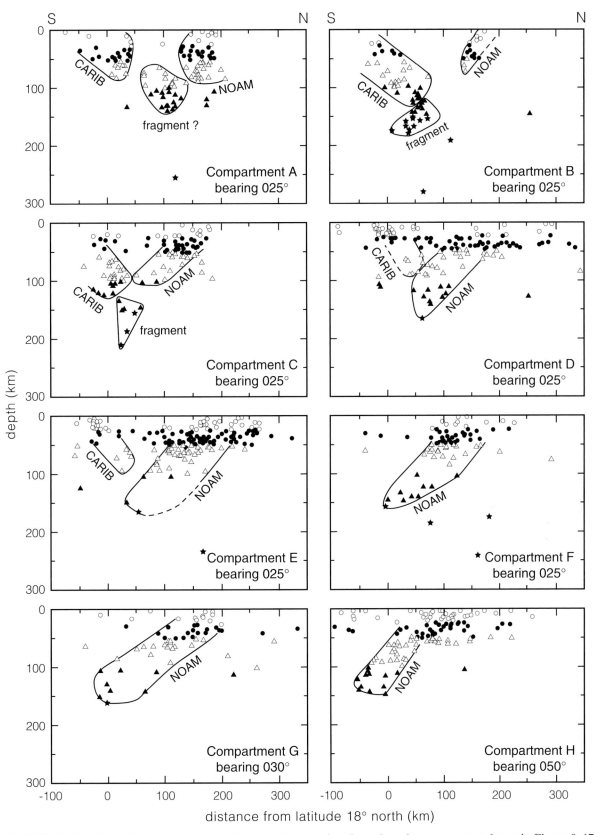

Figure 9–17. Plots of earthquake hypocenters projected to planes passing through each compartment shown in Figure 9–17. Bearings of the planes of projection are noted, and the rationale for choosing these compartments and projection planes is discussed in the text.

partments A through F). Compartment H (Figure 9–17) shows a well-defined zone of dipping earthquakes along a trend bearing 050° (dip direction is 230°). The boundaries between compartments G, H, and F are marked by some deep earthquakes in the range 50 to 150 km that are located farther out into NOAM crust (approximately northeast) than the average for the compartments. All of the isolated hypocenters that plot to the right and below the outlined slabs in compartments F, G, and H are located on compartment boundaries. Aside from these scattered hypocenters on slab boundaries in F and G, these two compartments display coherent SW-dipping groups of hypocenters, presumably representing the overridden NOAM. Through rigorous editing of this seismicity data (see caption, Figure 9–2) we have eliminated many doubtful points, but this approach has resulted in relatively few points for each interpreted slab; therefore we generally have arbitrarily shown the deep slab limit as normal to the dip (Figure 9–16). In the case of compartment G, though, a deep boundary for the slab oriented E-W seems identifiable on the map.

Compartments E through A also show southwest-dipping slabs of NOAM; the dip angle of the NOAM seismic zone is fairly constant at about 40° to 50°. In these compartments some earthquakes to the south also suggest the dipping of the CARIB beneath the PBZ. In compartments D and C there is a suggestion that NOAM and CARIB are in contact beneath Puerto Rico and the Mona Passage. The antecedents of the detached block in compartment A (Figure 9–17) and the structure of the NOAM block in A are unclear. This region is at the leading edge of the westward-moving flap of NOAM that is being shoved through the asthenosphere and, for that reason, some complex fragmentation may have occurred here.

Some important observations seem to be:

1   Compartments C, B, and (less clearly) A show discrete zones of seismicity at depth, presumably representing deep lithospheric fragments distinct from NOAM or CARIB.

2   The depth of the deepest extent of NOAM increases slightly and consistently from about 150 km to the east to 165 km in D, but further west the maximum depth of NOAM decreases abruptly to 103 km in C and 73 km in B.

3   The thickness of the NOAM slab, as defined by hypocenters, is constant at 60 to 70 km from the eastern compartment, west to compartment D, then thins abruptly to 45, then 30 km in C and B.

4   Like NOAM, CARIB, as defined by its seismicity, also thins toward its leading edge (toward the east in the case of CARIB). Its thickness is 60 to 65 km to the west, but decreases to about 40 km at its easternmost observation in compartment E.

5   CARIB appears to be in direct contact with NOAM in compartments C and D and with a deep lithosphere fragment in B.

6   In the eastern compartments, scattered deep earthquakes

occur primarily at compartment boundaries, and seismicity is less intense than to the west (or south).

TECTONIC MODEL

We will consider the seismicity patterns in light of the undoubted left-lateral transcurrent motion at the northern Caribbean Plate boundary. Clearly, as the NOAM slides westward, it is subducted on the eastern side of CARIB and the down-going slab of lithosphere causes flow and melting in the superjacent asthenosphere, generating the volcanic activity of the Lesser Antilles Island Arc. We propose that, along CARIB's northern boundary, a slab of lithosphere has been bent down at the northeast corner of the Caribbean Plate, and the slab simply slides westward beneath the Virgin Islands, Puerto Rico, and eastern Hispaniola as a downbent flap with no melting occurring (Figure 9–18). The dearth of volcanism in the eastern Greater Antilles probably results because NOAM is not being subducted to a significant extent and thus does not generate appropriate circulation in the asthenosphere to cause melting. Insofar as large-scale transcurrent motions are concerned, the PBZ and CARIB appear to be moving together, even though they are not actually coupled.

The distribution of earthquakes and patterns discerned in GLORIA appear to be consistent with this hypothetical model of an overridden, downbent, westward-moving slab of NOAM beneath the plate boundary zone (PBZ) region of lithospheric fragments, and furthermore, they suggest a comparable eastward-sliding slab of CARIB. The discrete zones of seismicity at depth in compartments C, B, and A (observation 1 in the previous analysis of seismicity patterns) probably represent detached fragments of lithosphere (as also suggested by McCann and Pennington 1990). Because the section of NOAM above them is exceptionally short (as indicated by its shallow basal depth – observation 2), we infer that the detached fragments were derived from the NOAM. The thinning of the NOAM seismicity zone in the western compartments (observation 3) may suggest that this western part of the overridden NOAM, which has been at depth longest, because it represents the leading edge of the overridden plate, may have begun to soften due to heating. The lower part of the lithosphere, which was warmest to begin with, would be the first to soften, whereas the upper part apparently still is sufficiently brittle to fracture and generate earthquakes. Alternatively, evolution of the mineralogical difference between upper lithosphere (crust) and lower lithosphere (upper mantle) may also influence this change in thickness of the earthquake layer by dehydration, basalt/eclogite transformation, and densification. If so, the process would seem to be far slower than in subduction zones. Thinning of the CARIB hypocenter pattern toward the east (4) may be explained similarly. The softening and probably the mineralogical changes of NOAM also would tend to weaken the plate and foster fragmentation in the western region that has been at depth longest. The CARIB

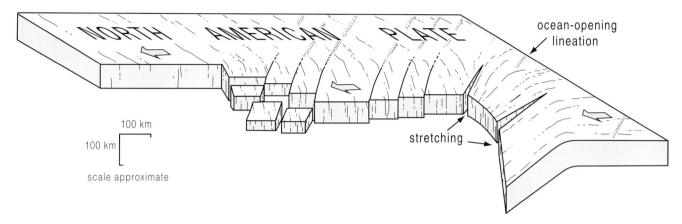

Figure 9–18. Perspective sketch of the North American Plate in the region of the northeastern Caribbean, viewing direction NNE. The lithospheres of the plate boundary zone and Caribbean Plate have been removed to afford a view of the North American Plate.

extends to its greatest depths in compartments B and C (about 130 km) and seems to be in contact with NOAM in C and D (observation 5) or the fragment of NOAM (in compartment B), which suggests that deep plate collision also may have contributed to the fragmentation of NOAM.

Complex stresses can be anticipated in NOAM near the northeast corner of CARIB, where CARIB is overriding NOAM. Along the *eastern* Caribbean subduction zone, NOAM is bent downward toward the west at near 45°, whereas along the *northern* plate boundary, the same plate is bent southward. The result is that the downbent part of the plate will tend to be pulled apart at the "turn" so that strong extensional stresses will be generated in the downbent slab (Figure 9–18). The deep earthquakes around the northeast corner of the CARIB (observation 6) probably represent the required fractures to allow this stretching of the NOAM plate. Perhaps this localized extension in a region of general compression helps to account for the dearth of earthquake activity in the turn area (note Figure 9–2), even though the directions of extension and compression would appear to be different. Horizontal extension in the downgoing slab due to this cause also might influence the apparent extensions in the Virgin Islands platform and Anegada Passage, but the existence and possible nature of coupling between the subducted North American Plate and overlying plate fragments in the PBZ is obscure. In this model the location of active extension in the downbent NOAM slab would have passed eastward from central Hispaniola to the Anegada Passage from Eocene to present, perhaps causing the fracturing of NOAM into the discrete slabs that we infer from the seismicity pattern. The fracturing may have followed the pattern of ocean-opening grain (trend of 025°) because this provided preexisting planes of weakness.

CARIB appears to be in collision with unbroken NOAM in compartments C and D (observation 5), where it appears that CARIB may be pushing down NOAM. The stress applied by the relatively eastward-sliding CARIB to the

NOAM may account for the subsidence of NOAM to form the Puerto Rico Trench. The cause of this trench, which has the largest free-air gravity anomaly in the world, has long been controversial (C. Bowin, oral communication 1990), because it exists where plate motion clearly is dominantly strike-slip, whereas subduction normal to a trench commonly is considered necessary to create such a feature. The maximum gravity anomaly of the trench occurs in compartments C and D (Bowin 1976; Westbrook 1990), as would be expected if deep plate collision were causing the stress.

There certainly was a time early in the development of the present Caribbean Plate when the downbent overridden flap of NOAM slid westward without contact with CARIB, and therefore when the interplate stresses that resulted in the Puerto Rico trench subsidence did not exist. This is clear from the gap between the area labeled "CARIB to be overridden" and the area labeled "NOAM to be overridden" that appears in the late Eocene reconstruction (Figure 9–19), a gap that had to be closed before interaction between NOAM and CARIB could start. Examination of the regional geology may indicate when development of the trench began and thus when this collision beneath the PBZ took place. Deposition before Pliocene time resulted in a thick layer of horizontally bedded limestone across almost all of Puerto Rico and the Virgin Islands during a long period of uniform subsidence from late Eocene (Lewis and Draper 1990). After about mid-Pliocene (about 3.3 Ma) the area experienced very rapid northward tilting and subsidence of the northern edge of the Puerto Rico-Virgin Islands carbonate platform and fragmentation of the platform in the Mona Passage west of Puerto Rico and Virgin Islands platform to the east. The subsidence tilted the previously horizontal shallow-water carbonate platform strata that are now on the northern insular slope of Puerto Rico to depths of more than 5,000 m. Although these strata occur on a microplate within the PBZ, their tilting clearly is part of the subsidence of the Puerto

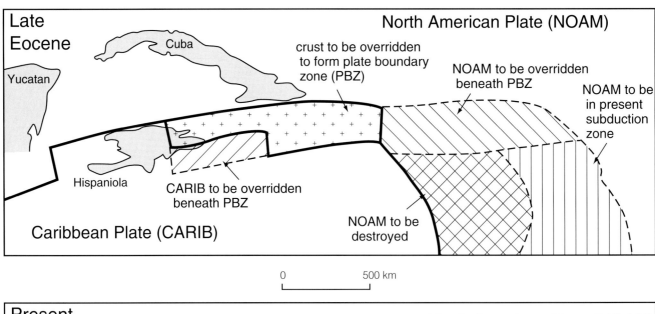

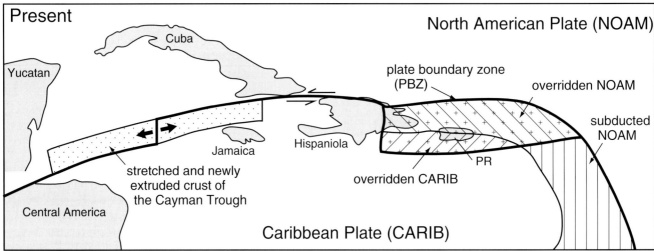

Figure 9–19. Present plate distribution (below) and Eocene reconstruction (above) of the Caribbean region.

Rico Trench. The timing of the initiation of this tilting in mid-Pliocene presumably identifies the age of deep plate collision. This tilting still is going on and results in massive deep-sea landslides that have created the amphitheaters observed in GLORIA.

PLATE MODEL

In order to reconstruct the arrangement of plates before initiation of the separate Caribbean Plate we must (1) slide the NOAM block back to the eastward, (2) remove the newly generated crust of the Cayman Trough, (3) fold back up the present overridden flaps and subducted slab, and (4) replace the NOAM crust that has been destroyed in the subduction zone. Such a reconstruction is presented in Figure 9–19, which also shows a sketch of the inferred present plate arrangements.

The amount of motion between the North American Plate and Caribbean Plate is about 1,200 km on the basis of the total opening of the Cayman Trough, which records this interplate motion (a minimum of 1,050 km according to Pindell and Barrett 1990). Therefore, the length of the overridden slab of NOAM (E-W direction) was ~ 1,200 km, as indicated in Figure 9–19. This distance is the same as the length of the zone of deep seismicity that defines the downbent flap west of the Lesser Antilles as it should be if both offsets were created by irregularities along a purely strike-slip boundary between rigid plates.

Please note that the location of Hispaniola shown for the Eocene (Figure 9–19) is its present position relative to the Caribbean Plate. Significant distortion of the PBZ would alter that location. The location of the island at the time of initiation of the present northern Caribbean Plate boundary is a major problem in Caribbean Plate reconstructions. Several authors (e.g., Calais and Mercier de Lépinay 1990b; Pindell and Barrett 1990) have attempted to align island-arc rocks in Cuba with similar rocks in Hispaniola in early Tertiary (and according to Calais and Mercier de Lépinay 1990b, they were together until early Miocene). Such a position allows for only 180 to 400 km of plate boundary offset along the eastern part of the northern CARIB boundary (Mann et al. 1991a) and only that amount of destruction of PBZ lithosphere at the Lesser Antilles Island Arc, a destruction that is necessary to allow space for the new crust formed in the Cayman Trough. However, there is evidence for about 1,200 km of differential plate motion at the Cayman Trough, which leaves the disposal of at least 800 km of lithosphere unaccounted for. If Hispaniola was immobile until Miocene relative to Cuba (which was part of the North American Plate), this destruction had to take place between the eastern end of the developing Cayman Trough and the western end of Hispaniola. Such destruction could occur by plate subduction or plate thickening within the PBZ. The former seems unlikely because evidence for a post-Eocene subduction zone crossing the PBZ should be obvious but is not apparent. Removal of lithosphere by plate thickening seems impossible, because present distance from the eastern end of the trough to Hispaniola is perhaps 200 km, and so an average crustal thickening in that region of five times would be necessary. Such thick crust certainly is not present. Transfer of the motion to the south side of the PBZ is not acceptable because the entire PBZ must move east to allow space for the new crust in the floor of the Cayman Trough (although we do admit that the last 400 km of Cayman Trough crust that has been formed [the central 400 km] actually is broader than older crust). A simpler solution may be to relax the requirement that Cuba and Hispaniola were adjacent when the new northern plate boundary was formed in Eocene time.

Summary and conclusions

Cayman Trough region

The GLORIA sidescan sonar reflection pattern in the Cayman Trough is dominated by backscatter from the north-trending ridges that form the grain of the floor of the trough. These are roughly parallel to, and presumably formed at, the spreading center, which forms part of the plate boundary. As indicated by seismicity, the main plate boundary extends west of the spreading center on the south side of the trough and east of the spreading center on the north side of the trough. At the spreading center, the rift valley appears as a 5- to 15-km-wide region of bright reflections. Triangular nodal basins containing little sediment cover appear at the north and south ends of the rift valley against the active transform faults (on the east side at the north end of the rift valley and west side at the south). The north wall of the trough contains a broad sediment-covered shelf about halfway from top to bottom, dividing the wall into upper

and lower escarpments. South of the north wall of the trough, a major ridge, probably formed of serpentinite, extends about 100 km east and west of the rift-transform intersection.

Western Hispaniola region

The insular slope of northwestern Hispaniola displays three distinct morphological/structural compartments. Steeply dipping (9° to 16°) insular slopes having evidence of mass movement and complex internal structure occur off northeast and northwest Haiti. The region between these compartments displays a much gentler slope (~ 4°), beneath which broad thrusted anticlines form a fold-controlled canyon system. The compartments probably are separated by tear faults, one of which is marked by a north-trending ridge that crosses the insular margin. The structure and morphology of the margin are that of a tectonic accretionary wedge. The shortening required to form such a feature presumably results from the effect of a restraining bend in the transcurrent plate boundary. The strike-slip motion, as indicated by seismicity, occurs near the coastline, south of the insular margin region that was imaged by GLORIA.

Sedimentation in the Hispaniola Basin is primarily from turbidity currents, and their flow is dominated by the tectonically formed morphology of the area. Many of the turbidity currents enter the basin from the insular slope of northwestern Hispaniola, which is being tectonically steepened, encouraging mass movement, and many turbidity flows are passed through the fold-controlled canyon system.

Puerto Rico region

GLORIA images indicate that the seafloor north of Puerto Rico has a set of ridges trending ~ 025°, interpreted to represent an ocean-opening grain of the oceanic crust. Just north of the Puerto Rico Trench, seismic profiles and GLORIA show a set of trench-parallel normal faults, probably formed by bending of the plate. South of the trench, lineaments trending 085° to 105° probably are traces of strike-slip faults formed by the transcurrent motion at the northern border of the Puerto Rico/Virgin Islands plate boundary zone microplate. On the northern insular slope of Puerto Rico lie 3.3 to 3.6-million-year-old shallow-water carbonate platform rocks that were formed flat lying but that now are tilted northward at 4° and occur at depths exceeding 5,000 m. Large amphitheater-shaped landslide scars are observed where these platform rocks have slid away due to gravity effects resulting from the tectonic tilting that they have undergone.

South of Puerto Rico is the Muertos Trough, which, like the Puerto Rico Trench, is marked at its oceanward (southern) side by trench-facing scarps caused by normal faulting at the bend of the oceanic plate (in this case the Caribbean Plate). Trough-parallel folds occur on the north side of the

Muertos Trough from south central Hispaniola to the eastern end of Puerto Rico. East of the compressional effects, the Virgin Islands region appears to be in extension. This suggests that the plate boundary zone, including the Virgin Islands, Puerto Rico (Masson and Scanlon 1991), and probably eastern Hispaniola, may represent a region that is rotating in a counterclockwise sense between the North American and Caribbean plates, perhaps as an edge-driven microplate (Schouten and Klitgord 1993). Alternatively, the eastern part of the plate boundary zone (perhaps the part east of the eastern Mona Passage) might be performing such a rotation while the western part (eastern Mona Passage to central Hispaniola) might be undergoing a southward wedging (escape tectonics) due to the shape of its northern boundary and the relatively westward movement of the NOAM.

Seismicity and a model for the northern Caribbean Plate boundary

Seismicity west of central Hispaniola can be explained as indicating almost pure strike-slip motion with a left offset (releasing bend) at the center of the Cayman Trough, which provides a location for opening and generation of new magnetic crust. Earthquakes are relatively shallow.

Seismicity in the plate boundary region east of central Hispaniola appears to define three sets of earthquakes, which we use to define three lithospheric units. One set occurs in lithosphere of the plate boundary zone (PBZ); a second set occurs in a zone that dips southward beneath the northern part of the PBZ and defines dipping North American Plate (NOAM) lithosphere; a third set dips northward beneath the southern part of the PBZ and defines Caribbean Plate (CARIB) lithosphere. The two trenches, the Puerto Rico Trench to the north of the PBZ and the Muertos Trough to the south, are the sites where the lithosphere is bent down to extend beneath the PBZ. Earthquake zones define the extent of brittle lithosphere, and, although they display dips, they do not imply subduction parallel to those dips. The plate motion can be explained as almost purely strike-slip. To the southeast of the region we consider here, in the Lesser Antilles Island Arc, the NOAM lithosphere dips westward and melts or softens at a depth of about 150 km. At the northeast corner of the CARIB, NOAM is bent down and overridden by the relatively eastward moving CARIB and associated PBZ, so that a southward-dipping flap of NOAM slides relatively westward along the northern boundary of the PBZ (PBZ and CARIB, although uncoupled, seem to be moving eastward at the same rate). At that northeast corner, a compound bend of NOAM is required, which requires stretching of the NOAM oceanic lithosphere. Apparently this stretching results in fracturing of NOAM along preexisting lines of weakness provided by the ocean-opening grain. This results in formation of a series of NNE-oriented (025°) blocks and, thus, the flap consists of a series of NNE-oriented slabs that are sliding to the west beneath the PBZ.

The presence of these slabs is indicated by a series of NNE offsets in the earthquake distribution.

Imaging of earthquake hypocenters shows not only the southward-dipping NOAM, but also, in the western part of the PBZ (west of the Virgin Islands), a northward-dipping flap of the CARIB. This is moving eastward relative to NOAM at the NOAM/CARIB rate of relative motion of perhaps 2 to 3 cm/year (Ladd et al. 1990). The latter motion has resulted in collision of the two flaps at depth beneath Puerto Rico and the Mona Passage, with CARIB wedging over NOAM. The result of this deep plate interaction probably is to force NOAM down (as a result of loading by CARIB) and to force CARIB up (by buoyancy forces). The Puerto Rico Trench is the deepest spot in the Atlantic and has the largest negative free-air gravity anomaly in the world, surprising considering its location in a strike-slip fault zone. This situation may be accounted for by the depression of NOAM due to interaction with CARIB at depth. Furthermore, the loading of NOAM and concomitant buoying up of CARIB generates a down-to-the-north rotation of the substructure of the PBZ beneath Puerto Rico that may account for the tilting of the Pliocene limestone platform. The onset of this tilting, about 3.3 Ma, may provide a date when the two plates collided at depth. The leading (western) edge of the NOAM flap seems to have become fragmented, possibly due to the collision or to gravity effects caused by densification resulting from phase changes.

A broad-scale horizontal counterclockwise rotation of the entire PBZ probably is indicated by the folding of sediments north of the Muertos Trough between the Beata Ridge and eastern Puerto Rico (Ladd et al. 1990) and the extension to the east of that point (Jany et al. 1990; Masson and Scanlon 1991). A detachment of the western end of the PBZ would be required, which may be provided by the Beata fault zone (Mann et al. 1991a; "eastern Beata fault zone," Lewis and Draper 1990). The entire PBZ may be rotating as a single block on the basis of this compression/extension pattern, but if so, it would be a highly elongate block, about five times as long (E-W) as its width (N-S). Obviously such a shape would be highly constrained in rotation. Possibly the extension in the Mona Canyon (note crosshatched area in Figure 9–12) is the site of fragmentation of the PBZ into more equidimensional blocks that will act more like roller bearings in the strike-slip zone. We might speculate that this division of the PBZ might extend southward to a location on the north wall of the Muertos Trough, where GLORIA shows a change in trend of the wall and its associated folding (Figure 9–15, about 67.3° W). A reentrant in the north wall of the Muertos Trough (Figure 9–1A) and a dearth of earthquakes along the same meridian (Figures 9–2 and 9–16) also mark this region of possible microplate fragmentation. The western block of the PBZ (formed of eastern Hispaniola and western Mona Passage) may be undergoing a southward wedging (escape tectonics) due to the shape of its northern boundary, which forms a restraining bend in conjunction with the relatively westward movement of the NOAM.

Final considerations

In conclusion, the northern boundary of the Caribbean Plate is characterized by strike-slip motion throughout its length, but overriding of slabs, extension, shortening, and rotation result from the geometry of the plate boundary. The geometric considerations include presence of a "corner" where one plate overrides another, as well as irregularities and fragmentation along the plate boundary. The seafloor effects and the large scale of these active tectonic features make them especially appropriate for analysis with the regional surveying capability of the GLORIA system. The tectonics of the region is extremely complex despite the basic simplicity of the plate motions, and therefore the solutions presented here represent only suggestions, which are likely to be controversial and certainly require further study.

Acknowledgments

We wish to thank all members of the seagoing scientific parties for the northern Caribbean GLORIA legs and especially Lindsay M. Parson and Douglas G. Masson of the Institute of Oceanographic Sciences. Valuable comments were made regarding the manuscript by reviewers Robert N. Oldale, Kim D. Klitgord, Elazar Uchupi, and John W. Ladd.

References

Austin, J. A., Jr. 1983. Overthrusting in a deep-water carbonate terrain. In *Seismic Expression of Structural Styles*, ed. A. W. Bally, pp. 3.4.2–167-72. Tulsa, OK: American Association of Petroleum Geologists Studies in Geology, Vol. 3, No. 15.

Bennetts, K. R., and Pilkey, O. H. 1976. Characteristics of three turbidites, Hispaniola-Caicos basin. *Geol. Soc. Am. Bull.* 87: 1291–300.

Biju-Duval, B., Bizon, G., Mascle, A., and Muller, C. 1983. Active margin processes: Field observations in southern Hispaniola. In *Studies in Continental Margin Geology*, eds. J. S. Watkins and C. L. Drake, pp. 325–44. American Association of Petroleum Geologists Memoir 34, Tulsa, OK.

Birch, F. S. 1986. Isostatic, thermal, and flexural models of the subsidence of the north coast of Puerto Rico. *Geology* 14: 427–9.

Bowin, C. 1975. The geology of Hispaniola. In *The Ocean Basins and Margins*, Vol. 3, eds. A. E. M. Nairn and F. G. Stehli, pp. 501–52. New York: Plenum.

Bowin, C. 1976. Caribbean gravity field and plate tectonics. *Geological Society of America Special Paper 169*, 79 p., Boulder, Colo.

Bracey, D. R., and Vogt, P. R. 1970. Plate tectonics in the Hispaniola area. *Geol. Soc. Am. Bull.* 81: 2855–60.

Bralower, T. J., Hutson, F., Mann, P., Iturralde-Vinent, M., and Sliter, W. V. 1993. Tectonics of oblique arc-continent collision in western Cuba 1: Stratigraphic constraints. *EOS.* 74(43): 546.

Burke, K., Cooper, C., Dewey, J. F., Mann, P., and Pindell, J. L. 1984. Caribbean tectonics and relative plate motions. In *GSA Memoir 162: The Caribbean-South American Plate Boundary and Regional Tectonics*, eds. W. E. Bonini, R. B. Hargraves and R. Shagam, pp. 31–63. Boulder, Colo.: GSA.

Burke, K. 1988. Tectonic evolution of the Caribbean. *Ann. Rev. Earth Plan. Sci.* 16: 201–30.

Byrne, D. B., Suarez, G., and McCann, W. R. 1985. Muertos Trough subduction – Microplate tectonics in the northern Caribbean. *Nature* 317: 420–1.

Caceres, D., Gordon, M. B., Mann, P., and Flores, R. 1993. Tectonics of oblique arc-continent collision in western Cuba 2: Structural constraints, *EOS.* 74(43): 546.

Calais, E., Mercier de Lépinay, B., Renard, V., and Tardy, M. 1989. Géometrie et régime tectonique de long d'une limite de plaques en coulissage: La frontiére nord Caraïbe de Cuba à Hispaniola, Grandes Antilles. *C. R. Acad. Sci. Paris.* 308, Sér. II: 131–5.

Calais, E., and Mercier de Lépinay, B. 1990a. A natural model of active transpressional tectonics. *Rev. Inst. Français du Pétrole.* 45: 147–60.

Calais, E., and Mercier de Lépinay, B. 1990b. Tectonique et paléogéographie de la côte sud de l'Oriente cubain; Nouvelles contraintes pour l'évolution géodynamique de la limite de plaques décrochante nord caraïbe de l'Eocene à l'Actuel. *C. R. Acad. Sci. Paris.* 310, Sér. II: 293–9.

Calais, E., and Mercier de Lépinay, B. 1991. From transtension to transpression along the northern Caribbean plate boundary off Cuba: Implications for the recent motion of the Caribbean plate. *Tectonophysics* 186: 329–50.

Calais, E., Bethoux, N., and Mercier de Lépinay, B. 1992. From transcurrent faulting to frontal subduction: A seismotectonic study of the northern Caribbean plate boundary from Cuba to Puerto Rico. *Tectonics* 11: 114–23.

Calais, E., and Mercier de Lépinay, B. 1993. Semi-quantitative modeling of strain and kinematics along the Caribbean/North America strike-slip plate boundary zone. *J. Geophys. Res.* p. 8293–8308.

Case, J. E., and Holcombe, T. L. 1980. Geologic-tectonic map of the Caribbean region. *U.S. Geological Survey Miscellaneous Investigations Map I-1100*, 3 sheets. Reston, Va.: USGS.

CAYTROUGH 1979. Geological and geophysical investigation of the Mid-Cayman Rise spreading center: Initial results and observations. In *Deep Drilling Results in the Atlantic Ocean: Ocean Crust:* eds. M. Talwani, C. G. Harrison, and D. E. Hayes, American Geophysical Union: Washington, D.C. Maurice Ewing Series 2, pp. 66–93.

Dillon, W. P., Vedder, J. G., and Graf, R. J. 1972. Structural profile of the northwestern Caribbean. *Earth Plan. Sci. Ltrs.* 17: 175–80.

Dillon, W. P., Scanlon, K. M., Austin, J. A., Jr., Edgar, N. T., Parson, L. M., and Ness, G. E. 1991. Morphology, structure and active tectonism of the insular margin off northwestern Hispaniola. In *Transactions of the 12th Caribbean Geological Conference*, St. Croix, USVI, eds. D. K. Larue and G. Draper, South Miami, Fla.: Miami Geological Society. pp. 133–42.

Dillon, W. P., Austin, J. A., Jr., Scanlon, K. M., Edgar, N. T., and Parson, L. M. 1992. Accretionary margin of north-western Hispaniola: Morphology, structure and development of part of the northern Caribbean plate boundary. *Mar. Petrol. Geol.* 9(1): 70–88.

Dillon, W. P., Edgar, N. T., Parson, L. M., Scanlon, K. M., Driscoll, G. R., and Jacobs, C. L. 1993. Magnetic anomaly map of the central Cayman Trough, northwestern Caribbean Sea. *U.S. Geological Survey Miscellaneous Field Studies Map MF 2083-B*. Reston, Va.: USGS.

Ditty, P. S., Harmon, C. J., Pilkey, O. H., Ball, M. M., and Richardson, E. S. 1977. Mixed terrigenous-carbonate sedimentation in the Hispaniola-Caicos turbidite basin. *Mar. Geol.* 24: 1–20.

Duncan, R. A., and Hargraves, R. B. 1984. Plate tectonic evolution of the Caribbean in the mantle reference frame. In *The Caribbean-South American Plate Boundary and Regional Tectonics*, eds. W. E. Bonini, R. B. Hargraves, and R. Shagam, pp. 81–93. Boulder, Colo.: Geological Society of America. Memoir 162,

Edgar, N. T. 1991. Structure and geologic development of the Cibao valley, northern Hispaniola. In *Geologic and Tectonic Development of the North American-Caribbean Plate Boundary in Hispaniola*, eds. P. Mann, G. Draper, and J. F. Lewis, pp. 281–99. Geological Society of America Special Paper 262, Boulder, Colo.

Edgar, N. T., Dillon, W. P., Parson, L. M., Scanlon, K. M., Jacobs, C. L., and Holcombe, T. L. 1991. GLORIA sidescan-sonar image and interpretation of the central Cayman Trough, northwestern Caribbean Sea. *U.S. Geological Survey Miscellaneous Field Investigations Map MF 2083-A*, 3 sheets. Reston, Va.: USGS.

Edgar, N. T., and Dillon, W. P. 1992. Comment on SeaMARC II mapping of transform faults in the Cayman trough, Caribbean Sea. *Geology* 4: 382–3.

EEZ-SCAN 85 Scientific Staff 1987. Atlas of the U.S. Exclusive Economic Zone, Eastern Caribbean. *U.S. Geological Survey Miscellaneous Investigations Series I-1864 B*, 58 p., scale 1:500,000. Reston, Va.: USGS.

Emery, K. O., and Milliman, J. D. 1980. Shallow-water limestones from slope off Grand Cayman Island. *J. Geol.* 88: 483–8.

Ewing, J. I., Antoine, J., and Ewing M. 1960. Geophysical measurements in the western Caribbean Sea and in the Gulf of Mexico. *J. Geophys. Res.* 65: 4087–126.

Fink, L. K., and Harrison, C. G. A. 1972. Palaeomagnetic investigations of selected lava units on Puerto Rico. In *Transactions of the 6th Caribbean Geological Conference*, Margarita, Venezuela, p. 379. Caracas, Venezuela: Impreso Por Cromotip.

Fox, P. J., and Gallo, D. G. 1986. The geology of North Atlantic transform plate boundaries and their aseismic extensions. In *The Geology of North America*, Vol. M, *The Western North Atlantic Region*, eds. P. R. Vogt and B. E. Tucholke, pp. 157–72. Boulder, Colo.: Geological Society of America.

Frankel, A., McCann, W. R., and Murphy, A. J. 1980. Observations from a seismic network in the Virgin Islands region: Tectonic structures and earthquake swarms. *J. Geophys. Res.* 85: 2669–78.

Garrison, L. E., Martin, R. G., Jr., Berryhill, H. L., Jr., Buell, M. W., Jr., Ensminger, H. R., and Perry, R. K. 1972. Preliminary tectonic map of the eastern Greater Antilles region. *U.S. Geological Survey Miscellaneous Investigations Map I-732*, scale 1:500,000. Reston, Va.: USGS.

Hall, S. A., and Westbrook, G. K. 1990. Magnetic anomaly map of the Caribbean area. In *The Geology of North America*, Vol. H, *The Caribbean Region*, eds. G. Dengo and J. E. Case, Plate 6. Boulder, Colo.: Geological Society of America.

Heezen, B. C., Nesteroff, W. D., Rawson, M., and Freeman-Lynde, R. P. 1985. Visual evidence for subduction in the western Puerto Rico Trench. *Géodynamique de Caraïbes Symposium*, Feb. 5–8, 1985, Paris: Editions Technip, p. 287–304.

Heubeck, C., and Mann, P. 1991. Geologic evaluation of plate kinematic models for the North American-Caribbean plate boundary zone. *Tectonophysics* 191: 1–26.

Hilde, T. W. C. 1983. Sediment subduction versus accretion around the Pacific. *Tectonophysics* 99: 381–97.

Holcombe, T. L., Vogt, P. R., Mathews, J. E., and Murchison, R. R. 1973. Evidence for sea-floor spreading in the Cayman Trough. *Earth Plan. Sci. Ltrs.* 20(3): 357–71.

Holcombe, T. L., and Sharman, G. F. 1983. Post-Miocene Cayman Trough evolution: A speculative model. *Geology* 11: 714–17.

Jacobs, C. L., Edgar, N. T., Parson, L. M., Dillon, W. P., Scanlon, K. M., and Holcombe, T. L. 1989. A revised bathymetry of the Mid Cayman Rise and central Cayman Trough using long range side scan sonar. Institute of Oceanographic Sciences, Deacon Laboratory, *Report No. 272*, 11 pp. Wormley, Surrey, U.K.: IOS.

Jany, I., Mauffret, A., Bouysse, P., Mascle, A., Mercier de Lépinay, B., Renard, V., and Stéphan, J.-F. 1987. Relève bathymetrique Seabeam et tectonique en décrochement au sud des Iles Vierges (Nord-Est Caraïbes). *C. R. Acad. Sci. Paris.* 304, Sér. II: 527–32.

Jany, I., Scanlon, K. M., and Mauffret, A. 1990. Geological interpretation of combined Seabeam, Gloria and seismic data from the Anegada Passage (Virgin Islands, North Caribbean). *Mar. Geophys. Res.* 12: 173–96.

Klitgord, K. L., and Schouten, H. 1986a. Magnetic lineations and fracture zones. In *The Geology of North America*, Vol. M, *The Western North Atlantic Region*, eds. P. R. Vogt and B. E. Tucholke, Plate 3. Boulder, Colo.: Geological Society of America.

Klitgord, K. L., and Schouten, H. 1986b. Plate kinematics of the central Atlantic. In *The Geology of North America*, Vol. M, *The Western North Atlantic Region*, eds. P. R. Vogt and B. E. Tucholke, pps. 351–78. Boulder, Colo.: Geological Society of America.

Ladd, J. W., Holcombe, T. L., Westbrook, G. K., and Edgar, N. T. 1990. Caribbean marine geology; Active margins of the plate boundary. In *The Geology of North America*, vol. H, *The Caribbean Region*, eds. G. Dengo and J. E. Case, pp. 261–90. Boulder, Colo.: Geological Society of America.

Ladd, J. W., Shih, T.-C., and Tsai, C. J. 1981. Cenozoic tectonics of central Hispaniola and adjacent Caribbean Sea. *Am. Assoc. Petrol. Geol. Bull.* 65: 466–89.

Ladd, J. W., and Watkins, J. S. 1978. Active margin structures within the north slope of the Muertos Trench. *Geol. en Mijnbouw.* 57: 255–60.

Ladd, J. W., and Watkins, J. S. 1979. Tectonic development of trench-arc complexes on the northern and southern margins of the Venezuelan Basin. In *Geological and Geophysical Investigations of Continental Margins*, eds. J. S. Watkins, L. Montadert, and P. W. Dickerson, pp. 363–71. Tulsa, Ok: Am. Assoc. Petroleum Geologists Memoir 29.

Ladd, J. W., Worzel, J. L., and Watkins, J. S. 1977. Multifold seismic reflection records from the northern Venezuela Basin and the north slope of the Muertos trough. In *Island Arcs, Deep Sea Trenches and Back-Arc Basins*, Maurice Ewing Series Vol. 1, pp. 41–56: Washington D.C.: American Geophysical Union.

Larue, D. K., Joyce, J., and Ryan, H. F. 1991. Neotectonics of the Puerto Rico Trench: Extensional tectonism and fore-arc subsidence. In *Transactions of the 12th Caribbean Geological Conference*, St. Croix, USVI, eds. D. K. Larue and G. Draper pp. 231–47. South Miami, Fla.: Miami Geological Society.

Lewis, J. F., and Draper, G. 1990. Geology and tectonic evolution of the northern Caribbean margin. In *The Geology of North America* Vol. H., *The Caribbean Region*, eds. G. Dengo and J. E. Case, pp. 77–140. Boulder, Colo.: Geological Society of America.

MacDonald, K. C., and Holcombe, T. L. 1978. Inversion of magnetic anomalies and sea-floor spreading in the Cayman Trough. *Earth Plan. Sci. Ltrs.* 40, p. 407–414.

Malfait, B. T., and Dinkelman, M. G. 1972. Circum-Caribbean tectonic and igneous activity and the evolution of the Caribbean plate. *Geol. Soc. Am. Bull.* 83(2): 251–72.

Mann, P., Hempton, M. R., Bradley, D. C., and Burke, K. 1983. Development of pull-apart basins. *J. Geol.* 91: 529–54.

Mann, P., and Burke, K. 1984. Neotectonics of the Caribbean. *Rev. Geophys. Space Phys.* 22: 309–62.

Mann, P., Burke, K., and Matumoto, T. 1984. Neotectonics of Hispaniola: Plate motion, sedimentation and seismicity at a restraining bend. *Earth Plan. Sci. Ltrs.* 70: 311–24.

Mann, P., Draper, G. Lewis, J. F. 1991a. An overview of the geologic and tectonic development of Hispaniola. In *Geologic and Tectonic Development of the North American-Caribbean Plate Boundary in Hispaniola,* eds. P. Mann, G. Draper, and J. F. Lewis, pp. 1–28. Geological Society of America Special Paper 262. Boulder, Colo.: GSA.

Mann, P., McLaughlin, P. P., and Cooper, A. 1991b. Geology of the Azua and Enriquillo basins, Dominican Republic; 2, Structure and tectonics. In *Geologic and Tectonic Development of the North American-Caribbean Plate Boundary in Hispaniola,* eds. P. Mann, G. Draper, and J. F. Lewis, pp. 3657–89. Geological Society of America Special Paper 262. Boulder, Colo.: GSA.

Masson, D. G., and Scanlon, K. M. 1991. Neotectonic setting of Puerto Rico. *Geol. Soc. Am. Bull.* 103(1): 144–54.

Mattson, P. H. 1984. Caribbean structural breaks and plate movements. In *The Caribbean-South American Plate Boundary and Regional Tectonics*, Memoir 162, eds. W. E. Bonini, R. B. Hargraves, and R. Shagam, pp. 131–52. Boulder, Colo.: Geological Society of America.

Mauffret, A., Jany, I., Mercier de Lépinay, B., Bouysse, P., Mascle, A., Renard, V., and Stéphan, J.-F. 1986. Relevé au sondeur multifaisceaux du bassin des Iles Vierges (extrémité orientale des Grandes Antilles): Role de l'extension et des décrochements. *C. R. Acad. Sci. Paris.* 303, Sér. II: 923–8.

Mauffret, A., and Jany, I. 1990. Collision et tectonique d'expulsion le long de la frontière Nord Caraïbe. *Ocean. Act.* Vol. Spécial: 1097–116.

McCann, W. R., and Sykes, L. R. 1984. Subduction of aseismic ridges beneath the Caribbean plate: Implications for the tectonics and seismic potential of the northeastern Caribbean. *J. Geophys. Res.* 89: 4493–519.

McCann, W. R., and Pennington, W. D. 1990. Seismicity, large earthquakes and the margin of the Caribbean Plate. In *The Geology of North America*, vol. H, *The Caribbean Region*, eds. G. Dengo and J. E. Case, pp. 291–306. Boulder, Colo.: Geological Society of America.

Molnar, P., and Sykes, L. R. 1969. Tectonics of the Caribbean and middle American regions from focal mechanisms and seismicity. *Geol. Soc. Am. Bull.* 80: 1639–84.

Molnar, P. 1977. Gravity anomalies and the origin of the Puerto Rico Trench. *Geophys. J. R. Astr. Soc.* 51:701–8.

Monroe, W. H. 1980. Geology of the Middle Tertiary Formations of Puerto Rico. *U.S. Geological Survey Professional Paper 953*. Washington, D.C.: U.S. Government Printing Office, 93 p.

Moussa, M. T., Seiglie, G. A., Meyerhoff, A. A., and Taner, I.

1987. The Quebradillas Limestone (Miocene-Pliocene), northern Puerto Rico, and tectonics of the northeastern Caribbean margin. *Geol. Soc. Am. Bull.* 99: 427–39.

Murphy, A. J., and McCann, W. R. 1979. Preliminary results from a new seismic network in the northeastern Caribbean. *Bull. Seismol. Soc. of Am.* 69: 1497–513.

Perfit, M. R. 1977. Petrology and geochemistry of mafic rocks from the Cayman trench: Evidence for spreading. *Geology* 5: 105–10.

Perfit, M. R., and Heezen, B. C. 1978. The geology and evolution of the Cayman Trench. *Geol. Soc. Am. Bull.* 89: 1155–74.

Pindell, J. L., and Dewey, J.F. 1982. Permo-Triassic reconstruction of western Pangea and the evolution of the Gulf of Mexico/Caribbean region. *Tectonics* 1: 179–212.

Pindell, J. L., and Barrett, S. F. 1990. Geological evolution of the Caribbean region: A plate-tectonic perspective. In *The Geology of North America*, Vol. H, *The Caribbean Region*, eds. G. Dengo, and J. E. Case, pp. 405–32. Boulder, Colo.: Geological Society of America.

Reid, J. A., and Plumley, P. W. 1991. Paleomagnetic evidence for late Miocene counterclockwise rotation of north coast carbonate sequence, Puerto Rico. *Geophys. Res. Ltrs.* 18: 565–8.

Rosencrantz, E., and Sclater, J. G. 1986. Depth and age in the Cayman Trough. *Earth Plan. Sci. Ltrs.* 79: 133–44.

Rosencrantz, E., Ross, M. I., and Sclater, J. G. 1988. Age and spreading history of the Cayman Trough as determined from depth, heat flow and magnetic anomalies. *J. Geophys. Res.* 93: 2141–57.

Rosencrantz, E., and Mann, P. 1991. SeaMARC II mapping of transform faults in the Cayman Trough, Caribbean Sea. *Geology* 19: 690–3.

Rosencrantz, E., and Mann, P. 1992. Reply on SeaMARC II mapping of transform faults in the Cayman Trough, Caribbean sea: *Geology* 4: 383–4.

Scanlon, K. M., Masson, D. G., and Rodriguez, R. W. 1988. GLORIA side-scan sonar survey of the EEZ of Puerto Rico and the U.S. Virgin Islands. *Transactions of the 11th Caribbean Geological Conference*, Barbados, 1986, pp. 32: 1–9. Bridgetown, Barbados: Ministry of Finance.

Schell, B. A., and Tarr, A. C. 1978. Plate tectonics of the northeastern Caribbean Sea region. *Geol. en Mijnbouw.* 57: 319–24.

Schouten, H., Klitgord, K. D., and Gallo, D. G. 1993. Edge-driven microplate kinematics. *J. Geophys. Res.* 98(B4): 6689–701.

Schneidermann, N., Beckmann, J. P., and Heezen, B. C. 1972. Shallow water carbonates from the Puerto Rico Trench Region. *Transactions of the 6th Caribbean Geological Conference Volume*, Margarita, Venezuela, pp. 423-5. Caracas, Venezuela: Universidad Central de Venezuela.

Schwab, W. C., Danforth, W. W., Scanlon, K. M., and Masson, D. G. 1991. A giant slope failure on the northern insular slope of Puerto Rico. *Mar. Geol.* 96: 237–46.

Speed, R. C., and Larue, D. K. 1991. Extension and transtension in the plate boundary zone of the northeastern Caribbean. *Geophys. Res. Ltrs.* 18: 573–6.

Stein, S., Wiens, D. A., and Engeln, J. F. 1986. Comment on "Subduction of aseismic ridges beneath the Caribbean Plate: Implications for the tectonics and seismic potential of the northeastern Caribbean" by W. R. McCann and L. R. Sykes. *J. Geophys. Res.* 91: 784–6.

Stein, S., DeMets, C., Gordon, R. G., Brodholt, J., Argus, D., Engeln, J. F., Lundgren, P., Stein, C., Wiens, D. A., and Woods, D. F. 1988. A test of alternative Caribbean Plate relative motion models. *J. Geophys. Res.* 93: 3041–50.

Sykes, L. R., McCann, W. R., and Kafka, A. L. 1982. Motion of the Caribbean plate during last 7 million years and implications for earlier Cenozoic movements. *J. Geophys. Res.* 87: 10656–76.

Vincenz, S. A., and Dasgupta, S. N. 1978. Paleomagnetic study of some Cretaceous and Tertiary rocks on Hispaniola. *Pure App. Geophys.* 116: 1200–10.

Vogt, P. R. 1986. Magnetic anomalies of the North Atlantic Ocean. In *The Geology of North America*, Vol. M. *The Western North Atlantic Region*, eds. P. R. Vogt and B. E. Tucholke, Plate 3. Boulder, Colo.: Geological Society of America.

Westbrook, G. K. 1990. Gravity anomaly map of the Caribbean region. In *The Geology of North America*, Vol. H, *The Caribbean Region*, eds. G. Dengo and J. E. Case, Plate 7. Boulder, Colo.: Geological Society of America.

IV U.S. West Coast EEZ

James V. Gardner and Michael E. Field

U.S. Geological Survey, Menlo Park, California

The initial USGS EEZ-SCAN surveys began in the Spring of 1984 in San Diego, California, and systematically surveyed northwards to the Canadian-U.S. border, then transited south to finish 100 days later in San Diego. These surveys mapped more than 750,000 km^2 of the U.S. West Coast EEZ with a series of tracklines spaced about 30 km apart. The mapping included digital GLORIA images, 160-in^3 airgun and 3.5-kHz seismic-reflection profiling, 10-kHz bathymetry, and magnetics data.

Studies of the Pacific EEZ in this volume are diversified in approach, topic, and area of investigation. They range in scale from huge areas, such as the reexamination of Monterey Fan, to small individual features such as the meandering channel north of Mendocino Fracture Zone. This diversity reflects the tectonic and sedimentary diversity that characterizes the Pacific margin.

The entire region is one of continued tectonic change, from the Miocene Plate interactions and subsequent transformation of the California margin to the formation and continued deformation of the young borderland basins of southern California to the presently active Cascadia Subduction Zone off northern California, Oregon, and Washington. The EEZ off California, Oregon, and Washington can be divided at $\sim 40°$ N into a northern province dominated by major tectonic ridges and fracture zones, with large sedimentary overprints, and a southern province dominated by a pattern of long-term sedimentation over an older terrain of seamounts (Figure IV–1). The northern province includes the Gorda-Juan de Fuca Ridge and associated Blanco and Mendocino Fracture Zones and the Cascadia Subduction Zone. Three large submarine fans – Nitnat, Astoria, and Gorda Fans, from north to south, respectively – splay out from the base of the subduction complex and blanket the relatively young oceanic crust with thick sediment deposits.

The southern province includes the Pioneer, Morro, and Murray Fracture Zones and the strike-slip regime of the Southern California continental borderland. Submarine fans

– Delgada, Farallone, Monterey, Arguello, and Conception – are the dominant sedimentological features, and their presence is a guide to the history of plate movements and changing sediment sources along the California margin. In general, sediment thickness in the southern province decreases towards the south to such an extent that many basement features are thinly covered off the southern California margin.

Carson et al. use an innovative technique of processing coregistered GLORIA images and bathymetry from Cascadia Subduction Zone to eliminate bathymetry-related backscatter and investigate the residual signal. Physical samples from the areas of high residual backscatter contain diagenetic carbonates formed by expelled fluids from previously mapped thrust faults of the subduction complex. Regional mapping of the backscatter residuals suggests that the high backscatter from the diagenetic carbonates allows one to map previously unmapped thrust faults in this region.

Cacchione, Drake, and Gardner use GLORIA and 3.5-kHz data to map a previously unmapped meandering channel on the deep-sea floor at the base of Gorda Escarpment of Mendocino Fracture Zone. Although rather small in scale, the channel is similar in geometry to other deep-sea meandering channels. Five turbidites, dated by ^{14}C, overlap in age with deposits described in the literature as related to great earthquakes on the Cascadia Subduction Zone.

Gardner et al. use GLORIA, TOBI (30-kHz deep-towed sidescan sonar), and 3.5-kHz seismic-reflection profiling to map the surface terrains and evolution of Monterey Fan. The new interpretation of the development of the fan includes the translation of the Pacific Plate, deflection of sediment pathways by basement highs, the buildup of the fan deposit, and the evential change of gradient and incipient fan-lobe erosion when deposition reached Murray Fracture Zone.

Lee et al. return to the theme of ground truthing GLORIA imagery where they compare measured backscatter and sediment properties from similar areas on Monterey and Mississippi Fans. Their study shows that conventional interpretations of sediment character from sidescan images is not always valid. The assumption that coarser sediment produces higher backscatter than does finer sediment is not axiomatic,

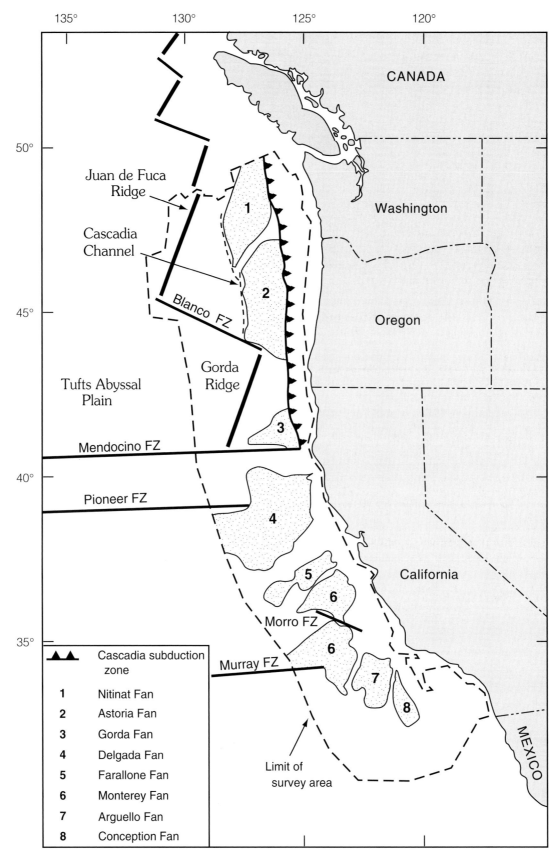

Figure IV–1. Major tectonic and sedimentological features of the area offshore California, Oregon, and Washington.

135° 130° 125° 120°

CANADA

50°

Juan de Fuca
Ridge

Washington

45°

Cascadia
Channel

Blanco FZ

Gorda
Ridge

Oregon

Tufts Abyssal
Plain

40°

Mendocino FZ

Pioneer FZ

California

35°

Morro FZ

Murray FZ

MEXICO

▲▲ Cascadia subduction zone

1 Nitinat Fan
2 Astoria Fan
3 Gorda Fan
4 Delgada Fan
5 Farallone Fan
6 Monterey Fan
7 Arguello Fan
8 Conception Fan

Limit of
survey area

166

and accurate interpretations must rely on physical sediment samples.

Lastly, Edwards, Field, and Kenyon use GLORIA images and 3.5-kHz profiling to investigate the backscatter signatures and processes of sediment transport and small-fan development in Santa Monica and San Pedro Basins of the inner Southern California Borderland. This study details how the central basin plains are fed by point-source channels that head in shallow water. Cores from various backscatter signatures show that coarse sediment is deposited beyond the lower-fan environment and that the backscatter characteristics are the result of volume inhomogenieties within the top few meters of sediment.

These studies are only a brief cross section of studies that are in progress from the California, Oregon, and Washington (COW) area. In addition to the EEZ-SCAN Atlas of the COW area, a series of papers and derivative maps already have been published (see reference list).

References

EEZ-SCAN 84 Scientific Staff 1986. Atlas of the exclusive Economic Zone, western conterminous United States. *U.S. Geological Survey Miscellaneous Investigations series I-1792*, 152p, scale 1:500,000. Reston, Va.: USGS.

EEZ-SCAN 84 Scientific Staff 1988. Physiography of the western U.S. Exclusive Economic Zone. *Geology,* 16: 131–4.

Field, M. E., Gardner, J. V., Drake, D. E., Cacchione, D. A. 1987. Tectonic morphology of offshore Eel River Basin, California. In *Tectonics, Sedimentation, and Evolution of the Eel River and Other Coastal Basins of Northern California,* eds. H. Schymiczek and R. Suchsland, pp. 41–8. San Joaquin Geological Society Miscellaneous Publication; No. 37, San Joaquin, Calif.

Gardner, J. V., Field, M. E., Lee, H., Edwards, B. E., Masson, D. G., Kenyon, N., and Kidd, R. B. 1991. Ground truthing 6.5-kHz sidescan sonographs: What are we really imaging?. *J. Geophys. Res.* 96: 5955–74.

Gardner, J. V. Cacchione, D. A., Drake, D. E., Edwards, B. D., Field, M. E., Hampton, M. A., Karl, H. A., Kenyon, N. H., Masson, D. G., McCulloch, D. S., and Grim, M. S. 1992a. Map showing sediment isopach in the deep-sea basins of the Pacific continental margin, Point Conception to Point Loma. Scale 1:1,000,000. *U.S. Geological Survey Miscellaneous Investigations Map I-2089-A.* Reston, Va.: USGS.

Gardner, J. V., Cacchione, D. A., Drake, D. E., Edwards, B. D., Field, M. E., Hampton, M. A., Karl, H. A., Kenyon, N. H., Masson, D. G., McCulloch, D. S., and Grim, M. S. 1992b. Map showing depth to basement in the deep-sea basins of the Pacific continental margin, Point Conception to Point Loma. Scale 1:1,000,000. *U.S. Geological Survey Miscellaneous Investigations Map I-2089-B.* Reston, Va.: USGS.

Gardner, J. V., Cacchione, D. A., Drake, D. E., Edwards, B. D., Field, M. E., Hampton, M. A., Karl, H. A., Kenyon, N. H., Masson, D. G., McCulloch, D. S., and Grim, M. S. 1993a. Map showing depth to basement in the deep-sea basins of the Pacific continental margin, Cape Mendocino to Point Conception. Scale 1:1,000,000. *U.S. Geological Survey Miscellaneous Investigations Map I-2090-A.* Reston, Va.: USGS.

Gardner, J. V., Cacchione, D. A., Drake, D. E., Edwards, B. D., Field, M. E., Hampton, M. A., Karl, H. A., Kenyon, N. H., Masson, D. G., McCulloch, D. S., and Grim, M. S. 1993b. Map showing depth to basement in the deep-sea basins of the Pacific continental margin, Cape Mendocino to Point Conception. Scale 1:1,000,000. *U.S. Geological Survey Miscellaneous Investigations Map I-2090-B.* Reston, Va.: USGS.

Gardner, J. V., Cacchione, D. A., Drake, D. E., Edwards, B. D., Field, M. E., Hampton, M. A., Karl, H. A., Kenyon, N. H., Masson, D. G., McCulloch, D. S., and Grim, M. S. 1993c. Map showing sediment isopachs in the deep-sea basins of the Pacific continental margin, Strait of Juan de Fuca to Cape Mendocino. Scale 1:1,000,000. *U.S. Geological Survey Miscellaneous Investigations Map I-2091-A.* Reston, Va.: USGS.

Gardner, J. V., Cacchione, D. A., Drake, D. E., Edwards, B. D., Field, M. E., Hampton, M. A., Karl, H. A., Kenyon, N. H., Masson, D. G., McCulloch, D. S., and Grim, M. S. 1993d. Map showing depth to basement in the deep-sea basins of the Pacific continental margin, Strait of Juan de Fuca to Cape Mendocino. Scale 1:1,000,000. *U.S. Geological Survey Miscellaneous Investigations Map I-2091-B.* Reston, Va.: USGS.

Hampton, M. A., Karl, H. A., and Kenyon, N. H. 1989. Sea-floor drainage features of Cascadia Basin and adjacent continental slope. *Mar. Geol.* 87:249–77.

10 Discrimination of fluid seeps on the convergent Oregon continental margin with GLORIA imagery

Bobb Carson,[1] Valerie Paskevich,[2] Erol Seke,[3] and Mark L. Holmes[4]

[1]Department of Earth and Environmental Sciences, Lehigh University, Bethlehem, Pennsylvania
[2]U.S. Geological Survey, Branch of Atlantic Marine Geology, Woods Hole, Massachusetts
[3]Department of Electrical Engineering and Computer Science, Lehigh University, Bethlehem, Pennsylvania
[4]U.S. Geological Survey, Branch of Pacific Marine Geology, University of Washington, Seattle, Washington

Abstract

Diagenetic carbonate cements and gas hydrates occur at or near the seafloor where pore fluids seep from continental margins. Because these deposits have acoustic impedances that differ significantly from those of hemipelagic deposits, they can be mapped with sidescan sonar if topographic effects (that affect backscattering angle and amplitude) are removed. We have made this topographic correction, using registered GLORIA and SeaBeam data, for a portion of the Oregon continental slope. Comparison of the processed images with local structure indicates that focused fluid expulsion is controlled by faults, but not all faults are active flow paths. Incipient thrust faults in Cascadia Basin just seaward of the base of the slope and an out-of-sequence thrust fault in the Pliocene section of the lower slope apparently channel fluids rapidly to the surface. Near-surface gas hydrates are inferred in the former location and known to precipitate very near the seafloor in the latter. In contrast, the frontal thrust fault that dips landward to the décollement at the toe of the slope shows no evidence of diagenesis associated with focused fluid flow. Instead, pore waters at the décollement may laterally migrate in this region to transverse strike-slip faults that define the northern and southern boundaries of the frontal thrust sheet. Preliminary flow measurements at vent sites, hydrogeologic tests, and thermal anomalies at Ocean Drilling Program (ODP) Site 892 suggest that near-surface formation of gas hydrates is associated with active, rapid fluid discharge. Carbonate cements may reflect slower seepage or relict deposits. If so, and if differences in backscatter amplitude from these contrasting deposits can be established, sidescan sonar imagery may be useful in mapping and evaluating variability in pore-water discharge on continental margins.

Introduction

Only in the past decade has it become apparent that significant quantities of pore fluids are expelled to the marine environment from both passive and active continental margins. Submersible exploration, which detailed chemosynthetic biological communities around submarine seeps (Paull et al. 1984; Suess et al. 1985; Kennicutt et al. 1985), provided initial recognition of this phenomenon. Subsequent field studies have measured the flow (Carson, Suess, and Strasser 1990; Foucher et al. 1992; Henry et al. 1992; Linke et al. 1994) and established the local transfer of inorganic and organic species from pore waters to the bottom-water regime (Han and Suess 1989; Gamo et al. 1992). Although questions remain about the volumetric fluid flux (Le Pichon et al. 1992), there is little doubt that expulsion on continental margins is an important hydrogeologic and geochemical process. A rough estimate of the global fluid budget suggests that discharge from continental margins contributes 100 km^3 of pore water to the ocean per year (Westbrook et al. 1987).

Flow may be driven by hydrostatic head (from groundwater recharge on land), by differential density within the pore fluids (induced by temperature, gas content, or fluid salinity), or by tectonic stresses that induce sediment compaction and dewatering. Whatever the driving mechanism, pore fluids contain biogenic and/or thermogenic hydrocarbons and inorganic ions that undergo diagenetic reactions at or near the seafloor as the fluids are exposed to the cold, oxidizing environment of deep seawater. Diagenetic deposits consist of carbonate cements (Ritger, Carson, and Suess 1987; Suess and Whiticar 1989; Kulm and Suess 1990; Sakai et al. 1992) and, in deep water, gas hydrate accumulations (Kvenvolden and Barnard 1983; Brooks et al. 1984, 1991; Hand, Katz, and Verma 1984; Roberts and Neurauter 1991).

The carbonate cements include calcite, dolomite, and aragonite, which occur as nodules (concretions), slabs, beds, and chimneys (Ritger et al. 1987; Kulm and Suess 1990; Sakai et al. 1992). Although nodules are deposited to several hundred meters below the seafloor (mbsf; Westbrook et al. 1994), massive deposits that commonly incorporate clamshells and worm tubes are characteristically deposited at or very near the sediment-water interface (Goedert and Squires 1990; Lalou et al. 1992).

Similarly, gas hydrate deposits that are thought to be normally concentrated at depths of 50 to 300 mbsf, above the bottom simulating reflector (BSR; Hyndman and Davis 1992), are known to occur at the seafloor where active fluid venting occurs (Brooks et al. 1985; Roberts et al. 1990; Roberts and Neurauter 1991; Carson et al. 1993). Methane hydrates are probably dominant, but other varieties also occur (Kvenvolden and Barnard 1983).

Surficial diagenetic deposits may be observed from submersibles and, if thick enough, imaged by multichannel seismic (MCS) systems (Roberts and Neurauter 1991). However, the deposits are often blanketed by hemipelagic sediment or are too thin (<10 to 15 m) to be resolved in MCS data. Hence, they are most successfully and efficiently imaged by sidescan sonar. In particular, the 6.5-kHz GLORIA system, which penetrates hemipelagic deposits to a depth of 3 to 10 m (Gardner et al. 1991), responds to these secondary precipitates in the shallow sulfate-reducing zone characteristic of submarine fluid seeps. Because these near-surface occurrences positively indicate fluid expulsion, the diagenetic deposits define the surface traces of faults, stratigraphic horizons, or diapirs of sufficiently high permeability to localize flow. The map distribution of these diagenetic deposits has been used to refine the position of breached hydrocarbon reservoirs (Brooks et al. 1985; Roberts and Neurauter 1991), indicate locations of subaerially charged aquifers (Allen et al. 1969; Friedman et al. 1971), and record the position of vents on convergent margins (Carson et al. 1991).

In this chapter, we confine our discussion to the active, accretionary margin off Oregon. Because flow-related diagenetic deposits on passive and active margins are apparently identical, the technique described is widely applicable and some of our conclusions have implications beyond the area considered here.

Geologic setting

The Cascadia margin forms the eastern boundary of Cascadia Basin (Figure 10–1), where up to 3.5 km of interlayered turbidites and hemipelagic deposits have accumulated (MacKay et al. 1992; Davis and Hyndman 1989; Spence et al. 1991). The margin is comprised of equivalent deposits accreted to the North American Plate since Eocene time (Hyndman et al. 1990). The upper continental slope is largely underlaid by incoherent reflectors, which are thought to reflect extensive small-scale folding and faulting of Pliocene and older deposits associated with their accretion (Westbrook et al. 1994). The lower continental slope, which incorporates Pleistocene deposits, is comprised of a series of short (~ 10 km along strike) anticlinal ridges formed by folding along both seaward- and landward-verging thrust faults (Figure 10–2; Carson et al. 1974). Accretion results in a column of clastic sediments >5 km thick beneath the

slope. Because convergence across much of the margin is oblique, the deformation has a shear component that is accommodated by strike-slip faults that transversely cut across the slope and shelf (Goldfinger et al. 1992). The deformation has stepped progressively westward with time (Carson et al. 1974), and the youngest faulting occurs along a series of blind thrusts that lie within Cascadia Basin, up to 6 km west of the base of the slope (protodeformation zone, Figure 10–1; MacKay et al. 1992).

Deformation and sedimentation have occurred coevally on this margin. Sediment dispersal by turbidity currents is controlled by a series of deep-sea channels (Hampton, Karl, and Kenyon 1989). However, transport paths across the slope have been periodically diverted by elevation of ridges through accretion-related folding and faulting. Turbidites are ponded behind the ridges, and slump deposits are common where the ridge flanks fail, probably due to reduction of internal friction by overpressured fluids (Figure 10–2; Orange and Breen 1992).

The particular area considered in this study (Figure 10–1) encompasses Cascadia Basin deposits near the base of the slope, the westernmost (first) ridge of the Oregon continental slope, and the adjacent landward (second) ridge. The first ridge was elevated <300,000 years before present (BP; Kulm and Fowler 1974), and late Pleistocene turbidites have filled a ponded slope basin behind it. The ridge is an anticlinal fault-bend fold (Moore et al. 1990), formed by movement along a landward-dipping frontal thrust fault that rises from the décollement (MacKay et al. 1992). A back-thrust that intersects the frontal thrust fault surfaces near the northern end of the first ridge (Figure 10–1; Moore et al. 1990). Sediment outcrops exposed on the second ridge are Pliocene in age, but radiolarian stratigraphy suggests that uplift occurred 1 to 2×10^6 years BP (L. D. Kulm, personal communication). The second ridge is cut by several landward-dipping out-of-sequence thrust faults, at least one of which (just west of the crest at 44°40′ N) is known to be hydrologically active (Linke et al. 1994; Westbrook et al. 1994).

Recognition of flow-related diagenetic deposits in GLORIA imagery

The Cascadia margin is dominated by clastic, terrestrially derived sediments that contain diagenetic carbonate and gas hydrate deposits at fluid-discharge sites. The carbonates have a significantly greater acoustic impedance (9 to 12 × 10^6 kg m^{-2} s^{-1}; H. P. Johnson, personal communication) than unconsolidated hemipelagic sediment (2 to 5 × 10^6 kg m^{-2} s^{-1}; from Christensen 1989). Reflection coefficients of carbonates at the seafloor vary from about 0.71 to 0.77 (versus ~0.06 for hemipelagic muds; Carson et al. 1991). Reflection coefficients of carbonates buried by mud (0.64 to 0.71) are still sufficiently large to give strong acoustic returns. Huggett and Somers (1988) have estimated backscat-

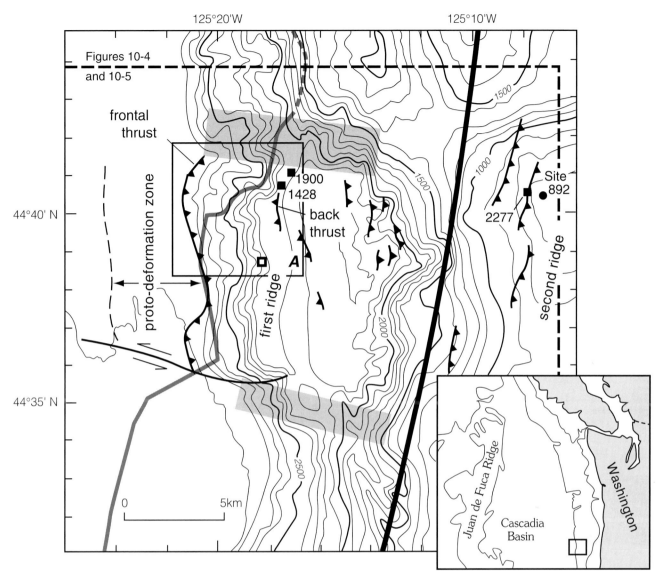

Figure 10–1. SeaBeam bathymetry of the study area (regionally located in inset) at the base of the Oregon continental slope. Position of the GLORIA track (apparent in Figures 10–3 and 10–4) is indicated by the heavy black line; the irregular gray line to the west plots the digital join between adjacent swaths (see Figure 10–3). Fault traces are taken from MacKay et al. (1992). Shaded areas indicate locations of near-surface fault splays associated with deep-seated strike-slip faults that define the northern and southern boundaries of the thrust sheet that incorporates the first ridge. Filled squares indicate positions of ALVIN dive sites where diagenetic carbonates were recovered. Both carbonates and gas hydrates occur at ODP Site 892 (filled circle). Area A outlines region detailed by Moore, Orange, and Kuhn (1990, Figure 10–7) where carbonate crusts were recorded only at ALVIN sites 1428 and 1900 and concretions occurred at the open square. Other locations traversed in area A were devoid of diagenetic carbonates. Boundary of Figures 10–4 and 10–5 is indicated by dashed line. Depth in meters.

ter strength from manganese nodules (20 dB above sediment returns) that exhibit contrasts in acoustic impedance with unconsolidated sediment and seawater similar to that of the carbonates considered here. Because the size, shape, and spatial distribution of carbonate deposits vary significantly (from individual nodules to extensive crusts and beds), we are unable to calculate monopole scattering from a single carbonate element and extrapolate that result to an entire vent site. The GLORIA data, however, suggest backscatter

contrast similar to that observed by Huggett and Somers (1988).

Pure hydrates have a compressional wave velocity of 3.3 to 3.8 km/s (Whalley 1980; Sloan 1990). Even in concentrations as low as 10% of the sediment pore volume, their occurrence increases the seismic velocity of surface sediments to >1.8 km/sec (Miller, Lee, and von Huene 1991). The resulting reflection coefficient (≥ 0.10; assuming bulk gas hydrate/sediment density = 1.0 to 1.2 g cm^{-3}) is two to

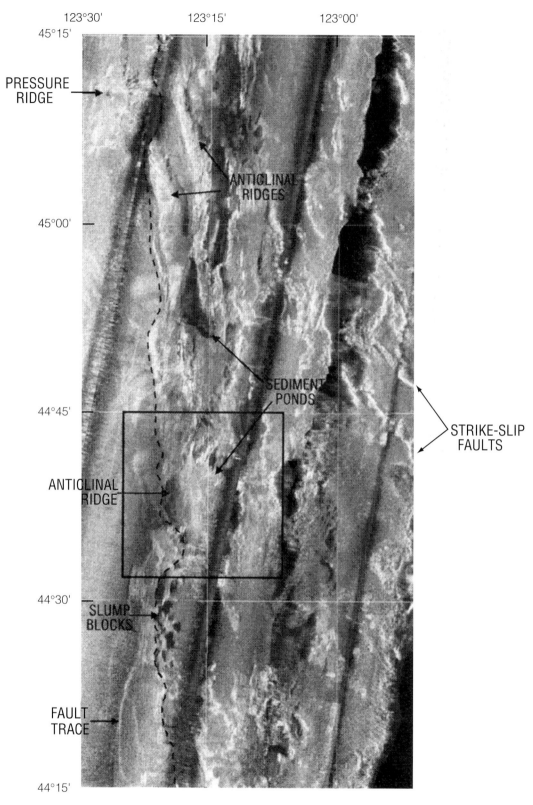

Figure 10–2. GLORIA mosaic of lower continental slope off Oregon, indicating position of topographic features associated with tectonic elements of the accretionary prism. The topographic base of the slope is indicated by the heavy dashed line. The rectangle defines the position of the study area detailed in Figures 10–1 and 10–3. From imagery published in EEZ-SCAN 84 Scientific Staff 1986.

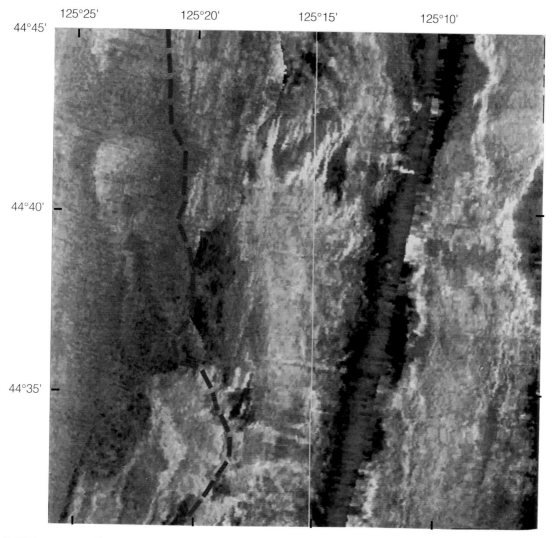

Figure 10–3. Histogram-equalized GLORIA image from a portion of the lower continental slope off Oregon (position indicated in Figures 10–1 and 10–2). The topographic base of the slope is indicated by the heavy dashed line. Note that higher-amplitude (lighter-colored) backscatter south of 45°35′ N extends across the continental slope – Cascadia Basin boundary, indicating nontopographic acoustic response.

four times greater than that of normal surficial muds. Buried gas hydrates are much less evident because they have reflection coefficients of 0.04 to 0.07. This latter case notwithstanding, the juxtaposition of surface and near-surface diagenetic deposits with acoustic characteristics distinct from the host detrital sediments implies that the secondary deposits might be differentiated by positive amplitude anomalies in sidescan sonar returns.

Sidescan sonar, however, also responds to variations in the backscatter angle (Wong and Chesterman 1968; Jackson et al. 1986), and the strong relief characteristic of most continental margins complicates interpretation. Indeed, in the complex topography of the Cascadia margin, topographic effects dominate (Figure 10–2; Hampton et al. 1989) and render lithologic interpretation with GLORIA imagery nearly impossible, although some compositional inferences

can be drawn in areas of constant bottom slope (Carson et al. 1991).

To separate the diagenetic (lithologic) and topographic components of the GLORIA images, we have undertaken a processing procedure detailed in Carson et al. (1994) and briefly outlined here. Initially, we produced digital mosaics of previously published GLORIA imagery from the Oregon margin (EEZ-SCAN 84 Scientific Staff 1986) on the USGS mini image processing system (Chavez 1984). The GLORIA mosaics were registered (rubber-sheeted) to SeaBeam bathymetry (Kulm et al. 1986) to provide a reference frame with robust navigation control. Visual registration of the two images is sufficiently imprecise that we cannot be certain that information from the same area of the seafloor is contained in every set of paired pixels, even though the SeaBeam data were combined and averaged to

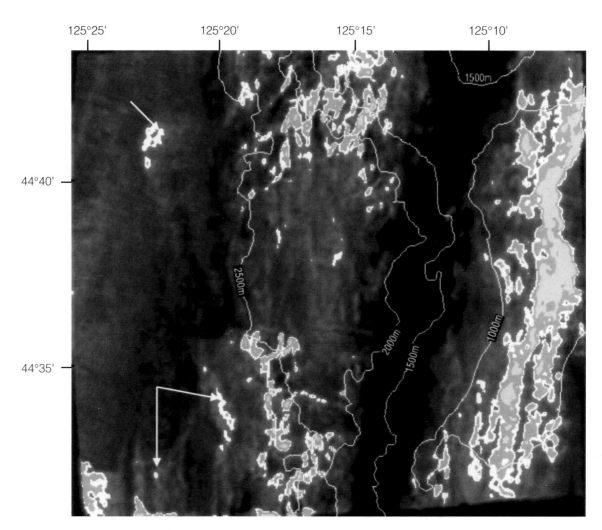

Figure 10–4. Residual GLORIA image from which topographic contributions have largely been removed. Backscatter values range from 0 to 255. Values between 190 and 225 are indicated by light gray; values between 170 and 190 are uniformly darker gray; values between 155 and 170 are white; values below 155 take on variable darker shades of gray or black. Arrows indicate locations of low-amplitude diagenetic deposits (gas hydrates?) in Cascadia Basin. Location of this image is indicated in Figure 10–1. Depth contours in meters.

produce pixels dimensionally equivalent (55 m along trackline) to the GLORIA pixels. Furthermore, noise in either data set, as evidenced by disparate amplitudes in adjacent pixels, complicates their comparison and registration. For these reasons, a three-pixel by three-pixel moving filter was applied to both the SeaBeam and GLORIA data to reduce local disparities. Only histogram equalization was applied to enhance the gray-scale spectrum of the original GLORIA image (Figure 10–3). To assess topographic effects in that image, we created an independent synthetic sidescan image from the SeaBeam bathymetry by calculating the angle of insonification for each pixel, using the geometry dictated by the position of the fish track and the bathymetry of the study area (Figure 10–1). An empirical function of backscattering strength versus angle of insonification (Jackson et al. 1986) was used to calculate amplitude variations in the synthetic

image. Because we had no a priori way of correlating the absolute backscatter intensity in the two images, they were iteratively subtracted as amplitudes in the synthetic image and were uniformly and sequentially reduced until topographic features could no longer be recognized on the final image. This residual image (Figure 10–4), from which topographic contributions were largely removed, shows marked variations in backscatter intensity, which we attribute to lithologic (diagenetic) contrasts. Because there are no topographic references recognizable on the final image, we draped that image over the bathymetry and show it in a perspective view (Figure 10–5).

The ability of the above procedure to delineate diagenetic deposits has been tested by submersible observations and ODP drilling. Near-surface carbonates and gas hydrates have been recovered only at locations characterized by high-

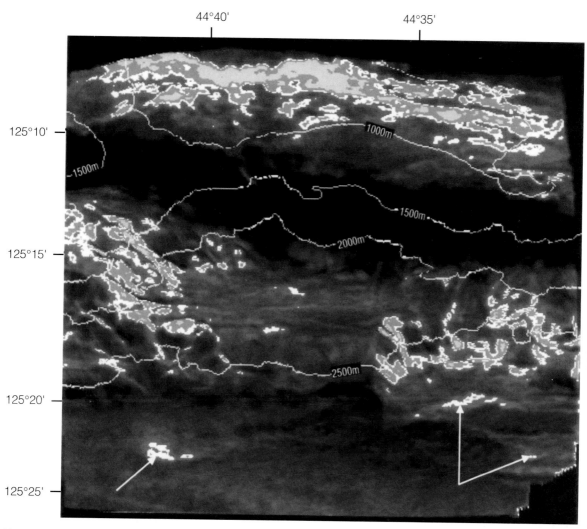

Figure 10–5. Residual GLORIA image with amplitude variation dominantly attributed to lithologic (diagenetic) variations. Image is draped over SeaBeam bathymetry. The oblique view is from the west, looking east from an elevation of 50°. Vertical exaggeration on bathymetry is 2X; depth contours in meters. Zone of no data across the lower portion of the second ridge is the fish track. Backscatter values as in Figure 10–4. Arrows indicate locations of low-amplitude diagenetic deposits (gas hydrates?) in Cascadia Basin as in Figure 10–4.

amplitude differences on the processed image. Conversely, in those portions of the study area where backscatter is low, no diagenetic deposits have been observed or sampled (Figure 10–1), although submersible operations necessarily covered a limited proportion of the seafloor.

Fluid seeps on the Oregon continental margin

Fluid discharge on this portion of the Oregon margin is controlled largely by faults. Diagenetic deposits are concentrated in three structurally distinct zones (second ridge, first ridge, and protodeformation zone) and have a backscatter pattern that is unique to each. Although thick horizons of

silty sand crop out on the western flanks of both the first and second ridges (Kulm et al. 1986; Moore et al. 1990), there is no evidence in the processed GLORIA imagery (Figures 10–4 and 10–5) of high-amplitude backscatter at these locations. We infer that these beds have not been affected by diagenetic deposition nor have they been important pore-water aquifers.

The most extensive diagenetic deposition occurs high on the second ridge (Figures 10–4 and 10–5) where a broad north-south, 38-km^2 band between 44°32–42′ N shows elevated residual-reflectance values (above 155). This large region is coincident with a landward-dipping thrust fault (MacKay et al. 1992), at least one portion of which is known to be hydrologically active. A carbonate bioherm (J. C. Moore, personal communication) that supports high levels of fluid expulsion (Linke et al. 1994) is located on the fault trace (ALVIN Site 2277, Figure 10–1); the GLORIA imagery indicates that there must be others. ODP drilling (Site 892; Figure 10–1) 300 m to the east of the bioherm revealed gas hydrate deposits 2 to 19 mbsf (Westbrook et al. 1994). Hence, the high-backscatter region on the second ridge may be due to both diagenetic carbonate and gas hydrate in the surface sediments.

The residual GLORIA image does not define a similar zone of high-amplitude backscatter where the frontal thrust fault emerges at the base of the first ridge. Submersible studies show that strong reflectance associated with the first ridge and its associated sediment pond is characterized by discontinuous patches of diagenetic deposits on the northern and southern margins of the anticlinal ridge, where deep-seated strike-slip faults bound the structure. Preliminary examination of multichannel seismic (MCS) lines that cross these left-lateral faults indicates that the fault zones extend at least to the décollement (G. F. Moore, personal communication), which occurs at a depth of about 2.5 km. The patchy distribution of high-amplitude backscatter may imply that fluids are transported upward along the faults and intercepted where erosional gullies cut the faults (Sample et al. 1993; Tobin et al. 1993) and/or that fluid discharge has not been sufficiently long-lived here to produce the areal continuity of diagenetic deposits that occur on the older second ridge. High backscatter along the base of the slope on the southern margin of the first ridge probably indicates intersection of the frontal thrust/décollement by the strike-slip fault zone and lateral flow to this seepage site from beneath the thrust sheet. The only high-amplitude backscatter zone clearly associated with the first ridge is located at its crest, on the northern end (44°40.5′ N, 125°18′ W), where it is associated with the back-thrust (Figure 10–1; MacKay et al. 1992). Numerous submersible dives to two locations (1428, 1900, Figure 10–1) confirm that the reflectors are diagenetic carbonate deposits (Kulm et al. 1986; Carson et al. 1990).

In contrast to the high-amplitude backscatter on the slope, apparent diagenetic lithification of sediments in the protodeformation zone seaward of the first ridge (Figure 10–1) shows lower residual amplitudes (generally <155, Figures 10–4 and 10–5) and a diffuse anastomosing areal distribution. We recognize two areas with anomalous backscatter amplitudes in these Cascadia Basin deposits (arrows, Figures 10–4 and 10–5): a relatively small region with maximum backscatter at 44°40.5′ N, 125°22.5′ W, and a much broader region south and west of the thrust sheet, with maximum backscatter at 44°33′ N, 125°20′ W. High-backscatter "stringers" that emanate from both of these regions may indicate subsurface (blind) thrust faults, because one of these tails evolves into a fault scarp (Figure 10–2) to the south of the area imaged in Figures 10–4 and 10–5. MCS profiles (MacKay et al. 1992) reveal numerous incipient faults in the protodeformation zone that are apparently dilated (Moore, Moore, and Cochrane 1993) and may control transport of

fluids to the surface sediments where diagenetic deposition occurs. Because residual amplitudes are distinctly lower in this subarea than on the slope, we suggest that the protodeformation zone diagenetic deposits dominantly consist of gas hydrates, although there have been no submersible dives or returned samples to confirm this conjecture.

Implications of GLORIA imagery for subduction-zone dewatering

One of the surprising findings of the GLORIA-defined seep distribution (Figures 10–4 and 10–5) is that although faults are the primary pore-water flow paths, apparently not all faults support flow. The frontal thrust beneath the first ridge and several landward-dipping faults that cut the second ridge give no acoustic evidence of flow-related diagenesis. It is conceivable that diagenetic deposits exist at these fault traces and have been covered by clastic deposits, although the occurrence of near-surface diagenesis in Cascadia Basin implies that active fluid discharge is able to penetrate unconsolidated sediments. A more likely possibility is that not all faults are hydrologically active. Even on the second ridge within the band of high backscattering, apparently not all of the faults are currently active flow zones. On the seismic profile across ODP Site 892 (Figure 10–1), the BSR is displaced toward the surface where it crosses the landward-dipping fault zone (Westbrook et al. 1994), because warm fluids moving up the fault destabilize hydrates at depth. However, the BSR shows no similar displacement south of 44°39′ N beneath the high-backscatter region (J. C. Moore, personal communication 1993), and we infer that flow that supported the diagenesis has ceased or declined to low levels.

The low-frequency GLORIA data reveal diagenetic deposits associated with both active and relict fluid seeps. Because all of the sites imaged in this study lie below the carbonate-compensation depth, one could expect that upon cessation of flow the diagenetic carbonates would undergo dissolution. Indeed, exposed carbonate deposits that have been recovered show clear evidence of dissolution rinds (Ritger et al. 1987). However, burial to even a few centimeters effectively isolates the cements from undersaturated ocean water, and it is unlikely that the absence of diagenetic carbonates at some fault traces is due to postdepositional dissolution. Similarly, postdepositional destabilization of surficial gas hydrates (induced by fluctuations in sea level or bottom-water temperature associated with glacial-interglacial variations; Revelle 1983) would be expected to affect all hydrate deposits in this area and cannot be invoked to account for their apparent association with fluid seeps.

Conceptually, faults are likely to fall into four classes: (1) tectonically and hydrologically active, with associated modern diagenetic deposits; (2) tectonically inactive but hydrologically active, with modern diagenetic deposits that could not be acoustically differentiated from (1); (3) tectonically and hydrologically inactive, although they might exhibit relict diagenetic deposits if the fault zone supported flow in the past; and (4) tectonically active but hydrologically unimportant, so that insignificant diagenesis occurs at the fault trace (at least as resolvable by sidescan sonar). This fourth case seems to describe the frontal thrust fault along which the first ridge has been elevated during the late Pleistocene. Either tectonic stress has prevented dilation and reduced permeability in this fault zone, fluids are being drained from the décollement along other flow paths, or both. Tobin et al. (1993) have suggested that fracture zone permeability is enhanced in strike-slip faults relative to thrust faults. We suspect that the left-lateral faults that define the northern and southern boundaries of the thrust sheet that includes the first ridge are the principal high-permeability conduits for fluids expelled from and beneath this structure, resulting in lateral flow from the décollement.

If flow from the prism does not occur along a plane orthogonal to the deformation front, then the two-dimensional numerical models presently used to estimate fluid budgets are inappropriate and inadequate. The GLORIA data from this small area on the Cascadia margin imply that three-dimensional models will be required, and furthermore, that field studies will be needed to define the distribution of hydrologically active fault zones to ensure application of appropriate boundary conditions in the models.

Although few measurements of flow rates have been made at submarine seeps, data collected in the present study area (Carson et al. 1990; Linke et al. 1994) show that mean rates of expulsion range from 2×10^{-6} m^3 m^{-2} s^{-1} (188 L m^{-2} d^{-1}) to 2×10^{-5} m^3 m^{-2} s^{-1} (1765 L m^{-2} d^{-1}) at vent sites. The lower flow rates come from site 1428 (Figure 10–1) on the first ridge, where diagenetic deposits exclusively consist of carbonate nodules and slabs. The higher flow was measured at site 2207 (Figure 10–1), where a massive carbonate bioherm occurs but also where ODP drilling subsequently sampled gas hydrates at 2 to 19 mbsf. These observations suggest that gas hydrates, which have a much lower reflection coefficient and should produce lower backscatter anomalies in the absence of associated carbonates, are brought to the surface only by relatively rapid fluid flow. Conversely, carbonate cementation (with associated strong positive backscatter anomalies) may accumulate at flow rates that are an order of magnitude lower. Both flow rates reflect movement along fault zones; dispersed intergranular discharge has not been measured, but is calculated to occur at much lower flow velocities (Davis et al. 1990). Our suggestion that hydrates are associated with most rapid discharge corroborates the qualitative correlation between seepage rate and type of diagenesis made previously by workers studying oil and gas seeps in the Gulf of Mexico (Roberts et al. 1990). If this correlation is valid, and if the low-amplitude anomalies observed in the protodeformation zone indicate gas hydrate deposition, then the GLORIA data suggest that some of the most rapid dewatering on the Oregon margin takes place in Cascadia Basin near the foot of the slope, perhaps in response to local consolidation and incipient thrust fault formation or to flow from the décollement that has bypassed the hydrologically inactive frontal thrust fault.

Summary

Submersible observations and samples collected in this study area indicate that GLORIA sidescan imagery, processed to remove topographic effects, can be used to map flow-related near-surface diagenetic deposits. This application is important because it is critical to detail the areal distribution of fluid-expulsion loci if we are to understand and model the water budget in accretionary prisms. This technique, however, may be used on passive as well as active margins and should provide an efficient means to define structural or stratigraphic control of fluid discharge on any continental margin. We speculate that the amplitude of GLORIA backscatter residuals may be related to rates of modern fluid expulsion, implying that sidescan imagery might be used to estimate flow rates. However, neither the correlation between flow rate and type of diagenesis nor our ability to discriminate acoustically between near-surface carbonate and gas hydrate deposits has yet been established.

Acknowledgments

This research was funded by a grant from the JOI/U.S. Science Support Program (JSC-8-90) through the National Science Foundation, to optimize site selection for drilling on the Oregon margin. D. Twichell assisted in processing the GLORIA and SeaBeam data. J. C. Moore, G. Cochrane, G. Moore, and M. MacKay generously provided access to, and interpretations of, multichannel seismic data across the study area. The final writing of this chapter occurred while Carson was on leave at the University of Wales, Aberystwyth. He gratefully acknowledges the hospitality of the staff and students there, and in particular the assistance of Alex Maltman. The final manuscript benefitted substantially from careful reviews by D. Twichell, D. Orange, and an anonymous reviewer.

References

Allen, R. C., Gavish, E., Friedman, G. M., and Sanders, J. E. 1969. Aragonite-cemented sandstone from outer continental shelf off Delaware Bay: Submarine lithification mechanism yields product resembling beachrock. *J. Sed. Petrol.* 39: 136–49.

Brooks, J. M., Kennicutt, M. C. II, Fay, R. R., MacDonald, T. J., and Sassen, R. 1984. Thermogenic gas hydrates in the Gulf of Mexico. *Science* 225: 409–11.

Brooks, J. M., Kennicutt, M. C. II, Bridigare, R. R., and Fay, R. A. 1985. Hydrate, oil seepage, and chemosynthetic ecosystems on the Gulf of Mexico shelf. *Trans. Am. Geophys. Union (EOS)* March:106.

Brooks, J. M., Field, M. E., and Kennicutt, M. C. II. 1991. Observations of gas hydrates in marine sediments, offshore northern California. *Mar. Geol.* 96: 103–9.

Carson, B., Yuan, J.-W., Myers, P. B., Jr., and Barnard, W. D. 1974. Initial deep-sea sediment deformation at the base of the Washington continental slope: A response to subduction. *Geology* 2: 561–4.

Carson, B., Suess, E., and Strasser, J. C. 1990. Fluid flow and mass flux determinations at vent sites on the Cascadia margin accretionary prism. *J. Geophys. Res.* 95: 8891–7.

Carson, B., Holmes, M. L., Umstattd, K., Strasser, J. C., and Johnson, H. P. 1991. Fluid expulsion from the Cascadia accretionary prism: Evidence from porosity distribution, direct measurements, and GLORIA imagery. *Trans. Royal Society of London, Series A* 335: 331–40.

Carson, B., Westbrook, G., Musgrave, R., and the Scientific Staff of ODP Leg 146 1993. Fluid flow and origin of BSR established in Cascadia accretionary wedge on ODP Leg 146. *EOS* 74(31): 337–9.

Carson, B., Seke, E., Paskevich, V., and Holmes, M. L. 1994. Fluid expulsion sites on the Cascadia accretionary prism: Mapping diagenetic deposits with processed GLORIA imagery. *J. Geophys. Res.* 99(B6): 11,959–69.

Chavez, P. C. 1984. U.S. Geological Survey mini image processing system (MIPS). *USGS Open-File Report 84-880,* 12 pp. Reston, Va.: USGS.

Christensen, N. I. 1989. Seismic velocities. In *CRC Handbook of Physical Properties of Rocks and Minerals.* pp. 429–526. Boca Raton, Fla., CRC Press.

Davis, E. E., and Hyndman, R. D., 1989. Accretion and recent deformation of sediments along the northern Cascadia subduction zone. *Geol. Soc. Am. Bull.* 101: 1465–80.

Davis, E. E., Hyndman, R. D., and Villinger, H. 1990. Rates of fluid expulsion across the northern Cascadia accretionary prism: Constraints from new heat flow and multichannel seismic data. *J. Geophys. Res.* 95: 8869–90.

EEZ-SCAN 84 Scientific Staff 1986. Atlas of the Exclusive Economic Zone, Western conterminous United States. *Miscellaneous Investigations Series No. I-1972.* Reston, Va.: U. S. Geological Survey.

Foucher, J.-P., Henry, P., Le Pichon, X., and Kobayashi, K. 1992. Time-variations of fluid expulsion velocities at the toe of the eastern Nankai accretionary complex. *Earth Plan. Sci. Ltrs.* 109: 373–82.

Friedman, G. M., Sanders, J. E., Gavish, E., and Allen, R. C. 1971. Marine lithification mechanism yields rock resembling beachrock. In *Carbonate Cements,* pp. 50–3. Baltimore and London: Johns Hopkins University Press.

Gamo, T., Sakai, H., Ishibashi, J., Shitashima, K., and Boulègue, J. 1992. Methane, ethane, and total inorganic carbon in fluid samples taken during the 1989 KAIKO-Nankai project. *Earth Plan. Sci. Ltrs.* 109: 383–90.

Gardner, J. V., Field, M. E., Lee, H., and Edwards, B. E. 1991. Ground-truthing 6.5-kHz side scan sonographs: What are we really imaging? *J. Geophys. Res.* 96(B4): 5955–74.

Goedert, J. L., and Squires, R. L. 1990. Eocene deep-sea communities in localized limestones formed by subduction-related methane seeps, southwestern Washington. *Geology* 18: 1182–5.

Goldfinger, C., Kulm, L. D., Yeats, R. S., Applegate, B., MacKay, M. E., and Moore, G. F. 1992. Transverse structural trends along the Oregon convergent margin: Implications for Cascadia earthquake potential and crustal rotations. *Geology* 20: 141–4.

Hampton, M. A., Karl, H. A., and Kenyon, N. H. 1989. Sea-floor drainage features of Cascadia Basin and the adjacent continental slope, northeast Pacific Ocean. *Mar. Geol.* 87: 249–72.

Han, M. W., and Suess, E. 1989. Subduction-induced pore fluid venting and the formation of authigenic carbonates along the Oregon/Washington continental margin: Implications for the global Ca cycle. *Palaeogeog. Palaeoclim. Palaeoecol.* 71: 119–36.

Hand, J. H., Katz, D. L. and Verma, V. K. 1984. Review of gas hydrates with implications for ocean sediments. In *Natural Gases in Marine Sediments,* pp. 179–94. New York: Plenum Press.

Henry, P., Foucher, J.-P., Le Pichon, X., Sibuet, M., Kobayashi, K., Tarits, P., Chamot-Rooke, N., Furuta, T., and Schultheiss, P. 1992. Interpretation of temperature measurements from the KAIKO-Nankai cruise: Modeling of fluid flow in clam colonies. *Earth Plan. Sci. Ltrs.* 109: 355–71.

Huggett, Q. J., and Somers, M. L. 1988. Possibilities of using the GLORIA system for manganese nodule assessment. *Mar. Geophys. Res.* 9: 255–64.

Hyndman, R. D., Yorath, C. J., Clowes, R. M., and Davis, E. E. 1990. The northern Cascadia subduction zone at Vancouver Island: Seismic structure and tectonic history. *Can. J. Earth Sci.* 27: 313–29.

Hyndman, R. D., and Davis, E. E. 1992. A mechanism for the formation of methane hydrate and sea floor bottom simulating reflectors by vertical fluid expulsion. *J. Geophys. Res.* 97: 7025–41.

Jackson, D. R., Baird, A. M., Crisp, J. J., and Thompson, P. A. G. 1986. High-frequency bottom backscatter measurements in shallow water. *J. Acoust. Soc. Am.* 80(4): 1188–99.

Kennicutt, M. C. II, Brooks, J. M., Bidigare, R. R., Fay, R. R., Wade, T. L., and McDonald, T. J. 1985. Vent-type taxa in a hydrocarbon seep on the Louisiana slope. *Nature* 317: 351–3.

Kulm, L. D., and Fowler, G. A. 1974. Oregon continental margin structure and stratigraphy: A test of the imbricate thrust model. In *The Geology of Continental Margins,* pp. 261–84. New York: Springer-Verlag.

Kulm, L. D., Suess, E., Moore, J. C., Carson, B., Lewis, B. T. R., Ritger, S. D., Kadko, D. C., Thornberg, T. M., Embley, R. W., Rugh, W. D., Massoth, G. J., Langseth, M. R., Cochrane, G. R., and Scamman, R. 1986. Oregon subduction zone: Venting, fauna, and carbonates. *Science* 231: 561–6.

Kulm, L. D., and Suess, E. 1990. Relationship between carbonate deposits and fluid venting: Oregon accretionary prism. *J. Geophys. Res.* 95: 8899–915.

Kvenvolden, K. A., and Barnard, L. A. 1983. Hydrates of natural gas in continental margins. In *Studies in Continental Margin Geology* pp. 631–40. Tulsa, Okla: American Association of Petroleum Geologists.

Lalou, C., Fontugne, M., Lallemand, S. E., and Lauriat-Rage, A. 1992. Calyptogena-cemented rocks and concretions from the eastern part of Nankai accretionary prism: Age and geochemistry of uranium. *Earth Plan. Sci. Ltrs.* 109: 419–29.

Le Pichon, X., Kobayashi, K., Cadet, J.-P., Ashi, J. Boulegue, J., Chamot-Rooke, N., Fiala-Medioni, A., Foucher, J. P., Furuta, T., Gamo, T., Henry, P., Iiyama, J. T., Lallemand, S. E., Lallemant, S. J. Ogawa, Y., Sakai, H., Segawa, J., Sibuet, M., Taira, A., Takeuchi, A., Tarits, P., and Toh, H. 1992. Fluid venting activity within the eastern Nankai Trough accretionary wedge: A summary of the 1989 Kaiko-Nankai results. *Earth Plan. Sci. Ltrs.* 109: 303–18.

Linke, P., Suess, E., Torres, M., Martens, V., Rugh, W. D., Ziebis, W., and Kulm, L. D. 1994. In situ measurements of fluid flow from cold seeps at active continental margins. *Deep Sea Res.* 41: 721–39.

MacKay, M. E., Moore, G. F., Cochrane, G. R., Moore, J. C., and Kulm, L. D. 1992. Landward vergence and oblique structural trends in the Oregon margin accretionary prism: Implications and effect on fluid flow. *Earth Plan. Sci.* 109: 477–91.

Miller, J. J., Lee, M. W., and von Huene, R. 1991. An analysis of a seismic reflection from the base of a gas hydrate zone, offshore Peru. *Am. Assoc. Petrol. Geol. Bull.* 75(5): 910–24.

Moore, J. C., Orange, D., and Kulm, L. D. 1990. Interrelationship of fluid venting and structural evolution: Alvin observations from the frontal accretionary prism. *J. Geophys. Res.* 95(B6): 8795–808.

Moore, J. C., Moore, G. F., and Cochrane, G. 1993. Reversed polarity reflections along faults: Indicators of channelized fluid flow? *Trans. Am. Geophys. Union (Eos):* 74(43): 579.

Orange, D. L., and Breen, N. A. 1992. The effects of fluid escape on accretionary wedges II: Seepage force, slope failure, headless submarine canyons, and vents. *J. Geophys. Res.* 97(6): 9277–95.

Paull, C. K., Hecker, B., Commeau, R., Freeman-Lynde, R. P., Neumann, C., Corse, W. P., Golubic, S., Hook, J. E., Sikes, E., and Curray, J. 1984. Biological communities at the Florida Escarpment resemble hydrothermal vent taxa. *Science* 226: 965–7.

Revelle, R. R. 1983. Methane hydrates in continental slope sediments and increasing atmospheric carbon dioxide. In *Changing Climate*, pp. 252–61. Washington, D.C.: National Academy Press.

Ritger, S., Carson, B., and Suess, E. 1987. Methane-derived authigenic carbonates formed by subduction-induced pore-water expulsion along the Oregon/Washington margin. *Geol. Soc. Am. Bull.* 98: 147–56.

Roberts, H. H., Carney, R., Larkin, J., and Sassen, R. 1990. Seafloor responses to hydrocarbon seeps, Louisiana continental slope. *Geo-Mar. Ltrs.* 10(4): 232–43.

Roberts, H. H., and Neurauter, T. W. 1991. Seismic expression and directly observed characteristics of seep-related features, northern Gulf of Mexico continental slope. *Am. Assoc. Petrol. Geol. Bull.* 75(9): 1537.

Sakai, H., Gamo, T., Ogawa, Y., and Boulegue, J. 1992. Stable isotopic ratios and origins of the carbonates associated with cold seepage at the eastern Nankai Trough. *Earth Plan. Sci. Ltrs.* 109: 391–404.

Sample, J. C., Reid, M. R., Tobin, H. J., and Moore, J. C. 1993. Carbonate cements indicate channelized fluid flow along a zone of vertical faults at the deformation front of the Cascadia accretionary wedge (northwest U.S. coast). *Geology* 21: 507–10.

Sloan, E. D. 1990. *Clathrate Hydrates of Natural Gases.* New York: Marcel Dekker, 641 pp.

Spence, G. D., Hyndman, R. D., Davis, E. E., and Yorath, C. J. 1991. Seismic structure of the northern Cascadia accretionary prism: Evidence from new multichannel seismic reflection data. *Continental lithosphere, deep seismic reflection.* Geodynamics Series. Washington D.C.: American Geophysical Union.

Suess, E., Carson, B., Ritger, S., Moore, J. C., Jones, M., Kulm, L. D., and Cochrane, G. 1985. Biological communities at vent sites along the subduction zones off Oregon. In *The Hydrothermal Vents of the Eastern Pacific: An Overview,* Bulletin of the Biological Society of Washington, Washington DC. pp. 475–84.

Suess, E., and Whiticar, M. J. 1989. Methane-derived CO2 in pore fluids expelled from the Oregon subduction zone. *Palaeogeog. Palaeoclim. Palaeoecol.* 71: 119–36.

Tobin, H. J., Moore, J. C., MacKay, M. E., Orange, D. L., and Kulm, L. D. 1993. Fluid flow along a strike-slip fault at the toe of the Oregon accretionary prism: Implications for the geometry of frontal accretion. *Geol. Soc. Am. Bull.* 105(5): 569–82.

Westbrook, G. K., Boulegue, J., Bowers, T. S., Cathles, L. M. III, Davis, E. E., Langseth, M. Leinen, M., Ohta, S., and Kastner, M. 1987. *Fluid circulation in the crust and the global geochemical budget. Report of the Second Conference on Scientific Ocean Drilling (COSOD II).* Strasbourg, France: European Science Foundation (ESF).

Westbrook, G. K., Carson, B., Musgrave, R., and Scientific Staff of ODP Leg 146 1994. *Initial Reports of the Ocean Drilling Program, Leg146.* College Station, Texas: Ocean Drilling Program.

Whalley, E. 1980. Speed of longitudinal sound in clathrate hydrates. *J. Geophys. Res.* 85: 2539–42.

Wong, H.-K., and Chesterman, W. D. 1968. Bottom backscattering near grazing incidence in shallow water. *J. Acoust. Soc. Am.* 44(6): 1713–18.

11 A meandering channel at the base of the Gorda Escarpment

David A. Cacchione, David E. Drake, and James V. Gardner

U.S. Geological Survey, Menlo Park, California

Abstract

Mendocino Channel lies at the base of Gorda Escarpment on the southern edge of Gorda deep-sea fan about 40 km off Cape Mendocino, California. The channel marks the seaward extension of the Mattole and Mendocino Submarine Canyons and contains several prominent meanders along its length. Based on water-gun and 3.5-kHz seismic records, the channel-levee geometry and structure are similar to deep-sea meandering channels reported elsewhere. Over the approximately 50-km section of the channel studied, channel sinuosity and channel slope are about 1.7 and 4.2 m/km, respectively. The valley slope is likely time-variable because of the tectonically induced movements of the southern Gorda Plate upon which the channel is situated. Changes to the valley slope will lead to changes in the meander development during future episodes of turbidity currents. Five (possibly six) turbidite units that have ^{14}C ages ranging from 960 yBP to 3,595 yBP were recovered in box cores from the channel floor. These ages overlap dates of great earthquakes that are estimated to have occurred along the region north of Cape Mendocino. Paleoseismic events are proposed to be the triggering mechanisms for the turbidity currents that deposited the turbidites in Mendocino Channel.

Introduction

With the increased use of swath surveying of the ocean floor, meandering (*senso latto*) channels have become common features found on the surface of many deep-sea fans. Meandering channels have recently been described from swath-mapping systems on the Amazon Fan (Flood and Damuth 1987; Damuth et al. 1983); Mississippi Fan (Twichell et al. 1991); Rhone Fan (O'Connell et al. 1991); Bengal Fan (Kenyon, personal communication 1992); Surveyor, Mukluk, Horizon, and unnamed fans in the Gulf of Alaska

(Bruns, Stevenson, and Dobson 1992; Carlson et al. this volume); and Monterey Fan (Gardner et al. this volume).

During the 1984 mapping of the EEZ off the west coast of the United States (Cacchione et al. 1984) with the GLORIA sidescan sonar, our attention was drawn to a meandering deep-sea channel situated off northern California on the southern edge of Gorda deep-sea fan (Figure 11–1; Chase et al. 1992; sheet 21 in EEZ-SCAN 84 Scientific Staff 1986). We refer to this feature as Mendocino Channel. A subsequent cruise during 1986 collected water-gun and high-resolution 3.5-kHz seismic profiles and core samples over a section of Mendocino Channel where high acoustic backscatter along the channel thalweg made the feature easily identifiable on the sidescan records. Here we present the results of the follow-on investigation and correlate the latter findings with the original GLORIA images of the meandering channel. We also suggest that radiocarbon-dated turbidites collected with box cores from the channel thalweg were emplaced following large paleoseismic events in this region.

Geological setting

The geographic relationship of the Mendocino Channel to nearby major physiographic seafloor features is shown in Figures 11–2 and 11–3. The Mendocino Transform Fault (MTF) at the base of the Mendocino Ridge is observed on GLORIA sonographs as a narrow linear feature that trends across the seafloor on an E-W strike, abutting Mendocino Channel in at least two places (Figure 11–3; and sheet 21 in EEZ-SCAN 84 Scientific Staff 1986). Right-lateral movement on MTF accommodates the relative motion between the Gorda Plate to the north and the Pacific Plate to the south. Masson, Caccione, and Drake (1988) analyzed the fabric of the basaltic intraplate ridges, magnetic-anomaly patterns, and earthquake data and proposed that since 5 Ma the Gorda Plate has moved obliquely to MFZ, leading to compression between the Gorda Plate and the Pacific Plate at this boundary. This concept agrees with the tectonic pat-

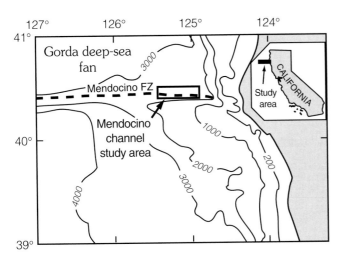

Figure 11–1. General location map showing the Mendocino Channel study area.

terns proposed earlier by Silver (1971) and Riddihough (1980), but disagrees with the interpretation of total strike-slip motion along the MTF (Wilson 1986). The Mendocino Triple Junction lies near Cape Mendocino (Figure 11–2; Clarke 1992) and is one of the earth's most geologically complex and seismically active regions. The seismicity in this region has been analyzed and discussed by Couch (1980) and more recently by Smith, Knapp, and McPherson (1993). Great earthquakes during the late Holocene have been inferred from paleoseismic studies to have occurred north of Cape Mendocino along the Cascadia Subduction Zone, and large-magnitude earthquakes in this region are predicted in the future (Clarke and Carver 1992; Smith et al. 1993).

Bathymetric charts indicate that Mendocino Channel is connected shoreward to Mendocino and Mattole submarine canyons, two significant features that incise the continental margin around Cape Mendocino (Figure 11–2). The complex nature of the acoustic backscatter caused by the highly variable bathymetry in the shallower sections of the continental margin in this region make it difficult to trace Mendocino Channel into water depths shallower than about 1,500 m in the GLORIA images (EEZ-SCAN 84 Scientific Staff 1986). However, the shoreward extension of Mendo-

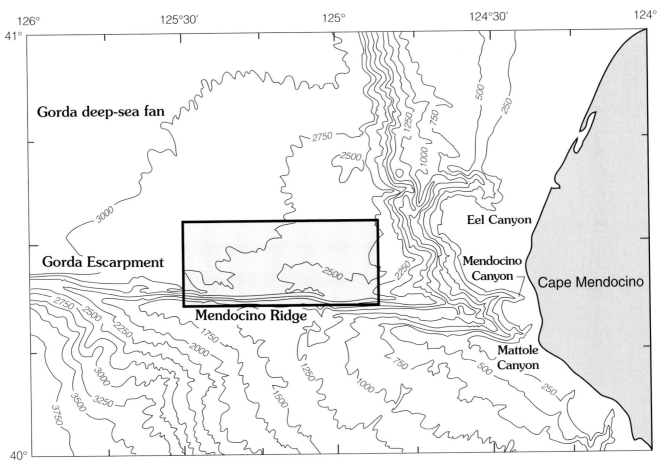

Figure 11–2. Bathymetric chart showing the study area (box) and prominent physiographic features. Bathymetric contours in meters (after NOAA 1969).

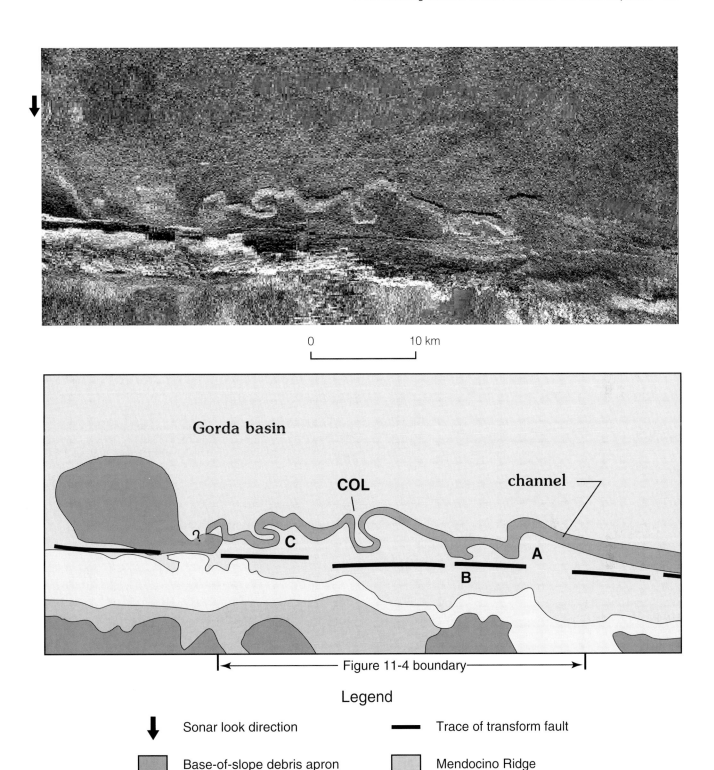

Figure 11–3. Enlargement of GLORIA sonograph and interpretation of the study area from the imagery showing the meander section of the Mendocino Channel. Mendocino Transform Fault appears as a lineament trending E-W in the image. Meander loops are labelled A, B, and C; the cutoff loop is labelled COL.

cino Channel into Mendocino and Mattole submarine canyons is obvious in the available bathymetry (NOAA chart NK 10-10 1969; also unpublished NOAA multibeam chart N400124W for Cape Mendocino region). The steep northern side of the Mendocino Ridge (Figure 11–2) forms a topo-

graphic barrier that prohibits southward development of the channel-levee system.

Mendocino Channel lies at the southern edge of Gorda deep-sea fan (Figures 11–1 and 11–2). The sedimentary history and structure of this fan have not been previously in-

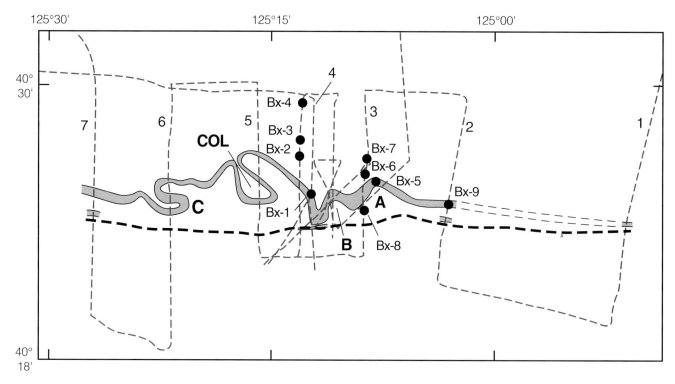

Figure 11–4. Diagram of the Mendocino Channel interpreted from water-gun and 3.5-kHz profiles, using the GLORIA sonographs as a guide. Ship tracklines (light dashed lines) that cross the channel are numbered. Box core locations are also shown (Bx-1–Bx-9). Heavy dashed line is the trace of the Mendocino Transform Fault. The meanders loops are labeled A, B, and C; the cutoff loop is labeled COL.

vestigated, but sandy turbidites found farther to the west in Escanaba Trough suggest that the turbidites may also comprise part of Gorda deep-sea fan. Based on mineralogical data from cores obtained during early Deep-Sea Drilling Project cruises, Vallier, Harold, and Girdley (1973) proposed that during Pleistocene low stands of sea level, sediments from the Columbia River and Klamath River drainage basins were transported by turbidity currents southwestward across Gorda Deep-sea Fan and into Escanaba Trough. The transport pathways of these flows are seen on GLORIA sonographs as a linear fabric in the surficial sediment cover on the fan (sheet 21 in EEZ-SCAN 84 Scientific Staff 1986). The lineated features in the GLORIA images are suggestive of thinly buried channels with a NE-SW trend. These features might represent the patterns of Pleistocene sediment movement during the turbidity-current episodes that brought materials into Escanaba Trough as proposed by Vallier et al. (1973). However, based on our results, it also appears that the southern edge of Gorda deep-sea fan contains the Mendocino channel-levee system that was constructed from Holocene turbidity currents emanating from submarine canyons on the northern California continental margin. The Mendocino Channel and its levees represent an additional important part of the construction of the southern end of Gorda deep-sea fan.

Experimental methods

An enlargement of the GLORIA image of the section of Mendocino Channel described in this study is shown in Figure 11–3. The meandering section of the channel along the southern boundary of Gorda deep-sea fan is clearly visible adjacent to MTF at the base of Mendocino Ridge (Figure 11–3). The locations of water-gun and 3.5-kHz profiles collected along several crossings of the meander section of the Mendocino Channel and nine box cores taken in the channel and on the adjacent levees are shown in Figure 11–4. The box cores were subsampled for sedimentological and geotechnical analyses, and a vertical slab was cut from each core for photography and x-radiography.

Routine shipboard navigation was by transit satellite and Loran-C. Detailed navigation for the seismic surveys and coring operations used a network of acoustic bottom transponders, but difficulties caused by false echoes and noise in the system permitted only sporadic use of the transponder information. The cores taken within the channel were all accurately located with the acoustic bottom technique, but because of the system problems, some of the cores from the levees had to be sited with the other navigation systems.

Channel characteristics

The portion of Mendocino Channel in this study extends to about 90 km west of Cape Mendocino, California (Figure 11–2). The path of the channel (Figure 11–4) was delineated from the 3.5-kHz profiles that cross the channel. We selected the 3.5-kHz and water-gun survey to define the channel geometry (Figure 11–4) and structure because of the consistent and accurate navigation and geographic positioning of the records. The GLORIA image of the channel (Figure 11–3) was used as a guide in the reconstruction. The channel section has three well-defined meanders and one nearly cut-off loop (COL in Figure 11–4) over the approximately 50 km of down-fan length. The characteristics of Mendocino Channel are similar in many respects to meandering channels on other deep-sea fans. In order to relate this feature to other meandering channels we draw comparisons of its acoustic backscatter, shape, and geometry to those described from the Amazon (Flood and Damuth 1987) and Rhone (O'Connell et al. 1991) Fans. Mendocino Channel has high acoustic backscatter (Figure 11–3) at the 6.5-kHz GLORIA frequency, similar to the rather high GLORIA backscatter from the meandering channels on the Amazon (Flood and Damuth 1987) and Rhone (O'Connell et al. 1991) Fans. The channel floor has high acoustic backscatter compared to the weaker backscatter of the surrounding levees and fan surface. We attribute this contrast in backscatter largely to the difference in sediment types, with sandy turbidites comprising the upper layers of the channel floor and predominantly muddy units covering the upper sections of the levees. The enlarged GLORIA image of the channel in Figure 11–3 also shows the shadows cast by the steep, relatively high channel walls on its north side. The channel loses its distinctive high backscatter at the far western side of the image where it appears to grade into a region of lower overall backscatter and an irregular backscatter texture (Figure 11–3 and EEZ-SCAN 84 Scientific Staff 1986). It also appears that a debris flow from the northern flank of the Mendocino Ridge transects the channel at the western end of the imaged section (Figure 11–3).

The water-gun data (Figure 11–5) show that the channel-levee complex is elevated 300 to 400 m relative to the seafloor to the north. The aggradation of the channel-levee system is characteristic of other such features on upper or middle fan regions (Normark 1978). Aggradation of the channel-levee deposits was reported for the Amazon and the Rhone Fans (Flood and Damuth 1987; O'Connell et al. 1991). However, part of the elevation of the Mendocino Channel-levee system is probably related to tectonic uplift of the Gorda Plate at its juncture with the Pacific Plate along MTF (Masson et al. 1988). Evidence of this uplift is seen in sedimentary units that are tilted up toward the south beneath the channel-levee deposits in the water-gun profiles

(profiles 5 and 6 in Figure 11–5). Also, the position of the channel valley headward of the meandering section appears to have remained approximately stationary for a prolonged period because no large lateral shifts are seen in the positions of subsurface channel reflectors in the watergun records (profiles 1 and 2 in Figure 11–5).

The deeply incised channel on the most landward trackline crossing (line 01 in Figures 11–4 and 11–5) has a channel water depth of about 2,300 m, and is bounded by levees that rise about 320 m (north side) and 290 m (south side) above the thalweg. Table 11–1 summarizes the channel depth and levee heights obtained from the 3.5-kHz records. The channel depth systematically increases down-fan to about 2,745 m on the farthest western crossing (line 07 in Table 11–1). The levees generally become lower and less pronounced down-fan (Table 11–1), similar to the channel levees on other fans. At the westernmost crossing of the channel (line 07 in Figure 11–4), the levees are barely discernible and rise less than 10 m above the channel floor. The levees are typically asymmetric, a characteristic found on other channel-levee systems, with the southern levees less pronounced and generally lower, particularly on crossings where the channel is close to Gorda Escarpment (Figure 11–5). Nearly thirty years ago, Menard (1955) noted that the levees on the right sides of channels, as viewed down-canyon, were higher than those on the left sides. He proposed that turbidity currents moving down-canyon would preferentially overflow the right bank due to the Coriolis effect, thereby depositing more material on that levee. This explanation for the asymmetry of levees surrounding fan channels has been accepted by others (Shepard and Dill 1966).

The down-fan valley length for the meandering section of Mendocino Channel (between lines 02 and 07 in Figure 11–4) is about 35 km; the valley slope (channel drop measured as a straight line down the regional slope) over this section is about 8.2 m/km. This valley slope is comparable to that estimated for meandering channels on the upper fan sections of Rhone (7.3 m/km; O'Connell et al. 1991) and Amazon (8 to 12 m/km; Flood and Damuth 1987) Fans, but much steeper than gradients on lower fan channels on the Mississippi (Twichell et al. 1991) and Monterey Fans (Gardner et al. this volume). The channel slope (channel drop measured along the channel thalweg) of Mendocino Channel is about 4.3 m/km, which is within the range of channel slopes found on the upper fan channels of the Amazon system (4 to 8 m/km; Flood and Damuth 1987). The sinuosity of the channel (ratio of channel length to valley length) along the section between lines 2 and 7 in Figure 11–4 is 1.7. Values of sinuosity greater than 1.5 classify the channel as a meandering type (Leopold and Wolman 1960). The value of 1.7 compares favorably with a maximum sinuosity of 1.7 on upper Amazon Fan (Flood and Damuth 1987), but is less than 2.3 for one of the older channels reported on Rhone Fan (O'Connell et al. 1991).

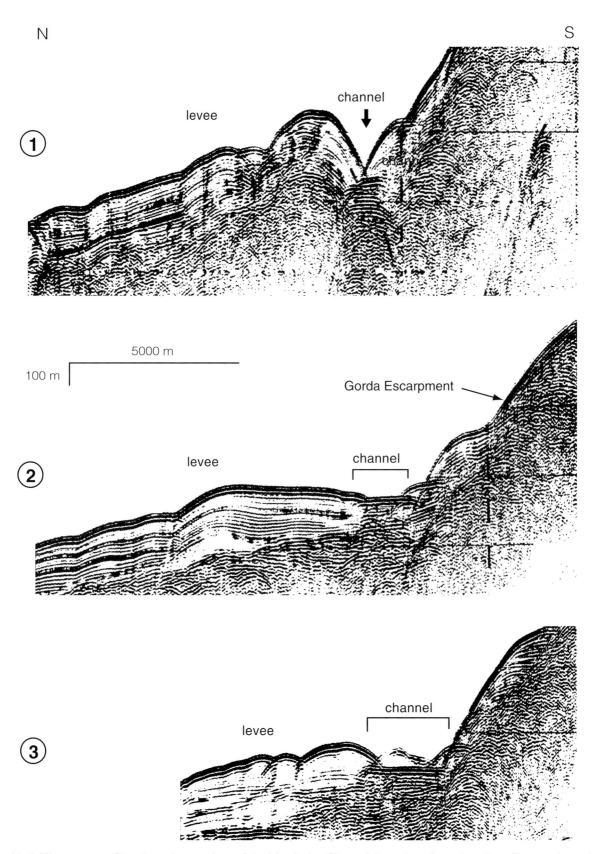

Figure 11–5. Water-gun profiles along the crossings of the Mendocino Channel. Location of numbered profiles are shown in Figure 11–4. Channel axis locations are verified with the 3.5-kHz records. Gorda Escarpment is the steep slope on the south side, as labeled on profile 2.

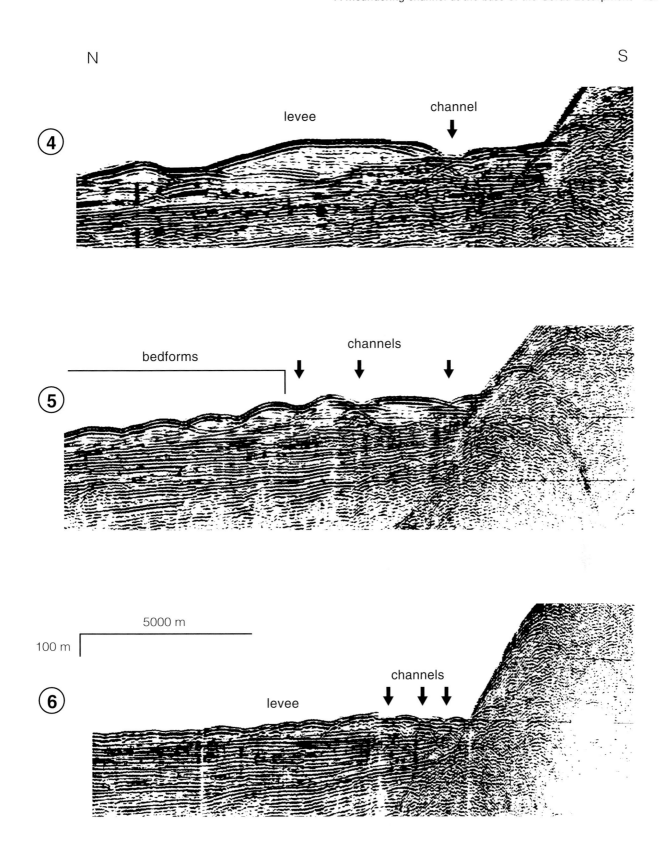

The meanders of Mendocino Channel can be characterized by meander length (L), channel width (W), and radius of curvature (R). Table 11–2 contains these measures for each of the three Mendocino Channel meanders and the cut-off loop (Figure 11–4) and also shows the range of values of these parameters for the Amazon Fan Channels (Flood and Damuth 1987). The channel width was the most difficult characteristic to evaluate from the available data. Chan-

Table 11–1. *Mendocino Channel characteristics*

Line crossing from Figure 11–4	Channel depth, m	Levee height, m	
		N levee	S levee
01	2,300	320	290
02	2,456	70	70
03	2,530	120	120
05	2,620	55	50
06	2,685	37	23
07	2,745	<10	<5

Table 11–2. *Meander geometric parameters of the Mendocino Channel*

	Meander*				Amazon† Range	Cut-off Loop*
	a	b	c	Average		
Length, km	3.89	2.96	1.85	2.90	2.0–15.0	4.07
Width, km	1.11	0.74	0.37	0.74	0.3–1.4	1.11
Radius, km	1.11	0.93	0.74	0.93	0.5–7.0	0.74

*See Figure 11–4 for location of meanders.
†Range of values for Amazon channel meanders (Flood and Damuth 1987).

nel width was measured both as the width of the highly reflective axial zone on the GLORIA image and from the interpreted diagram of the channel (Figure 11–4). Comparisons of the values obtained from these two techniques suggest that the errors in estimates of channel width are no more than 50 m.

Turbidite units

Locations of box cores taken within the channel-levee complex are shown in Figure 11–4. The three cores from the channel (Bx-1, Bx-5, and Bx-9) contain sandy turbidites separated by zones of olive-grey hemipelagic mud. As can be seen in the stratigraphic sections of the box cores (Figure 11–6), cores Bx-1 and Bx-5 have five distinct sandy turbidites and Bx-9 has four turbidites. A thin sandy layer at the base of Bx-1 might represent a sixth turbidite bed, but the sample was too disturbed and meager to permit confidence in making the identification. Each turbidite is characterized by an upward-fining sandy layer that contains plant and woody debris. Thin peat layers were found within the sandy sections of turbidite units in Bx-1 and Bx-9 (Figure 11–6). The mean grain size of the turbidites is that of coarse silt, although measurable amounts of very fine to fine sand are present in the lower sections of most units. Generally the bottom contacts between the turbidites and the

hemipelagic mud are sharp and irregular, indicating erosion before emplacement, whereas the upper contacts between the turbidites and the overlying muddy zones are smooth to wavy but distinct (Figure 11–6). The topmost turbidite in each box core is overlain by a thin mud layer about 3 to 5 cm thick, suggesting the last turbidity current event occurred on the order of several hundreds of years ago. At a rate of sedimentation of about 10 cm/kyr (Menard 1964), the topmost mud unit would have been deposited about 300 to 500 years ago. These dates are likely minima because of the potential loss of mud from the top of the box cores during sampling. High percentages of benthic foraminifers *Elphidium* and *Buccella* in the turbidites from Bx-1 and Bx-9 indicate displacement from the inner shelf into deep water, presumably by the turbidity currents.

Cores from the levees generally have thin sand layers intercalated within thicker zones of hemipelagic olive-grey mud (Figure 11–6). The muddy zones often contain sand-filled burrows scattered throughout. The sand layers are not graded and the contacts between the mud and sand units are generally conformable and sharp. Two- to three-meter-long gravity cores were also collected from the levees and showed essentially the same sediment properties and sequences as the box cores from this region.

Bulk samples of organic-rich sediment from the turbidites in Bx-1 were dated by conventional ^{14}C dating (Figure 11–4). Approximately 2 cm of core length was used for each of the samples. Dates from the five turbidites in Bx-1 are shown in Figure 11–6 and Table 11–3. The dates are all mid to late Holocene, ranging from 970 ±80 yBP for the topmost turbidite to 3,595 ±220 yBP for the bottom unit. This age information will be used in the next section to investigate the possible link between the turbidity currents that deposited the turbidites and large paleoseismic events that have occurred in this tectonically active region.

Discussion

The turbidite sequences in the box cores suggest five and possibly six turbidity currents travelled through this channel system over the past 3,500 yrs. Even though separated by a layer of hemipelagic mud, the turbidites dated at 3,215 and 3,275 yBP (Figure 11–6) are possibly from the same event, given their dating errors. Adams (1990) examined sequences of turbidites from gravity cores collected in the Cascadia Channel and along the Oregon-Washington continental margin and concluded that each of the thirteen turbidites he found is related to a great Cascadia Subduction Zone earthquake (magnitudes equal to or greater than 8.5) that occurred in this region since the Mount Mazama ash deposited about 6,850 yBP. His analysis of the number of turbidites and of the pelagic mud intervals that were deposited between the turbidites suggests the great earthquakes occurred on average every 590 ±170 years. He argued that, because thir-

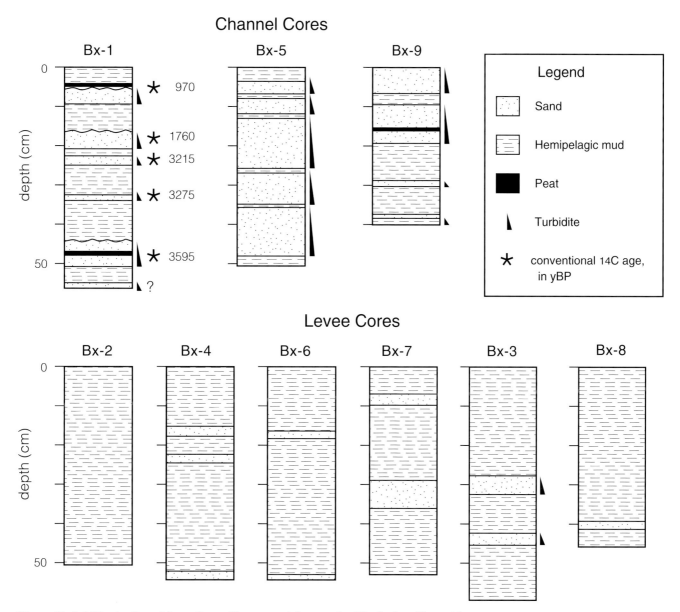

Figure 11–6. Lithostratigraphic sections of box cores taken on the Mendocino Channel-levee system.

teen turbidite sequences were also found at other widely separated locations on the Oregon-Washington margin, simultaneous self-triggering or localized mass failure were unlikely causes for the turbidity currents. He reasoned that synchronous effects of great subduction-zone earthquakes would be experienced throughout the Pacific Northwest. Adams (1990) concluded that where great earthquakes were likely to have occurred, the turbidite record would be directly related to the timing of the paleoseismicity.

Along the southern section of Cascadia Subduction Zone where Mendocino Channel is located, great earthquakes are likely (Heaton and Kanamori 1984; Dengler, McPherson, and Carver 1992). Evidence for earthquake-related effects and timing of the events from the region of Cape Mendo-

cino (Figure 11–2) comes from a recent study of tectonics and paleoseismicity of the southern part of the Cascadia Subduction Zone by Clarke and Carver (1992). Their analysis is largely based on apparently synchronous fault offsets of stratigraphic features by as much as 5 to 7 m, sudden subsidence of marsh peats and fossil forests, and uplift of coastal marine terraces during the past 4,800 yrs. Clarke and Carver (1992) argued that the large earthquakes that caused these dislocations were associated with tectonic subduction of the Gorda Plate and had moment magnitudes of 8.4 or greater. Table 11–3 summarizes the dates and nature of the evidence for the great earthquakes found by Clarke and Carver (1992), together with dates from the turbidites in Bx-1 from Mendocino Channel. Within the limits of the errors, the ^{14}C dates

Table 11–3. *Age of turbidites (box core 1) and of coastal seismic events*

Box core 1 ages, (yBP from ^{14}C dating)	Coastal seismic events* (yBP, approximate ages)	Nature of data for coastal seismic events
970 ± 80	960–1,257	Uplifted wave-cut terrace north of Humboldt Bay.
1,760 ± 140	1,450–1,650	Submerged peaty marsh deposits below modern intertidal zone.
3,215 ± 90	3,100–3,700	Raised beaches on north side of Cape Mendocino with fossil mollusk shells in growth position.
3,275 ± 190	Same as above	Same as above.
3,595 ± 220	2,149–4,802	Fluvial flood plain deposits displaced vertically about 18 m across thrust fault.

*Seismic events and analyses are presented in Clarke and Carver (1992).

for the turbidites in Bx-1 all overlap the ages of the large paleoseismic events reported by Clarke and Carver (1992) and Valentine et al. (1992). Although this finding does not prove that the triggering mechanism for the turbidity currents that emplaced the Mendocino Channel turbidites were the proposed great earthquakes, the close correspondence in the timing of the earthquakes and the ages of the turbidites provides strong circumstantial evidence that the turbidites resulted from the seismic events. This correlation also supports the conclusions of Clarke and Carver (1992) regarding the occurrences of great earthquakes along the southern section of the Cascadia Subduction Zone in the time periods proposed. Furthermore, sandy turbidites of late Holocene age in the northeastern Pacific Ocean have generally been linked to earthquakes as triggering mechanisms for the turbidity currents (Adams 1990).

Finally, we find that geometric characteristics of meanders in Mendocino Channel are similar to those found for the Amazon and Rhone channel-levee systems. The relationships between meander wavelength and channel width or radius of curvature fall within the range of those found for Amazon meanders (Table 11–2) and fit the curves that relate these parameters for river meanders. Flood and Damuth (1987) show that meander wavelength plotted against channel width and radius of curvature for the Amazon data closely follows the plots of these parameters for large rivers that were initially derived by Leopold and Wolman (1960). The relationships between changes in valley slope and changes in meander geometry for large rivers were originally derived by Schumm (1977) and are reviewed by Flood and Damuth (1987). It was proposed that if valley slope were to increase beyond a threshold point, which is specific to each channel system, then the channel would in-

crease its meandering to regain or maintain its channel slope. Schumm (1977) argued that for river systems, this increased meandering would be in equilibrium with flow velocities just capable of transporting the sediment loads.

In the specific case of Mendocino Channel, there is a high likelihood that the valley slope has varied during the Holocene and will continue to change in time because of the interplate tectonic collision at the nearby MTF and because of channel blockages caused by debris flows from the adjacent Mendocino Ridge. Although major channel migration through time is not observed in the available seismic records, there is some indication that channels may have been abandoned (see lines 02 and 07 in Figure 11–4). In addition, GLORIA imaged a channel section abutting the MTF with anomalously high backscatter (Figure 11–3) that might represent an older channel route. Continued tilting of the seafloor in this region resulting from tectonic activity will likely alter the valley slope and induce changes in the meander development within Mendocino Channel. Additionally, debris flows like that shown in Figure 11–3, possibly triggered by large seismic events along the MTF, could block the channel and induce further meander development.

Conclusions

A detailed geological investigation of a section of Mendocino Channel, originally observed on GLORIA sonographs, reveals meanders that are similar in geometry and structure to those found on other deep-sea channel systems (Flood and Damuth 1987; O'Connell et al. 1991). Data from water-gun and 3.5-kHz seismic records were used to determine a channel slope and sinuosity of 4.3 and 1.7 m/km, respectively.

These values are similar to those determined for the Amazon upper fan channels. The relationships between meander wavelength and channel width or radius of curvature are also typical of those for the Amazon channels and for large rivers.

Within the limits of conventional ^{14}C age determinations, age dates for the turbidites from the channel floor overlap ages of large earthquakes in the southern Cascadia Subduction Zone that were determined from onland studies of late Holocene paleoseismicity in this region. Modern seismicity of Cascadia Subduction Zone and along the MTF has been well documented (e.g., Smith et al. 1993), and earthquakes with moment magnitudes in excess of 8.4 are thought to have occurred in the recent past in the region (Clarke and Carver 1992; Valentine et al. 1992). We have not demonstrated positive proof that the turbidity currents that deposited the Mendocino channel turbidites were triggered by great earthquakes, but the correspondence of the ^{14}C ages of the turbidites and the ages of major earthquake-induced geological effects along the coast, and the general agreement that great earthquakes have occurred in this region, provide strong circumstantial evidence to support this hypothesis.

The concept that meander development is related to valley and channel slope (Schumm 1977) suggests that future turbidity currents that flow through Mendocino Channel will likely continue to produce changes in the meander geometry. This forecast follows from the indications that regional uplift and tilt of the southern Gorda Plate is continuing along the MTF (Masson et al. 1988). The uplift will contribute to a change in the valley slope of the channel and possibly trigger alterations to the meander geometry by future turbidity currents. In addition, changes in the course of the channel might be induced by debris avalanches and flows that originate from the northern face of Mendocino Ridge and block the channel path.

Acknowledgments

The authors gratefully acknowledge Homa Lee and Robert Kayen of the U.S. Geological Survey for their capable and untiring assistance during the research cruise on the R/V *Samuel P. Lee* to collect the seismic records and core samples. We also thank Douglas Masson of the Institute of Oceanographic Sciences, Great Britain, for his collaboration during the collection of the GLORIA data. We acknowledge the efforts by Paula Quintero of the U.S. Geological Survey for providing the identifications of the microfossils in the core samples. We are particularly thankful to Samuel Clarke of the U.S. Geological Survey for his assistance during the GLORIA cruise and especially for his sound and illuminating review of the manuscript. He and Paul Carlson, also with the U.S. Geological Survey, provided many helpful suggestions to improve the quality of the manuscript.

References

Adams, J. 1990. Paleoseismicity of the Cascadia subduction zone: Evidence from turbidites off the Oregon-Washington margin. *Tectonics* 9: 569–83.

Bruns, T. R., Stevenson, A. J., and Dobson, M. R. 1992. GLORIA investigation of the Exclusive Economic Zone in the Gulf of Alaska and off southeast Alaska: R/V *Farnella* cruise F7-89-GA, June 14–July 13, 1989. *U.S. Geological Survey Open-File Report 92-317,* 16p. Reston, Va.: USGS.

Cacchione, D. A., Drake, D. E., Clarke, S. H., and Masson, D. G. 1984. Physiography of the sea-floor off northern California and southern Oregon (abs.), *Trans. Am. Geophys. Union (EOS)* 65(45): 1083.

Chase, T. E., Wilde, P., Normark, W. R., Evenden, G. I., Miller, C. P., Seekins, B. A., Young, J. D., Grim, M. S., and Lief, C. J. 1992. Map showing bottom topography of the Pacific continental margin, Cape Mendocino to Point Conception. *U.S.G.S. Miscellaneous Investigations Series I-2090-C.* Reston, Va.: USGS.

Clarke, S. H., Jr. 1992. Geology of the Eel River basin and adjacent region: Implications for late Cenozoic tectonics of the southern Cascadia subduction zone and Mendocino triple junction. *Am. Assoc. Petrol. Geol. Bull.* 76: 199–224.

Clarke, S. H., Jr., and Carver, G. A. 1992. Late Holocene tectonics and paleoseismicity, southern Cascadia subduction zone. *Science* 255: 188–92.

Couch, R. 1980. Seismicity and crustal structure near the north end of the San Andreas fault system. In *Studies of the San Andreas Fault Zone in Northern California,* R. Streitz and R. Sherburne, pp. 131–7. Special Report 140. Sacramento, CA: California Division of Mines and Geology.

Damuth, J. E., Kolla, V., Flood, R. D., Kowsmann, R. O., Monteiro, M. C., Gorini, M. A., Palma, J. J., and Belderson, R. H. 1983. Distributary channel meandering and bifurcation patterns on the Amazon deep-sea fan as revealed by long-range side-scan sonar (GLORIA). *Geology* 11: 94–8.

Dengler, L., McPherson, R. and Carver, G. 1992. Historic seismicity and potential source areas of large earthquakes in North Coast California. In *Pacific Cell, Friends of the Pleistocene Guidebook for the Field Trip to Northern Coastal California,* pp. 112–8. R. M. Burke and G. A. Carver (eds.) Humboldt State University, Arcata, CA.

EEZ-SCAN 84 Scientific Staff 1986. Atlas of the Exclusive Economic Zone, Western Conterminous United States. *U.S.G.S. Miscellaneous Investigations Series I-1792,* 152 pages. Reston, Va.: USGS.

Flood, R. D., and Damuth, J. E. 1987. Quantitative characteristics of sinuous distributary channels on the Amazon deep-sea fan. *Geol. Soc. Am. Bull.* 98: 728–38.

Heaton, T. H., and Kanamori, H. 1984. Seismic potential associated with subduction in the northwestern United States. *Bull. Seismol. Soc. Am.* 75: 933–42.

Leopold, L. B., and Wolman, M. G. 1960. River meanders. *Geol. Soc. Am. Bull.,* 71: 769–94.

Masson, D. G., Cacchione, D. A., and Drake, D. E. 1988. Tectonic evolution of Gorda Ridge inferred from sidescan sonar images. *Mar. Geophys. Res.* 10: 191–204.

Menard, H. W. 1955. Deep-sea channels, topography, and sedimentation. *Am. Assoc. Petrol. Geol. Bull.* 39: 236–55.

Menard, H. W. 1964. *Marine Geology of the Pacific.* New York: McGraw-Hill, 271 p.

NOAA (National Oceanic and Atmospheric Administration) 1969. *National Ocean Survey Bathymetric Chart NK 10-10.* Washington, D.C.: NOAA.

Normark, W. R. 1978. Fan valleys, channels and depositional lobes on modern submarine fans: Characters for recognition of sandy turbidite environments. *Am. Assoc. Petrol. Geol. Bull.* 62: 912–31.

O'Connell, S., Normark, W. R., Ryan, W. B. F., and Kenyon, N. H. 1991. An entrenched thalweg channel on the Rhone fan: An interpretation from a SeaBeam and SeaMARC I survey. In *From Shoreline to Abyss: Contributions in Marine Geology in Honor of Francis Parker Shepard,* ed. R. H. Osborne, pp. 259–70. Society of Economic Paleontologists and Mineralogists Special Pub. No. 46. Tulsa, OK.

Riddihough, R. P. 1980. Gorda plate motions from magnetic anomaly analysis. *Earth Plan. Sci. Ltrs.* 51: 163–70.

Schumm, S. A. 1977. *The Fluvial System.* New York: John Wiley & Sons, 338 p.

Shepard, F. P., and Dill, R. F. 1966. Submarine Canyons and Other Sea Valleys. Chicago: Rand McNally, 381 p.

Silver, E. A. 1971. Tectonics of the Mendocino Triple Junction. *Geol. Soc. Am. Bull.* 82: 2965–78.

Smith, S. W. Knapp, J. S., and McPherson, R. C. 1993. Seismicity of the Gorda plate, structure of the continental margin, and an eastward jump of the Mendocino triple junction. *J. Geophys. Res.* 98: 8153–71.

Twichell, D. C., Kenyon, N. H. Parson, L. M., and McGregor, B. A. 1991. Depositional patterns of the Mississippi fan surface: Evidence from GLORIA-II and high resolution seismic profiles. In *Seismic Facies and Sedimentary Processes of Submarine Fans and Turbidite Systems.* eds. P. Weimer and M. H. Link, pp. 349–63. New York: Springer-Verlag.

Valentine, D., Vick, G. S., Carver, G. A., and Manhart, C. S. 1992. Late Holocene stratigraphy and paleoseismicity, Humboldt Bay, California: In *Pacific Cell, Friends of the Pleistocene Guidebook for the Field Trip to Northern Coastal California,* eds. R. M. Burke, and G. A. Carver, pp. 182–7. Arcata, CA: Humboldt State University

Vallier, T. L., Harold, P. J., and Girdley, W. A. 1973. Provenances and dispersal patterns of turbidite sand in Escanaba Trough, Northeastern Pacific Ocean. *Mar. Geol.* 15: 67–87.

Wilson, D. S. 1986. A kinematic model for the Gorda deformation zone as a diffuse southern boundary of the Juan de Fuca plate. *J. Geophys. Res.* 91: 10259–69.

12 The morphology, processes, and evolution of Monterey Fan: A revisit

James V. Gardner,[1] Robert G. Bohannon,[1] Michael E. Field,[1] and Douglas G. Masson[2]

[1]U.S. Geological Survey, Menlo Park, California

[2]Institute of Oceanographic Sciences, Southampton, United Kingdom

Abstract

Long-range (GLORIA) and mid-range (TOBI) sidescan imagery and seismic-reflection profiles have revealed the surface morphology and architecture of the complete Monterey Fan. The fan has not developed a classic wedge shape because it has been blocked for much of its history by Morro Fracture Zone. The barrier has caused the fan to develop an upper-fan and lower-fan sequence that are distinctly different from one another. The upper-fan sequence is characterized by Monterey and Ascension Channels and associated Monterey Channel-levee system. The lower-fan sequence is characterized by depositional lobes of the Ascension, Monterey, and Sur-Parkington-Lucia systems, with the Monterey depositional lobe being the youngest. Presently, the Monterey depositional lobe is being downcut because the system has reached a new, lower base level in the Murray Fracture Zone.

A five-step evolution of Monterey Fan is presented, starting with initial fan deposition in the Late Miocene, about 5.5 Ma. This first stage was one of filling bathymetric lows in the oceanic basement in what was to become the upper-fan segment. The second stage involved filling the bathymetric low on the north side of Morro Fracture Zone, and probably not much sediment was transported beyond the fracture zone. The third stage witnessed sediment being transported around both ends of Morro Fracture Zone and initial sedimentation on the lower-fan segment. During the fourth stage Ascension Channel was diverted into Monterey Channel, thereby cutting off sedimentation to the Ascension depositional lobe. The fifth stage occurred when Monterey Canyon captured Ascension Canyon on the upper-fan segment and then the entire Monterey sedimentation system found a new base level in the Murray Fracture Zone and began to downcut the Monterey depositional lobe.

This chapter is not subject to US copyright.

Introduction

Submarine fans attracted the attention of geologists in the early stages of marine geology, not only because of their use as a modern analog of ancient hydrocarbon reservoirs, but also because they often contain intriguing features. Large erosional submarine canyons commonly evolve down-fan into constructional meandering channels that are bordered by levee systems and end in depositional lobes. These characteristic fan features, as well as many others, were deduced from poorly navigated first-generation echograms and single-channel seismic-reflection profiles and a sparse collection of rather randomly collected cores and bottom samples. However, technologies developed in the 1970s and 1980s, especially digital sidescan sonars and Global Positioning System (GPS) navigation, provided the tools necessary to map the surface of a complete fan system accurately and systematically and, when integrated with seismic-reflection profiles, to develop a comprehensive view of the architecture and history of an individual fan.

In 1984 the U.S. Geological Survey mapped the entire U.S. West Coast Exclusive Economic Zone using GLORIA long-range sidescan sonar, digital seismic-reflection profiling, 3.5- and 10-kHz high-resolution profiling, and magnetics. A portion of this survey covered Monterey Fan (Figure 12–1A). Here we interpret this large database of sidescan sonar imagery and seismic data and additional new data from Monterey Fan to extend and elucidate the morphologies, processes, and development of this large depositional feature.

Data set and methods

The data sets used for this study include the following: (1) digitally reprocessed and mosaicked GLORIA sidescan images from the EEZ-SCAN cruises (EEZ-SCAN 84 Scientific Staff 1986); (2) two-channel digital 160-in^3 airgun seismic-reflection profiles (EEZ-SCAN 84 Scientific Staff 1986), 3.5-kHz high-resolution profiles simultaneously col-

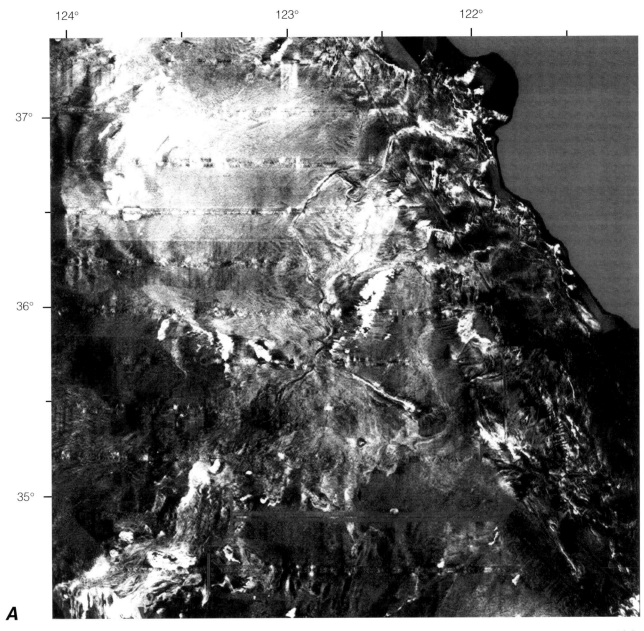

Figure 12–1. A, Digital GLORIA mosaic of Monterey Fan (modified from EEZ-SCAN 84 (1986). Lighter areas represent higher backscatter.

lected with the GLORIA imagery and seismic data, and 3.5-kHz profiles from local surveys before the EEZ-SCAN cruises as well as subsequent cruises in 1989 and 1990 (Gardner et al. 1991a); and (3) TOBI (Murton et al. 1992) deep-towed 30-kHz digital sidescan sonar images collected in 1990 (Gardner et al. 1991b; 1991c; Masson et al. 1995). Line spacings for the EEZ-SCAN survey are less than 30 km (Figure 12–1B), and the survey was navigated with GPS and Loran-C. The TOBI sidescan survey was navigated by GPS and Loran-C as well as with a seafloor acoustic-transponder net that allowed us to locate the fish accurately (±5 m) while it was within the net. TOBI surveys were con-

ducted to provide complete coverage of local areas identified on the GLORIA imagery.

The GLORIA sidescan sonographs represent our most extensive data set. All the sonographs in this report have been reprocessed from the original raw data and have, in addition, been digitally mosaicked using the schemes outlined by Chavez (1986) and Chavez et al. (this volume). The GLORIA mosaics were first interpreted by assigning purely descriptive sidescan-backscatter codes such as bright-speckled, rather than any genetic names, such as debris-flow deposits, to each of the identified backscatter patterns. The entire surface of Monterey Fan was mapped in this way. The

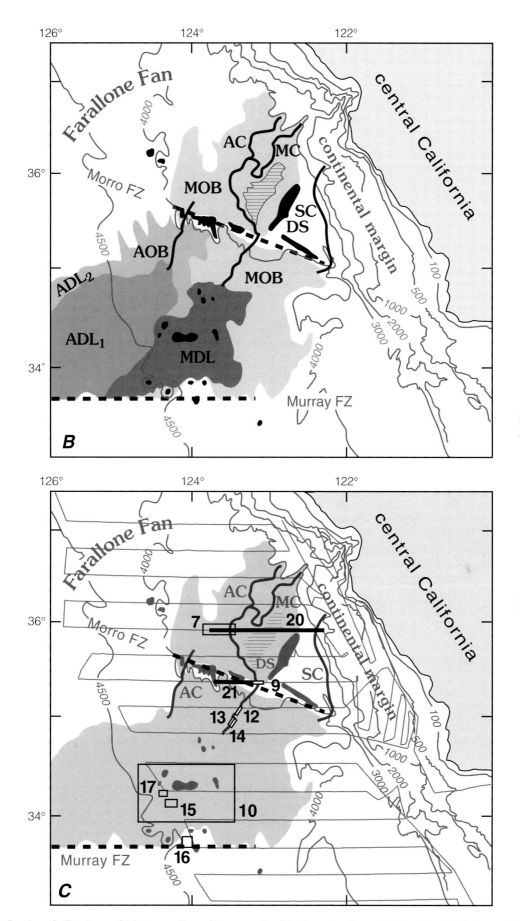

Figure 12–1. *Continued.* (B), Area of Monterey Fan with generalized bathymetry (in meters) and major features. DS = Davidson Seamount, FZ = Fracture Zone, AC = Ascension Channel, ADL = Ascension depositional lobe, AOB = Ascension overbank deposits, MC = Monterey Channel, MDL = Monterey depositional lobe, MOB = Monterey overbank deposits, SC = Sur-Parkington-Lucia Channel. Heavy dashed line is trend of fracture zone, cross-hatched area is Sur Slide. Dark-shaded enclosed features are seamounts. C, Locations of survey tracks, and seismic and imagery in subsequent figures.

second phase of mapping involved annotating the ship's tracklines with interpretations from the seismic-reflection and 3.5-kHz profiles for relief features (channels, levees, basement outcrops, etc.) and reflection characteristics (strong surface, no subbottom, mushy, etc.) following the methodology of Damuth (1975). These two different interpretations were combined and overlaid on the sonograph mosaic, and a final interpretation was made using geological intuition and traditional marine-geological conventions to generate a terrain map. The terrain map forms the basis for the interpretation of the surface morphology of the fan.

Terminology

Any attempt to describe or discuss a submarine fan runs headlong into a tangled web of various fan nomenclatures. For the past thirty years or more, numerous workers on submarine fans, whether interpreting rocks as ancient fan deposits or interpreting seismic data or cores from modern fans, have defined criteria for subenvironments of a particular fan (for example Normark 1970a; Mutti and Ricci Lucchi 1972; Walker 1978; Stow 1981; Bouma 1983/1984; Mutti and Normark 1987; Damuth et al. 1988; Shanmugam and Moiola 1988; Weimer and Link 1991; and many others). The names and criteria used by various authors often conflict with those used by others and, adding to the confusion, there is a growing realization that fans are far more spatially and temporally complex than the models have been able to emulate. Although many efforts and successes have been made in understanding fan morphology and origin, these efforts are not completely applicable to large submarine fans on the floors of ocean basins, so we here make an effort to avoid placing Monterey Fan into any particular fan model.

Even with more than thirty years of colorful literature behind us, we are still faced with the dilemma faced by those earlier workers: how to meaningfully subdivide Monterey Fan with descriptive names and discuss the features and morphology that make one area of the fan different from adjacent areas without getting caught in the terminology trap. Terms such as depositional lobe, channel, gully, levee, channel-levee system, distributary, tributary, and so on are widely used descriptive terms in fan studies, and we reluctantly employ them here. However, we remind the reader that these terms have had varied use and meanings in the literature. We define the terms we use as they apply to Monterey Fan so that others who may want to compare our observations with those from other fans can do so, albeit inevitably with some difficulty. We have endeavored to make our definitions descriptive and straightforward.

Previous studies

One of the earliest studies of Monterey Fan was by Shepard (1966) who collected a series of bathymetric profiles

across the Monterey Channel meander, a pronounced loop in the trend of the channel that occurs about a third of the way down-fan (Figure 12–1A). Later studies (Normark 1970a, 1970b, 1978; Hess and Normark 1976; Normark et al. 1983/1984) concentrated on the sediment history and morphology of the upper half of the fan, and they included efforts to define the extent of the fan deposits based on seismic-reflection profiles. These studies describe the capture of Ascension Channel by Monterey Channel and the abandonment of Monterey East Channel (Normark 1970b). A small portion of a field of bedforms west of Ascension Channel was interpreted as migrating sediment waves by Normark et al. (1980). Normark et al. (1983/1984) also interpreted the role played by Morro Fracture Zone (called Chumash Fracture Zone by them) in the growth of Monterey Fan.

Studies of Monterey Fan for the past decade have concentrated on understanding various phenomena and features at particular locations. These include investigations of features within Monterey Canyon and the Monterey Channel meander from submersibles (Embley et al. 1990), Sur Slide from GLORIA imagery and cores (Normark and Gutmacher 1988), and using cores, high-resolution sidescan sonar, and 3.5-kHz profiles to describe the depositional lobe of the fan (Gardner et al. 1991a, 1991b, 1991c).

Geological evolution

Monterey Fan covers about 100,000 km^2, and has an average thickness of about 0.5 km, giving a volume of fan sediments in excess of 50,000 km^3. The fan rests on oceanic crust of the Pacific Plate and it is bounded on the east by the slope of the central California margin, on the south by Murray Fracture Zone, and on the north by Farallone Fan (Figure 12–1). Monterey Fan thins and eventually merges with the basin plain on the west, more than 400 km from the shoreline. The fan has developed on a broad expanse of seafloor with considerable relief, including major fracture zones, basement ridges, and seamounts. The continental margin to the east of the fan is cut by a series of strike-slip faults that have moved the fan progressively northward relative to the adjacent land (Atwater 1989; McCulloch 1989; Greene, Clarke, and Kennedy 1991). Because the regional basement and tectonics have had an important impact on the development of Monterey Fan, a thorough review of them follows.

AGE OF THE SUBFAN CRUST

Monterey Fan lies west of Davidson Seamount, which marks an ancient segment of the Farallon/Pacific spreading ridge, which was abandoned at 19 Ma (Lonsdale 1991) leaving a small relict portion of the Farallon Plate to its east (Figure 12–2). Davidson Seamount is a large linear high that has acted as a barrier to transport of sediment derived on the continent. The oceanic crust beneath the fan to the southwest of the seamount formed 20 to 33 my ago as dated by magnetic anom-

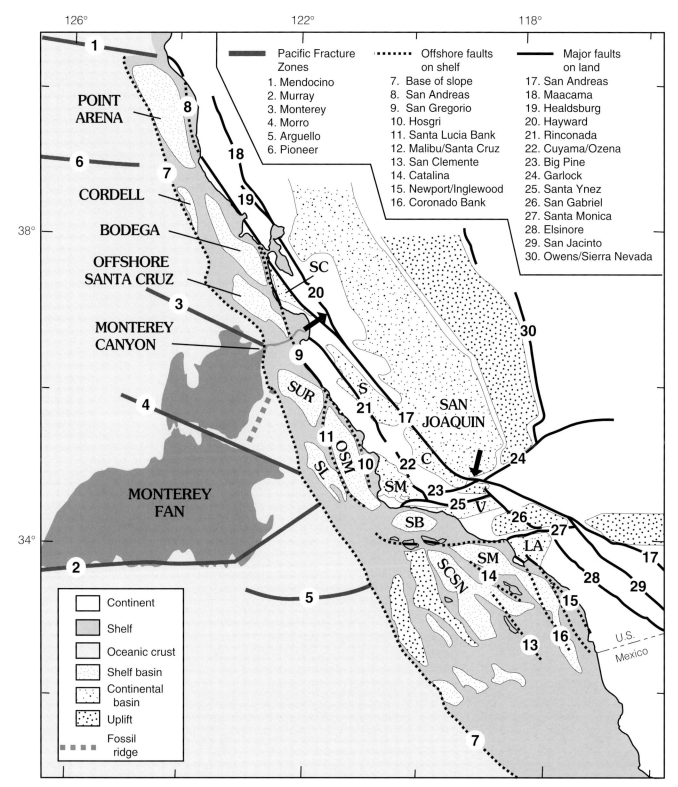

Figure 12–2. Map showing present position of Monterey Fan relative to (1) largest fracture zones in Pacific Plate, (2) major faults with known or suspected right-lateral slip in coastal and offshore California, and (3) basins on land and on the shelf that contain thick accumulations of sediment correlative in age to Monterey Fan deposits. Two arrows mark the maximum Neogene offset on the San Andreas Fault (Stanley 1987; Irwin 1990). The northern arrow on the west side of the fault indicates the projected intersection of Monterey Canyon with the fault trace. The southern arrow shows the pre-Miocene position of Monterey Canyon relative to uplifts on east side of the San Andreas Fault. Faults and fracture zones are numbered and listed on diagram. Basins are named or abbreviated as follows: SC = Santa Cruz, OSM = Offshore Santa Maria, SL = Santa Lucia, S = Salinas, C = Caliente, SM = Santa Maria, SB = Santa Barbara, V = Ventura, LA = Los Angeles, SM = Santa Monica, SCSN = Santa Cruz/San Nicolas.

alies 6 through 12 (Atwater and Severingus 1989). The youngest subfan crust, anomaly 6 is adjacent to the west side of the Davidson Seamount on the north side of the Morro Fracture Zone. The crust beneath the far southwestern reaches of the fan formed during anomaly 12.

The oldest continentally derived sediments in the Monterey Fan must be younger than 20 Ma in age because the Pacific-Farallon spreading ridge was active between the Mendocino and Murray Fracture Zones during the 20- to 30-Ma time period (Stock and Molnar 1988). The active ridge probably was a physiographic barrier that confined continentally derived sediment to the subducting Farallon Plate to its east. By 26 to 20 Ma the Farallon Plate had locally fragmented into the Monterey Plate north of Morro Fracture Zone and the Arguello Plate to its south (Lonsdale 1991); however, there is no evidence that this fragmentation resulted in breaches in the high ground of the spreading and fracture-zone system through which sediment could have passed to the west. Consequently, the only sedimentation at the fan site during the 30- to 20-Ma time period was probably pelagic. The Pacific and North American Plates were in contact between the Mendocino and Murray Fracture Zones after 25 to 19 Ma (Atwater 1970; Atwater and Molnar 1973; Engebretson, Cox, and Gordon 1985; Stock and Molnar 1988; Atwater and Severingus 1989; and Lonsdale 1991), so clastic sediment might have reached the fan site any time after that.

NORTHWARD TRANSLATION OF THE MONTEREY FAN SITE RELATIVE TO NORTH AMERICAN SOURCES

Plate reconstructions (Atwater 1970; Atwater and Molnar 1973; Engebretson et al. 1985; Stock and Molnar 1988; and Atwater and Severingus 1989) conclusively indicate that the part of the Pacific Plate that includes Monterey Fan was 8.5° to 10° south of its present location 33 my ago when the oldest oceanic crust beneath the fan was formed. The site of Monterey Fan has migrated 950 to 1250 km northward, parallel to the California coast, to its present position, because seafloor spreading ceased at 19 to 20 Ma. About 345 to 370 km of this northward translation can be accounted for by Neogene displacement on the San Andreas Fault system (Huffman 1972; Powell 1981; Stanley 1987). If Huffman's (1972) and Stanley's analyses (1987) are correct, then at least 249 to 259 km of that displacement occurred in the last 10 my (Figure 12–3). Most of the differential Pacific/North American translation must have occurred offshore, to the west of the San Andreas system. We infer that a large part of the differential translation occurred at the base of the continental slope at the site of the ancient subduction system before 10 Ma.

BASEMENT POSITION AND POTENTIAL
SEDIMENT SOURCES

Before describing the morphology and development of the fan, it is also important to review the potential sources of

sediment that were available while the oceanic crust was in a position to receive fan sedimentation. The potential sources and positions of the crust are shown in reconstructions of 20, 11, 5.5, and 0 Ma (Figure 12–3).

20 Ma. The site of Monterey Fan was offshore of Los Angeles Basin, outboard of the California Continental Borderland and was adjacent to the southern end of San Joaquin Basin across the San Andreas Fault system 20 my ago (Figure 12–3). It is unlikely that much sediment was transported across the shelf during this time because the basin-ridge physiography that characterizes the borderland province probably started to develop about 20 my ago (Vedder 1987) and the numerous basins in that province would have trapped most of the sediment on the shelf. Vedder (1987) described a large area of uplift and erosion on the outer continental borderland, but that feature did not develop until the Late Miocene. By that time, Monterey Fan was too far to the north to receive sediment derived from this uplift.

11 Ma. The 11-Ma model places Monterey Fan adjacent to the Santa Maria Basin, slightly offset from the southern end of San Joaquin Basin (Figure 12–3). There probably was not much clastic deposition at the Monterey Fan site during the period between 20 and 5.5 Ma because Monterey Canyon was too far to the north of the future site of the fan head to provide sediment to the fan. Also, the fan was moving northward past numerous basins on the continental shelf, which probably trapped most incoming sediment. The Sur, Santa Lucia, Santa Maria, and Santa Barbara Basins all trapped large volumes of clastic sediment during this time period and there is no evidence that any of them completely filled to overflowing to allow sediment to spill into the Pacific Basin.

5.5 Ma. Significant volumes of clastic sediment began to reach Monterey Fan around 5.5 Ma. Uplift of the Sierra Nevada Range was under way (Huber 1981) and large volumes of sediment were shed into the southern part of the San Joaquin Basin at the end of the Miocene and beginning of the Pliocene (Goodman and Malin 1992). The Sierra Nevada offers the largest and most proximal potential source during this time period. Uplift of the central California Coast Ranges had not begun in the Late Miocene and Early Pliocene. This region was the site of widespread deposition of shallow marine and fluvial deposits. In addition, large areas in the vicinity of the present Transverse Ranges were also covered with sediment well into the Late Miocene (Jennings 1977). A Sierran source requires complete filling and overflow of the southern San Joaquin Basin and northwestward transport of the sediment parallel to the San Andreas Fault to the heads of Monterey and Sur Canyons (Figure 12–3), either of which could have channeled sediment to the abyssal ocean floor. At present, Sur Canyon funnels only sediment that is locally derived in the Santa Lucia Range,

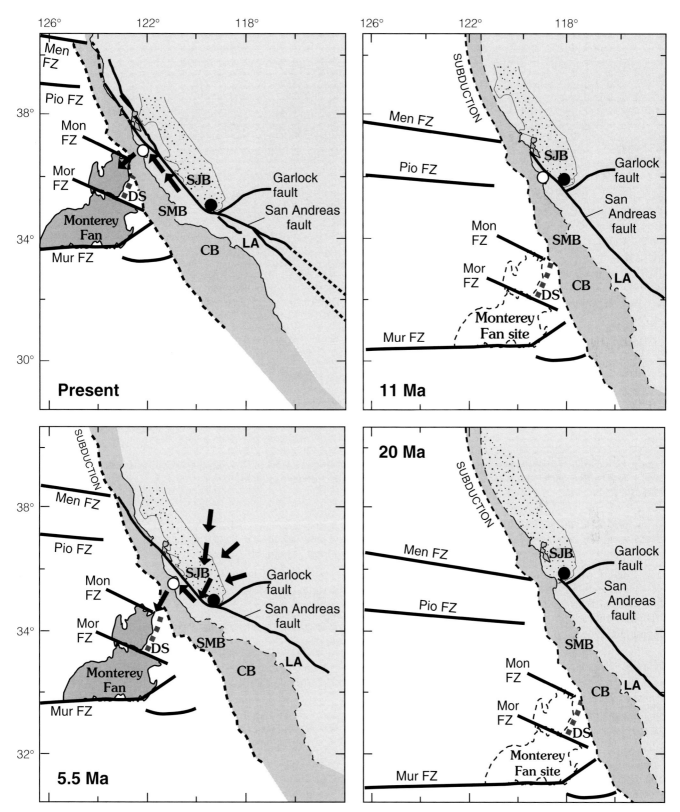

Figure 12–3. Tectonic reconstructions of the area of Monterey Fan and adjacent continent at 20, 11, 5.5, and 0 Ma. Fracture zones are abbreviated Mur FZ = Murray Fracture Zone, Mor FZ = Morro Fracture Zone, Mon FZ = Monterey Fracture Zone, Pio FZ = Pioneer Fracture Zone, Men FZ = Mendocino Fracture Zone; basins are abbreviated CB = California Borderland Basins, SMB = Santa Maria Basin, SJB = San Jose Basin; DS = Davidson Seamount; and sediment transport paths are indicated with arrows.

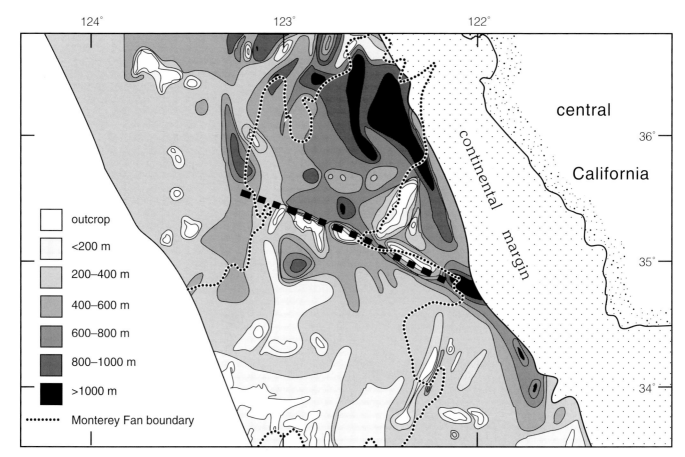

Figure 12–4. Isopach map of total sediment thickness of region of Monterey Fan reflecting the varied basement topography. Heavy dashed line represents Morro Fracture Zone. Adapted from Gardner et al. (1992).

but it may have been more important as a channel for regionally derived sediment 5.5 my ago before the uplift of the mountains. Because there are no large submarine canyons between the Big Sur area and Arguello Canyon to the south of Santa Maria Basin (Figure 12–2), there was no direct route for sediment through the Sur, Santa Lucia, and offshore Santa Maria Basins, which were located between the fan and its Sierran source at that time.

Present.   Monterey Fan is presently aligned with and fed by Monterey and Sur Canyons (Figure 12–3). Monterey Canyon heads in Monterey Bay where sediment is brought into the bay by the Salinas River and by littoral drift. The Salinas River drains a large area of the central coast ranges to the south and channels sediment northwestward parallel to the San Andreas Fault. The Salinas drainage basin is confined to the coast ranges whose recent uplift has isolated Monterey Fan from more distal sources, such as the Sierra Nevada. Sur Canyon heads at the steep coastal region along the west side of the Santa Lucia Range, which was uplifted in the last few million years. Ascension Canyon must be supplied sediment by littoral drift, but only during eustatic lows in sea level. None of the sediment shed off the present

Sierra Nevada Range reaches Monterey Fan; all that sediment enters the Pacific Ocean through San Francisco Bay and is fed to Farallone Fan.

Morphology of Monterey Fan

Major geomorphic features

The major geomorphic features that comprise Monterey Fan include (1) the continental margin with submarine canyons and related gullies; (2) an upper-fan segment with several channels, Sur Slide, and a large bedform field; and (3) a lower-fan segment with a series of overlapping depositional lobes, several channels, isolated seamounts, and a complicated system of tributaries and distributaries (Figure 12–1C). Morro Fracture Zone separates the upper-fan segment from the lower-fan segment, and Murray Fracture Zone marks the southward limit of the fan. For much of the fan's history, its development was blocked by Morro Fracture Zone (Figure 12–4), a deep (>700 m) fracture-zone basin and related basement high (more than 2000 m above regional basement level). One consequence of all this disruption to normal progradation of the fan is that terms such as upper fan, mid-

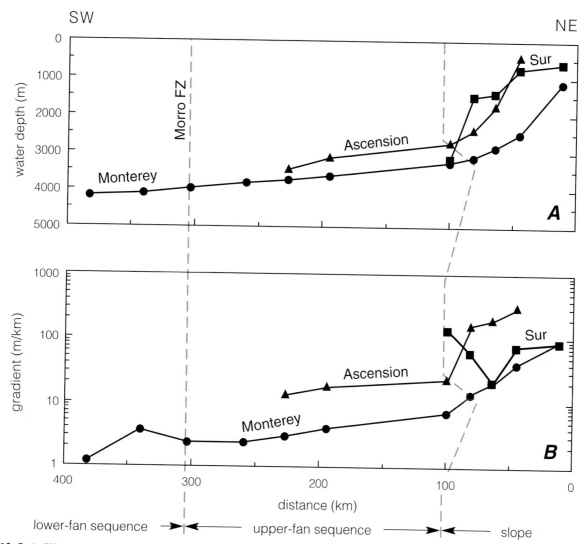

Figure 12–5. A, Water depths of axes of Monterey, Sur, and Ascension Canyons and channel floors measured on 3.5-kHz profiles. B, Channel gradients calculated from channel lengths measured on GLORIA images.

dle fan, and lower fan, with all their variously interpreted genetic and morphologic connotations, are not particularly useful in discussions about Monterey Fan.

THE CONTINENTAL MARGIN

The continental margin (shelf and slope) off central California has a very narrow (<2 to ~10 km) and relatively featureless shelf and a steep (7° to 11°) and very irregular slope. The margin extends from the shoreline to depths of 3000 m. Monterey Bay is a 40 km (N-S) by 10 km (E-W) reentrant of the coastline that forms a shallow, relatively broad portion of shelf. The shelf break north of Monterey Bay is incised by numerous submarine canyons, the most important of which is Ascension Canyon. The entire shelf in the area of Monterey Bay is incised by the large Monterey Canyon and its Soquel and Carmel branches, and by Sur Canyon. Farther south Partington, Lucia, and several unnamed canyons incise the shelf break.

The floor of Monterey Canyon varies in width between 0.2 and 2.5 km and has gradients that decrease down-canyon from 94 to 20 m/km (Figure 12–5). Ascension Canyon is somewhat narrower (<1 km) and steeper with gradients that gradually decrease from 311 to 167 m/km. Sur Canyon, whose width varies between 0.2 and 2.5 km, has gradients that range from 136 to 20 m/km but in a very irregular pattern. The variable widths of the canyons are caused by variation in the relative erodability of the rock types (siltstone, sandstone, limestone, granodiorite, and metamorphic rocks) that the canyons incise (Martin and Emery 1967).

UPPER-FAN SEGMENT OF MONTEREY FAN

Monterey Fan can be divided into upper-fan and lower-fan segments with Morro Fracture Zone as the dividing line. We use the term upper-fan segment to represent the area of the fan between the slope and Morro Fracture Zone (Figure 12–1C). This segment of the fan is in water depths between

W   E

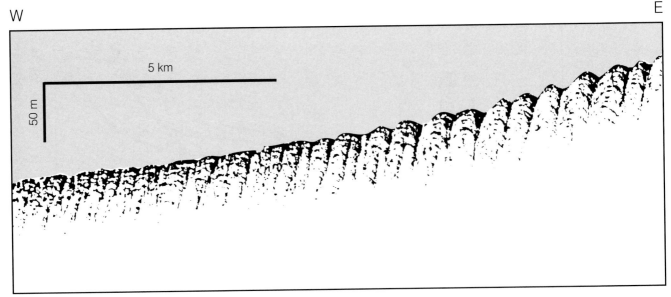

Figure 12–6. 3.5-kHz profile showing extensive regular bedforms developed adjacent to Ascension Channel. Location of profile shown in Figure 12–7. These have been previously interpreted as migrating sediment waves created by overbank flow of turbidity currents (Normark et al. 1980).

about 3000 m and 3900 m. The upper-fan segment is divided into two distinct provinces. The first province is a vast gently sloping region west of Ascension-Monterey Channels that appears on sonar images as a region of intermediate microscale speckled backscatter. The 3.5-kHz profiles of this area show irregular to hummocky seafloor with parallel to subparallel subbottom reflectors that indicate large-scale bedforms (Figure 12–6). Sediment from this bedform area merges to the northwest with bright mesoscale ropey backscatter from sediments of Farallone Fan, a separate small submarine fan to the north. Although this difference in backscatter from the two fans is visible on the images as a distinct contact (Figure 12–1A), the interfingering of the two systems cannot be identified on either 3.5-kHz or seismic profiles. The smaller of the two provinces of the upper-fan segment is located between Monterey Channel and Davidson Seamount and appears on the GLORIA imagery as intermediate microscale speckled backscatter. The 3.5-kHz profiles show sediment blanketing irregular seafloor topography, but regular, large-scale bedforms do not occur.

Monterey Channel traverses the upper-fan segment of the fan with thalweg gradients from 7 m/km near the slope to 2.4 m/km just north of Morro Fracture Zone (Figure 12–5B). Monterey Channel has captured Ascension Channel in the upper portions of the upper-fan segment (Normark 1970b; Hess and Normark 1976; Greene, Stubblefield, and Theberge 1989). Upstream from the capture, Ascension Channel has a gradient of 17 to 13 m/km, which is somewhat steeper than Monterey Channel, but has a similar rate of change progressing down-fan (Figure 12–5B). Monterey Channel is deeply incised into the upper-fan segment, has a

prominent levee on its northwestern side, and has terraces on its walls. The channel depth has a maximum relief of more than 500 m just southwest of the slope, but relief rapidly decreases downfan to 65 m just north of Morro Fracture Zone. The channel has a channel-floor width of about 1.5 km until it approaches Morro Fracture Zone, where the channel floor narrows to less than 100-m wide. Monterey Channel has a gentle channel sinuosity of 1.6, almost identical to that of the youngest channel on the midregion of Mississippi Fan (Kastens and Shor 1985, 1986; Twichell et al. 1992) and many of the channels on Amazon Fan (Damuth et al. 1988). This sinuosity value places Monterey Channel in the class of meandering channels of fluvial systems (Leopold and Wolman 1957). The relief of Ascension Channel is less than 100 m except above its junction with Monterey Channel, where it is incised more than 200 m deep. Ascension Channel lacks an obvious constructional levee. The remnants of the original course of Monterey Channel down-channel from the point of capture of Ascension Channel is a distinct hanging channel (Monterey East Fan Valley of Normark 1970b).

The floors of all the channels of the upper-fan segment have dark microrelief, smooth backscatter on GLORIA images (Figure 12–1A). The low backscatter of the channel is similar to that found in the Equatorial Atlantic midocean channel (Belderson and Kenyon 1980) and the Bering Channel (Bering Sea EEZ-SCAN Scientific Staff 1991; Karl et al. this volume), but is in contrast to high backscatter of channels from Amazon Fan (Damuth et al. 1983), Mississippi Fan (Twichell et al. 1991), and Mendocino Channel (Cacchione et al. this volume). The differences in backscat-

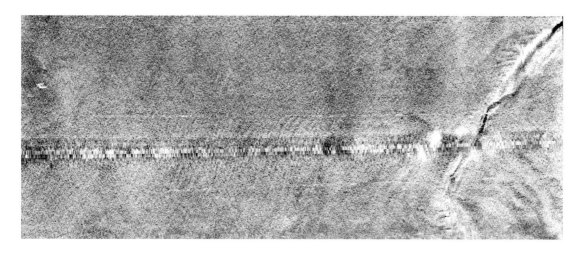

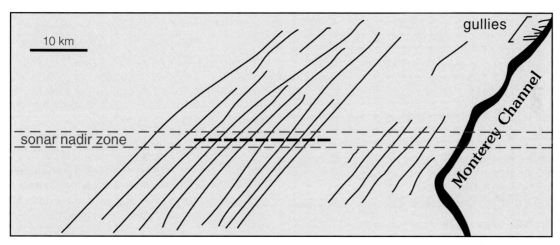

Figure 12–7. GLORIA image and interpretation showing bedforms in Figure 12–6. Lighter areas on sonograph represent higher backscatter. Location of image shown as boxed area labeled 7 on Figure 12–1C.

ter result from differences in acoustic impedance of the upper few meters of the channel fill, but do not necessarily indicate primarily sand in the high-backscatter channels and primarily mud in the low-backscatter channels (Gardner et al. 1991a).

The 1200-km² field of bedforms within the upper-fan segment of the fan is confined to the area west of Monterey Channel. The 3.5-kHz profiles show these bedforms are generally asymmetrical with the steep side oriented upslope, suggesting upslope migration (Figure 12–6). Individual bedforms can be followed on the imagery for several 10s of km and trend subparallel to the fan channel (Figure 12–7). The bedforms have wave heights (measured from top of crest to bottom of trough) ranging from less than 2 to 10 m, but the average is about 3 m. Wavelengths range from less than 200 to 2800 m (corrected for trackline azimuth versus bedform trends from the imagery) with an average of about 400 m (Figure 12–8). The areal distribution of these migrating bedforms demonstrates that sediments have banked against the north side of Morro Fracture Zone and have been deflected

to the west and, to some extent, to the north. Cores from the bedforms collected predominately silt and mud with thin sand and sandy silt interbeds that apparently cannot be correlated between cores (Normark et al. 1985; Mutti and Normark 1987). Based on a study of a small area of these bedforms, Normark et al. (1980) suggested that they are migrating sediment waves that were built by thick low-velocity, low-concentration turbidity flows that overflowed the Monterey channel-levee system.

East of the bedform field, fan deposits are relatively flat-lying, except for the large western levee of Monterey Channel and Sur Slide. Sur Slide (Figure 12–1), originally studied by Normark and Gutmacher (1988), covers an area of about 5600 km² and extends downslope west of the channel, as shown on the imagery by the area of bright microscale speckled backscatter of debris-flow deposits (Figure 12–1A). Locally, this material partially fills Monterey Channel (Figure 12–9). Normark and Gutmacher (1989) speculate the failure is mid to late Holocene in age and was triggered by an earthquake.

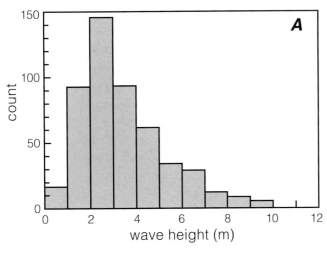

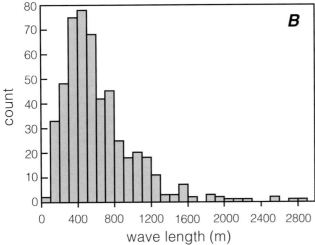

Figure 12–8. Statistics of the wave heights (A) and wave lengths (B) of bedforms, measured from 3.5-kHz profiles (*n* = 509).

The eastern part of the upper-fan segment is dominated by Davidson Seamount, which rises more than 2000 m above the fan surface, and Morro Fracture Zone, which has relief of up to 850 m above the fan surface. These features have been effective barriers to fan sedimentation to the east and south, respectively. After fan deposition filled the depression north of Morro Fracture Zone and elevated the seafloor to the level of the fracture zone, Monterey Channel (at 35°22′ N, 123°08′ W) and Ascension Channel (at 35°38′ N, 124°00′ W) breached Morro Fracture Zone and initiated sedimentation on the lower-fan segment of Monterey Fan.

LOWER-FAN SEGMENT OF MONTEREY FAN

The lower-fan segment of the fan is in water depths of from 4,000 to 4,600 m and is composed of several large, relatively smooth provinces each incised by a series of shallow (<20-m-deep) channels and is interpreted as a region of relatively large depositional lobes. Scattered isolated seamounts of various sizes occur throughout the lower-fan segment and have contributed to the control of sedimentation pathways (Figure 12–10A). The lower-fan segment of Monterey Fan terminates to the south at Murray Fracture Zone and to the west as an overlap on sediments of a nonfan basin facies.

The acoustic backscatter of the lower-fan segment is distinctly different from the upper-fan segment. The lower-fan segment is composed of three areas with intricate patterns of bright to intermediate mottled microscale speckled to ropey backscatter overlying dark microscale smooth backscatter. The seismic-reflection profiles show virtually no differences across these backscatter boundaries, although there is a distinct change on the 3.5-kHz record (Figure 12–10B). The dark microscale smooth backscatter facies (Figure 12–10A) represents basin-sediment facies that has not been

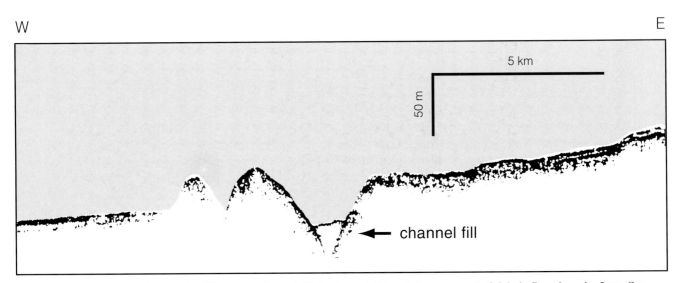

Figure 12–9. 3.5-kHz profile crossing Monterey Channel. Note channel fill mainly composed of debris-flow deposits from Sur Slide. Location of profile shown as eastern open end of line labelled 9 on Figure 12–1C.

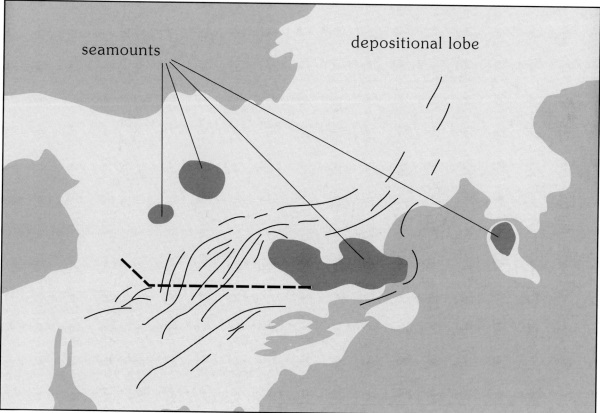

seamounts

depositional lobe

A

Figure 12–10. A, GLORIA image and interpretation of most recent depositional lobe of Monterey Channel. Lighter areas on sonograph represent higher backscatter, lines on interpretation indicate distributary channels interpreted from the GLORIA imagery. Location is box labeled 10 on Figure 12–1C.

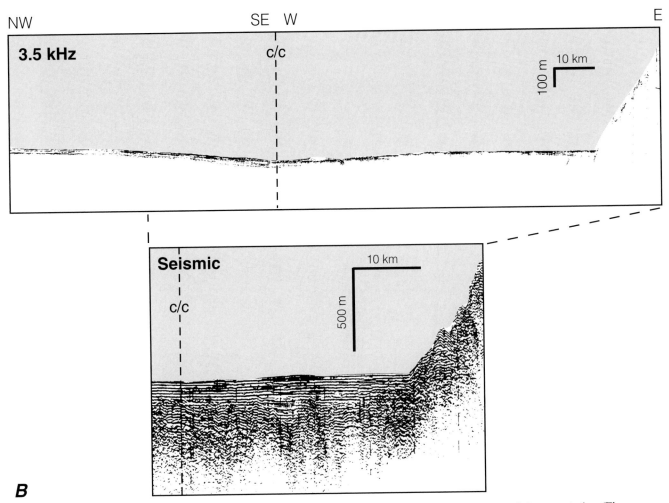

Figure 12–10. *Continued.* **B, Airgun-seismic and 3.5-kHz profiles shown as dashed line on sonograph interpretation (Figure 12–11A). Note difference in subbottom penetration on either side of course change (c/c) on 3.5-kHz profile but no difference on seismic-reflection profile. Also note that the distributary channels are too small to be resolved by 3.5-kHz system.**

covered by more than a few meters of fan deposits. The higher backscatter patterns are related to the various facies of the lower-fan segment of Monterey Fan that are thicker than a few meters (Gardner et al. 1991a).

Two major channels cross the lower-fan segment: One is the continuation of Monterey Channel, and the other is a remnant of Ascension Channel (Figures 12–1 and 12–11). In addition, Sur-Parkington-Lucia Channel is located east of the upper-fan segment, but abruptly turns south at the base-of-slope and has fed sediment around the eastern end of Morro Fracture Zone to the lower-fan segment (Figure 12–1). The portion of Monterey Channel in the lower-fan segment has no topographic levees (e.g., Figure 12–10B) and gradients that decrease downslope from 3.7 to 1.2 m/km (Figure 12–5B). Monterey Channel shows a distinct increase in channel gradient at Morro Fracture Zone with gradients of 2.4 m/km just north of Morro Fracture Zone (in the upper-fan segment) and 3.7 m/km just south of the fracture zone (in the lower-fan segment).

Monterey Channel is clearly defined in the GLORIA imagery in the northern half of the lower-fan segment; however, in the middle of the lower-fan segment the channel appears only in relatively short (~10 km) isolated sections (Figure 12–1A). This appearance and disappearance is similar to that described by Damuth et al. (1988) from GLORIA images of Amazon Fan. The backscatter from the floors of some of the Amazon channels are generally lower than the backscatter from the adjacent levee areas; this difference in backscatter is similar to that seen on the GLORIA images of Monterey Channel. Box cores from the axis of Monterey Channel in this area collected 10 to 40 cm of clay to clayey silt overlying coarse- to fine-grained, ungraded sand. Piston cores from the adjacent levees recovered thick sequences of silty clay with thin (<5 cm) silt to very fine-grained sand laminae.

A survey with the 30-kHz TOBI sidescan sonar of a portion of Monterey Channel (Gardner et al. 1991b, 1991c; Masson et al. 1995) confirms that some segments of the

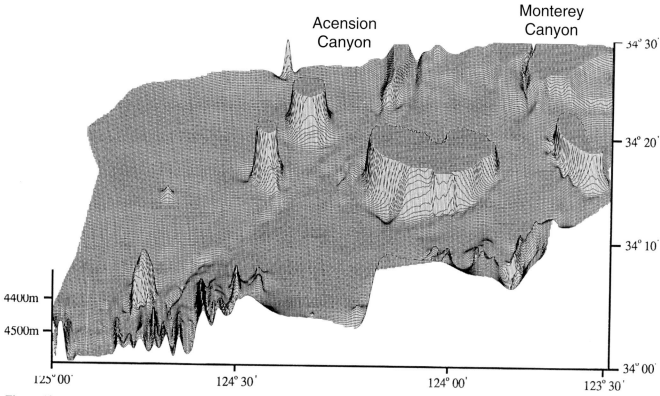

Figure 12–11. Perspective diagram of bathymetry of the most recent depositional lobe. This view is inclined 60° and has a vertical exaggeration of 150×. The tops of the seamounts have been arbitrarily truncated to make the illustration a manageable size. The solid lines represent the main distributary channels across the surface of the depositional lobe. This diagram encompasses the bottom half of area shown on Figure 12–10.

channel are well defined for short distances, but others are poorly defined. Channel walls and four terrace levels are well developed for part of the reach (Figure 12–12), then disappear as the channel broadens. Both 3.5-kHz and deep-towed 7-kHz profiles across the channel show that the channel depth (relief) varies from about 50 m (well developed) to less than 2 m (poorly developed) along a distance of less than 15 km. The floor of the well-defined portion of the channel has a series of stepped failures, or cascades, with downstream fields of transverse bedforms associated with each one (Figure 12–12). The cascades are concave down-channel and of undetermined heights, and we interpret them to be zones of active headward erosion of the channel. The bedforms apparently represent material collapsed from the cascade that is being transported down-channel.

Slumps and gullies are clearly evident on the margins of the channel (Figures 12–13 and 12–14). The outer bank shows the greatest amount of mass wasting, probably in response to undercutting of the channel walls and migration of the channel. One section of Monterey Channel imaged with the 30-kHz sidescan shows a cutoff where channel migration has stranded a portion of the channel and formed a new segment (Figure 12–14). The TOBI images of this portion of Monterey Channel suggest rather frequent prolonged flow, presumably by dilute low-density flows of the type discussed by Damuth et al. (1988) for Amazon Fan and Normark and Piper (1991) rather than episodic turbidity currents.

A large area of bright to intermediate mottled microscale speckled backscatter occurs over an area of at least 50,000 km^2 and represents an unburied portion of deposits of an older Ascension depositional lobe. The exposed portion of a depositional lobe of the Ascension system beyond the overbank deposits (Figure 12–1C) covers more than 25,000 km^2 and is characterized by bright to intermediate mottled microscale ropey to interwoven backscatter. The 3.5-kHz profile across the depositional lobe of the Ascension system shows no differences in character from the overbank region, yet it is well defined on the imagery. The less well-defined, and somewhat subdued, appearance of the backscatter for both the overbank and depositional lobe of the Ascension system suggests that the Ascension system may have been inactive for a long time, but determinations of relative age from GLORIA imagery are equivocal at best.

The area of the highest backscatter, one of bright microscale ropey to interwoven character, is clearly defined on

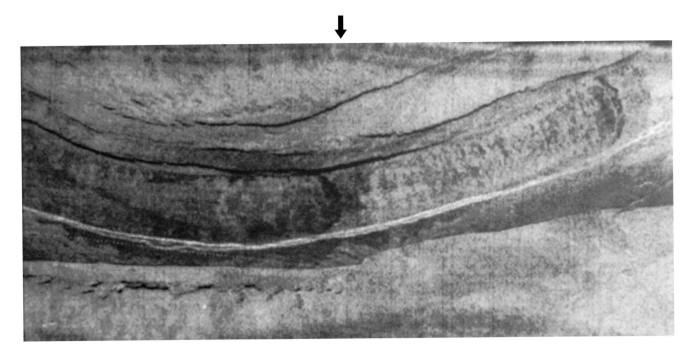

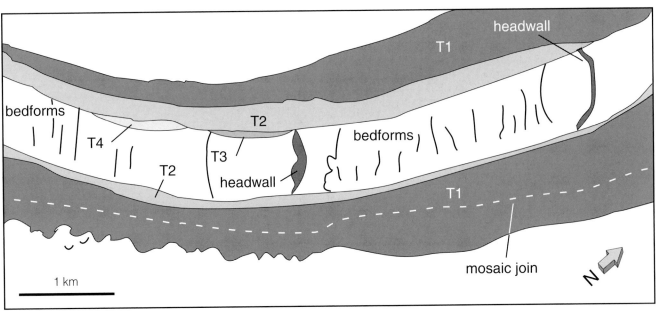

Figure 12–12. TOBI sidescan sonar image and interpretation of a segment of Monterey Channel on lower-fan segment of fan. T1 through T4 represent different levels of terraces. The headwalls of the stepped failures appear as acoustic shadows. Arrow on sonograph shows direction of insonification. Lighter area on sonograph represents higher backscatter. Location of image shown as white box labeled 12 on Figure 12–1C.

the GLORIA imagery (Figure 12–10A), yet again with no change in seismic-reflection character on the 3.5-kHz profiles across the boundary (Figure 12–10B). This area represents the most recent depositional lobe of the Monterey Channel system. Monterey Channel cannot be followed directly into the depositional lobe; rather, it appears on GLORIA sonographs to break up into a series of poorly defined smaller somewhat braided distributary channels (Figure 12–10A).

The Monterey depositional lobe covers an area in excess of 25,000 km², similar in size to that of the Ascension depositional lobe. Both GLORIA and 30-kHz TOBI sidescan imagery show that the depositional lobe contains a system of distributary channels that splay out across its surface (Figures 12–10A and 12–15). The TOBI images show these channels as relatively straight to slightly sinuous rather than meandering, as parallel to subparallel to one another, and

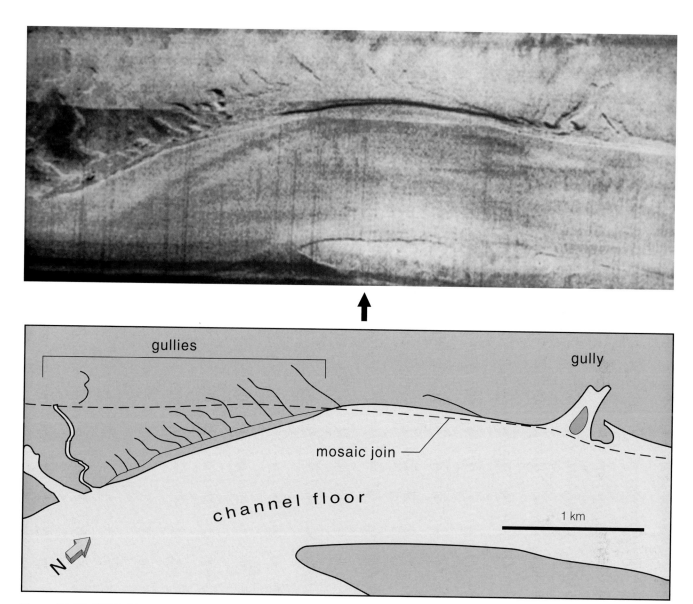

Figure 12–13. TOBI sidescan sonar image and interpretation of a segment of Monterey Channel on upper lower-fan segment. Note gullying on outer bank of channel curve. Lighter areas on sonograph represent higher backscatter. Arrow on sonograph shows direction of insonification. Location of image shown as white box labeled 13 on Figure 12–1C.

they are not braided. The distributary channels constrict in the area confined between two seamounts (Figure 12–10A), and, in places, they form a distinct crosscutting pattern suggesting successive phases of distributary activity (Figure 12–15). The distributary channels are less than 100 m wide with less than 2 m of relief and have low backscatter on TOBI 30-kHz images, similar in many aspects to the distal-lobe channels described by Twichell et al. (1992) from Mississippi Fan. The distributary channels are incised into the depositional lobe, appear to be actively downcutting, and have reached the area of Murray Fracture Zone that is deeper than the depositional lobe. The channel's courses trend through the lows between the northern foothills of Murray

Fracture Zone and ultimately into the east-west-trending bathymetric low of the fracture zone. The flows through the distributary channels between the foothills have been strong enough to undercut sediment that mantles the hills and cause large slabs of sediment to slump away (Figure 12–16).

The imagery suggests that the Monterey system has recently found a new base level in the area of Murray Fracture Zone and the entire system is now adjusting to a new equilibrium profile. The erosion of the distributary channels of the depositional lobe and the undercutting of the sediment mantling the small bathymetric highs adjacent to the fracture zone support this conclusion.

Cores collected from the most recent depositional lobe of

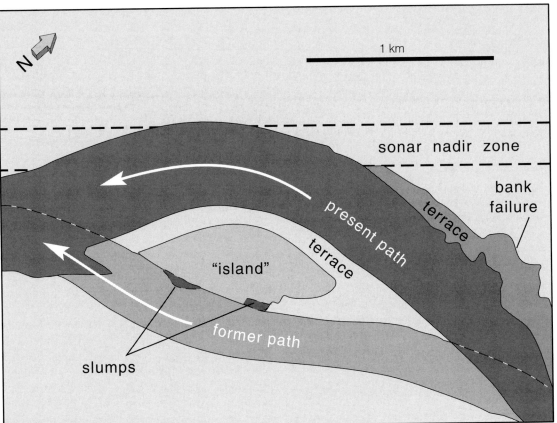

Figure 12–14. TOBI sidescan sonar image and interpretation of a segment of Monterey Channel on upper lower-fan segment where the channel has abandoned a former course cut off as it migrated. Also note large zone of failure along bank. Arrows on sonograph show direction of insonification. Lighter areas on sonograph represent higher backscatter. Location of image shown as white box labeled 14 on Figure 12–1C.

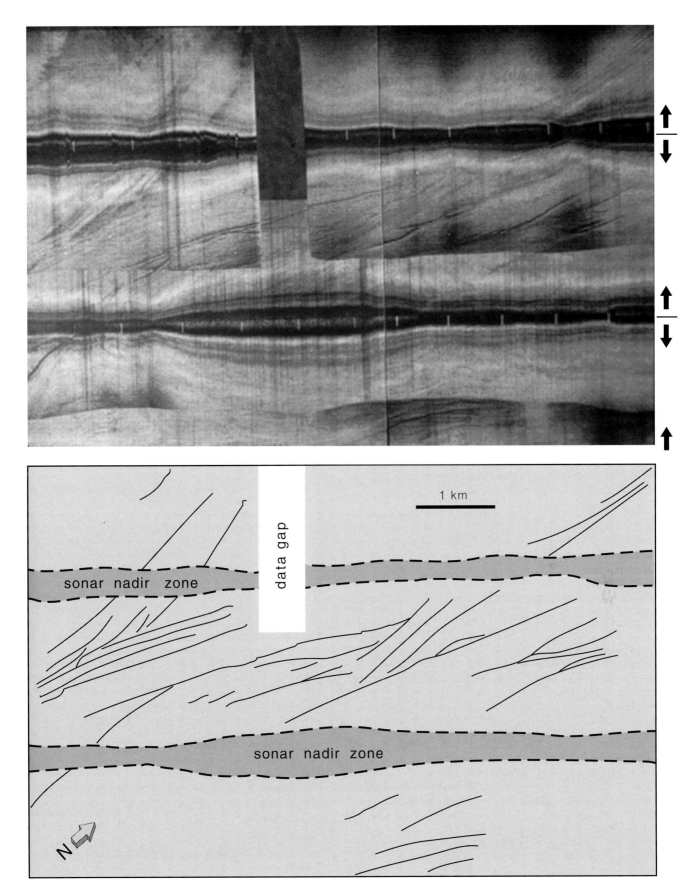

Figure 12–15. TOBI sidescan sonar image mosaic and interpretation of small distributary channels on the lower end of the most recent depositional lobe of Monterey Channel. Lines on interpretation represent incised and crosscutting channels. Lighter areas on sonograph represent higher backscatter. Arrows on sonograph show direction of insonification. Location of image shown as box labeled 15 on Figure 12–1C.

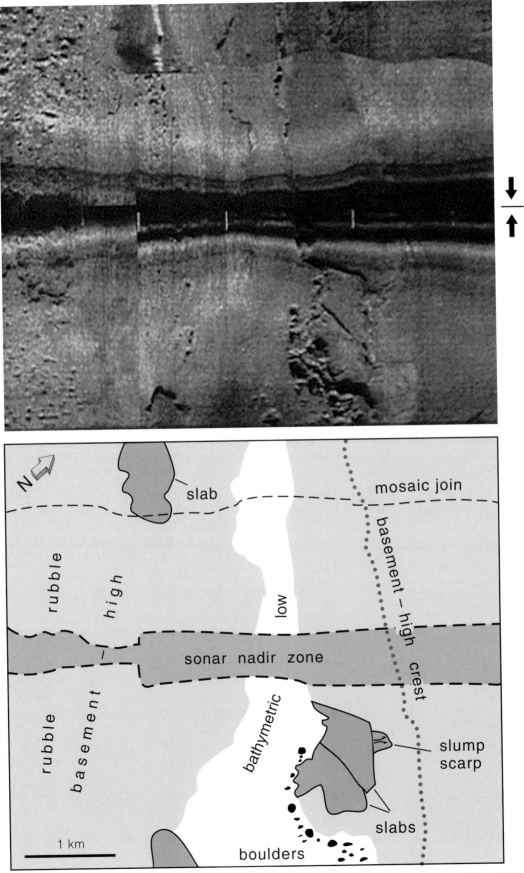

Figure 12–16. TOBI sidescan sonar image and interpretation of lowermost end of the most recent depositional lobe of Monterey Channel. Location of image shown as box labeled 16 on Figure 12–1C, just north of Murray Fracture Zone. Note slabs of sediment eroding from basement highs by undercutting caused by flows moving down the bathymetric low. Arrows on sonograph show direction of insonification. Lighter areas on sonograph represent higher backscatter.

Monterey Fan show that the highest backscatter area is composed of silty clay with minor interbeds of sand, whereas the areas of low backscatter are composed predominately of sand (Gardner et al. 1991a). This correlation of sediment facies versus backscatter is similar to that found on Mississippi Fan (Twichell et al. 1992; Nelson et al. 1992). Available ^{14}C ages (Gardner et al. 1991a) show that sandy turbidites, commonly containing terrigenous plant debris, were deposited on the most recent depositional lobe as recently as 6,500 years ago.

Another similarity that the most recent depositional lobe has with the depositional lobes of Mississippi Fan is the lack of correlation of coarse beds from core to core, even in closely spaced cores (Twichell et al. 1992; Nelson et al. 1992). This lack of lateral continuity of beds suggests that sand and gravel deposition on Monterey Fan occurs in overlapping discontinuous lenses of mass-transport flows and not in laterally extensive sheet deposits.

The most dramatic of the enigmatic features found on the most recent depositional lobe are large (many >20 m) boulders. These boulders occur both as isolated individual boulders and as fields of boulders (Figures 12–17 and 12–18). The fields of boulders show a rough alignment with lineations in backscatter intensity that we interpret as sediment-transport flow patterns. There is no decrease in the number of boulders with distance from the two closest seamounts, suggesting the boulders may be from a more-distant source. The boulders might have been rafted down channel by large mass-transport flows. Similar large boulders occur on the continental rise off the east coast of the U.S. (Robb 1991) and also were interpreted to have been emplaced by debris flows. These observations, together with the cores, suggest that mass-transport events may have been prevalent on the lower-fan segment.

The evolution of Monterey Fan

The different scales of resolution of the various systems used in this compilation cause much of the difficulty in interpreting the lithostratigraphic architecture of Monterey Fan and placing it in the same context as the surface morphology. Processed GLORIA imagery has a *horizontal* pixel resolution of more than 50 m and integrates the backscatter from a sediment volume within the upper 10 m of sediment. Processed TOBI imagery has a *horizontal* pixel resolution of 3 m and integrates backscatter from the upper 1 m of sediment. In contrast, the seismic-reflection profiler has a *vertical* resolution of about 20 m and the 3.5-kHz profiler has a *vertical* resolution of about 1 m. The immediate consequence of this mismatch in dimensional resolutions is that the imagery can resolve a sometimes bewildering amount of surface and near-surface details, whereas the airgun seismic data show only the large-scale integrated changes in acoustical properties with depth. The complexities revealed on the

surface of the fan by the sidescan sonar data strongly suggest that the evolution of the fan is vastly more complex than is revealed by the seismic-reflection profiles.

The line spacing of our seismic data is about 30 km, except in small areas where detailed surveys were run. However, these detailed surveys are concentrated on the Monterey depositional lobe. One consequence of this rather wide line spacing is that our interpretation of the growth history of the fan must be somewhat speculative. Guided by these realizations, we have strived to be cautious in our interpretations.

We define five fundamental stages in the development of Monterey Fan, most of which significantly differ from the most recently published scenario (Normark et al. 1983/1984, 1985). During the first stage, sediment was supplied to the region north of Morro Fracture Zone by a series of margin canyons and shallow distributary-channel systems that filled individual bathymetric lows in the basement (Figure 12–19A). The presence of these small systems is deduced from the ponded sediment that fills basement lows observed on seismic-reflection profiles. We see no evidence of major channel-levee systems in the deepest 0.5 km of sediment on the seismic records during this stage (Figure 12–20), suggesting that each system was small (too small to resolve on the airgun seismic system) and rather local in effect.

Eventually the second stage evolved, in which three large separate channel-levee systems developed across the small growing ramp of fan sediment (the early upper-fan segment) and terminated against the northern side of Morro Fracture Zone (Figure 12–19B). The ancestral Ascension Channel migrated across the western region and, from seismic records (Figure 12–20), appears to have been a meandering channel-levee system. The seismic-reflection profile across Ascension Channel shows a broad area of high-amplitude reflectors (HARs) interpreted as channel deposits in the subsurface (Damuth et al. 1983, 1988; Kastens and Shor 1985; Manley and Flood 1988) that eventually break into two distinct channels. Actually, these two apparent channels probably represent a single meandering channel crossed by the seismic line at a large meander loop. The early Monterey Channel appears to have begun at about the same time as the early Ascension Channel but, surprisingly, its location has not changed much since it developed (Figure 12–20). The early Monterey Channel was a much more entrenched aggradational channel-levee system and was much less sinuous or meandering than Ascension Channel. The differing nature of the two channel morphologies suggests that the regional bathymetry of the fan during this early stage of development was quite varied; Ascension Channel was migrating across a flat low-gradient plain, whereas Monterey Channel was traversing across a steeper region. The early Sur-Partington-Lucia Channel is ill defined on the seismic profiles and is only suggested by a thick zone of HARs that might represent a chaotic channel deposit at its base-of-slope location (Figure 12–20).

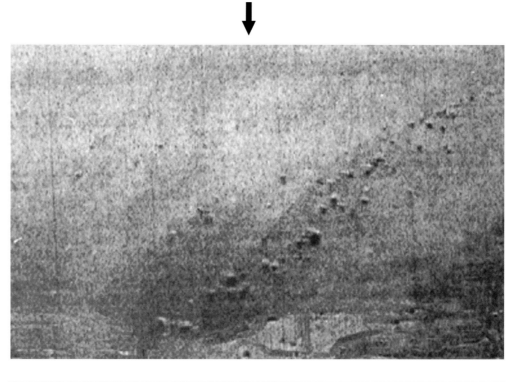

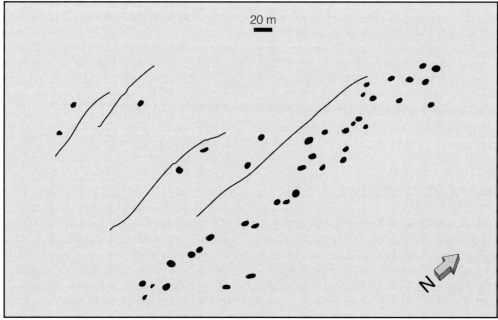

Figure 12–17. TOBI sidescan sonar image and interpretation of boulders resting on surface of most recent depositional lobe of Monterey Channel. Lines represent interpreted sediment-transport flow patterns. Arrow on sonograph shows direction of insonification. Lighter areas on sonograph represent higher backscatter. Location of image shown as box labeled 17 on Figure 12–1C and on Figure 12–18.

The third stage in the development of Monterey Fan involved transport of sediment around the two ends of Morro Fracture Zone to initiate sedimentation on the lower-fan segment (Figure 12–19C). The continued progradation of the

Sur-Parkington-Lucia Channel system carried sediment around the southeast end of Morro Fracture Zone, whereas Ascension Channel continued to prograde across the upper-fan segment and extended fan sedimentation to the south-

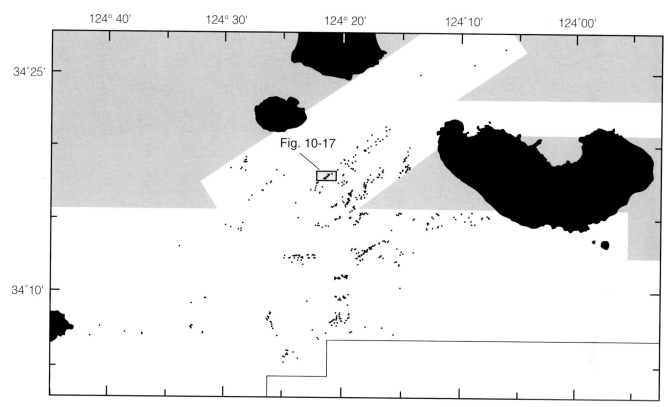

Figure 12–18. Location of all boulders identified on TOBI images. A total of 492 boulders were located. White area is region insonified with TOBI system, and small box is location of Figure 12–17.

west. Monterey Channel also continued to be active during this stage, but deposition remained confined behind Morro Fracture Zone. This stage represents a period of almost complete burial of the basement north of Morro Fracture Zone and the building of a thick wedge of sediment out from the continental margin to the fracture zone.

The fourth stage of development occurred when the established course of Ascension Canyon was diverted in the middle of the upper-fan segment and its course joined with Monterey Channel (Figure 12–19D). The diversion could have been caused by avulsion as the channel migrated across the fan's surface, but the seismic-reflection profile across this area shows an acoustically transparent unit directly above Ascension Channel that is interpreted as a large debris-flow deposit (Figure 12–20). The position of this debris-flow deposit suggests that it may have caused the diversion. The diversion of Ascension Channel caused the lower reach on the lower-fan segment to be cut off and the depocenter shifted to the central area of the upper-fan segment against Morro Fracture Zone. The merging of the two channel systems provided an increased sediment supply that finally built the sediment pile high enough against Morro Fracture Zone to breach the lowest pass. A seismic-reflection profile striking obliquely across Morro Fracture Zone clearly shows the banked sediment on the northeast side of the basement ridge and the drop in bathymetry to the southwest (Figure 12–21). Eventually,

Monterey Channel extended through the pass in the summit region and then built out across the Ascension and Sur-Parkington-Lucia depositional bodies and prograded farther to the south (Figure 12–19D).

The fifth stage in the evolution of Monterey Fan was the progradation of lower-fan sedimentation to the south, burying basement topography and eventually reaching the bathymetric lows of Murray Fracture Zone (Figure 12–19E). During this stage the upper reach of Monterey Channel formed the large meander loop (Shepard 1966) and captured Ascension Channel (Normark 1970b). The most recent depositional lobe of Monterey Channel is clearly observed on sidescan imagery, but it must be quite young because it is too thin (<10 m) to show as a distinct unit on 3.5-kHz or seismic-reflection records.

Conclusions and unresolved fundamental problems

Although Monterey Fan has many commonly observed fan elements (channel-levee systems, depositional lobes, etc.) it is by no means a typical submarine fan, if indeed a typical fan exists. Monterey Fan has developed adjacent to a tectonically active margin and undoubtedly has been affected by numerous large-magnitude earthquakes. The fan was

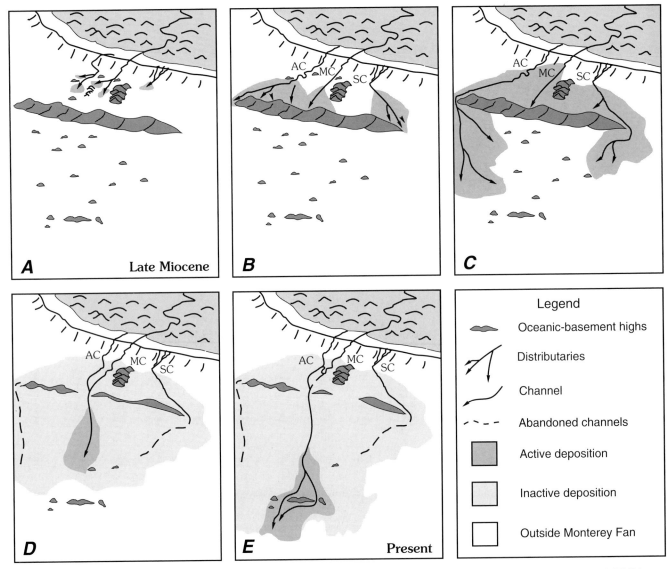

Figure 12–19. Conceptual drawings showing evolution and growth pattern of Monterey Fan. AC is Ascension Channel, MC is Monterey Channel, and SC is Sur-Partington-Lucia Channel system.

built on a basement of high relief, and was blocked from simple downslope progradation by large seamounts and fracture zones. Although initial fan sedimentation may have occurred as early as Early Miocene, rapid fan development probably did not occur until the Late Miocene, because of the accessibility of sediment sources and the amount of basement topography that had to be filled.

Three major channel systems have fed Monterey Fan through its evolution. Ascension and Monterey Channels fed sediment to the upper-fan segment of the fan, and later in the development, the Sur-Parkington-Lucia Channel system reached the growing submarine fan. The latest stage in the evolution of Monterey Fan was the breaching of Morro Fracture Zone by Monterey Channel and the progradation of fan sediments south of the fracture zone. Once the distributary

channels of the latest depositional lobe of Monterey Fan reached Murray Fracture Zone, the entire system began an adjustment to the new base level, and a period of erosion and channel downcutting began. This stage appears from the TOBI images to be continuing today.

There are many dilemmas in the history of Monterey Fan that will not be resolved until extensive ground-truth cruises and a drilling campaign are conducted. There is a critical need to recover complete lithostratigraphic sequences through the fan sediments on a relatively tight spatial grid. The following are some of the principle questions left unanswered:

(a) What is the age of initial Monterey Fan deposition on the upper-fan segment? Did initial fan deposition begin in the Late Miocene during major uplift of the Sierra Nevada or later during uplift of the Coast Ranges?

W

E

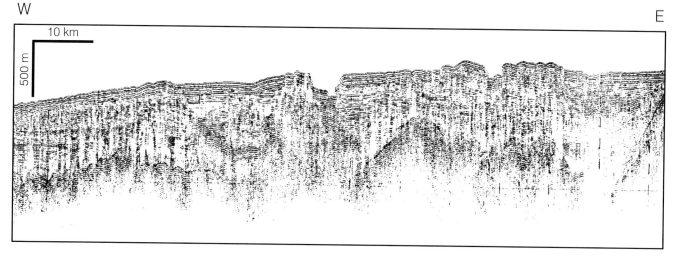

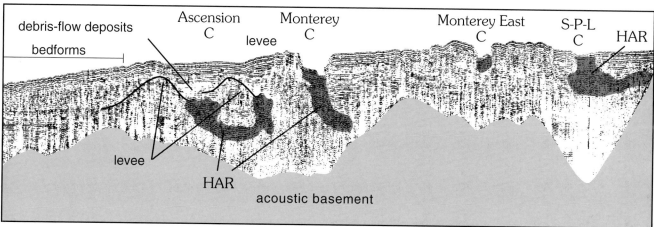

Figure 12–20. Seismic-reflection profile and interpretation of line crossing middle of upper-fan segment of Monterey Fan showing well-developed Ascension and Monterey Channel-levee systems. HAR is high-amplitude reflectors interpreted as channel deposits. Location of profile shown as black line labeled 20 on Figure 12–1C.

(b) What is the source(s) of the sediment comprising the deeper parts of Monterey Fan? How volumetrically important is Sierra Nevada – derived sediment relative to Coast Range – derived sediment?

(c) What is the age of initial fan deposition south of Morro Fracture Zone on the lower-fan segment?

(d) What is the age of the capture of Ascension Channel by Monterey Channel?

(e) What is the origin, age, and process that rafted the large boulders to the depositional lobe?

(f) What are the submarine processes that form and modify Monterey Channel on the lower-fan segment and caused geomorphologies so similar to subareal analogs?

(g) What role, if any, did eustatic fluctuations of sea level play in the fan sedimentation, relative to periodic large-magnitude seismic events?

These and many more questions await higher-resolution remote-sensing tools and sampling by extensive drilling.

Acknowledgments

We first want to acknowledge the collaborative science agreement between the U.S. Geological Survey and the Institute of Oceanographic Sciences, U.K., that made this study possible. Also, the sidescan sonar innovations of GLORIA and TOBI are a direct result of the efforts of J. Revie, M. L. Somers, J. Campbell, D. Bishop, A. Harris, C. G. Flewellen, N. Millard, I. P. Rouse, E. Darlington, A. Gray, and their colleagues at IOS. To them we owe a major debt of gratitude for designing, building, and maintaining these superb instruments. We also wish to acknowledge P. Chavez, J. Anderson, S. Sides, L. Bellisime, and J. Bowell of the USGS, Flagstaff, Arizona, who developed and perfected the image-processing techniques for GLORIA and TOBI. Without them, we would still be using scissors and a magnifying lens. And lastly, but really most importantly, we wish to thank the Captains and crews of R/V *Farnella,* for their dedication in collecting all these data. Cap-

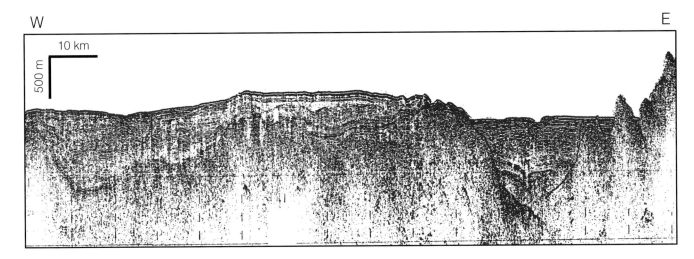

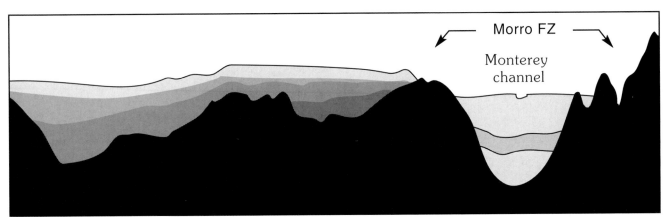

Figure 12–21. Seismic-reflection profile and interpretation of line obliquely crossing Morro Fracture Zone and showing the upper lower-fan segment of Monterey Fan. Location of profile shown as black line labeled 21 on Figure 12–1C.

tains Hadgraft, Cannon, and Nichols and their men were always there when we needed them.

We also acknowledge the thorough and constructive reviews of earlier versions of the manuscript by Steve Eittreim, Dave Twichell, John Damuth, and Rob Kidd.

References

Atwater, T. 1970. Implications of plate tectonics for the Cenozoic tectonic evolution of western North America. *Geol. Soc. Am. Bull.* 81(12): 3513–35.

Atwater, T., and Molnar, P. 1973. Relative motion of the Pacific and North American plates deduced from sea-floor spreading in the Atlantic, Indian, and South Pacific oceans. In *Conference on Tectonic Problems of the San Andreas Fault System: Proceedings,* Vol. 13, *Geological Science,* pp. 136–48. Palo Alto, Calif.: Stanford University.

Atwater, T. 1989. Plate tectonic history of the northeast Pacific and western North America. In *The Geology of North America,* Vol. N, *The Eastern Pacific Ocean and Hawaii* eds. E. L. Winterer, D. M. Hussong and R. W. Decker. pp. 21–72. Boulder, Colo.: Geological Society of America.

Atwater, T., and Severingus, J. 1989. Tectonic maps of the northeast Pacific. In *The eastern Pacific Ocean and Hawaii, The Geology of North America,* vol. N, eds. E. L. Winterer, D. M. Hussong and R. W. Decker. pp. 15–20. Boulder, Colo.: Geological Society of America.

Belderson, R. H., and Kenyon, N. H. 1980. The Equatorial Atlantic mid-ocean canyon seen on a sonograph. *Mar. Geol.* 34: M77–M81.

Bering Sea EEZ-SCAN Scientific Staff 1991. Atlas of the U.S. Exclusive Economic Zone, Bering Sea. *U.S. Geological Survey Miscellaneous Investigations Series I-2053,* 145p., scale 1:500,000. Reston, Va.: USGS.

Bouma, A. H. ed. 1983/1984. *Geo-Mar. Letters,* v. 3, 224 p.

Chavez, P. S. 1986. Processing techniques for digital sonar images from GLORIA. *Photogram. Engrg. Rem. Sens.* 52: 1133–45.

Damuth, J. E. 1975. Echo character of the western equatorial Atlantic floor and its relationship to the dispersal and distribution of terrigenous sediments. *Mar. Geol.* 18: 17–45.

Damuth, J. E., Kolla, V., Flood, R. D., Kowsmann, R. O., Monteiro, M., Gorini, M. A., Palma, J. J., and Belderson, R. H. 1983. Distributary channel meandering and bifurcation patterns on the Amazon deep-sea fan as revealed by long-range side-scan sonar (GLORIA). *Geology* 11: 94–8.

Damuth, J. E., Flood, R. D., Kowsmann, R. O., Belderson, R. H., and Gorini, M. A. 1988. Anatomy and growth pattern of

Amazon deep-sea fan as revealed by long-range side-scan sonar (GLORIA) and high-resolution seismic studies. *Am. Assoc. Petrol. Geol. Bull.* 72: 885–911.

EEZ-SCAN 84 Scientific Staff 1986. Atlas of the Exclusive Economic Zone, western conterminous United States. *U.S. Geological Survey Miscellaneous Investigation Series I-1792.* 152p., scale 1:500,000. Reston, Va.: USGS.

Embley, R. W., Eittreim, S. L., McHugh, C. H., Normark, W. R., Rau, G. H., Hecker, B., DeBevour, A. E., Greene, H. G., Ryan, W. B. F., Harrold, C., and Baxter, C. 1990. Geological setting of chemosynthetic communities in the Monterey Fan Valley System. *Deep-Sea Res.* 37: 1651–67.

Engebretson, D. C., Cox, A., and Gordon, R. G. 1985. Relative motions between oceanic and continental plates in the Pacific basin. *Geological Society of America, Special Paper 206,* 59 p. Boulder, Colo.: GSA.

Gardner, J. V., Field, M. E., Lee, H., Edwards, B. E., Masson, D. G., Kenyon, N., and Kidd, R. B. 1991a. Ground-truthing 6.5-kHz side-scan sonographs: What are we really imaging? *J. Geophys. Res.* 96: 5955–74.

Gardner, J. V., Field, M. E., Barber, J. H., Edwards, B. D., and Masson, D. G. 1991b. Erosion and deposition on Monterey Fan: New evidence for long- and mid-range side-scan sonar. Abstracts with program SEPM meeting. Portland, OR: Society of Economic Paleontologists and Mineralogists. Tulsa, OK.

Gardner, J. V., Field, M. E., Barber, J. H., and Masson, D. G. 1991c. Recent major erosional and depositional processes on Monterey Fans: Evidence from mid-range side-scan sonar (abs). *Geological Society of America 1991 Annual Mtg, Abstracts and Programs, A385.* Boulder, Colo.: GSA.

Gardner, J. V., Cacchione, D. A., Drake, D. E., Edwards, B. D., Field, M. E., Hampton, M. A., Karl, H. A., McCulloch, D. S., Masson, D. G., Kenyon, N. H., and Grim, M. 1992. Map showing sediment isopachs in the deep-sea basins on the Pacific continental margin, Point Conception to Point Arena, Scale 1:1,000,000. *U.S. Geological Survey Miscellaneous Investigations Series Map I-2090-C.* Reston, Va.: USGS.

Goodman, E. D., and Malin, P. E. 1992. Evolution of the southern San Joaquin Basin and mid-Tertiary "transitional" tectonics, central California. *Tectonics* 11(3): 478–98.

Greene, H. G., Stubblefield, W. L., and Theberge, A. E., Jr. 1989. Geology of Monterey submarine canyon system and adjacent areas offshore central California: Results of NOAA SeaBeam survey, descriptive report for the Surveyor cruise. *U.S. Geological Survey Open-File Report 89-221.* 33p. Reston, Va.: USGS.

Greene, H. G., Clarke, S. H., and Kennedy, M. P. 1991. Tectonic evolution of submarine canyons along the California continental margin, In *From shoreline to abyss: Contributions in marine geology in honor of Francis Parker Shepard,* ed. R. H. Osborne, pp. 231–48. SEPM Special Paper 46. Society of Economic Paleontologists and Mineralogists, Tulsa, OK

Hess, G. R., and Normark, W. R. 1976. Holocene sedimentation history of the major fan valleys of Monterey Fan. *Mar. Geol.* 22: 233–51.

Huber, M. K. 1981. Amount and timing of late Cenozoic uplift and tilt of the central Sierra Nevada, California – Evidence from the upper San Joaquin River Basin. *United States Geological Survey Professional Paper 1197,* 28 p. Reston, Va.: USGS.

Huffman, O. F. 1972. Lateral Displacement of Upper Miocene Rocks and the Neogene History of Offset along the San Andreas Fault in Central California. *Geol. Soc. Am. Bull.* 83(10): 2913–46.

Irwin, W. P. 1990. Geology and plate tectonic development. R. E. Wallace, ed., The San Andreas Fault System, California. U.

S. Geological Survey Professional Paper 1515, 61–80. Reston, Va: USGS.

Jennings, C. W. 1977. Geologic map of California: California Division of Mines and Geology, *California Geologic Data Map Series, Map 2,* Scale 1:750,000. Sacramento, CA.

Kastens, K. A., and Shor, A. N. 1985. Depositional processes of a meandering channel on the Mississippi Fan. *Bull. Am. Assoc. Petrol. Geol.* 79: 190–202.

Kastens, K. A., and Shor, A. N. 1986. Evolution of a channel meander on the Mississippi deep-sea fan. *Mar. Geol.* 71: 165–75.

Leopold, L. B., and Wolman, M. G. 1957. River channel patterns: Braided, meandering, and straight. *U. S. Geological Survey Professional Paper 282-B,* pp. 39–85. Reston, Va.: USGS.

Lonsdale, P. F. 1991. Structural patterns of the Pacific floor offshore of Peninsular California, In *Gulf and Peninsula Provinces of the Californias.* American Association of Petroleum Geologist-Memoir 43, 87–125. Tulsa, OK: AAPG.

Manley, P. I., and Flood, R. D. 1988. Cyclic sediment deposition within Amazon deep-sea fan. *Am. Assoc. Petrol. Geol. Bull.* 72: 912–25.

Martin, B. D., and Emery, K. O. 1967. Geology of Montery Canyon, California. *Bull. Am. Assoc. Petrol. Geol.* 51: 2281–304.

Masson, D. G., Kenyon, N. H., Gardner, J. V., and Field, M. E. 1995. Monterey Fan: Channel and overbank morphology. In *An Atlas of Deep Water Systems: Turbidity System Architectural Styles,* eds. K. Pickering et al., p. 74–9. London.

McCulloch, D. S. 1989. Evolution of the offshore central California margin. *The eastern Pacific Ocean and Hawaii,* In *The Geology of North America,* Vol. N. eds. E. L. Winterer, D. M. Hussong, and R. W. Decker, pp. 373–82. Boulder, Colo.: Geological Society of America.

Murton, B. J., Rouse, I. P., Millard, N. W., and Flewellen, C. G. 1992. Multisensor, deep-towed instrument explores ocean floor. *EOS, Trans. Am. Geophys. Union* 73: 225–8.

Mutti, E., and Ricci Lucchi, F. 1972. Le torbiditi dell'Apennino settentrionale: Introduzione all'analisi Di facies. *Mem. Soc. Geol. Italy* 11: p. 61-199. (1978 English translation by T. H. Nilsen, *Int. Geol. Rev.* 20: 125–6.

Mutti, E., and Normark, W. R. 1987. Comparing examples of modern and ancient turbidity systems: Problems and concepts. In *Marine Clastic Sedimentology,* London: Graham and Trotman. eds. J. L. Leggett and G. G. Zuffa, pp. 1–38.

Nelson, C. H., Twichell, D., Schwab, W., Lee, H. J., and Kenyon, N. H. 1992. Upper Pleistocene turbidite sand beds and chaotic silt beds in the channelized, distal, outer-fan lobes of the Mississippi fan. *Geology* 20: 693–6.

Nilsen, T. H. 1984. Offset along the San Andreas Fault of Eocene strata from the San Juan Bautista area and western San Emigdio Mountains, California. *Geol. Soc. Am. Bull.* 95(5): 599–609.

Normark, W. R. 1970a. Growth patterns of deep-sea fans. *Bull. Am. Assoc. Petrol. Geol.* 54: 2170–95.

Normark, W. R. 1970b. Channel piracy on Monterey deep-sea fan. *Deep-Sea Res.* 17: 837–46.

Normark, W. R. 1978. Fan Valleys, channels, and depositional lobes on modern submarine fans: Characters for recognition of sandy turbidite environments. *Bull. Am. Petrol. Geol.* 62: 912–31.

Normark, W. R., Hess, G. R., Stow, D. A. V., and Bowen, A. J. 1980. Sediment waves on the Monterey Fan levee: A preliminary physical interpretation. *Mar. Geol.* 37: 1–18.

Normark, W. R., Gutmacher, C. E., Chase, T. E., and Wilde, P. 1983/1984. Monterey Fan: Growth pattern control by basin morphology and changing sea levels. *Geo-Mar. Ltrs.* 3: 93–100.

Normark, W. R., Gutmacher, C. E., Chase, T. E., and Wilde, P. 1985. Monterey Fan, Pacific Ocean. In *Submarine fans and related turbidite systems,* eds. A. H. Bouma, W. R. Normark, and N. E. Barnes, pp. 79–86. New York: Springer-Verlag.

Normark, W. R., and Gutmacher, C. E. 1988. Sur submarine slide, Monterey Fan, central California. *Sedimentology* 35: 629–47.

Normark, W. R., and Gutmacher, C. E. 1989. Major submarine fans of the California continental rise. In *The Geology of North America,* In *The Eastern Pacific Ocean and Hawaii,* E. L. Winterer, D. M. Hussong, and R. W. Decker, pp. 373–82. Boulder, Colo.: Geological Society of America.

Normark, W. R., and Piper, D. J. W. 1991. Initiation processes and flow evolution of turbidity currents: Implications for the depositional record, In *From Shoreline to Abyss: Contributions in Marine Geology in Honor of Francis Parker Shepard,* ed. R. H. Osborne, pp. 207–30. SEPM Special Paper 46. Tulsa, OK: Society of Economic Paleontologists and Minerologists.

Powell, R. E. 1981. Geology of the crystalline basement complex, eastern Transverse Ranges, southern California: Constraints on regional tectonic interpretations. Pasadena, California Institute of Technology, Ph.D. thesis, 441 p.

Robb, J. M. 1991. Multibeam bathymetry and GLORIA images show complex morphology and sedimentation patterns of the continental rise offshore New York, *Abstracts with Program,* First Annual Theme Meeting, SEPM meeting, Portland, Ore., p. 29.

Shanmugam, G., and Moiola, R. J. 1988. Submarine fans: Characteristics, models, classification, and reservoir potential. *Earth-Science Rev.* 24: 383–428.

Shepard, F. P. 1966. Meander in valley crossing a deep-ocean fan. *Science* 154: 385–6.

Stanley, R. G. 1987. New estimates of displacement along the San Andreas fault in central California based on paleobathymetry and paleogeography. *Geology* 15: 171–4.

Stock, J., and Molnar, P. 1988. Uncertainties and implications of the Late Cretaceous and Tertiary position of North America relative to the Farallone, Kula, and Pacific plates. *Tectonics,* 7(6): 1339–84.

Stow, D. A. V. 1981. Laurentian Fan: Morphology, sediments, processes and growth pattern. *Am. Assoc. Petrol. Geol. Bull.* 65: 375–93.

Twichell, D. C., Kenyon, N. H., Parson, L. M., and McGregor, B. A. 1991. Depositional patterns of the Mississippi fan surface: Evidence from GLORIA II and high-resolution seismic profiles. P. Weimer and M. H. Link, In *Seismic facies and sedimentary processes of Submarine Fans and Turbidite Systems,* pp. 349–63. New York: Springer-Verlag.

Twichell, D., Schwab, W., Nelson, C. H., Kenyon, N. H., and Lee, H. J. 1992. Characteristics of a sandy depositional lobe on the outer Mississippi fan from SeaMARC 1A side-scan sonar images. *Geology* 20: 689–92.

Vedder, J. G. 1987. Regional geology and petroleum potential of the Southern California borderland. In *Geology and Resource Potential of the Continental Margin of Western North America and Adjacent Ocean Basins. Beaufort Sea to Baja California,* Circum Pacific Council for Energy and Mineral Resources, Earth Science Series, vol. 6, eds. D. W. Scholl, A. Grantz, J. G. Vedder. pp. 403–47. Houston, TX.

Walker, R. G. 1978. Deep water sandstone facies and ancient submarine fans: Models for exploration for stratigraphic traps. *Bull. Am. Assoc. Petrol. Geol.* 62: 932–66.

Weimer, P., and Link, M. H. eds. 1991. *Seismic facies and sedimentary processes of submarine fans and turbidite systems.* New York: Springer-Verlag, 447 p.

13 Ground-truth studies of West Coast and Gulf of Mexico submarine fans

Homa J. Lee,[1] Robert E. Kayen,[1] Brian D. Edwards,[1] Michael E. Field,[1] James V. Gardner,[1] William C. Schwab,[2] and David C. Twichell[2]

[1]U.S. Geological Survey, Menlo Park, California
[2]U.S. Geological Survey, Woods Hole, Massachusetts

Introduction

The use of sidescan sonar technology has greatly expanded in recent years. One impediment to interpreting sidescan sonar images, which are a representation of the amount of sound backscattered from the seafloor, is the incomplete understanding of the physical meaning of acoustic backscatter intensity variations. Ground-truth studies can help us to understand the causes of variations in backscatter. We need to measure physical and geometric properties of seafloor sediment and correlate them with variations in sidescan sonar acoustic backscatter. We present in this paper comparative ground-truth studies of two deep-sea fan depositional lobes. We show that sediment lithology influences sidescan sonar images, but that the relation between backscatter intensity and sediment grain size is not uniquely defined.

Some of the seafloor characteristics that are potential causes of variations in acoustic backscatter intensity are surface roughness, variations in sediment composition, grazing angle of insonification, and seafloor slope, including topographic variability (Urick 1983). The influence of each of these and the subbottom depth range over which sediment compositional variations are important will vary with the characteristics of the sidescan system, including frequency, pulse length, bandwidth, time-varying gains, and footprint size. The number of variables needs to be kept to a minimum in order to simplify a ground-truth study. The distal parts of deep-sea fans are good locations for such studies because they tend to have nearly horizontal seafloor surfaces. Thus, the topographic variability – bottom slope effect can be ignored. Also, deep-sea fans commonly include sandy deposits emplaced far from shore by turbidity currents. The physical contrast between these sandy deposits and surrounding finer-grained hemipelagic deposits is strong and should provide an excellent opportunity to isolate and evaluate the influence of sediment composition on backscatter intensity.

According to Lyons, Anderson, and Dwan (in press), the field of acoustic energy scattered by the seafloor can consist of two general components: scattering from interface roughness and scattering from volume inhomogeneities. Figure 13–1 illustrates the situation in which scattering from either the sediment-water interface or a shallow subbottom interface dominates the field of scattered energy. For each of the four cases shown, the hachured area represents the relative amount of sound energy scattered from the bottom as a function of angle. For a conventional sidescan sonar configuration, the backscattered sound recorded on the sonograph is represented by the part of the hachured area that intersects the incoming ray. That is, the recorded backscattered sound is the part that returns back to the receiver along the path of the incoming ray. The factors that determine backscatter intensity are (1) contrasts in acoustic impedance across interfaces (sediment-water interface and subsequent subbottom interfaces) and (2) interface roughnesses. Acoustic impedance is a physical property of a sound-transmitting medium and is equal to the product of saturated bulk density (ρ) and compressional wave sound velocity (c) in the medium. Sandy sediment typically has a higher acoustic impedance than does muddy sediment (Hamilton 1970).

For a relatively smooth bottom with high impedance contrast (Figure 13–1A), a large part of the incoming sound energy is reflected away from the receiver along a ray path with an angle of reflection equal to the angle of incidence. Some sound is refracted into the bottom in a direction indicated by Snell's Law. In contrast to the energy scattered in the forward (specular) direction, a lesser amount of sound energy is backscattered along the incoming ray path to be recorded at the receiver. If the bottom is relatively rough and has a high impedance contrast (Figure 13–1B), then more sound is backscattered along the incoming ray path to be recorded at the receiver. If the bottom has a low impedance contrast, as is common in marine sediment (Figures 13–1C and D), then proportionally more sound energy is refracted into the sediment and less is scattered or reflected into the water column than for high impedance contrast. Of this lesser amount of sound that is scattered or reflected from the seafloor, more sound is backscattered to be recorded at the receiver if the bottom is relatively rough (Figure 13–1D).

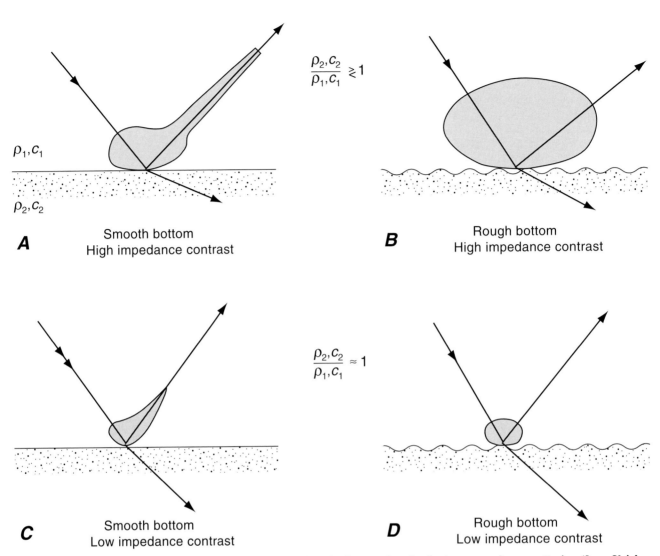

Figure 13–1. Acoustic backscatter from the seafloor if interface backscattering dominates over volume scattering (from Urick 1983); ρ is sediment density, c is sediment compressional wave velocity.

In summary, if interface scattering dominates and if the bottom roughnesses are similar, then a bottom with high acoustic impedance (e.g., a typical sand) will produce greater acoustic backscatter than a bottom with low acoustic impedance (e.g. a typical mud). A rough sandy surface will also produce more backscatter than a smooth muddy surface. In the case of a smooth sandy surface juxtaposed against a rough muddy surface, the relative level of acoustic backscatter can be determined only through quantitative knowledge of the roughness and impedance of each facies. When volume scattering from internal inhomogeneities becomes important, the relative level of acoustic backscatter can be determined only from quantitative knowledge of the seafloor internal heterogeneity. Conventional interpretations of sidescan sonar imagery often ignore these last two cases. Rather, an ad hoc assumption is commonly made that sandy sediment produces more acoustic backscatter than muddy sediment.

Approach

We conducted ground-truth investigations of the distal depositional lobes of two deep-sea fans, Mississippi Fan in the Gulf of Mexico and Monterey Fan in the eastern Pacific Ocean. In both cases, we have complete coverage of long-range GLORIA sidescan sonar data (EEZ-SCAN 84 Scientific Staff 1986; EEZ-SCAN 85 Scientific Staff 1987). We used the GLORIA images to help identify juxtaposed high- and low-backscatter areas on what is interpreted in each case as the most recent depositional lobe.

The GLORIA images of Mississippi Fan (Figures 13–2 and 13–3) were used as a reconnaissance tool to select an area for higher-resolution sidescan sonar studies using the SeaMARC IA system (Twichell et al. 1992). We used the SeaMARC IA images, part of which are shown in Figure 13–4, to select locations for piston and gravity coring within

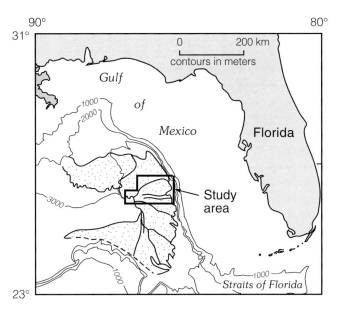

Figure 13–2. Location map – Mississippi Fan study area.

the depositional lobes and in presumed basin-plain (hemipelagic) sediment. Both the SeaMARC IA tracklines and the cores were navigated relative to the same bottom-transponder acoustic net; consequently, the location of the cores relative to the image is accurate. We consider seventeen of these coring stations in this study. These cores (20 to 650 cm long) were collected in the vicinity of three recently active sublobes termed the young sublobe, the intermediate sublobe, and the old sublobe (Figure 13–3). Figure 13–4 shows the locations of some of the cores in the intermediate sublobe relative to the sidescan imagery. Descriptions of these cores were given by Nelson et al. (1992).

Although we use the quantitative characteristics from the SeaMARC IA images in our ground-truth study, the qualitative characteristics of both the GLORIA images and the SeaMARC IA images are similar. That is, both show depositional lobes as high backscatter areas and the surrounding basin plains as low backscatter regions. The major difference in the images is the detail apparent along the margins of the lobes; the higher-frequency (30 kHz) SeaMARC IA images show much clearer and more complex boundaries and allow better definition of core locations than do the 6.5-kHz GLORIA images. Because the GLORIA and SeaMARC IA images were collected at different times, using different navigation systems, there will always be a question of the detailed coregistration of the two data sets. Several of the cores were collected on the margin of the depositional lobe (Figure 13–4), and we cannot be positive that they are not just beyond the limit of the depositional lobe on the basin facies. Consequently, we consider our ground-truth study of the SeaMARC IA images to be a qualitative ground-truth study of the GLORIA images.

The GLORIA mosaic of Monterey Fan (Figures 13–5 and

13–6) was used to locate coring sites on its most recent depositional lobe (Gardner et al. 1991b). The fan is characterized by generally high acoustic backscatter. However, a digitate low-backscatter area (Figures 13–5, 13–6, and 13–7) occurs within the fan and is termed the fingers area. We performed a detailed study over the fingers area and the surrounding high-backscatter region using bottom photography; 3.5-kHz subbottom profiling; and piston, gravity, and box coring (Gardner et al. 1991b). Analyses of the cores were tabulated by Ramirez-Bader, Gardner, and Field (1991); results from fourteen of the box cores (20 to 60 cm long) are included in this paper. A later cruise surveyed the fingers area using the TOBI 30-kHz deep-towed sidescan sonar (Gardner et al. 1991a).

We measured downcore variations in density (calculated from water content assuming 100% saturation) for all cores taken in both fan systems at a 2- to 10-cm spacing. Compressional-wave sound velocities were measured using a through-the-liner velocimeter developed by the U.S. Geological Survey. Acoustic impedance was calculated as the product of density and sound velocity.

Results

Mississippi Fan

Cores near the edges of Mississippi Fan distal lobes show a clear association between the occurrence of sediment with a relatively high acoustic impedance and high backscatter on the SeaMARC IA mosaic. Cores from the high-backscatter sublobe areas have sandy turbidites or chaotic silt beds buried under 9 to 230 cm of hemipelagic mud (Figure 13–8). The chaotic silt beds are interpreted as debris-flow deposits (Nelson et al. 1991). Mean thickness of hemipelagic mud (Table 13–1) is consistent with a designation of the sublobe as young (9 cm), intermediate (76 cm), or old (230 cm) that was inferred from crosscutting relations seen on the SeaMARC IA image. The muds overlying the sands or debris-flow deposits have properties similar to those present in the surrounding basin plain (Table 13–1). Accordingly, we suggest that the subsurface sands and debris-flow deposits produce the high backscatter on the sidescan mosaic. Such a relation between sand presence and high acoustic backscatter implies either the dominance of interface backscattering or an association of the presence of subsurface sands and chaotic silts with enhanced volume scattering. Also, the results suggest that the 30-kHz signal of SeaMARC IA penetrates at least 2.3 m into the muddy sediment.

The sediment from the low-backscatter basin plain areas beyond the acoustically defined lobes and sublobes is characteristically a nearly uniform hemipelagic gray clay with occasional thin silt laminae in the uppermost 4 to 6 m (Figure 13–8 shows a characteristic impedance profile). The sediment has density values similar to those of basin-plain sediment

Figure 13–3. A part of the GLORIA images from the Gulf of Mexico showing depositional lobes on the outer Mississippi Fan. Lighter areas represent higher backscatter.

elsewhere in the world (Hamilton 1970). The hemipelagic sediment displays low backscatter on the sidescan mosaic.

To develop a quantitative comparison of sidescan sonograph characteristics with sediment character, we first obtained representative values of the processed sidescan eight-bit backscatter intensity (DN) over an eleven- by eleven-pixel grid (137 × 137 m area on the SeaMARC IA images) surrounding each core location and averaged to obtain a representative value. During processing, backscatter values are modified to account for distance from nadir and other geometric and regional effects. In general, although not quantitatively precise, higher values of DN represent a greater backscatter of sound. For a quantitative comparison, Figure 13–9A shows this representative mean value of DN plotted versus the mean acoustic impedance for each core considered in the basin plain, beyond the acoustically defined lobes. We assume the mean acoustic impedance is representative of the seafloor surface, because the basin-plain sediment is homogeneous and has little subbottom variation (e.g., see Figure 13–8).

Figure 13–4. SeaMARC IA imagery of part of the intermediate sublobe of the Mississippi Fan study area. Lighter areas represent higher backscatter. Locations of some of the cores considered in the present study are shown by dots.

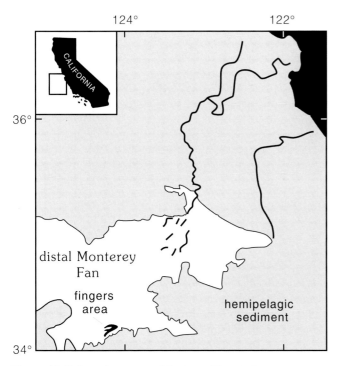

Figure 13–5. Location map – Monterey Fan study area.

Also shown on Figure 13–9A are the representative mean DN values and the mean acoustic impedance of the uppermost sand bed for cores from the young sublobe. We assume the acoustic impedance for the upper sand bed is representative of the seafloor surface because only a thin (<10 cm) veneer of mud overlies it. Impedance contrast across interfaces (rather than acoustic impedance itself) is the critical parameter influencing interface scattering. Because bottom seawater impedance is nearly constant over this area, we can consider the representative acoustic impedance of the seafloor to be directly related to, and a measure of, the acoustic impedance contrast across the seafloor. Although the mean DN – mean acoustic impedance data have considerable scatter, the correlation shown in Figure 13–9A is consistent with interface backscattering being dominant. That is, in a general sense, the higher the acoustic impedance contrast across an interface, the greater the amount of backscatter in this area.

For each of the three Mississippi Fan sublobes considered, the mean acoustic impedance of the uppermost sand or chaotic silt is roughly the same, although the thickness of overlying mud varies greatly (Table 13–1). Most likely,

124°30'
35°00'

123°00'

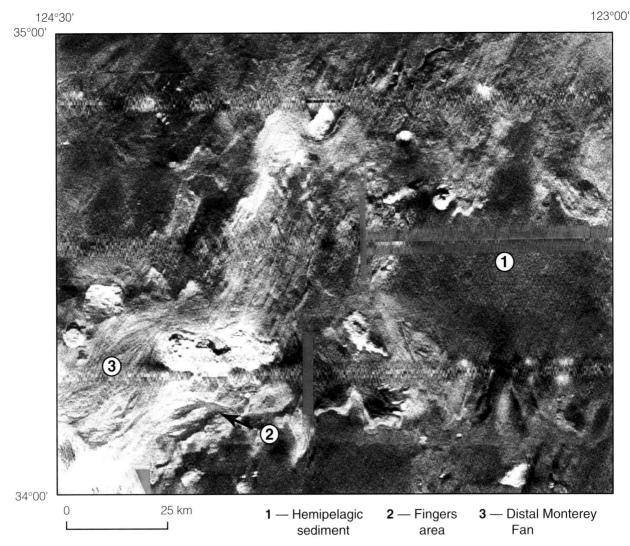

34°00'

0   25 km

1 — Hemipelagic   **2** — Fingers   **3** — Distal Monterey
sediment   area   Fan

Figure 13–6. A part of the GLORIA image from the California margin showing depositional lobes on the distal Monterey Fan.

acoustic backscatter from each of the sublobes varies as a result of sound attenuation in the overlying mud units. In Figure 13–9B we plot representative DN values versus thicknesses of the overlying mud units (equivalent to depth to top of sandy turbidite – chaotic silt facies). Again, the data show scatter (linear regression correlation coefficient of only 0.49), but the general trend is consistent: thicker surficial mud units tend to produce less backscatter, probably because sound scattered from the first turbidite – chaotic silt facies is attenuated in passing through the mud, both in the transmitted and backscatter paths.

Perhaps only coincidentally, if the DN, acoustic impedance, and mud thickness values from Figure 13–9 are averaged for each of the three lobes and the basin plain, a relatively good trend emerges (Figure 13–10). A large separation is apparent in Figure 13–10A between basin-plain (hemipelagic) sediment and the sandy turbidite – chaotic silt facies for the young sublobe. Interestingly, a nearly linear

relation between mean DN and mean surficial mud thickness is apparent for the three sublobes in Figure 13–10B.

Monterey Fan

All seven of the box cores from the low-backscatter fingers area of Monterey Fan have a similar lithostratigraphy: approximately 10 cm of silty clay overlying a thick (44 cm or more) unit of graded sand (see Figures 13–11 and 13–12 for typical lithologies and acoustic impedance profiles). The sand has density and sound velocity characteristics that are remarkably similar to those of sandy turbidites from the Mississippi Fan (Table 13–1). The only obvious difference between the Monterey and Mississippi sands is that the Monterey sands contain less mud (16% <63 μm vs. 48% <63 μm for the Mississippi sands).

The depositional lobe deposits outside the fingers area are varied and contain hemipelagic mud, clay, and silt clasts,

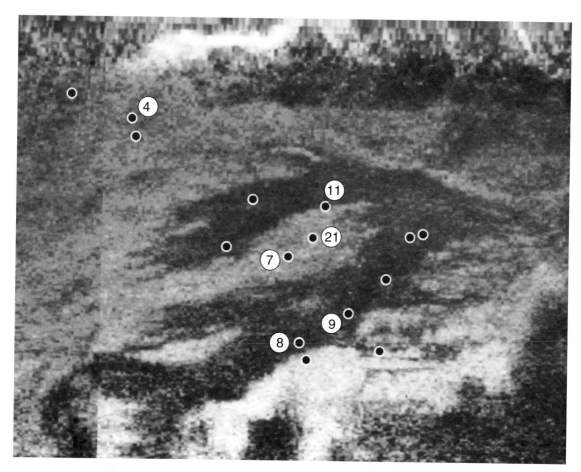

Figure 13–7. GLORIA imagery, fingers area and immediate vicinity; Monterey Fan. Locations of cores considered in the present study are shown by dots.

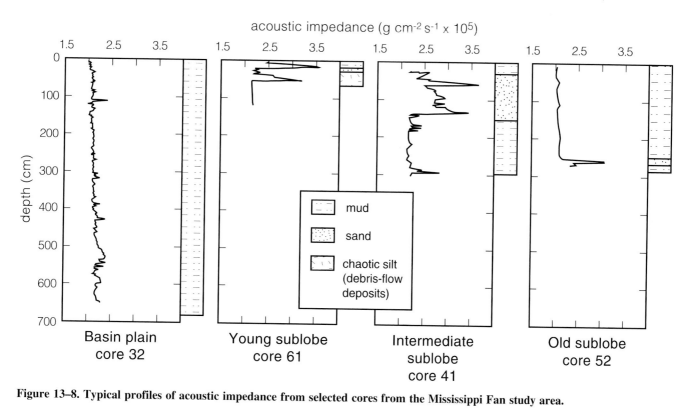

Figure 13–8. Typical profiles of acoustic impedance from selected cores from the Mississippi Fan study area.

Table 13–1. *Average properties of cores from Monterey and Mississippi Fan study areas*

	Monterey Fan	Mississippi Fan
Acoustically defined fingers area (Monterey) or lobe areas (Mississippi)		
Surficial hemipelagic mud (mud above first subbottom facies interface) Thickness (cm)	10	9 (young sublobe) 76 (intermediate sublobe) 230 (old sublobe)
Sound velocity (km/s)	1.55	1.51
Density (g/cm^3)	1.43	1.46
Acoustic impedance (g cm^{-2}s^{-1} $\times$ 10^5)	2.21	2.21
Sediment subjacent to hemipelagic mud (below first subbottom facies interface; sandy turbidites and chaotic silts) Mean thickness (cm)	>44	48 (range of 10–100)
Sound velocity (km/s)	1.74	1.64
Density (g/cm^3)	1.84	1.93
Acoustic impedance (g cm^{-2}s^{-1} $\times$ 10^5)	3.20	3.17
Mean % sand	84	52
High backscatter area surrounding fingers; distal Monterey Fan (entire cores)		
Sound velocity (km/s)	1.59	
Density (g/cm^3)	1.52	
Acoustic impedance (g cm^{-2}s^{-1} $\times$ 10^5)	2.41	
Hemipelagic sediment beyond fan deposits (entire cores)		
Sound velocity (km/s)	1.50	1.51
Density (g/cm^3)	1.44	1.40
Acoustic impedance (g cm^{-2}s^{-1} $\times$ 10^5)	2.16	2.11

apparent debris-flow materials, and some sandy turbidites (Figures 13–11 and 13–12). The mean acoustic impedance of the sediment outside the fingers area is low (2.41 g/cm^2/s $\times$ 10^5) relative to that of the sand from the fingers area (3.2 g/cm^2/s $\times$ 10^5). The sand-rich fingers area has lower backscatter than the relatively higher backscatter of the surrounding deposits on GLORIA imagery (Figure 13–7). In addition, the fingers do not show up at all on the 30-kHz TOBI sidescan sonar images (Gardner et al. 1991a). These results argue against interface scattering as the dominant backscatter mode for this area. If interface scattering were dominant, the sand-rich, high acoustic impedance deposits would have high backscatter and the lower acoustic impedance deposits outside the fingers would have lower backscatter. In fact, Lyons et al. (in press) showed that volume scattering in the heterogeneous sediment beyond the fingers is very significant and most likely the cause of the high backscatter of these deposits.

The abyssal-hill hemipelagic sediment beyond the distal fan deposits has a similar acoustic impedance (2.16 g/cm^2/s $\times$ 10^5) to that of the basin-plain sediment beyond the Mississippi sublobes and produces comparably low acoustic backscatter in contrast with the neighboring deep-sea fan.

Figure 13–13 shows these results quantitatively in the form of mean acoustic impedance versus processed DN value from the GLORIA image (GLORIA DN not directly comparable with the DN values from the SeaMARC IA). This figure emphasizes the anomalous nature of the fingers area. With some data scatter, the sandy sediment from the fingers area typically has a higher mean acoustic impedance and a lower amount of backscatter than the distal fan sediment surrounding the fingers area.

Not only is the backscatter from the fingers area low relative to the surrounding distal fan sediment, but it is also low relative to that of the hemipelagic sediment that lies be-

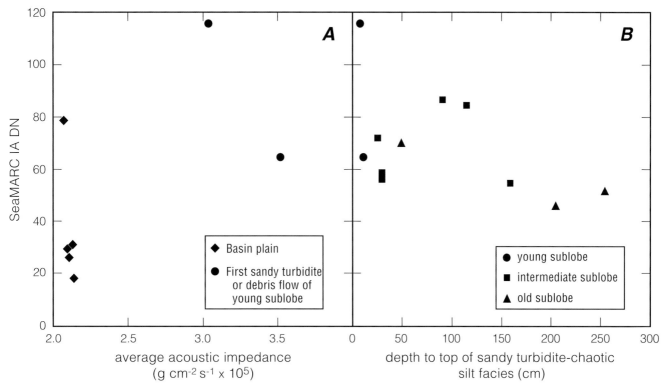

Figure 13–9. Sediment characteristics versus processed SeaMARC IA DN, a measure of intensity of backscatter; full suite of cores, Mississippi Fan study area.

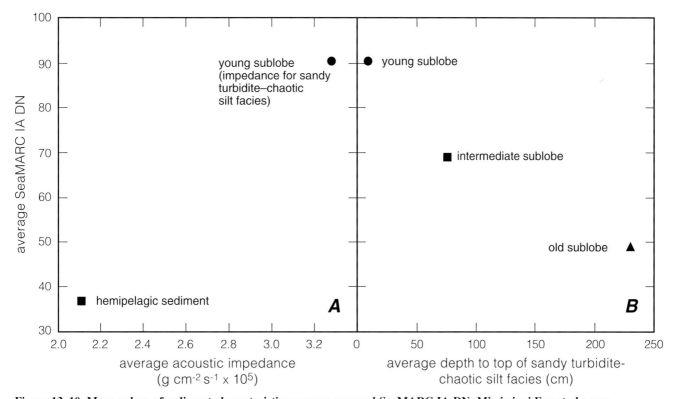

Figure 13–10. Mean values of sediment characteristics versus processed SeaMARC IA DN; Mississippi Fan study area.

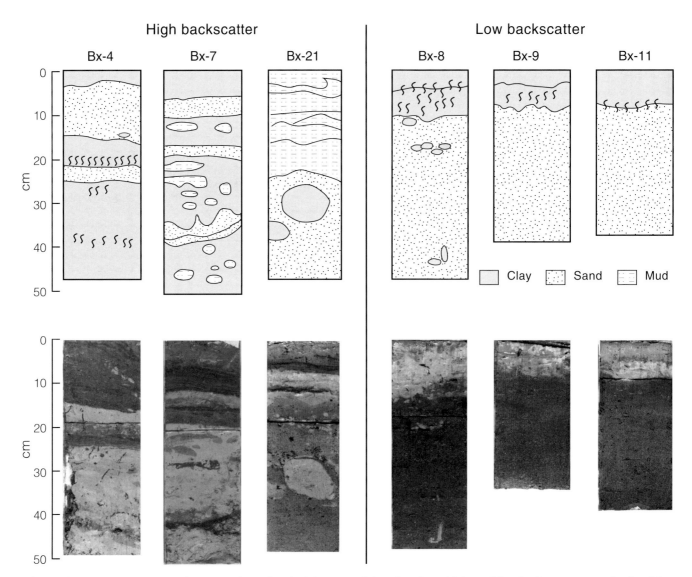

Figure 13–11. Photographs and interpretations of typical box core slabs taken from high- and low backscatter areas in the vicinity of the fingers area; Monterey Fan.

yond the fan. This effect cannot be shown well by evaluating processed DN values because radiometric corrections applied to data files during processing are impacted by the absolute DN values contained within each data file. Consequently, data values of neighboring points skew the resultant processed DN values. Areas containing hemipelagic sediment typically lie in different data files from those showing the fingers. To make a quantitative comparison of data from the fingers, surrounding distal fan, and surrounding hemipelagic sediment, we must apply data that have not been subjected to radiometric corrections. These uncorrected or raw acoustic data were taken from subregions of the imagery at a constant distance from nadir to avoid water-column and grazing-angle influences. Figure 13–14 shows these mean values of raw DN plotted versus mean values of acoustic impedance for the fingers area, high-backscatter

area surrounding fingers, and a core from the hemipelagic sediment beyond the fan deposits. This figure confirms that the sandy sediment from the fingers area has the highest acoustic impedance (at least over the depth range that was sampled) and the lowest amount of backscatter, even lower than that of the hemipelagic sediment that surrounds the fan.

Discussion

Some of the results of this study agree with the assumption often used by marine geologists in interpreting sidescan sonar imagery that sandy sediment produces higher backscatter in comparison with muddy sediment. For example, the depositional lobes of Mississippi Fan and the overall distal Monterey Fan show high backscatter compared with the sur-

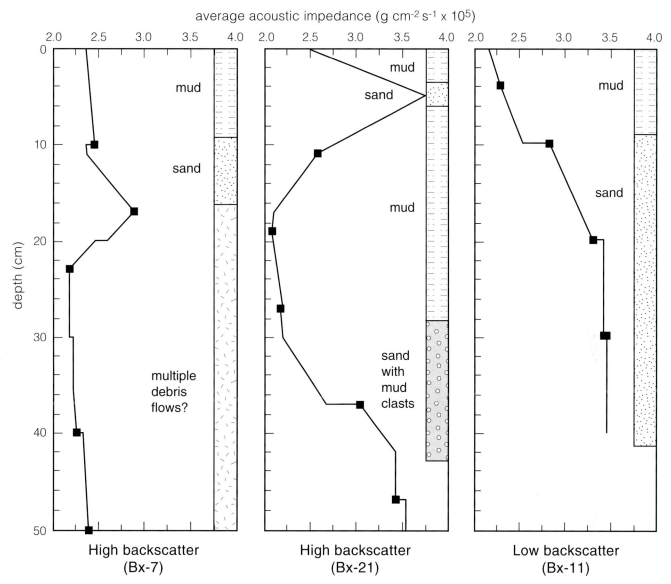

Figure 13–12. Acoustic impedance profiles from selected sediment cores taken from the vicinity of the fingers area; Monterey Fan.

rounding hemipelagic sediment. In contrast with this assumption is the response of the fingers area on Monterey Fan. Gardner et al. (1991b) presented a numerical model of sound reflection and refraction for this area and could not explain the causes of the low backscatter returned from the fingers. In addition, the sedimentology and physical properties of the Mississippi sublobe sands are almost identical to the fingers sands but produce an opposite backscatter response.

One partial explanation for this response lies in the geologic complexity of fan deposits that surround and underlie the sandy deposit of the fingers area. The deposits are highly heterogeneous and likely contain internal inhomogeneities such as debris-flow clasts that could serve as good volume scatterers of sound (Lyons et al. in press). Also, the properties shown in Table 13–1 and Figure 13–13 were measured on box core samples that were only 20 to 60 cm long. Because the GLORIA system has a frequency of 6.5 kHz, the signal penetrated meters into the seafloor and may have backscattered from more deeply buried sediment having a high acoustic impedance (Gardner et al. 1991b). Accordingly, the fan deposits may be much better scatterers of sound than suggested by the low measured mean acoustic impedance of box cores, and the very high backscatter (DN values in Figures 13–13 and 13–14) from the fan deposits may be explained in this way. If so, a partial explanation for the low-backscatter appearance of the fingers would be contrast with the very high backscatter from the distal fan.

Contrast with an area of very high backscatter is not the full explanation for the low backscatter of the fingers. The unprocessed GLORIA DN values (at a common distance

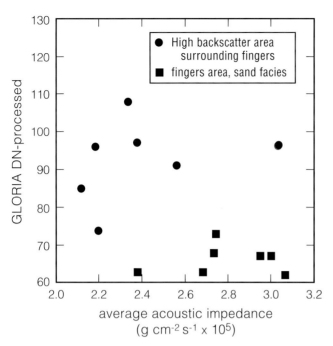

Figure 13–13. Sediment characteristics versus processed GLORIA DN; full suite of cores, Monterey Fan study area.

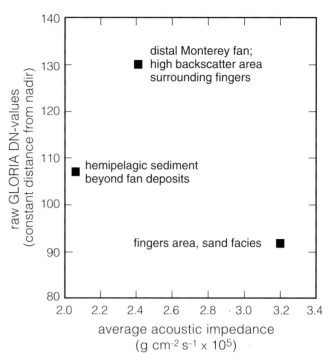

Figure 13–14. Sediment characteristics versus raw GLORIA DN values at a constant distance from nadir; Monterey Fan study area.

from nadir) show that the fingers area produces even less backscatter than the hemipelagic sediment surrounding the distal fan (Figure 13–14). Accordingly, this is a clear example of a sandy deposit producing less backscatter than a mud. Attributes that could cause such low backscatter are (1) a smooth sand surface (below the Rayleigh roughness criteria) that would reflect most sound away from the ship rather than scatter it back (e.g., a situation like that of Figure 13–1A) and (2) a significant thickness of coarse, homogeneous sand that would have low backscattering within the sand and would attenuate sound passing through the sand (Hamilton 1972) and prevent backscattering from deeper reflectors or volume scatterers. We observed that the sand from the fingers has a lower fines content and possibly is thicker than the sand from the Mississippi Fan. Both of these factors could lead to greater attenuation.

Conclusions

Sidescan sonar is a valuable tool for mapping sediment facies variations in topographically simple environments. However, the thinking by many that coarser sediment always produces greater acoustic backscatter is incorrect. In some cases, an inverse correlation between sediment grain size and backscatter can exist, as was seen in the Monterey Fan fingers area. Clearly, those interpreting sidescan images should be cautious in believing that coarser sediment textures necessarily result in higher backscatter. Ground truthing by sampling sediment is required to confirm inter-

pretations derived from sidescan-sonar imagery until better theories are developed.

The scattered data in Figures 13–9 and 13–13 show that DN values obtained from GLORIA and SeaMARC IA images relate in a rough sense to the acoustic impedance values from associated cores. These acoustic impedance values relate, in turn, to sediment lithology and other characteristics and properties. The scatter of data is too great, however, to suggest that pixels or groups of pixels can be used to predict specific sediment character at a particular seafloor location. In general, however, regional patterns of acoustic backscatter relate to regional sediment characteristics.

References

EEZ-SCAN 84 Scientific Staff 1986. Atlas of the Exclusive Economic Zone, western conterminous United States. *U.S. Geological Survey Miscellaneous Investigations Series I-11792*, 152 p.

EEZ-SCAN 85 Scientific Staff 1987. Atlas of the Exclusive Economic Zone, Gulf of Mexico and eastern Caribbean areas. *U.S. Geological Survey Miscellaneous Investigations Series I-1864-A*, 104 p.

Gardner, J. V., Field, M. E., Lee, H. J., Barber, J. H., and Masson, D. G. 1991a. Recent major erosional and depositional processes on Monterey Fan: Evidence from mid-range sidescan sonar (abs). *Geological Society of America 1991 Annual Mtg., Abstracts and Program*, p. A385. Boulder, Colo.: GSA.

Gardner, J. V., Field, M. E., Lee, H. J., Edwards, B. D., Masson, D. G., Kenyon, N. H., and Kidd, R. B. 1991b. Ground-truthing 6.5 kHz side scan sonographs: What are we really imaging? *J. of Geophys. Res.* 96: 5955–74.

Hamilton, E. L. 1970. Sound velocity and related properties of marine sediments, north Pacific. *J. Geophys. Res.* 75:4423–46.

Hamilton, E. L. 1972. Compressional-wave attenuation in marine sediments. *Geophysics* 37: 620–46.

Lyons, A. P., Anderson, A. L., and Dwan, F. S. in press. Acoustic scattering from the seafloor; Modeling and data comparisons. *J. Acoust. Soc. Am.*

Nelson, C. H., Twichell, D. C., Schwab, W. C., Lee, H. J., and Kenyon, N. H. 1992. Upper Pleistocene turbidite sand beds and chaotic silt beds in the channelized, distal outer-fan lobes of the Mississippi fan. *Geology* 20: 693–6.

Ramirez-Bader, L., Gardner, J. V., and Field, M. E. 1991. Core descriptions, grain size, and carbon analyses data of cores collected on cruises F5-87-SC, F1-88-SC and F3-89-SC from the Monterey Fan, off central California. *U. S. Geological Survey Open-File Report 91-625.* Reston, Va.: USGS.

Twichell, D. C., Schwab, W. C., Nelson, C. H., Kenyon, N. H., and Lee, H. J. 1992. Characteristics on a sandy depositional lobe on the outer Mississippi fan from SeaMARC IA side-scan-sonar images. *Geology* 20: 689–92.

Urick, R. J. 1983. *Principles of Underwater Sound*, Ed. 3, 423 p. New York: McGraw-Hill.

14 Morphology of small submarine fans, inner California continental borderland

Brian D. Edwards,[1] Michael E. Field,[1] and Neil H. Kenyon[2]

[1]U.S. Geological Survey, Menlo Park, California
[2]Institute of Oceanographic Sciences, Southampton, United Kingdom

Abstract

Long-range sidescan sonographs from the GLORIA sidescan sonar system provide a new perspective on the morphology and sediment distribution of small active submarine fans in the Santa Monica and San Pedro Basins of the California Continental Borderland. These sonographs, combined with 3.5-kHz seismic-reflection profiles, depict elongate submarine fan systems characterized by intermediate acoustic backscatter in the middle fan region and low backscatter in the distal reaches where the lower fan feeds onto the high-backscatter central basin plain. The fans are fed by low-backscatter channels that originate in shallow water at the northwest corner of each basin. These channels subsequently branch downstream into a system of smaller channels and lineations that extend to the tips of the distalmost deposits. In these distal reaches, fan deposition occurs in low-relief (1 to 2 m), tapering, low-backscatter fingerlike distributaries that extend to the high-backscatter central basin. The low-backscatter fingers are lens-shaped in cross section. Topographic lows occurring between adjacent fingers apparently direct the transport of subsequent flows with resulting shifts of the depocenters over time.

Core samples show that turbidity currents have deposited coarse sediment beyond the mid- and lower-fan environments and onto the western part of the flat central floor of both basins. The cores, combined with bottom photographs and high-resolution seismic-reflection profiles, indicate that patterns observed in sedimented areas on the GLORIA mosaic are caused largely by scattering from volume inhomogeneities and subbottom interfaces of sediment layers within the upper few meters of the sediment column. Neither variations in surface-sediment facies (upper few centimeters) nor small-scale surface features (on the order of centimeters) appear to affect GLORIA backscatter intensity in these sediment-dominated areas. Rather, thick masses of sand in the fingerlike distributaries likely are responsible for the low backscatter. In contrast, laterally extensive sand layers, deposited by larger-volume turbidity currents that flowed across the distal basin floor, appear to produce either high or low backscatter depending on the relative thickness of the sands and the nature of the interbedded turbidite and hemipelagic muds.

Introduction

The California Continental Borderland is a tectonically active province that evolved throughout the Neogene as part of a broad transform plate boundary between the North American and Pacific Plates (Shepard and Emery 1941; Atwater 1970; Crowell 1974; Howell 1976). During this time, lateral shearing and related folding along east-west and northwest-southeast fractures have produced a series of basins separated by irregular topographic highs (sills and islands) that extend from Point Conception, California, to Isla Cedros, Baja California, Mexico. Santa Monica and San Pedro Basins, located immediately offshore of the Los Angeles metropolitan area, occupy the central, inner part of this borderland (Figure 14–1).

The dominant sediment input to these basins is a combination of detritus from adjacent coastal streams combined with pelagic sediment from a core of nutrient-rich upwelling water centered south of the Northern Channel Islands (Gorsline 1980). Of these inputs, 70% to 80% of the total flux (Schwalbach and Gorsline 1985) is derived from intermittent rivers that drain the adjacent mountainous semiarid regions. Early studies by Gorsline and Emery (1959) described the general character of basin infilling that produced a basin floor of ponded sediment with a very smooth and nearly horizontal surface (<1:600). The nearly continuous rain of hemipelagic sedimentation that characterizes the inner borderland is supplemented by the episodic deposition of sands and coarse silts as turbidity currents flow into the basins through the small canyons and channels that head nearshore and incise the continental shelf and slope. During the present eustatic high sea level, basin-wide turbidites have been deposited at intervals of 100 to 400 years (Gorsline

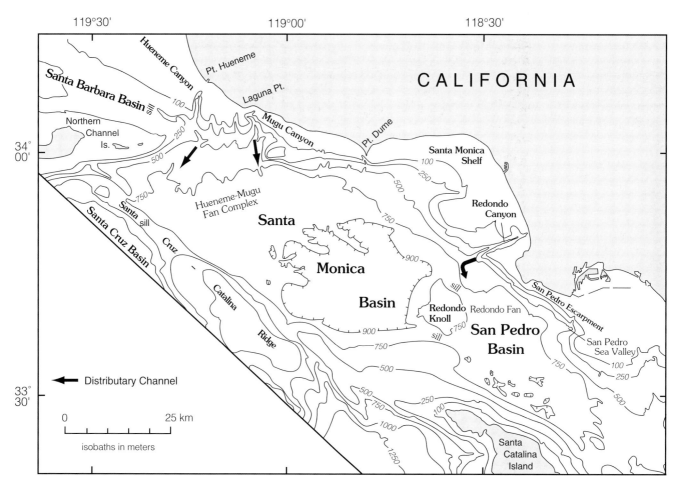

Figure 14–1. Physiographic map of Santa Monica and San Pedro Basins showing bathymetry, major morphologic features, and major distributary systems.

and Emery 1959; Malouta, Gorsline, and Thornton 1981; Reynolds 1987). In contrast, during eustatic low sea levels of glacial periods, the rate of terrigenous contribution to the basin increased several fold (Gorsline and Prensky 1975; Pao 1977) through a significantly higher frequency of turbidity-current events. The resulting basin-fill deposits consist of green hemipelagic muds interbedded with sand and silt layers that range in thickness from partings to layers at least 1 m thick.

Fan sedimentation within Santa Monica Basin has been the most important depositional process over the past 2.5 my, followed by downslope mass movements along the mainland margin and settling from plumes of suspended sediment centered off Pt. Dume (Field and Edwards 1980; Gorsline 1980; Teng and Gorsline 1989). Two fan systems dominate Santa Monica Basin: the Hueneme Fan fed by Hueneme Canyon, and the Mugu Fan fed by Mugu Canyon. Both fans head in the northwestern part of the basin and coalesce to form the Hueneme-Mugu Fan complex, which comprises more than 50% of the basin area. The fan complex covers 750 km² and, in the thickest parts, contains more than 3 km of Cenozoic sediment (Teng and Gorsline 1989;

Schwalbach, Edwards, and Gorsline in press). Hence, the complex is about 0.25% the size of the Mississippi Fan (Twichell et al. 1991) and is similar in size, setting, and general morphology to the well-studied Navy Fan (Normark and Piper 1972; Normark, Piper, and Hess 1979) located farther south in the borderland. A large (140 km²) area of mass movement in the proximal part of Hueneme and Mugu Fan complex separates the upper mid-fan regions of the two fans (Nardin 1983).

Sedimentation in the adjacent San Pedro Basin is dominated by a single submarine fan, the Redondo Fan, fed by the Redondo Canyon. The fan is smaller than the Hueneme-Mugu Fan complex and covers only about 200 km² (Normark 1974; Nardin 1983). The direction of fan growth has been controlled, in part, by the Redondo Knoll, which deflects sediment southeastward from the mouth of Redondo Canyon and into San Pedro Basin (Haner 1971; Nardin 1983). A small inactive fan at the mouth of the San Pedro sea valley has coalesced with the Redondo Fan and this combined system is bounded to the north by a large (100 km²) area of mass movement that originated on the San Pedro Escarpment (Figure 14–1).

Seismic-reflection studies show that these fan systems are characterized by a complicated high-relief mounded seismic facies in the upper fan, broad low-relief mounded facies in the mid-fan, and onlap basin-fill facies in the lower fan (Nardin 1983). Locally, the restricted nature of these small borderland basins confines fan development and at some locations renders the lower-fan and basin-plain seismic facies indistinguishable from each other based on seismic-reflection data alone (Teng and Gorsline 1989).

Although the morphology and depositional histories of these fan systems are generally well established, the detail of the surface channels and distal reaches of these systems have not been clearly identified. Our intent in this chapter is (1) to describe the sedimentary features of the fans within two of the small confined basins of the inner California Continental Borderland by combining GLORIA imagery with high-resolution seismic-reflection data and published sedimentological data and (2) to evaluate factors controlling GLORIA backscatter intensity.

Methods

We used the GLORIA II (Geological LOng-Range Inclined Asdic) sidescan sonar during the summer of 1984 to conduct a reconnaissance-scale survey of the West Coast Exclusive Economic Zone of the United States including the tectonically active inner California Continental Borderland (Figures 14–1 and 14–2). GLORIA II, a shallow-towed digital chirp sidescan sonar developed by the Institute of Oceanographic Sciences (IOS) of the United Kingdom operates at a central frequency of about 6.5-kHz with a wavelength of about 20 cm (Somers et al. 1978; Somers this volume). The relatively fast towing speeds (typically 8 kn) made possible by the shallow tow depth (about 50 m) allows rapid insonification of large areas of the seafloor. Concerns of potential vehicle impact with the seafloor prevent towing GLORIA in shallow water (<300 meters). Accordingly, some regions, notably shelf and upper-slope areas, were not surveyed. In addition, the rugged, irregular topography and shallow waters of the borderland required unusually close trackline spacing during acquisition of the GLORIA data (Figure 14–2A).

Variations in signal strength backscattered from the seafloor are recorded by GLORIA as digital values that range from 0 to 255. After computer processing to correct for geometric and radiometric distortions, the system has a resolution of approximately 50 m by 50 m (Chavez 1986). High backscatter is displayed as bright areas on the images, whereas low backscatter and acoustic shadows are displayed as dark areas on the images. Additionally, minor variations in processing for radiometric distortions can result in tonal differences between computer files that make up the mosaic.

More than 900 trackline-km of digital GLORIA sonar data

were collected in the inner California continental borderland concomitantly with 10-kHz bathymetric, 3.5-kHz seismic-reflection, and 160-in^3 airgun data during the 1984 field season. Additional uniboom and 3.5-kHz seismic-reflection data collected during a 1981 cruise have been incorporated into the data set. Navigation for these cruises was by Loran-C in hyperbolic mode combined with independent transit satellite fixes. Navigational accuracy is estimated to be about ±200 m. Researchers at the University of Southern California provided sedimentologic information from box cores, piston cores, and bottom photographs. Tracklines for both the 1981 seismic-reflection and 1984 GLORIA surveys are shown in Figure 14–2A.

The GLORIA mosaic

Hueneme-Mugu Fans

The Hueneme and Mugu Fans are displayed as elongate sediment systems that extend from the northwest margin of the Santa Monica Basin onto the central basin floor (Figures 14–2B and 14–3). The Hueneme Fan covers approximately 500 km^2 and is clearly defined for a distance of about 65 km from the canyon headwall into the central basin. To the east, the upper- and mid-fan regions of the Hueneme Fan blend into the Mugu Fan; the submarine mass-movement deposit identified by Nardin (1983) cannot be resolved. The lower mid-fan and lower fan regions of Hueneme and Mugu Fans are distinct. To the west the Hueneme Fan is bounded by the basin slope and sill separating Santa Monica Basin from Santa Cruz Basin (the Santa Cruz – Catalina Ridge). In comparison, the 40-km-long Mugu Fan, bounded on the east by the mainland slope and the much smaller Dume and Santa Monica Fans, covers about half as much area (350 km^2) as the Hueneme Fan.

Due to shallow water and the confined nature of the borderland, only limited parts of the upper-fan region were imaged during the survey. Accordingly, the upper-fan region is not displayed well in the data. The mid-fan region has a generally uniform intermediate backscatter character except where crossed by feeder channels. The uncharacteristically high backscatter tone seen in a part of the western mid-fan region of Hueneme Fan occurs due to a computer processing error (Figure 14–3A). In contrast to the nondescript intermediate-backscatter tone of the mid-fan region, the feeder channels for both the Hueneme and Mugu Fans are seen as low-backscatter curvilinear features. These channels become less distinct in the lower mid-fan regions of both Hueneme and Mugu Fans where, in the distal reaches, they degrade into high- and low-backscatter linear streakings that are about 1 to 2 pixels (50 to 100 m) wide (Figures 14–2B and 14–3). Coincident with this channel degradation, the Hueneme Fan curves eastward, probably in response to topographic steering of flows by the steep wall of the sill be-

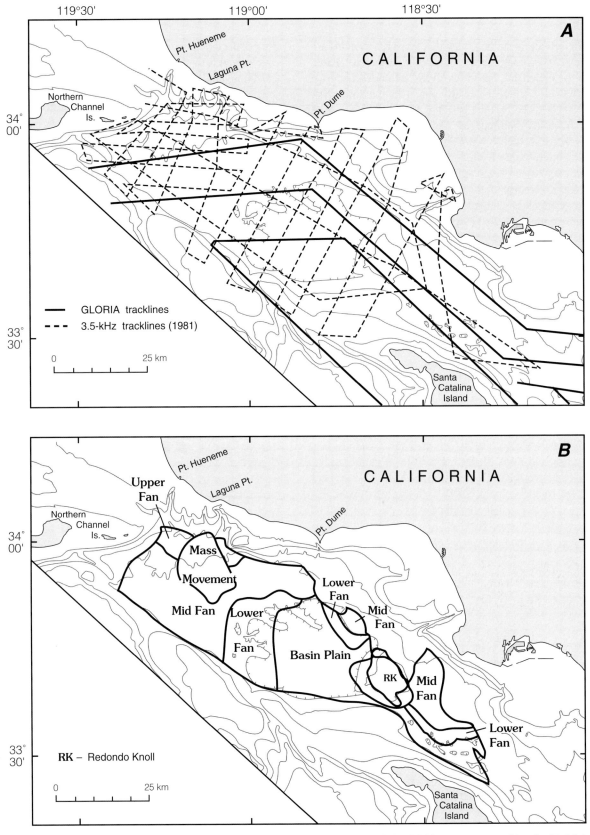

Figure 14–2. A, Trackline chart of the study area. Heavy lines represent tracks of GLORIA coverage combined with high-resolution seismic-reflection data. Thin lines show tracks of ancillary (1981) high-resolution seismic-reflection data. B, Zonation of late Quaternary depositional systems in Santa Monica and San Pedro Basins based on seismic-reflection facies analysis. Modified from Nardin (1983).

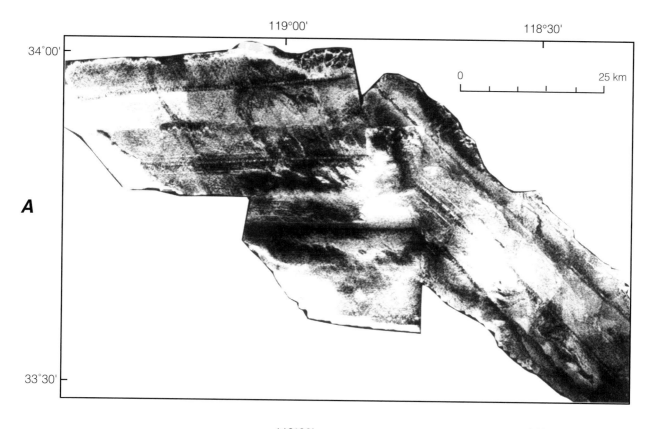

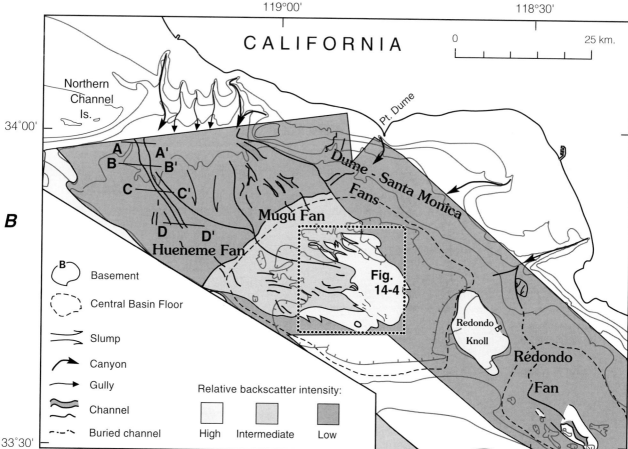

Figure 14–3. A, Digitally mosaicked processed GLORIA sonograph. B, Line drawing interpretation of major features in Santa Monica and San Pedro Basins. Note the development of a branching distributary system on the upper mid-fan and the tapering digitate form of the distal fan deposits (dark, fingerlike region in the central basin). Note also that both bedrock outcrops (e.g., Redondo Knoll) and ponded basin-plain sediment can give high-backscatter returns.

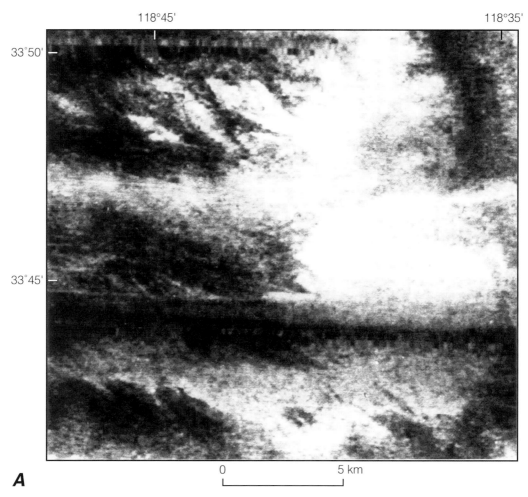

118°45'   118°35'

33°50'

33°45'

0   5 km

A

Figure 14–4. A, Enlarged sonograph of the central Santa Monica Basin.

tween Santa Monica and Santa Cruz Basins. Such deflection of flows by steep walls is a common feature of turbidity-current transport and fan growth in confined elongate basins (Hsü, Kelts, and Valentine 1980; Siegenthaler, Hsü, and Kelboth 1984; Kneller et al. 1991).

The lower-fan region of both fans is characterized by bifurcating low-backscatter fingers that are continuous with narrow (50 to 100 m wide) high- and low-backscatter streaking patterns (feeder channels?). These fingers taper to sharp points that terminate in the central area of Santa Monica Basin. Apparent overlapping of fingers in both the Hueneme and Mugu Fan systems is suggestive of lateral migration and temporal shifting of the feeder channels. Where the fingers terminate, the flat central basin area is characterized by contrasting high backscatter (Figures 14–3 and 14–4). This high-backscatter region extends 10 to 15 km basinward from the tips of the fingers and occupies much of the basin plain defined by Nardin (1983). The distal boundary of the high-backscatter region typically has a more lobate form although some fingering is characteristic of the southern boundary. Small high-backscatter pockets occur where the

low-backscatter fingers bifurcate and overlap, leaving the impression that the sediment represented by the low-backscatter fingers stratigraphically overlies the high-backscatter region. The remainder of the basin-plain environment is characterized by intermediate backscatter tones.

Redondo Fan

The Redondo Fan is smaller than either the Hueneme or Mugu Fans. The main distributary channel for the fan extends from the mouth of Redondo Canyon, trends southwest toward Redondo Knoll, then bifurcates, one channel (perhaps older) turning westward toward the eastern part of Santa Monica Basin and the other branching southeastward into San Pedro Basin (Haner 1971). The westward-trending branch produces no resolvable features on the GLORIA data from Santa Monica Basin. The southeastern branch of Redondo Channel, however, is somewhat sinuous along its length but becomes obscured on the middle-fan surface, possibly by a large region of sediment failure that originated on the steep San Pedro Escarpment (Gorsline et al. 1984). The

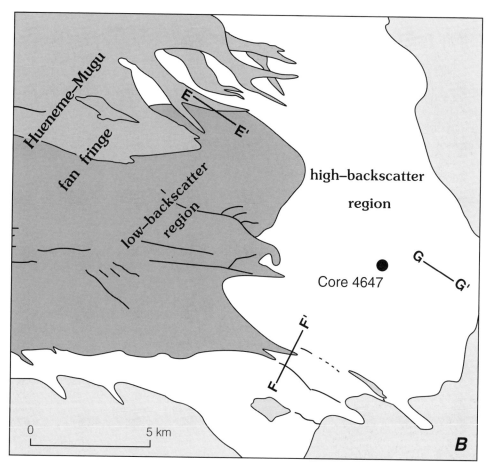

Figure 14–4. *Continued.* **B,** Line drawing detailing the boundary between distal Hueneme-Mugu Fan Complex deposits and central basin sediment. Note the numerous elongate channel-like deposits that lead to the digitate branching tips of the distal fan. Here, the low-backscatter quality of the distal fan deposits is in marked contrast with the high-backscatter character of the central basin sediment.

channel meanders south and east on the lower mid-fan and lower fan within San Pedro Basin and then, much like the Hueneme-Mugu Fan complex, terminates on the adjacent high-backscatter basin plain in a 1-km-wide low-backscatter digitate feature (Figure 14–3).

Miscellany

Although incompletely imaged because of shallow water depths, much of the mainland slope in Santa Monica Basin is characterized by slope-normal linear features that likely correlate with gullies leading into the basin from the shelf edge. A small (8 km²) slope failure is located just to the east of Redondo Canyon (Figure 14–3A and B) and is distinguished from the adjacent slope sediment by a bounding low-backscatter line that likely represents the headwall scarp. The bounding Santa Cruz-Catalina Ridge to the south and west is characterized by high backscatter. Another irregularly shaped high-backscatter region lies between Santa Monica and San Pedro Basins. This feature correlates with

Redondo Knoll — a large bedrock high composed of Miocene sedimentary and volcanic rocks and pre-late Cretaceous metamorphic rocks (Vedder et al. 1986) that rises 350 m above the basin floor.

Controls on backscatter intensity

Tonal variations observed in the GLORIA imagery are controlled by the intensity of acoustic energy backscattered from the seafloor as well as by subsequent computer processing of data files. Earlier interpretations of sidescan sonar data from relatively flat regions of sediment-covered seafloor typically assumed that backscatter intensity is controlled by (1) local variations in high-frequency bed roughness and (2) acoustic impedance contrasts at the sediment-water interface (Kenyon, Belderson, and Stride 1975; Roberts, Kidd, and Masson 1984; EEZ-SCAN 84 Scientific Staff 1986; Gardner and Kidd 1987). More recent studies, however, show that, as with seismic-reflection profiling, the sidescan sonar acoustic energy penetrates into the sediment and this penetration can have a profound effect on the backscatter response. Thus, in general, the strength of backscatter intensity is controlled by a combination of three fundamental parameters: (1) the angle of incidence of the incoming wave to the seafloor, (2) the volume inhomogeneities within the bed, and (3) changes in acoustic properties across the sediment-water interface and across shallow subbottom sediment interfaces (Gardner et al. 1991; Lyons, Anderson, and Dwan 1994; Lee et al. this volume).

Within parts of the study area, slope angle (which relates to the angle of incidence) of the acoustic target can dominate and control backscatter intensity. For example, in the inner borderland, the characteristic high backscatter of topographically rugged areas at basin margins and interbasin highs (e.g., Santa Cruz – Catalina Ridge and Redondo Knoll) probably results from steep slopes combined with thin or absent sediment cover over rough bedrock. In contrast, in sedimented areas (such as mid-fan channel deposits, distal-fan deposits, and central-basin floors), distinct differences in acoustic backscatter are observed that cannot be caused by slope angle, variations in incidence angle, or bed roughness. Here, GLORIA backscatter differences probably result from variations of one or a combination of factors such as the following: (1) small-scale roughness of the seafloor, (2) seafloor and shallow subbottom interfaces having impedance contrasts, or (3) sediment heterogeneity within the subbottom. We used high-resolution seismic-reflection profiles, bottom photographs, and core samples to evaluate the importance of these variables.

High-resolution (3.5-kHz and uniboom) seismic-reflection profiles provide evidence that neither slope angle nor angle of incidence exerts a major control on backscatter intensity observed from these highly sedimented environments. For example, the profiles show the upper mid-fan to be a region of complex topography characterized by leveed fan valleys, particularly well developed on the western flank (Figure 14–5). Characteristically, feeder channels, as much as 1500 m wide with 200 m of relief, become less incised downfan as the longitudinal gradient diminishes. Concurrently, the well-developed west-side levee decreases in size until the lower mid-fan is characterized by a low-relief, broadly mounded region that merges gently with the flatter lower fan (Figure 14–5). The detail of these changes, seen in the seismic-reflection data, however, is not revealed in the GLORIA data. Rather, the variations in relief of the levee and levee-flank areas produce only nondescript intermediate-backscatter tones. These intermediate-backscatter deposits, however, are acoustically distinct from the low-backscatter character of the distributary channel floors that cross the mid-fan region (compare Figure 14–5 with Figure 14–3).

Likewise, many of the sedimented topographic features on the mid-fan, such as steep levee crests (Figure 14–5A) and sediment waves (Figure 14–5C) seen in the seismic-reflection data, and mass failure deposits reported in the liter-

ature (Nardin 1983), are not observed in the GLORIA data. Typically, high-resolution seismic-reflection records of the levees and levee-flank areas are characterized by sharp bottom echoes with continuous parallel to subparallel subbottom reflectors. This acoustic signature, which correlates with the muted intermediate-backscatter tones in the GLORIA data, is in contrast to the prolonged bottom-echo response characteristic of the low-backscatter channel floors. Such a prolonged bottom-echo character commonly is attributed to concentrations of coarse-grained sediment (Damuth 1980), an interpretation consistent with the suggestion that the channels in Santa Monica Basin are conduits for the transport of coarse-grained sediment from the adjacent Ventura shelf to the deep basin floor (Gorsline and Emery 1959; Malouta et al. 1981; Schwalbach et al. in press). Apparently, the textural variation between the coarse-grained channel floor and the finer-grained levee deposits provides an acoustic contrast that is resolved by GLORIA and displays the curvilinear pathways whereas the topographic variations (steep levee crests, sediment waves) within the uniform soft levee deposits do not provide a sufficiently strong target to be resolved by GLORIA.

The GLORIA mosaic shows distinct linear streaks of high and low backscatter as the gradient decreases within the lower fan that again is suggestive of sediment pathways (Figures 14–3 and 14–4). Few channels in this area are evident at the seafloor on high-resolution profiles although channels are seen in the shallow subsurface (Figure 14–6). High-resolution seismic profiles across the streaks show semiprolonged bottom echoes with few discontinuous subbottom reflectors. Facies differences, or variations in the state of consolidation relative to the adjacent sediment, are the probable cause of the low- and high-backscatter streaking observed on the GLORIA mosaic.

The interfingering of low-backscatter digitate patterns on the GLORIA mosaic with the high-backscatter areas on the central-basin plain shows that Hueneme-Mugu Fan complex deposits extend beyond the fan boundaries as defined by seismic-reflection systems (compare Figures 14–2B and 14–3). High-resolution seismic-reflection profiles collected across the distal ends of the low-backscatter fingers (Figure 14–7) reveal poorly defined narrow (100- to 400-m-wide) low-relief mounds (1 to 2 m high) overlying horizontally stratified basin-plain deposits. The high-resolution data showing the low-relief mounds are not compelling by themselves, but, viewed in conjunction with the GLORIA mosaic, document the build-up of distal fan deposits on top of the more flat-lying basin-plain sediment. Apparently, these mounded deposits help steer subsequent flows, and sediment infilling between the mounds may be another possible cause of the linear streak patterns observed on the images. Because of the small scale of these lenticular deposits and the resolution capability of the shipboard seismic and navigation systems, we cannot accurately correlate each of these low-relief mounds and intervening channels with individual streaks seen on the GLORIA mosaic.

Upper fan

West

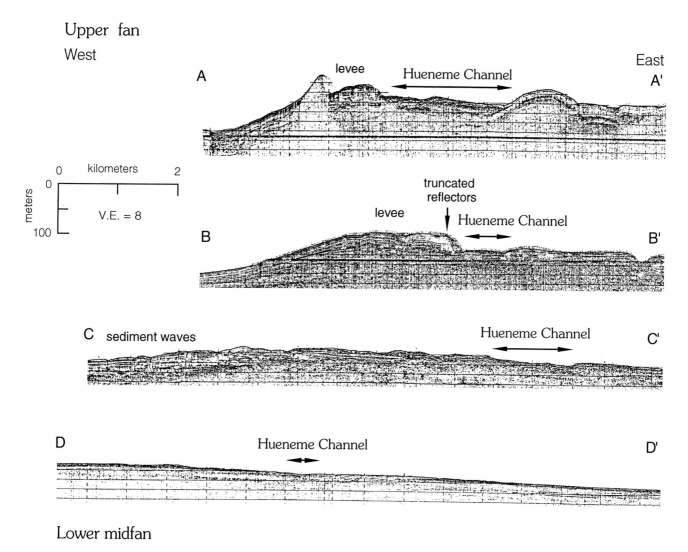

Lower midfan

Figure 14–5. High-resolution (uniboom) seismic-reflection profiles across the Hueneme Fan. Note the well-developed truncated reflectors characteristic of the levee environment as contrasted with the poor penetration along the sand-rich channel axis in the upper mid-fan (profiles A-A′ and B-B′). Channel relief diminishes rapidly downfan (profiles A-A′ to D-D′), and a sediment-wave field is developed on the flank of the levee system at profile C-C′. Location of profiles is given on Figure 14–3B.

The area of high backscatter on GLORIA records extends across the deepest part of Santa Monica Basin until the region where the floor starts to shoal near the far side of the basin. The backscatter intensity decreases at that point to intermediate values. High-resolution seismic-reflection data from the central basin show a prominent, high-amplitude reflector that correlates with the high-backscatter part of the central basin (Figure 14–8). The reflector is either exposed at the seafloor or becomes buried as the edges of the basin are approached. To the west, the reflector is covered by an acoustically transparent unit 1 to 2 m thick yet still correlates with high-backscatter returns on the GLORIA sonographs. As this overlying unit thickens further westward, low-backscatter returns predominate.

We used data available in the literature to evaluate the control of surface texture on backscatter intensity. More than 130 box cores have been collected in Santa Monica Basin with a concentration in shelf and ridge areas (Figure 6 of Schwalbach et al. in press). Cores from the basins were collected on a widely spaced grid pattern, and hence are not sited ideally with respect to the GLORIA imagery. Textural characteristics of surface (0 to 2 cm) samples from these cores (Gorsline 1980; Malouta et al. 1981; Reynolds 1987; Schwalbach et al. in press) correlate in only a general way with the GLORIA mosaic. Isopleths of mean grain size bow basinward along fan-channel axes indicating the coarser nature of the channel deposits (Figure 6 of Schwalbach et al. in press) and correlating with the low-backscatter tones of the channels in the GLORIA mosaic. In the central basin, where GLORIA shows distinct low-backscatter fingers jux-

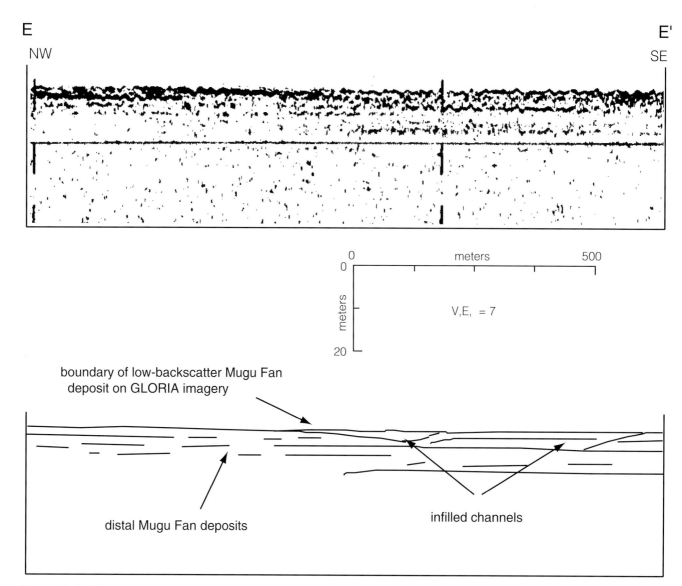

Figure 14–6. High-resolution (3.5 kHz) seismic-reflection profile across the distal end of the Mugu Fan. Note the poor penetration, semiprolonged bottom echoes, and discontinuous subbottom reflectors suggesting a sand-rich environment. Infilled channels are common. Location given on Figure 14–4B.

taposed against the high-backscatter deposits, mean grain size of the surface sediment shows little variation and has no apparent correlation with the GLORIA backscatter patterns. This lack of correlation indicates that textural variations in the surface sediment (upper few centimeters) are not the major modulator of backscatter intensity.

Similarly, in the central basin area, variation in small-scale roughness at the seafloor appears to contribute little to the observed differences in backscatter intensity. Dozens of bottom photographs taken of the lower fan and basin plain in Santa Monica Basin show this region to be monotonously uniform across acoustic boundaries with surface relief less than 5 cm (Malouta et al. 1981; Gorsline, oral communication 1989). Therefore, variations in sediment properties ei-

ther at the seafloor surface or within the sediment mass must be the major modulators of backscatter intensity.

To determine the possible downcore sedimentologic controls on backscatter intensity, we considered the limited number of box, hydroplastic, and piston cores described in the literature. Cores collected from the lower fan (low-backscatter intensity) typically contain about 9 cm of laminated hemipelagic muds overlying stacked graded sands (e.g., Figure 14–9, core DOE3; Gorsline, written communication 1988). In contrast, a 4-m-long piston core (core 4647, Figure 14–9) collected from the high-backscatter part of the central Santa Monica Basin contains about 90 cm of uniform hemipelagic mud overlying a sequence of 10- to 15-cm-thick uniform to graded sand layers interbedded with

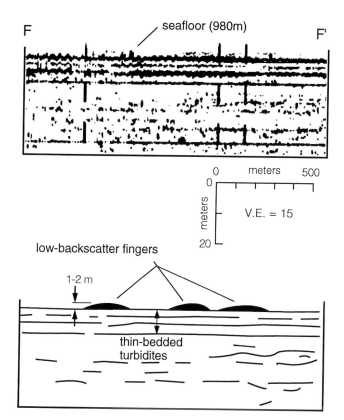

Figure 14–7. Enlarged high-resolution (3.5-kHz) seismic-reflection profile collected where the fan fringe deposits of the Hueneme system rest on thin-bedded turbidites of the basin plain. Note the low relief and associated depressions created by these lens-shaped deposits. Location of profile given on Figure 14–4B.

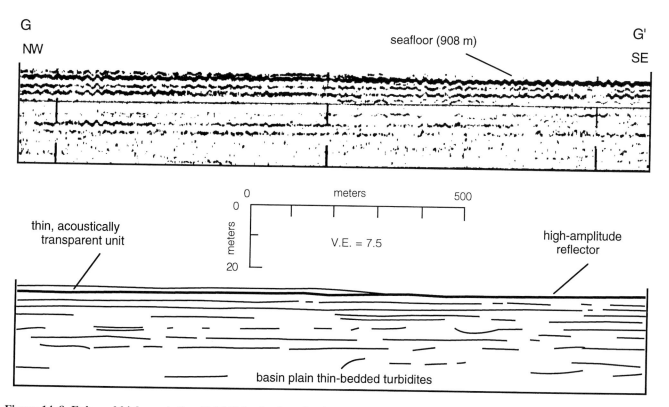

Figure 14–8. Enlarged high-resolution (3.5-kHz) seismic-reflection profile collected within the high backscatter of the central Santa Monica Basin plain. Note the generally well-bedded sequence of subbottom reflectors and the shallow-seated high-amplitude reflector that is overlaid by a thin acoustically transparent unit. Location of profile given on Figure 14–4B.

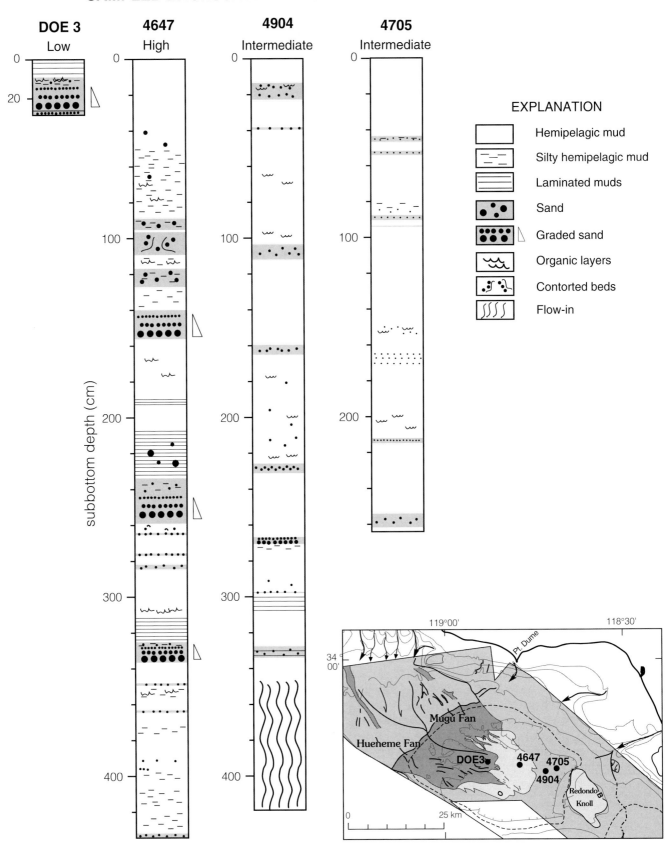

Figure 14–9. Generalized interpretations of selected box and piston cores recovered from Santa Monica Basin (after Gorsline 1958; and Gorsline, written communication 1989). Note the change in thickness of sands and interbedded muds along a transect from the low-backscatter distal fan-fringe deposits (core DOE 3), across the high-backscatter central basin region (core 4647), and onto the adjacent basin-plain sequence (cores 4904 and 4705).

hemipelagic mud. The 90-cm-thick surface-mud layer likely correlates with the acoustically transparent layer that overlies the high-amplitude reflector on seismic-reflection records (Figures 14–4B and 14–8), and the interbedded sand interval provides the impedance contrast that is recorded as the high-amplitude reflector. The sand layers thin and become separated by greater thicknesses of hemipelagic mud (cores 4904 and 4705, Figure 14–9) farther into the basin correlating with a decrease in backscatter intensity on the GLORIA records.

Discussion

The combination of high-resolution seismic-reflection profiles, bottom photographs, and stratigraphic evidence from cores of these inner borderland basins indicates that facies changes within the shallow subsurface coupled with bedding sequences and layer thickness are the important controls on GLORIA backscatter intensity. For example, the core data from channel thalwegs of the mid-fan and on the lower fan suggest that low-backscatter returns correlate with near-surface abundance of thick turbidite sands. Core DOE 3 (Figure 14–9) recovered from the low backscatter part of the distal Hueneme Fan recovered 9 cm of laminated hemipelagic mud overlying more than 20 cm of graded sand. The corer was unable to penetrate more deeply into the sediment, presumably due to thickness of the sand unit. Gardner et al. (1991) and Lee et al. (this volume), in a study of GLORIA backscatter patterns on Monterey Fan, noted a similar correlation between very low backscatter returns and thick near-surface sand beds. Their studies postulate that the low-backscatter character observed from the Monterey Fan deposits results from a combination of (1) a smooth, high impedance contrast surface causing strong signal reflection rather than backscatter and (2) a significant thickness of sand that effectively attenuates the passage of sound. Although core control from Santa Monica Basin is sparse, we infer a similar relationship between sand-rich deposits in the channel axes and the distal fan deposits to control GLORIA backscatter intensity in these inner borderland basins.

Farther east, the central-basin areas are characterized by either high or intermediate backscatter (Figures 14–3 and 14–4). These backscatter intensities appear to correlate with (1) subbottom depth to sand layers, (2) thickness and number of sand layers, and (3) thickness of mud interbeds. The high-backscatter region loosely correlates with stacked sands having a thickness of about 20 cm (core 4647, Figure 14–9) that are covered by about 1 m of hemipelagic mud. Many of the sands are graded or have contorted beds. The intensity of the GLORIA backscatter decreases as the sand layers thin farther into the basin (cores 4904 and 4705, Figure 14–9) and become separated by greater thicknesses of hemipelagic mud. Although these data are equivocal, varying thicknesses and continuity of interbedded sand and mud

layers appear to correlate with backscatter intensity, thin sand and thick mud interbeds giving lower backscatter returns but not as low as that produced by the thick sands of the lower-fan region.

Apparently, in this hemipelagic turbidite-basin setting, acoustic energy from the 6.5-kHz GLORIA source penetrates into the bed and scattering involves sediment in the shallow (upper few meters) subsurface rather than being controlled by the uppermost few centimeters of the substrate. Thus, volume scattering and subbottom interface scattering are the major factors controlling backscatter intensities; small-scale relief of the seafloor and sediment character in the upper few centimeters appear to be unimportant. Studies by Clague et al. (1988), Gardner et al. (1991), Lyons et al. (1994) and Lee et al. (this volume) support our view that acoustic energy from the GLORIA system can penetrate into the sediment column and that patterns appearing on the sonographs can result from geological conditions below (>1 m) the seafloor.

Conclusions

The areal view of the inner California borderland provided by the GLORIA mosaic gives us a new perspective on sedimentation in confined basins. Well-developed leveed feeder channels on the upper fan of the Hueneme-Mugu Fan complex lead to a system of low-backscatter curvilinear channels and branching distributaries on the mid-fan that contrast with the general intermediate backscatter of the mid-fan levee deposits. The lower-fan and basin-plain depositional environments, however, record a more striking variation in backscatter intensity. Characteristically, the lower fan returns a low backscatter in response to unchannelized sands that bypassed the mid-fan region. Streaks of high and low backscatter within these deposits delineate paths of sediment flows that cannot be resolved by shipboard high-resolution profiling systems. The sediment flows are shown as a series of branching fingerlike, low-backscatter, sand-rich distributaries overlying a high-backscatter field of the central basin. These distributary deposits produce low-relief mounds separated by depressions that allow subsequent flows to shift laterally along the periphery of the fan. The resulting distal sand-rich deposits are lens-shaped, exhibit poor lateral continuity, and are surrounded by muddy turbidites and hemipelagic mud.

Deposition in the central basin results from hemipelagic sedimentation and occasional large-volume turbidity currents that transport and deposit laterally extensive sand layers across most of the basin floor. Extensive sand layers produce a high-backscatter response at GLORIA frequencies when they are about 20-cm thick (close to the GLORIA wavelength), are coarse grained, contain contorted beds, and are interbedded with muddy turbidites and hemipelagic mud. A low-backscatter response becomes characteristic as the sand beds thin and become intercalated with thicker mud units.

Thus, volume and subbottom interface scattering of the acoustic source related to sand-mud interbed thickness modulates the GLORIA backscatter response in these flat, sedimented, central-basin floor settings and can result in either low- or high-backscatter returns. More rugged areas, however, such as the unsedimented Redondo Knoll or insular basin flanks, produce a high-backscatter return. Thus, backscatter intensity at a given incident frequency can vary rapidly in response to local changes in sediment character and bed roughness.

Acknowledgments

We thank the captains, crews, technicians, and scientific parties of the M/V *Farnella* and the R/V *S. P. Lee* for their support during collection of the data. D. S. Gorsline kindly provided core data and bottom photographs from the Santa Monica Basin.

References

Atwater, T. 1970. Implications of plate tectonics for the Cenozoic tectonic evolution of western North America. *Geol. Soc. Am. Bull.* 81: 3513–36.

Chavez, P. S., Jr. 1986. Processing techniques for digital sonar images from GLORIA. *Photogram. Engrg. and Rem. Sens.* 52: 1133–45.

Clague, D. A., Moore, J. G., Torresan, M., Holcomb, R., and Lipman, P. 1988. Shipboard report for Hawaii GLORIA ground-truth cruise F2-88-HW, 25 Feb.–9 March, 1988. *U.S. Geological Survey Open-File Report 88-292,* 54 pp. Reston, Va.: USGS.

Crowell, J. C. 1974. Origin of late Cenozoic basins in southern California. In *Tectonics and Sedimentation,* ed. W. R. Dickinson, pp. 190–204. *Society of Economic Paleontology Mineralogists Special Publication 22.* Tulsa, OK: SEPM.

Damuth, J. E. 1980. Use of high-frequency (3.5-12 kHz) echograms in the study of near-bottom sedimentation processes in the deep-sea: A review. *Mar. Geol.* 38: 51–76.

EEZ-SCAN 84 Scientific Staff 1986. Atlas of the Exclusive Economic Zone, western conterminous United States. *U.S. Geological Survey Miscellaneous Investigations Series I-1792,* 152 pp., scale 1:500,000. Reston, Va.: USGS.

Field, M. E., and Edwards, B. D. 1980. Slopes of the southern California continental borderland: A regime of mass transport. In *Quaternary Depositional Environments of the Pacific Coast,* vol. 4, eds. M. E. Field, A. H. Bouma, I. P. Colburn, R. G. Douglas, and J. C. Ingle, pp. 169–84. Soc. Econ. Paleontol, Mineral. Pac. Sec., Pacific Coast Paleog. Symposium. Society of Economic Paleontologists and Mineralogists, Pacific Section, Pacific Coast Paleogeography Symposium, Tulsa, OK.

Gardner, J. V., and Kidd, R. B. 1987. Sedimentary processes on the northwestern Iberian continental margin viewed by long-range sidescan sonar and seismic data. *J. Sed. Petrol.* 57: 397–407.

Gardner, J. V., Field, M. E., Lee, H., Edwards, B. D. 1991. Ground-truthing 6.5-kHz side scan sonographs: What are we really imaging? *J. Geophys. Res.* 96: 5955–74.

Gorsline, D. S. 1958. Marine geology of San Pedro and Santa Monica Basins and vicinity, California. *Ph.D. dissertation,* University of Southern California, Los Angeles. (Unpublished.)

Gorsline, D. S., and Emery, K. O. 1959. Turbidity-current deposits in the San Pedro and Santa Monica Basins off southern California. *Geol. Soc. Am. Bull.,* 70: 279–90.

Gorsline, D. S., and Prensky, S. E. 1975. Paleoclimatic inferences for late Pleistocene and Holocene from California Continental Borderland basin sediments. In *Quaternary Studies,* pp. 147–54. Royal Society of Wellington: New Zealand.

Gorsline, D. S. 1980. Depositional patterns of hemipelagic Holocene sediments in borderland basins on an active margin. In *Quaternary Depositional Environments of the Pacific Coast,* Vol. 4, eds. M. E. Field, A. H. Bouma, I. P. Colburn, R. G. Douglas, and J. C. Ingle, pp. 185–200. Soc. Econ. Paleontol. Mineral. Pac. Sec., Pacific Coast Paleog. Symposium. Society of Economic Paleontologists and Mineralogists, Pacific Section, Pacific Coast Paleogeography Symposium, Tulsa, OK.

Gorsline, D. S., Kolpack, R. L., Karl, H. A., Drake, D. E., Fleischer, P., Thornton, S. E., Schwalbach, J. R., and Savrda, C. E. 1984. Studies of fine-grained sediment transport processes and products in the California Continental Borderland. *Fine-Grained Sediments: Deep-Water Processes and Facies,* eds. D. A. V. Stow and D. J. W. Piper, pp. 395–415. Geological Society Special Publication No. 15. Oxford: Blackwell Scientific Publications.

Haner, B. E. 1971. Morphology and sediments of Redondo submarine fan, southern California. *Geol. Soc. Am. Bull.* 82: 2413–32.

Howell, D. G. ed. 1976. Aspects of the geologic history of the California continental borderland. *American Association of Petrology Geologists Miscellaneous Publication 24,* 561 pp. Tulsa, OK: AAPG.

Hsü, K. J., Kelts, K., and Valentine, J. W. 1980. Resedimented facies in Ventura Basin, California, and model of longitudinal transport of turbidity currents. *Am. Assoc. Petrol. Geol. Bull.* 64: 1034–51.

Kenyon, N. H., Belderson, R. H., and Stride, A. H. 1975. Plan views of active faults and other features on the lower Nile cone. *Geol. Soc. Am. Bull.* 86: 1733–9.

Kneller, B., Edwards, D., McCaffrey, W., and Moore, R. 1991. Oblique reflection of turbidity currents. *Geology* 14: 250–2.

Lyons, A. P., Anderson, A. L., and Dwan, F. S. 1994. Acoustic scattering from the seafloor: Modeling and data comparison. *J. Acous. Soc. Am.* 95: 2441–51.

Malouta, D. N., Gorsline, D. S., and Thornton, S. E. 1981. Processes and rates of recent (Holocene) basin filling in an active transform margin: Santa Monica Basin, California continental borderland. *J. Sed. Petrol.* 51: 1077–96.

Nardin, T. R. 1983. Late Quaternary depositional systems and sea level change – Santa Monica and San Pedro basins, California continental borderland. *Am. Assoc. Petrol. Geol. Bull.,* 67: 1104–24.

Normark, W. R., and Piper, D. J. W. 1972. Sediments and growth pattern of Navy deep-sea fan, San Clemente Basin, California borderland. *J. Geol.* 80: 198–223.

Normark, W. R. 1974. Submarine canyons and fan valleys: Factors affecting growth patterns of deep-sea fans. In *Modern and Ancient Geosynclinal Sedimentation,* eds. R. H. Dott and R. H. Shaver, pp. 56–68. Special Publication 19. Society of Economic Paleontologists and Mineralogists. Tulsa, OK: SEPM.

Normark, W. R., Piper, D. J. W., and Hess, G. R. 1979. Distributary channels, sand lobes, and mesotopography of Navy submarine fan, California borderland, with applications to ancient fan sediments. *Sedimentology* 26: 749–74.

Pao, G. A. 1977. Sedimentary history of California continental borderland basins as indicated by organic carbon content. Unpublished M.S. thesis, University of Southern California, Los Angeles, 112 p.

Reynolds, S. 1987. A recent turbidity current event, Hueneme Fan, California: Reconstruction of flow properties. *Sedimentology* 34: 129–37.

Roberts, D. G., Kidd, R. B., and Masson, D. G. 1984. Long-range sonar observations of sedimentary facies patterns in and around Amirante Trench, western Indian ocean. *J. Geol. Soc. London,* 141: 975–83.

Schwalbach, J. R., and Gorsline, D. S. 1985. Holocene sediment budgets for the basins of the California continental borderland. *J. Sed. Petrol.* 55: 829–42.

Schwalbach, J. R., Edwards, B. D., and Gorsline, D. S. in press. Depositional features produced by a series of events at high sea level: Contemporary channel-levee systems in active borderland basin plains. In *Sedimentary Geology,* Special Issue, eds. T. Shiki, S. Chough, and G. Einsele.

Shepard, F. P., and Emery, K. O. 1941. Submarine topography off the California coast: Canyons and tectonic interpretation. *Geological Society of America Special Paper 31,* 171 p. Boulder, Colo.: GSA.

Siegenthaler, C., Hsü, K. J., and Kleboth, P. 1984. Longitudinal transport of turbidity currents – a model study of Horgen events. *Sedimentology* 31: 187–93.

Somers, J. L., Carson, R. M., Revie, J. A., Edge, R. H., Barrow, B. J., and Andrews, A. G. 1978. GLORIA II – an improved long range sidescan sonar. In *Oceanology International 1978,* pp. 16–24. London: BPS Publications.

Teng, L. S., and Gorsline, D. S. 1989. Late Cenozoic sedimentation in California continental borderland basins as revealed by seismic facies analysis. *Geol. Soc. Am. Bull.* 101: 27–41.

Twichell, D. C., Kenyon, N. H., Parson, L. M., and McGregor, B. A. 1991. Depositional patterns of the Mississippi Fan surface: Evidence from GLORIA II and high-resolution seismic profiles. In *Seismic Facies and Sedimentary Processes of Submarine Fans and Turbidite Systems,* eds. P. Weimer and M. H. Link, Chapter 19, pp. 349–63. New York: Springer-Verlag.

Vedder, J. G., Greene, H. G., Clarke, S. H., and Kennedy, M. P. 1986. Geologic map of the mid-Southern California continental margin – Map No. 2A. In *California Continental Margin Geologic Map Series Mid-Southern California Continental Margin Area 2 of 7,* eds. H. G. Greene and M. P. Kennedy, sheet 1 of 4. Division of Mines and Geology, State of California. Scale 1:250,000. Sacramento, CA

V Alaskan EEZ

Herman A. Karl and Paul R. Carlson
U.S. Geological Survey, Menlo Park, California

Introduction

About 2.7×10^6 km^2 of the seafloor of the Bering Sea and North Pacific Ocean were mapped using the GLORIA long-range sidescan sonar system as part of the EEZ-SCAN program of the U.S. Geological Survey (Figure V–1). The areas surveyed include the following: (1) the Bering Sea region, which is all of the Aleutian Basin and Bowers Basin deeper than 200 m east of the U.S. – U.S.S.R. Convention Line of 1867; (2) the Aleutian Arc region, which is the seafloor south of the Aleutian Islands from the insular slope out to the 200 n.m. (370 km) boundary of the U.S. Exclusive Economic Zone (EEZ) and from the 1867 Convention Line eastward to Unimak Pass; and (3) the Gulf of Alaska region, which is that area within 200 nautical miles of the EEZ eastward from Unimak Pass to Dixon Entrance. The Alaskan EEZ survey spanned four years (1986 through 1989) and required twelve cruises each (with one exception) approximately a month in duration. The results of the Bering Sea survey have been published in an atlas (Bering Sea EEZ-SCAN Scientific Staff 1991).

Surveys of the Alaskan EEZ began in 1986 with three cruises to the Bering Sea (Cooper et al. 1986; Gardner et al. 1986; Carlson et al. 1987). A fourth cruise was required in 1987 to complete the area (Cooper et al. 1987). Four cruises, two in 1987 and two in 1988, were required to survey the Aleutian Arc region (Karl et al. 1987; Edwards et al. 1988; Stevenson and Scholl, personal communication, 1988; Vallier et al. 1989). Surveys of the Gulf of Alaska region began in 1986 with a short cruise (Carlson, Bruns, and Fisher 1990a); mapping of this region resumed in 1988 with one cruise and concluded in 1989 with two cruises (Carlson et al. 1990b; Bruns, Stevenson, and Dobson 1992).

The GLORIA mosaics provide an unprecedented view of the regional geologic and geomorphic provinces – outer plate, trench, inner wall, insular slope, fore-arc terrace, volcanic ridge, and back-arc basin – which characterize an active island arc/subduction zone system. Interpretations of

tectonic and sedimentologic patterns and processes are derived not only from the GLORIA mosaic, but also from high-resolution and dual-channel airgun seismic-reflection, gravity, and magnetic data. In the following section we provide a brief overall summary of the contrasting tectonic and sedimentologic styles across this vast expanse of seafloor.

Overview of styles of sedimentation and tectonism

The Bering Sea region comprising a back-arc basin, in contrast to the Aleutian Arc, is tectonically quiet. The Beringian margin, along the eastern edge of the back-arc basin, is the site of a former subduction zone where the Pacific (Kula) Plate was subducted beneath the North American Plate (Scholl et al. 1987). In early Tertiary time, the Aleutian Island Arc began to form, and subduction shifted from the Beringian margin to the Aleutian Trench (Scholl et al. 1975). Remnants of active subduction in the Bering Sea include Bowers Ridge, which is an inactive volcanic arc flanked by a filled trench. The most striking aspects of the Bering Sea region are the enormous canyons and the products of mass movement of sediment (see Karl, Carlson, and Gardner this volume). On the northern side of the Aleutian Ridge, large volumes of volcaniclastic sediment shed from the island arc are conveyed to the Bowers and Aleutian Basins by mass transport down the insular slopes. To the northeast a large percentage of the Beringian margin is failing by mass wasting. Mass wasting processes not only distribute sediment to the basin floor, but also are the principal control on the morphology of the basin margins. The enormous canyons that incise the Beringian margin are principally the result of mass wasting.

Sediment reaches the Bering Sea Basin floor by turbidity currents via three channels that issue from Bering, Umnak, and Pochnoi submarine canyons. Bering Channel, the largest of the three, extends for more than 500 km and terminates in a large fan, informally called Bering Fan, in the deepest part of the eastern Aleutian Basin. Bering Fan is a very low

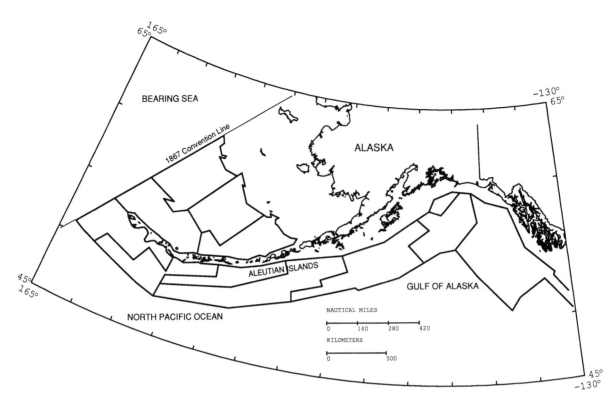

Figure V–1. Index map of the region offshore Alaska showing the twelve EEZ-SCAN survey areas.

relief feature with a complex system of distributaries, indicating that most transport and deposition on the fan is by sheet-flow. Debris shed from the Aleutian Ridge and funneled down the flanking canyons of the ridge appears to accumulate near Bering Fan as well, and this area is thereby the primary locus of modern deposition in the eastern Bering Sea (see Karl et al. this volume).

Major features in the North Pacific seaward from the Aleutian Island Arc are a canyon-incised insular slope, irregular and discontinuous fore-arc basins, the Aleutian Trench, an accretionary prism, and the outer plate. Sedimentation patterns on the outer plate reflect pelagic and hemipelagic depositional processes that are typical of deep ocean environments. Near the Convention Line of 1867, seismic-reflection profiles demonstrate that Meiji tongue sediment overlies well-stratified deposits as old as late (?) Cretaceous on both the Pacific Plate and a remnant of the Kula Paleo-Plate (see Scholl et al. 1977 for discussion of Meiji sediment). Farther east, south of Umnak and Unimak Islands, distal Zodiac Fan deposits mask the underlying basement topography (Stevenson, Scholl, and Vallier 1983). In the Zodiac Fan area the GLORIA images show the flat and smooth topography of an abyssal plain in striking contrast to the fractured abyssal hill patterns common in the western and central part of the survey area. Tectonic elements within and near the active subduction complex include a flat trench floor, landward-verging thrusts in the accretionary prism, and faults on

the seaward slope of the trench that are oriented subparallel to the direction of plate convergence. Tectonic elements that are unique to the adjacent North Pacific include Stalemate Ridge, a fragment of the Kula Paleo-Plate, and the extinct Kula-Pacific Ridge (see Vallier et al. this volume). Stalemate Ridge is a fossil transform fault (Londsdale 1988). The collision of Stalemate Ridge with the Aleutian Trench has raised the trench floor as much as 1400 m.

Across the trench on the arc margin, structurally-controlled submarine canyons head at the shelf of the summit platform, but most do not continue to the trench floor. Instead, these canyons terminate in fore-arc basins that are filled with thick accumulations of sediment (see Dobson, Karl, and Vallier this volume). At the western end of the island arc, subtle patterns in the backscatter intensity on the sonographs suggest pathways of sediment dispersal from the sediment-filled fore-arc basins to the trench floor. The trench floor slopes from east to west from the Alaska Peninsula to longitude 180° where the slope direction of the axial gradient reverses. No channels are discernible on the trench floor. However, at its eastern end, the trench is open to input of sediment from the Alaska mainland, and turbidity currents generated in the Gulf of Alaska are free to flow westward along the trench axis. In contrast, sediment is restricted from entering the trench longitudinally at the western end of the Aleutian chain because Stalemate Ridge intersects the trench, blocking sediment input from the west.

The GLORIA images show that transverse input of sediment to the trench increases west of about 180°.

The Gulf of Alaska (GoA) extends from Unimak Pass (165° W) to Dixon Entrance (133° W) along the Canada-Alaska Boundary below southeast Alaska. This 2300-km stretch of the GoA EEZ extends from a classic subduction margin in the west to a transpressive margin in the north-central to an active strike-slip margin in the east. Because of the complexity of this active tectonic boundary between the Pacific and North American Plates, some very different morphologic and sedimentologic features have been noted on the GLORIA mosaics (see Carlson et al. this volume).

Tectonic elements in the GoA include the Queen Charlotte – Fairweather Fault, which slices along the upper continental slope off southeast Alaska, forming the boundary between the Pacific and North American Plates. This fault, equivalent in length and other dimensions to the better-known San Andreas Fault, has dramatically offset small slope canyons. As the Pacific Plate travels north along this boundary at a rate of about 6 cm/yr, fault motion beheads numerous canyons that drain from the heavily glaciated southeast Alaska coastal mountains. This seafloor conveyor belt system thus collects sediment from numerous sources along its journey north, which contributes to the evolution of large sediment bodies forming at abyssal depths on the Pacific Plate. Although these large deep-sea fans are best shown in the vertical dimension of seismic-reflection profiles, they are characterized by very extensive deep-sea channels, from south to north – Mukluk, Horizon, Chirikof, and Surveyor Channels. GLORIA imagery reveals that the slope is devoid of large canyons, but rather is incised by many small gully systems, some of which merge into small canyons. These small slope gully and canyon systems in turn merge into large channel systems that carry the glacially derived sediment onto the abyssal plain, forming the large fan bodies that underlie the Alaskan Abyssal Plain.

Alaska EEZ chapters

Four chapters in this section discuss in detail the sedimentation and tectonic styles that characterize each of the three regions briefly described in this introduction. The chapter by Karl et al., "Aleutian Basin of the Bering Sea: Styles of Sedimentation and Canyon Development," discusses the processes responsible for the formation of the enormous canyons that incise the Beringian margin and the pathways and mechanisms by which sediment is transported to the deep ocean floor. Dobson et al. in the chapter entitled "Sedimentation along the Fore-arc Region of the Aleutian Island Arc, Alaska" examine depositional systems associated with intraoceanic convergent margins. The contribution by Vallier et al., "Geology of the Kula Paleo-Plate, North Pacific Ocean," describes the tectonic elements and subduction history of the Pacific and Kula Plates. Carlson et al. in the chap-

ter "Sediment Pathways in Gulf of Alaska from Beach to Abyssal Plain" discuss the paths of sediment transport from the glaciated Alaska coastal region across the shelves, principally through the large troughs that extend to the outer shelf to the deep ocean floor.

References

Bering Sea EEZ-SCAN Scientific Staff 1991. Atlas of the U.S. Exclusive Economic Zone, Bering Sea: *U.S. Geological Survey Miscellaneous Investigations Series I-2053*, 145 p., scale 1:500,000. Reston, Va.: USGS.

Bruns, T. R., Stevenson, A. J., and Dobson, M. R. 1992. Gloria investigation of the Exclusive Economic Zone in the Gulf of Alaska and off Southeast Alaska: M/V *Farnella* cruise F7-89-GA, June 14–July 13, 1989. *U.S. Geological Survey Open-File Report 92-317,* 16 p. Reston, Va.: USGS.

Carlson, P. R., Marlow, M. S., Parson, L. M., and Somers, M. L. 1987. GLORIA investigations of the Exclusive Economic Zone in the deep Bering Sea, M/V *Farnella* cruise F2-86-BS. 5 July to 4 August 1986. *U.S. Geological Survey Open-File Report 87-72,* 15 p. Reston, Va.: USGS.

Carlson, P. R., Bruns, T. R., and Fisher, M. A. 1990a. Development of slope valleys in the glaciomarine environment of a complex subduction zone, Northern Gulf of Alaska. In *Glaciomarine Environments: Process and Sediments,* eds. J. A. Dowdesell and J. D. Scourse, pp. 139-53. Geological Society of America Special Publication No. 53. Boulder, Colo.: GSA.

Carlson, P. R., Mann, D. M., Huggett, Q. J., and Bishop, D. 1990b. GLORIA investigation of the Exclusive Economic Zone in the Gulf of Alaska: M/V *Farnella* cruise F6-89-GA. *U.S. Geological Survey Open-File Report 90-71,* 17 p. Reston, Va.: USGS.

Cooper, A. K., Stevenson, A. J., Kenyon, N. H., and Bishop, D. 1986. GLORIA study of the Exclusive Economic Zone off Alaska – southern Bering Sea: Initial report for cruise F4-86-BS, 3 September to 30 September 1986. *U.S. Geological Survey Open-File Report 86-596,* 11 p. Reston, Va.: USGS.

Cooper, A. K., Hunter, R. E., Huggett, Q., and Harris, M. 1987. GLORIA study of the EEZ off Alaska – southern Bering Sea, Bowers Basin, initial report for cruise F1-87-BS. *U.S. Geological Survey Open-File Report 87-579,* 20 p. Reston, Va.: USGS.

Edwards, B. D., Bohannon, Robert, G., Dobson, M. R., and Campbell, J. M. 1988. Cruise report for cruise F3-87-AA, GLORIA survey of the west-central Aleutian Arc and adjacent North Pacific Ocean. *U.S. Geological Open-File Report 88-0227,* 9 p. Reston, Va.: USGS.

Gardner, J. V., Karl, H. A., Huggett, Q., and Harris, M. 1986. Cruise report for EEZ-SCAN 86 cruise F3-86-BS, Zhemchug Canyon and the central Aleutian Basin, Bering Sea, August 6 through September 1, 1986. *U.S. Geological Survey Open-File Report 86-597,* 9 p. Reston, Va.: USGS.

Karl, H. A., Vallier, T. L., Masson, D., and Bishop, D. 1987. Cruise report of EEZ-SCAN cruise F2-87-AA, western Aleutian arc and adjacent to North Pacific, July 28 through August 26, 1987. *U.S. Geological Survey Open-File Report 87-643,* 5 p. Reston, Va.: USGS.

Lonsdale, P. 1988. Palogene history of the Kula Plate: Offshore evidence and onshore implications. *Geol. Soc. Am. Bull.* 100: 733–54.

Scholl, D. W., Buffington, E. C., and Marlow, M. S. 1975. Plate tectonics and the structural evolution of the Aleutian-Bering Sea region. In *Contributions to the Geology of the Bering Sea Basin and Adjacent Regions*, ed. R. B. Forbes, pp. 1–32. Geological Society of America Special Paper 151. Boulder, Colo.: GSA.

Scholl, D. W., Hein, J. S., Marlow, M. S., and Buffington, E. C. 1977. Meiji sediment tongue: North Pacific evidence for limited movement between the Pacific and North American plates. *Geol. Soc. Am. Bull.* 88:1567–76.

Scholl, D. W., Vallier, T. L., and Stevenson, A. J. 1987. Geologic evolution and petroleum potential of the Aleutian Ridge. In *Geology and Resources of Western North America*, eds. D.

W. Scholl, A. Grantz, and J. Vedder, pp. 123–56. vol. 6. Houston, Tex.: Circum-Pacific Council for Energy and Mineral Resources.

Stevenson, A. J., Scholl, D. W., and Vallier, T. L. 1983. Tectonic and geologic implications of the Zodiac fan, Aleutian Abyssal Plain, northeast Pacific. *Geol. Soc. Am. Bull.* 94: 259–73.

Vallier, T., Karl, H. A., Masson, D. G., and Cherriman, J. 1989. Cruise report for EEZ-SCAN cruise F8-88-AA: Eastern Aleutian Arc oceanic plate and western Alaska peninsula margin between 179° and 158° west longitude, north Pacific Ocean. *U.S. Geological Survey Open-File Report 89-497*, 4 p. Reston, Va.: USGS.

15 Sediment pathways in Gulf of Alaska from beach to abyssal plain

Paul R. Carlson,[1] Andrew J. Stevenson,[1] Terry R. Bruns,[1] Dennis M. Mann,[1] and Quentin Huggett[2]

[1]U.S. Geological Survey, Menlo Park, California

[2]Institute of Oceanographic Science, Southampton, United Kingdom

Abstract

In the northern Gulf of Alaska large glacially formed sea valleys extend to the edge of the continental shelf and form conduits for transport of suspended sediment. GLORIA sidescan sonar imagery shows that the adjacent continental slope is devoid of large submarine canyons. A variety of mechanisms act to transport sediment to the base of the slope. From Kodiak to Middleton Island, discontinuous actively growing deformation structures trap or divert sediment transported downslope into the Aleutian Trench. In those areas where channelways continue to the trench, we observe trellised and dendritic patterns of erosion on the GLORIA imagery. Some gullies merge to form small canyons that extend to the relatively shallow (4000 m) northern part of the Aleutian Trench where small fans are being constructed from debris flow and/or turbidity-current transported sediment. On the continental slope seaward of the Malaspina Glacier, small canyons and extensive gullies merge to form three turbidity-current channels, which in turn merge to form Surveyor Channel. About 350 km from the base of the slope, Surveyor Channel crosses the structural barrier formed by the Kodiak-Bowie Seamount chain and heads south another 150 km where it bends northwesterly, perhaps influenced by the oceanic basement relief of the Patton Seamounts. The now deeply entrenched channel is up to 6 km wide and 500 m deep and continues northwesterly for 200 km where it empties into the Aleutian Trench, some 700 km from the Yakutat margin. Immediately south of Surveyor Channel, GLORIA imagery reveals evidence of two other channels, one active, one inactive. Both channels meander through gaps in the seamount chain. The active one, Chirikof Channel, splays out in fan form just south of the seamount chain. The inactive channel eventually bends northwesterly and may have conveyed turbidity currents to the Aleutian Trench before the advent of the modern, active Surveyor Channel.

Introduction

Between 1986 and 1989, four cruises used GLORIA (Geological LOng-Range Inclined Asdic), a side-looking sonar (Somers et al. 1978; Swinbanks 1986) to image the Gulf of Alaska EEZ (Exclusive Economic Zone) (Figure 15–1). A mosaic of sidescan sonar images was produced that shows the geologic and morphologic features of the seafloor and can be used for an initial assessment and evaluation of the economic potential of the Alaskan EEZ (Bering Sea EEZ-SCAN Scientific Staff 1991).

The GLORIA sonar array is towed at eight to ten knots, 300 m behind the vessel at a depth of 40 to 50 m. The instrument consists of two rows of thirty transducers each on either side of the instrument. The transducers send a 4-s burst of energy at 30-s intervals with frequencies of 6.3 and 6.7 kHz and 100-Hz bandwidth. Incoming digitally recorded signals are computer processed both onboard and later onshore to correct for distortions due to slant-range and changes in ship's speed.

In addition to GLORIA data, two-channel seismic-reflection profiles, 3.5-kHz high-resolution profiles, 10-kHz bathymetric profiles, and magnetic and gravity potential-field measurements were collected. Navigation incorporated Global Positioning System (GPS), satellite, and rho-rho loran positioning.

This report documents and illustrates, from GLORIA imagery, seismic-reflection profiles, and sparse seafloor samples, the character and configuration of slope, rise, and abyssal plain sediment pathways with special emphasis on the Surveyor Channel system. We discuss the role of slope and rise gullies and connected deep-sea channels as conduits of sediment from the gullied continental margin of the northeast Gulf of Alaska across the Alaskan Abyssal Plain to the Aleutian Trench.

Regional setting

Tectonics

The continental margin of the northern Gulf of Alaska is underlaid by Paleozoic- through Cenozoic-age rocks within

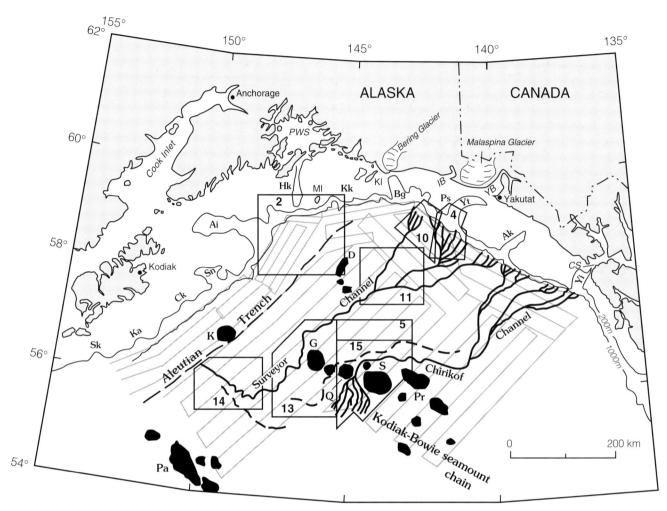

Figure 15–1. Trackline map of GLORIA coverage in northern Gulf of Alaska overlaid by generalized sketch showing relations among continental margin, deep-sea channels, Aleutian Trench, and seamounts (K = Kodiak; G = Giacomini; Q = Quinn; S = Surveyor; Pr = Pratt; and Pa = Patton). Sea Valleys from W to E: Sk = Sitkinak Trough; Ka = Kiliuda Trough; Ck = Chiniak Trough; Sn = Stevenson Trough; Ai = Amatuli Trough; Hk = Hinchinbrook Sea Valley; Kk = Kayak Trough; Bg = Bering Trough; Ps = Pamplona Spur and Trough; Yt = Yakutat Valley; Ak = Alsek Valley; Yi = Yakobi Valley. Other features: PWS = Prince William Sound; MI = Middleton Island; KI = Kayak Island; IB = Icy Bay; YB = Yakutat Bay; CS = Cross Sound. Numbers in boxes show locations of other figures in this chapter.

faultbounded tectonostratigraphic terranes that were accreted to North America in late Mesozoic and Cenozoic time (Jones et al. 1986). Plate motion between the Pacific and North American Plates currently is accommodated by strike-slip motion along the Queen Charlotte – Fairweather fault zone (Bruns and Carlson 1987) and by subduction along the Aleutian Trench (Vallier et al. this volume). In the transition region between the strike-slip and subduction regimes, the small Yakutat terrane is moving with the Pacific Plate and colliding with and subducting beneath southern Alaska (Bruns 1983, 1985; Plafker 1987).

The region adjacent to the northern Gulf of Alaska is structurally complex, with rapid uplift of the coastal mountains resulting from plate collision. This mountain building has a profound affect on both river and glacial transport of sediment to the Gulf of Alaska.

The continental shelf is underlaid by a fold and thrust belt west of Pamplona Spur (Figure 15–1) and by a relatively undeformed basin to the east (Bruns 1985). Rocks beneath the shelf include up to 5 km of Eocene and Oligocene strata overlaid by up to 5 km of lower Miocene through Quaternary strata of the Yakataga Formation, a unit strongly affected by glacial processes and consisting mostly of glacial-marine pebbly mudstone, glacial till, and marine mudstone (Plafker and Addicott 1976).

Structure along the slope from southwest of Middleton Island to Cross Sound changes markedly from west to east (Bruns and Schwab 1983; Bruns 1985). Southwest of Middleton Island, the slope is characterized by large, laterally extensive, shelf-break parallel anticlines and synclines (Figures 15–2 and 15–3). At the base of the slope, an accretionary wedge becomes more pronounced northeast to south-

Figure 15–2. Mosaic of GLORIA images in northern Gulf of Alaska showing continental slope morphology seaward of Middleton Island. Note disrupted compressional fold ridges to left half of slope and systems of gullies merging to form small canyons incised in the margin at top right of mosaic. Elongate isolated feature in the lower right is Dall Seamount, with relief of about 1000 m, rising from northern part of Alaskan Abyssal Plain. White is high backscatter on this and subsequent GLORIA figures. See Figure 15–1 for location.

west. Between Middleton Island and Pamplona Spur, a large anticlinal fold forms a ridge (Khitrov Ridge) at the base of the slope (Figure 15–3) which splays northeast into multiple anticlines and continues northeastward across the shelf as the Pamplona zone fold and thrust belt. Between the Pamplona zone and Fairweather Ground, on the south side of Yakutat Valley, there is little evidence of compressional deformation (Bruns 1985). Fairweather Ground is a large struc-

tural high on the shelf south of Alsek Valley. The base of the slope contains a trough or basin that is filled with up to 6 km of sediment including two Pliocene or younger sediment fans off Yakutat and Alsek valleys, each about 2.5 km thick (Bruns 1985).

The continental slope directly seaward of Middleton Island shows little structure in contrast to the pronounced ridges developed southwest of the island (Figures 15–2 and

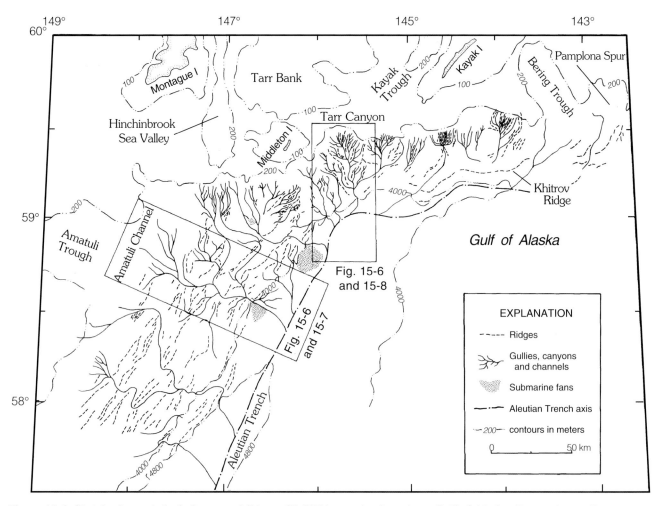

Figure 15–3. Sketch of morphologic features visible on GLORIA mosaic of northern Gulf of Alaska. Rectangles outline two very different types of sediment pathways incised into this portion of the continental slope (see Figures 15–6 through 15–8 for enlargements of these pathways). Bathymetry from Atwood et al. (1981).

15–3). Bruns (1985) suggests that the variation in structure from the large folds and accretionary structures southwest of Middleton Island to a region of little structure adjacent to Middleton Island may be the result of the Yakutat terrane passing through the subduction zone. Subduction of the terrane may remove previously developed structures, and after passage of the Yakutat terrane, new subduction and accretion-related structures may develop as the Pacific Plate continues to subduct beneath the North American Plate.

Morphology

The continental shelf in the northern Gulf of Alaska ranges in width from 20 to 80 km and has an average gradient of about 0.25° from shore to the 200 m shelf break (Atwood et al. 1981). Six prominent sea valleys cut the shelf (Figure 15–4) between Cross Sound and Prince William Sound and five valleys or troughs cut the shelf from Middleton Island to the southwest of Kodiak Island (Figure 15–1). All were

cut by tongues of glacier ice that crossed the shelf during Pleistocene sea-level low stands (Thrasher 1979; Carlson et al. 1982; Hampton et al. 1986). Characteristics of the sea valleys, many of which are located seaward of large glaciers, include the following: a subbottom erosional surface incised into the lithified strata of the shelf; U-shaped cross sections; concave longitudinal sections that commonly shoal at their seaward end; till-like sediment along the walls of the valleys and at the outer shelf and upper slope adjacent to the troughs; seismic stratigraphy and bottom samples indicative of glacially derived strata; and disconnected depressions on the modern valley floor (Carlson et al. 1982).

The continental slope in the northern Gulf of Alaska varies in width from 25 to 100 km and varies in gradient from about five to ten degrees. The slope is about 100 km wide off Amatuli Trough and less than 30 km wide off Bering Trough (Figure 15–1). The base of the slope in the western Gulf coincides with the landward wall of the Aleutian Trench. Westward along the central Aleutian Arc, the

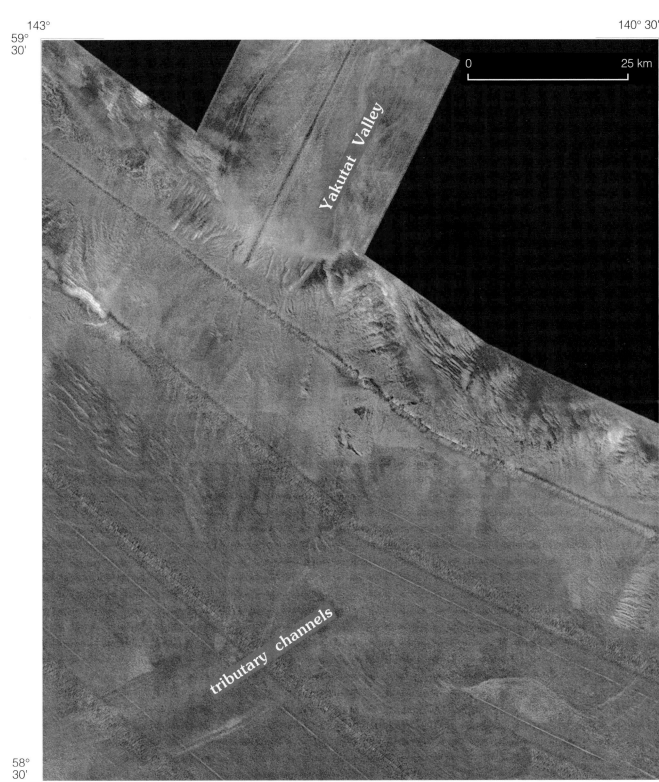

143°
59°
30'

140° 30'

0   25 km

Yakutat Valley

tributary channels

58°
30'

Figure 15–4. Mosaic of GLORIA images showing extensive gullying of continental slope near one prominent shelf sea valley, Yakutat Valley, seaward of massive Malaspina Piedmont Glacier and large glaciers that filled Icy and Yakutat bays within the last 100 and 700 years, respectively. At upper middle part of image, northeastward-oriented segment portrays outer part of Yakutat Valley, a 110-km-long shelf valley that extends from mouth of Yakutat Bay to edge of continental shelf. See Figure 15–1 for location.

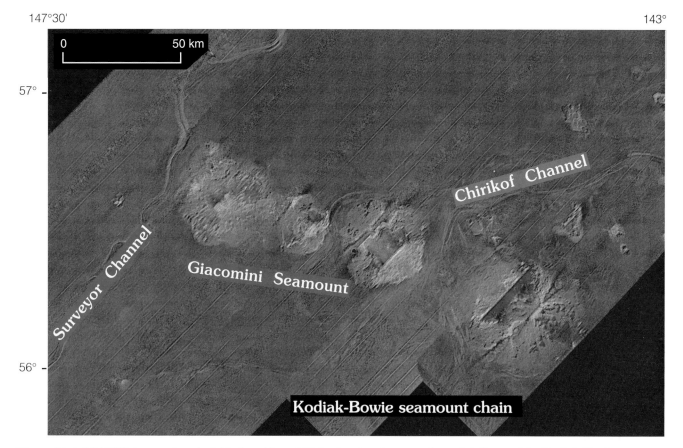

Figure 15–5. GLORIA image of part of Alaskan Abyssal Plain showing deep-sea channels influenced by Kodiak-Bowie Seamount Chain causing Surveyor Channel to wind around Giacomini Seamount. In lower part of image, Chirikof Channel also can be seen to be influenced by structural barrier posed by seamounts. See Figure 15–1 for location.

trench floor exceeds water depths of 7000 m; however, in the northern Gulf of Alaska the trench floor shoals to less than 4000 m. The trench floor has an axial gradient of 0.2° toward the west at its northeastern end, a lesser gradient than the shelf.

The Alaskan Abyssal Plain covers an area of about 27,000,000 km^2 between the Patton Seamount at its southwest boundary and the base of the Alaska continental slope extending from Kodiak Island to the Canadian boundary. The northern 45% is underlaid by the Surveyor Fan which contains a volume of sediment (pelagic and turbidite deposits) exceeding 175,000 km^3 (Stevenson and Embley 1987).

Sediment

Patterns of surficial sediment distribution on the shelf combined with high-resolution seismic-reflection profiles provide a good picture of the Quaternary depositional environment (Carlson et al. 1977; Molnia and Carlson 1978, 1980). Sand accumulating in the nearshore zone grades offshore into silty sand and then to clayey silt by mid-shelf depths. This Holocene glacial-marine deposit wedges out near the outer shelf where an underlying thin Pleistocene (?) diamict

consisting of gravelly sand, muddy gravel, and pebbly mud crops out. The diamict appears to mantle the outer shelf and upper slope in much of the northern Gulf of Alaska.

Complex sediment distribution patterns emerge where tectonic activity has caused positive relief features such as banks and islands and where glacial advances created negative relief features such as sea valleys. Winnowing of fine sediment from the tops of banks and islands creates cobble and boulder lag deposits such as those present today on banks north of Middleton Island (Carlson 1989). Ancient lag deposits are exposed in Middleton Island's uplifted terraces (Eyles, Eyles, and Miall 1985). Eyles (1987) interpreted some Yakataga Formation sand and gravel facies on Middleton Island as fill deposited by sediment gravity flows within a submarine channel complex. This channel system perhaps was cut in an outer shelf – upper slope environment by sediment gravity flows generated by meltwater streams draining from a large shelfal ice sheet.

The modern glacially derived sea valleys (Figure 15–4) have concentrations of relict coarse sediment adjacent to their mouths on the outer shelf and upper slope (Sharma 1979; Carlson et al. 1982), probable reworked end moraine deposits. The sea valleys are pathways for currents and sediments to traverse the shelf and are also modern sediment traps. Seismic-reflection profiles show thick accumulations of modern sediment (>100 m) in some of the sea valleys (Carlson et al. 1982; Hampton 1983).

Although much is known about the shelf sediment, the information about slope sediment in this area is sparse. Data from twelve shelf edge and upper slope samples, between Amatuli Trough and Hinchinbrook Sea Valley (Sharma 1979), show quantities of gravel ranging from trace to more than 25%, and all samples have less than 40% clay. These gravelly, sandy silts and silty sands, whether glacial marine pebbly mud, contact till, or outwash delta sediment, owe their origin to some glacially related processes. The sediment was probably deposited at or beyond the shelf edge during a recent glacial stage when sea level was at or near the shelf edge.

Gravity cores collected by the USGS from the slope in the northern Gulf recovered lenses of coarse clastics or, in some cases, bottomed in gravelly sand and sandy or silty mud. Some samples include glacial marine mud with dropstones. Other samples apparently were derived from mass movement of glacial contact tills that were deposited by ice tongues at the mouths of sea valleys and were then carried to the base of the slope by earthquake- or gravity-triggered slides and slumps.

Cores from the Alaskan Abyssal Plain are also relatively sparse. Both DSDP Hole 178 drilled near Surveyor Channel and Hole 180 in the Aleutian Trench contained ice-rafted pebbles and thin silt and sand turbidite layers (Kulm et al. 1973). Hole 178, located on the abyssal plain between Surveyor Channel and the Aleutian Trench, penetrated nearly 780 m of sediment with the upper 270 m (middle Pliocene to Holocene) containing ice-rafted erratics. This glacially influenced section is underlaid by nearly 500 m of interbedded diatomaceous mud and thin, poorly sorted silt to sand turbidites. Hole 180, drilled to 470 m depth on a depositional levee within the Aleutian Trench, bottomed in upper Pleistocene sediment. This section of rapid sediment accumulation (1500 ± 300 m/my), consisting primarily of interbedded graded silt turbidites and mud, was interpreted by DSDP scientists, based on occurrences of ice-rafted erratics, to contain two glacial and one interglacial interval. ODP Leg 145 also drilled in the northeast Pacific, encountered significant numbers of glacial dropstones at 2.6 Ma, indicating onset of major glaciation (Rea et al. 1993). They also report dropstones at 4.3 Ma, signaling glaciers at sea level in the early Pliocene.

Piston cores up to 15 m in length were collected at thirteen sites on the Alaskan Abyssal Plain including three within Surveyor Channel (Ness and Kulm 1973). These cores contained sediment characteristic of turbidity-current deposits including numerous fine lenses of silt or sand. Some of the cores contained graded sand and silt units. The three channel-axis cores contained relatively thick (0.5 to 5 m) sand layers; two had sand at the tops of the cores. The presence of both ice-rafted and turbidite deposits in the piston and drill-hole cores from the Alaskan Abyssal Plain indicate continued glacial and turbidity-current activity from Pliocene to modern time.

GLORIA imagery

Mosaics created from GLORIA images display prominent geomorphic features of the Gulf of Alaska continental slope and rise, the Aleutian Trench, and the Alaskan Abyssal Plain. GLORIA, seismic-reflection, and bathymetric data show that the continental slope is incised by relatively small canyons and cut by intensive gullying (Figures 15–2 and 15–4). In spite of the large glacially formed sea valleys that extend to the edge of the continental shelf (Figure 15–4, and Carlson et al. 1982; Hampton 1983), the slope is not dissected by large submarine canyons that are so common to many slopes throughout the world (and seen on images in other chapters in this volume).

The major features imaged on the northern part of the abyssal plain are the Kodiak-Bowie Seamounts (Figure 15–5), which extend in a northwesterly trend across the surveyed area; the Surveyor channel system, which extends across the Alaskan Abyssal Plain from the Yakutat continental margin to the Aleutian Trench; the Chirikof channel system, which extends across the abyssal plain from the margin between the Alsek and Yakobi Valleys to beyond the seamount chain; and an older partially filled channel between Surveyor and Chirikof Channels (Figures 15–1 and 15–5). Although the seamount chain and the Surveyor channel system have been known for many years (Menard 1964; Ness and Kulm 1973; Stevenson and Embley 1987), the GLORIA imagery permits continuous and more precise mapping of the deep-sea channels, seamounts, and surrounding abyssal features.

The Kodiak-Bowie Seamount chain consists of more than thirty seamounts that range in size from 50 km in long dimension (for example, Giacomini, Pratt, and Surveyor Seamounts; Figures 15–1 and 15–5) to small seamounts less than 5 km in diameter, including some with visible craters. GLORIA imagery allows us to map all of the seamounts in detail and to assess their effects on the routes of the deep-sea channels. The sidescan imagery shows the most northerly sea-mounts and allows us to visualize their effects on the routes of the deep-sea channels (Figure 15–5).

This section focuses on two segments of the mosaic that illustrate types of sediment pathways that are incised in the continental slope to abyssal environments. The first segment is between Kodiak Island and Bering Trough and the second between Bering Trough and Yakobi Valley.

Kodiak Island to Bering Trough

Along the slope between Kodiak Island and Pamplona Spur, actively growing discontinuous anticlines interrupt the

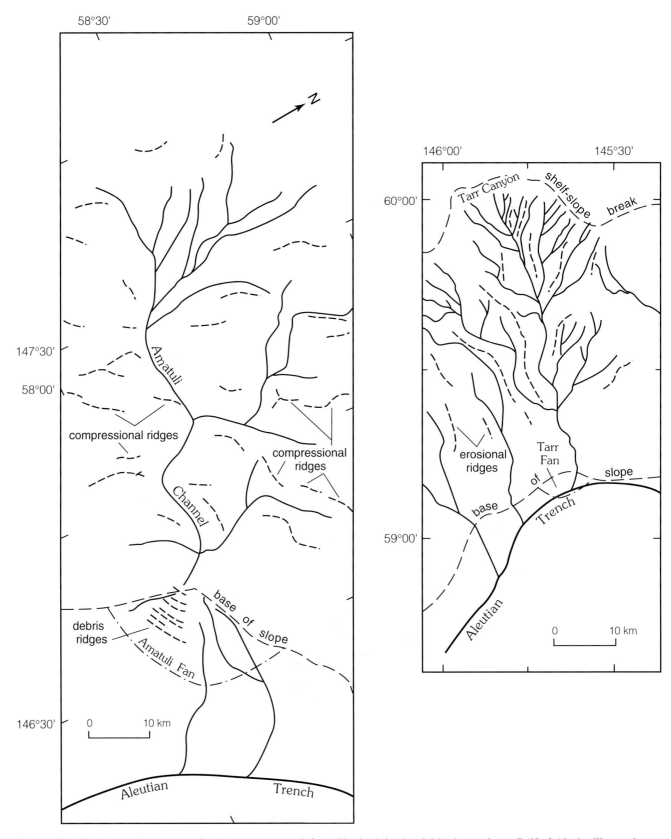

Figure 15–6. Sketches of two types of drainage patterns (*left*, trellised; *right*, dendritic) in northern Gulf of Alaska illustrating variations in slope morphology and effect on type of sediment pathways developed. Figures 15–7 and 15–8 show part of GLORIA imagery used to make these sketches.

147°36'   146°30'
59°48'

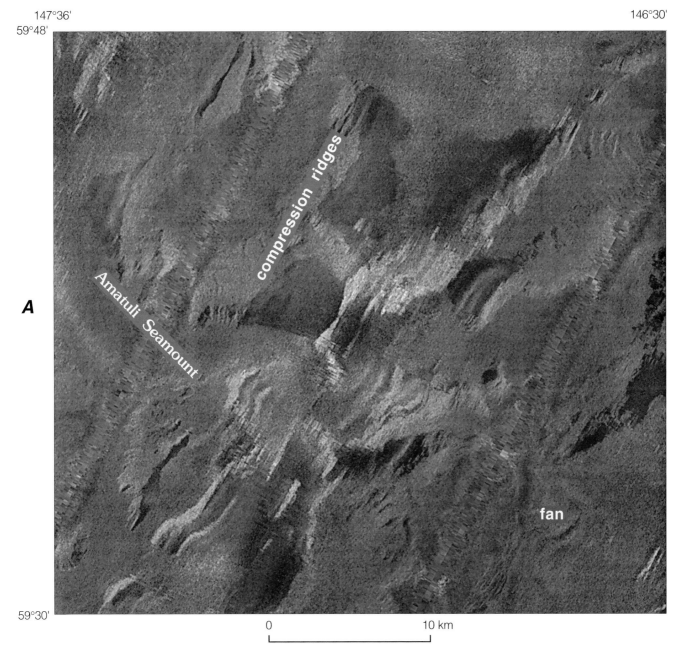

A

59°30'

0   10 km

Figure 15–7. Digitally mosaicked segments of GLORIA imagery showing parts of lower course of Amatuli Channel. A, Winding path controlled by compression ridges. Note debris fan at center of lower right quarter of image.

downslope transport of sediment into the Aleutian Trench (Figures 15–2 and 15–3). These subduction margin ridges, some as long as 35 km and with more than 1000 m of relief, result in the ponding of Quaternary sediment to thicknesses of several hundred meters in the interslope basins (Carlson, Bruns, and Fisher 1990).

Southwest of Hinchinbrook Sea Valley, large ridges dominate the slope (Figures 15–3, 15–6, and 15–7A). These ridges vary in length from less than 5 km to greater than 35 km and in width from 2 to 9 km. The ridges vary in relief from

200 m to more than 1000 m and are asymmetrical in transverse section with slopes that vary from ten to thirty degrees. The ridges are generally steeper on the seaward side, a result of both fold vergence and of ponded sediment onlapping the ridges from the landward side. The ridges obstruct the downslope drainage, causing detours of the flow patterns parallel to the ridges. In some places the channels can be traced around the ridges, and the flow may extend in a circuitous pattern from the shelf break all the way to the Aleutian Trench (Figures 15–3, 15–6, and 15–7A). The Amatuli channel system

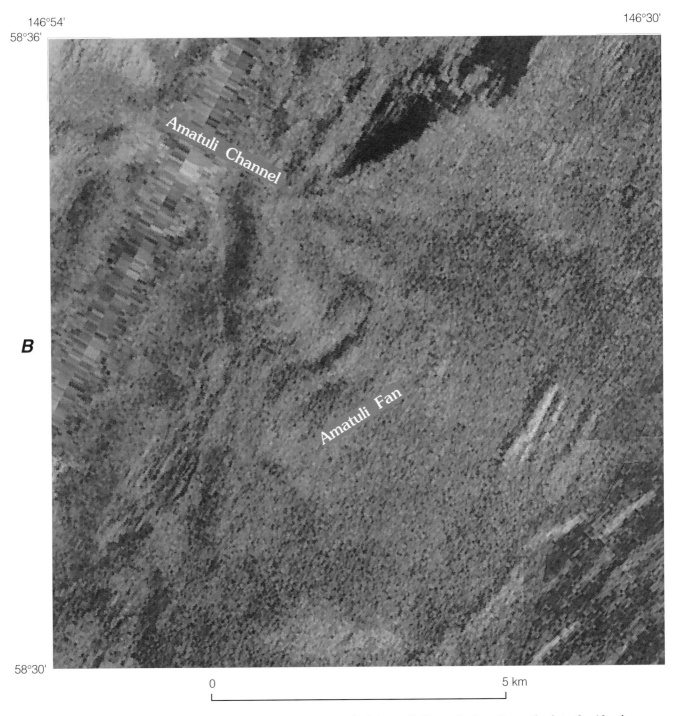

146°54'
58°36'

146°30'

B

Amatuli Channel

Amatuli Fan

58°30'

0   5 km

Figure 15–7. *Continued.* **B, Enlarged view of debris fan at lower end of Amatuli Channel where it empties into the Aleutian Trench. See Figure 15–6 for location.**

has formed seaward of Amatuli Trough. Similar trellised drainage patterns have been mapped from the GLORIA mosaic further to the southwest (Figure 15–3). Amatuli Channel is about 105 km long from the shelf break at 200 m to the base of the slope at 4200 m where it empties onto Amatuli Fan, a small deposit on the trench floor (Figures 15–6 and 15–7B).

The gradient of the main thalweg of the channel is about 2.2°. The width of the channel floor varies from 1 to 7 km and the channel width from rim to rim is as much as 12 km.

Well-developed dendritic drainage patterns are present south and southeast of Middleton Island and Tarr Bank where gullies merge to form small canyons that are incised

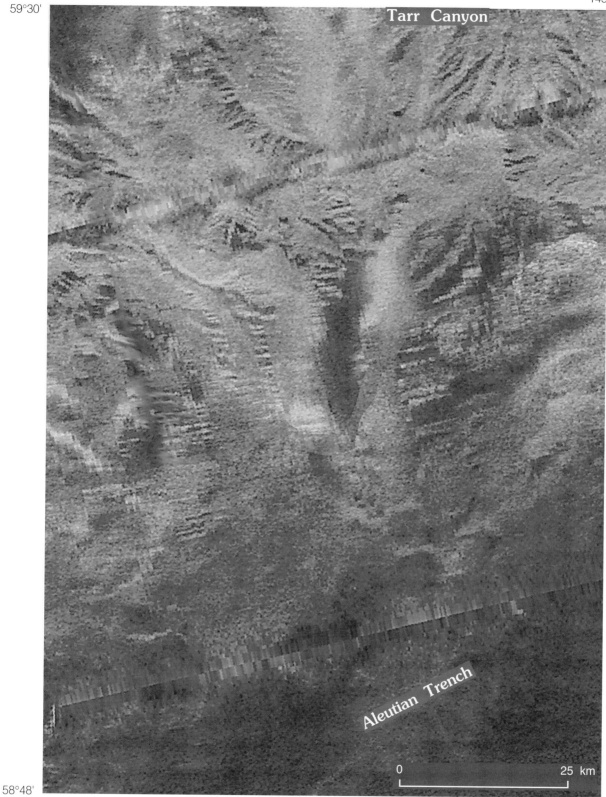

Figure 15–8. Digitally mosaicked segment of GLORIA imagery showing intensely gullied dendritic head of Tarr Canyon. See Figure 15–6 for location.

265

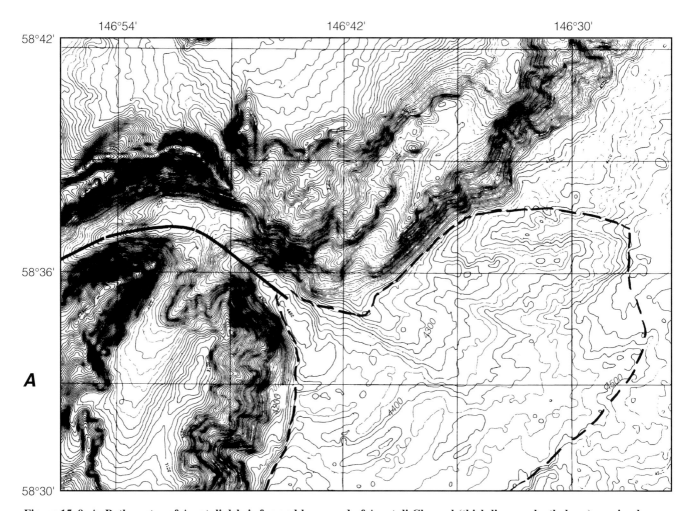

Figure 15–9. A, Bathymetry of Amatuli debris fan and lower end of Amatuli Channel (thick line marks thalweg) curving between compressional fold ridges on lower slope southwest of Middleton Island. Bathymetry from NOAA SeaBeam data gathered by NOAA ship *Surveyor* in 1986. See Figure 15–3 for location.

into the lower slope and empty into the Aleutian Trench (Figures 15–3, 15–6, and 15–8). The most prominent and largest dendritic gulley system (Tarr Canyon) heads at the shelf break near Tarr Bank southeast of Middleton Island (Figures 15–6 and 15–8). Numerous gullies merge into tributary canyons and eventually at 2900 m into one canyon that extends to the floor of the Aleutian Trench at a depth of about 4400 m. The overall length of Tarr Canyon is about 48 km with a thalweg gradient of about 4.6°. The width of the canyon floor varies from 0.5 to 1 km, and the canyon from rim to rim is as wide as 5 km.

On the lower slope, between Kayak Island and Bering Trough, Khitrov Ridge is greater than 50 km in length and 1 to 5 km in width, and its flanks slope 17 to 21° seaward and 7 to 12° landward. The ridge (Figure 15–3) is cut at about 144.5° W longitude by a low pass or saddle with relief of about 25 m. This pass would be overtopped by large, thick turbidity currents that could flow down the adjacent valley. The pervasive dendritic patterns that char-

acterize the slope to the west are not as evident immediately seaward of Bering Trough (Figure 15–3). The smoothness of the slope suggests a covering by sediment, perhaps dumped there when ice tongues occupied the shelf valleys.

At the base of the slope in the landward part of the Aleutian Trench, small submarine fans (areas 410 and 440 km^2) are forming at the mouths of at least two of the slope channel systems (Figures 15–3 and 15–9A and B). For comparison, the area of Monterey Fan, off central California, is about 68,000 km^2. The SeaBeam bathymetry (Figures 15–9A and B) of these fans shows subtle fan channels near the north edge of each fan. These channels appear to be about 12 km long and 0.5 km wide. The most prominent fan on the GLORIA imagery is located at the mouth of Amatuli Channel at a depth of about 4200 m (Figures 15–3 and 15–9A). This fan is portrayed on the sidescan imagery as consisting of a series of ridges that are similar in appearance to bedforms, but are more likely slide or slump struc-

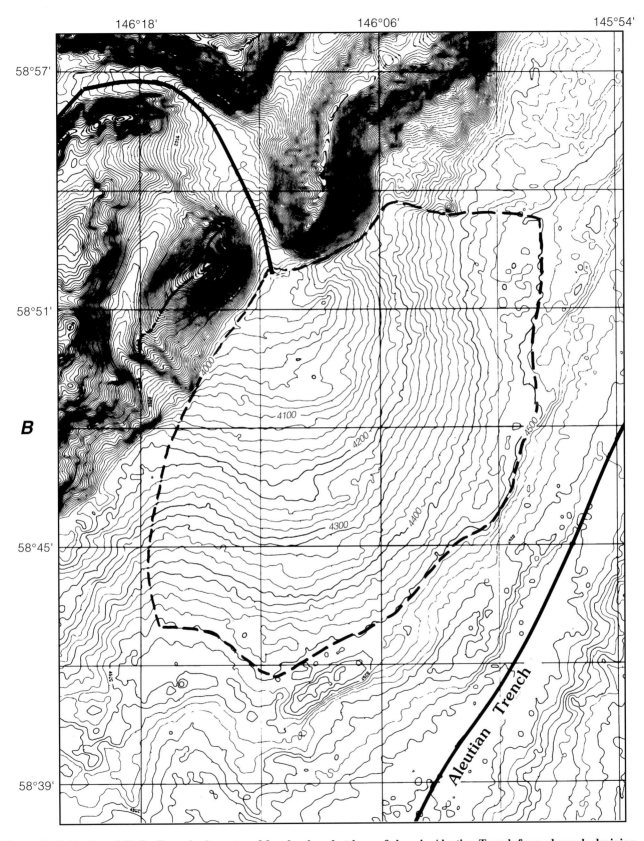

Figure 15–9. *Continued*. **B, SeaBeam bathymetry of fan developed at base of slope in Aleutian Trench from channels draining Hinchinbrook Sea Valley. See Figure 15–3 for location.**

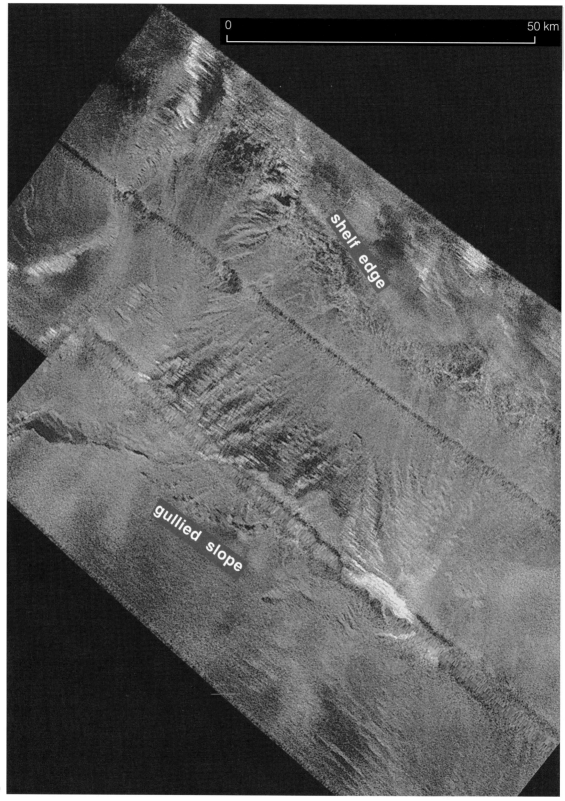

143°30'   142°00'

0   50 km

shelf edge

gullied slope

58°24'

Figure 15–10. GLORIA mosaic of extensively gullied continental slope between Bering Trough and Yakutat Valley. See Figure 15–1 for location.

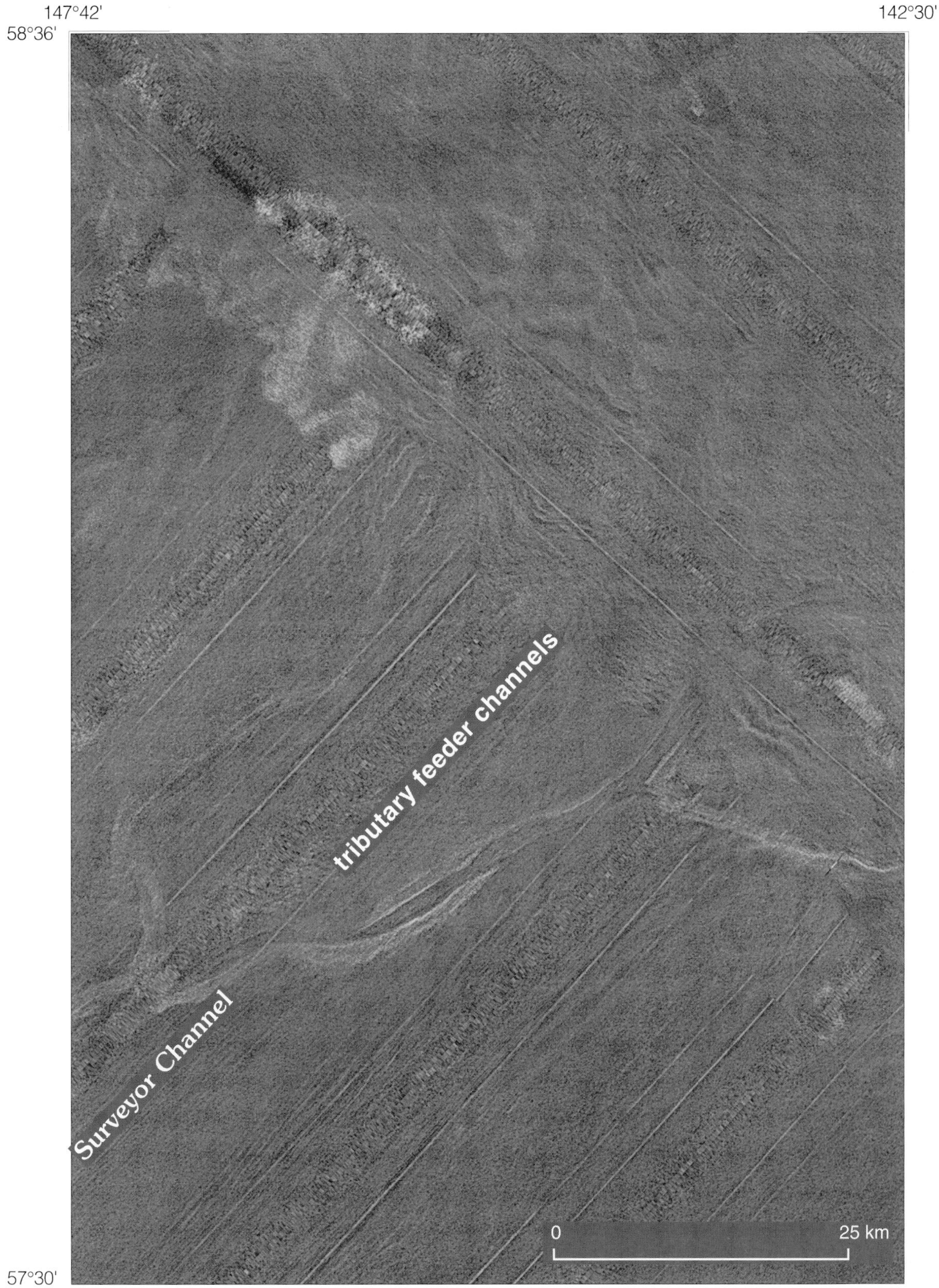

147°42'

142°30'

58°36'

Surveyor Channel

tributary feeder channels

0 — 25 km

57°30'

Figure 15–11. GLORIA mosaic of deep-sea channels merging on lower slope of Yakutat margin. At lower left of image is beginning of Surveyor Channel. See Figure 15–1 for location of image.

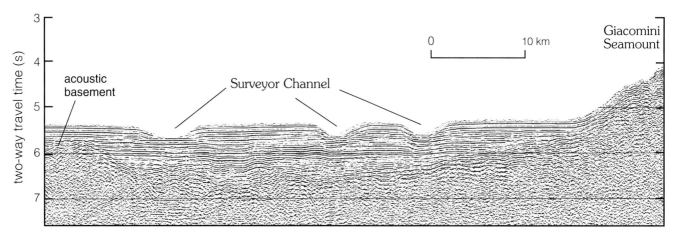

Figure 15–12. Two-channel seismic-reflection line 17-89 across meandering Surveyor Channel west of Giacomini Seamount with stacking and band-pass filter applied. Two closely spaced channel crossings are parts of tight meander apparent on Figure 15–13 southwest of Giacomini Seamount.

tures that have formed a fanlike deposit of debris elements. This fan has a length of 12 km and a width of about 15 km. The gradient down fan is 0.6°. The larger fan is located at the inner wall of the trench below the channel system that heads near Hinchinbrook Sea Valley (Figure 15–9B). This fan has a length of 15 km and a width of about 20 km. The gradient down fan is about 1.9°.

Cores collected from the trench floor sediment hundreds of kilometers west of the small fans contain graded sand layers attributed to turbidity-current deposition (Johnson 1986; Underwood 1986). Some of these turbidity-current deposits may have been generated from slides or debris flows such as those making up Amatuli Fan.

Bering Trough to Yakobi Island

Between the Bering Trough and Yakobi Valley, areas of intensively gullied slopes are common (Figures 15–4 and 15–10). The gullies often merge to form a small dendritic canyon system that extends to abyssal depths (Figure 15–1). Between Middleton Island and Pamplona Spur, these systems extend directly to the northeastern end of the Aleutian Trench (Figure 15–3). Southeast of Pamplona Spur, the dendritic systems reach the base of the slope where they feed extensive deep-sea channel systems (Carlson et al. 1990; Bruns, Stevenson, and Dobson 1992) (Figure 15–1).

No similar small local fans are visible on the GLORIA imagery along the base of slope between Bering Trough and Yakobi Valley. However, based on seismic-reflection data, Bruns (1985) reported two buildups of Pliocene-Quaternary age sediment that could be deep-sea fans off the Alsek and Yakutat Valleys. These sedimentary deposits are each about 2.5 km thick. The GLORIA imagery does, however, show that the gullies incised in the upper slope beyond the shelf

valleys coalesce on the lower slope and rise (Figures 15–1, 15–10, and 15–11) eventually to form large deep-sea channel systems that cross the abyssal plain.

The Surveyor channel system begins at the base of the gullied slope (Figures 15–4 and 15–10) between Bering Trough and Alsek Valley, offshore from the large Malaspina and Bering Glaciers and the glacially formed Icy and Yakutat Bays (Figure 15–1). Here, numerous small slope gullies merge into three channels, probably formed by turbidity currents, which in turn merge about 150 km south of the margin to form Surveyor Channel (Figures 15–1 and 15–11). The resulting well-defined Surveyor Channel, with pronounced meanders (Figures 15–12 and 15–13), extends about 700 km from the margin to the Aleutian Trench. About 350 km from the margin, the channel encounters the structural barrier formed by the Kodiak-Bowie Seamount chain. The channel here is about 5 km wide and has walls 170 m high and a gradient of about 2.5 m/km (Table 15–1). The channel crosses the barrier to the northwest of Giacomini Seamount (Figures 15–1, 15–5, and 15–13) and heads south for another 150 km where it bends northwesterly, following the oceanic basement low between the Kodiak-Bowie and Patton Seamounts. The now deeply entrenched channel (Figure 15–14 and Table 15–1) is up to 6 km wide (2 km wide floor), 500 m deep, with a gradient as great as 5.6 m/km, and continues this northwesterly trend for about 200 km where it empties into the Aleutian Trench, some 700 km from the Yakutat margin. The cause of the deep incision of the lower end of Surveyor Channel is due to the large bulge adjacent to the southern wall of the trench. The bulge is the result of the flexing of the ocean crust as the Pacific Plate is subducted beneath the North American Plate at the Aleutian Trench.

South and southeast of Surveyor Channel (Figures 15–1

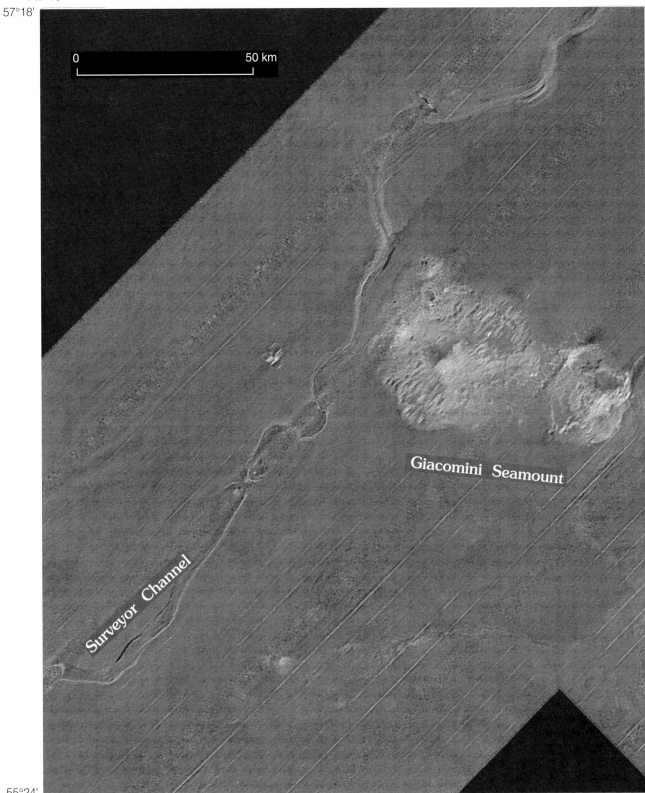

0   50 km

Giacomini   Seamount

Surveyor Channel

Figure 15–13. Enlargement of part of mosaic of Figure 15–5. This shows a pronounced meander in Surveyor Channel after it was diverted around Giacomini Seamount. Gradient in this part of channel changes rather dramatically from about 1 m/km on each side of seamount chain to about 2.5 m/km near seamount to as much as 6 m/km west of seamount chain to trench. Figure 15–12 shows two-channel airgun profile across meander. Location of image in Figure 15–1.

Table 15–1. *Measurements of channel depth, relief, width, and gradient for Main Surveyor Channel (5), Yakutat Branch, middle (Y), Icy Bay Branch, north (I), and Alsek Branch, south (A)*

Crossing/Number	Depth	Relief	Width	Distance	Gradient
Main Surveyor Channel					
S 1	5,160 m	60 m	Aleutian Trench Floor	30 km	4.3 m/km
S 2	5,030 m	385 m	6,798 m	34 km	4.9 m/km
S 3	4,865 m	465 m	6,798 m	45 km	5.6 m/km
S 4	4,615 m	332 m	6,180 m	57 km	4.0 m/km
S 5	4,385 m	250 m	Oblique	70 km	2.0 m/km
S 6	4,242 m	192 m	7,416 m	22 km	0.8 m/km
S 7	4,225 m	200 m	7,416 m	8.5 km	2.4 m/km
S 8	4,205 m	220 m	6,180 m	72 km	1.0 m/km
S 9	4,130 m	167 m	Oblique	19 km	0.7 m/km
S 10	4,117 m	174 m	Oblique	137 km	1.1 m/km
S 11	3,965 m	135 m	Oblique	19 km	0.4 m/km
S 12	3,958 m	165 m	Oblique		
Average	3,958–5,160 m	Average of 10 crossings: 252 m	Average width: 6,798 m	Total distance: 513.5 km	Average gradient: 2.47 m/km
Icy Bay Branch (north)					
I 1	3,640 m	75 m	9,880 m	S13–I1 109 km	S13–I1 2.9 m/km
I 2	3,542 m	40 m	18,540 m	38 km	2.6 m/km
I 3	3,400 m	50 m	12,978 m	39 km	3.6 m/km
I 4	2,705 m	25 m	Several channels 100–300 m wide	39 km	18.0 m/km
Average	3,330 m	48 m	Average width 13,800 m	Total distance 226 m	Average gradient 6.8 m/km
Yakutat Branch (middle)					
Y 1	3,860 m	120 m	3,090 m	S13–Y1 44.7 km	S13–Y1 2.4 m/km
Y 2	3,850 m	100 m	6,798 m	18.0 km	0.6 m/km
Y 3	3,755 m	50 m	13,000 m	21.8 km	4.4 m/km
Y 4	3,647 m	50 m	27,000 m	22.8 km	4.7 m/km
Average	3,778 m	80 m	Average width ~5,000 m	Total distance 107 km	Average gradient 3.0 m/km
Alsek Branch (south)					
A 1	3,816 m	125 m	1,854 m	Y1–A1 28.9 km	Y1–A1 1.5 m/km
A 2	3,755 m	65 m	3,090 m	20.8 km	2.9 m/km
A 3	3,681 m	90 m	4,944 m	21.9 km	3.4 m/km
Average	3,751 m	93 m	Average width 3,296 m	Total distance 23.9 km	Average gradient 2.6 m/km

and 15–5), two other channels are shown by GLORIA imagery and seismic data. One channel (unnamed) is partially filled and is older than Surveyor Channel based on overlapping levee deposits. The older channel is most readily traced by using seismic reflection data in conjunction with the GLORIA data. This older channel meanders through a gap in the seamount chain between Quinn and Giacomini Seamounts (Figure 15–5) and eventually bends northwesterly. This channel may have carried turbidity currents to the Aleutian Trench prior to the advent of the modern active Surveyor Channel or may be the former course of Chirikof Channel.

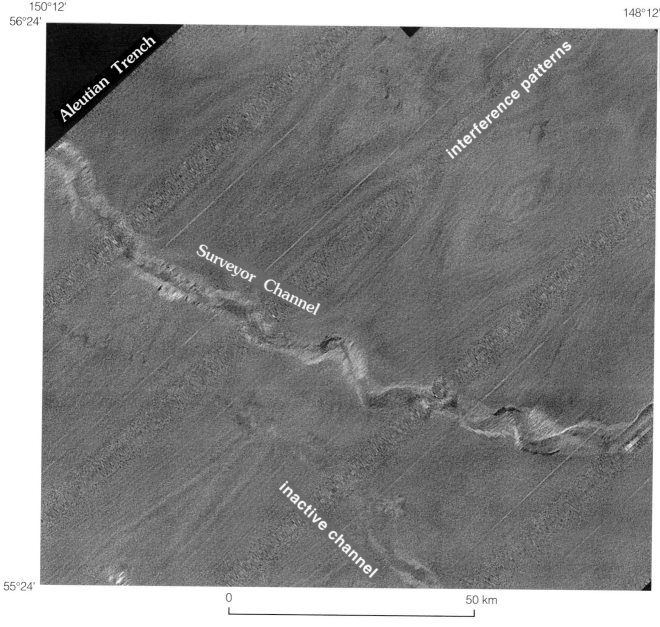

Figure 15–14. GLORIA mosaic of deeply entrenched portion of Surveyor Channel along part that heads north and into Aleutian Trench. Channel relief along this stretch varies from 330 to 500 m. Complex patterns on abyssal plain on both sides of channel and subparallel to tracklines are called interference fringes by Huggett et al. (1992). They suggest these patterns are a thin-layer interference effect caused by sound reflecting off a shallow subsurface layer resulting in constructive or destructive interference by backscattered sound. See Figure 15–1 for location of image.

The second channel, still active, we have informally named the Chirikof channel system (Bruns et al. 1989). The Chirikof Channel bends around Pratt Seamount and then winds through a gap between Surveyor and Quinn Seamounts (Figures 15–1 and 15–15) where it splays into a fanlike form and covers an area of about 7200 km², on the abyssal plain southwest of the seamount chain (Figure 15–1). This channel, which originates at the base of the continental margin between Alsek and Yakobi Valleys (Bruns et al. 1989), is about 570 km long and varies from 2 to 3 km wide with 25 to 50 m of relief and a gradient of about 2 m/km (Table 15-2).

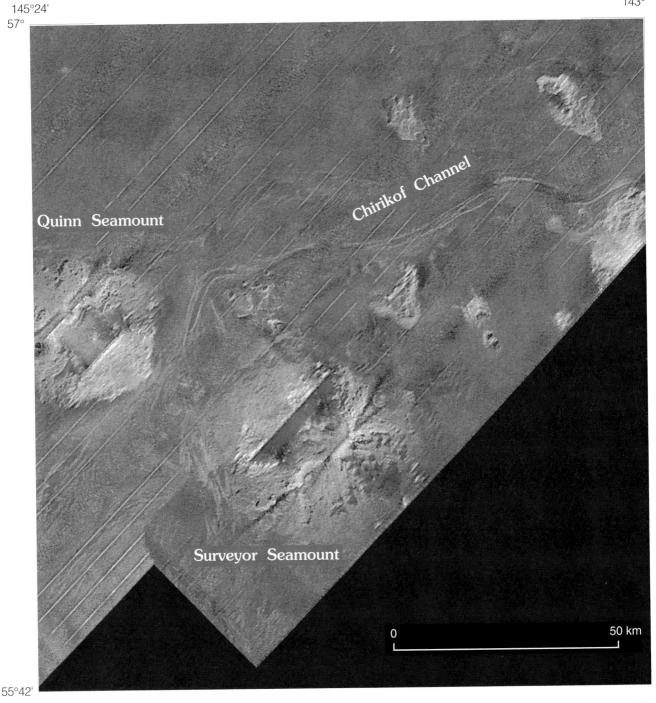

Figure 15–15. GLORIA image showing Chirikof Channel extending through a gap between Surveyor and Quinn Seamounts. After passing through gap, channel splays into a fanlike shape on abyssal plain. See Figure 15–1 for image location.

Summary and conclusions

GLORIA has provided new perspectives on the deep-sea floor in the Gulf of Alaska EEZ. The wide swath (up to 45 km total width per pass) of GLORIA allowed us to survey the entire Gulf of Alaska EEZ from the shelf break to the 200 n mi (370 km) boundary, an area of about 760,000 km², in about 3.5 months' time. The major features observed include (1) continental margin deformation features and their effects on a variety of types of continental slope sediment transport pathways, and (2) extensive abyssal plain submarine channel systems and their interaction with seamounts

Table 15–2. *Measurements of channel depth, relief, width, and gradient for Main Chirikof Channel (C), Fairweather Branch, north (F), and Yakobi Branch, south (B).*

Crossing/Number	Depth	Relief	Width	Distance	Gradient
Main Chirikof Channel					
C 1	3,925 m	20 m	1,800 m		
C 2	3,875 m	45 m	5,562 m	31.9 km	1.6 m/km
C 3	3,830 m	25 m	4,635 m	35.9 km	1.3 m/km
C 4	3,800 m	60 m	7,416 m	26 km	.9 m/km
C 5	3,751 m	40 m	5,562 m	22 km	2.2 m/km
C 6	3,717 m	30 m	7,516 m	24 km	1.4 m/km
C 7	3,547 m	60 m	3,708 m	48 km	3.5 m/km
Average	3,777 m	40 m	5,171 m	187.8	1.8 m/km
Fairweather Branch (north)					
F 1	3,368 m	40 m	5,635 m	C7–F1 46 km	C7–F1 3.9 m/km
F 2	3,260 m 3,279 m	50 m	Many small channels 50–100 m wide	18 km	5.4 m/km
F 3	3,285 m	135 m	16,000 m	15 km	1.7 m/km
F 4	2,220 m	250 m	5,562 m	15 km	71 m/km*
Average	3,082 m	119 m		100 km	3.7 m/km
Yakobi Branch (south)					
B 1	3,350 m	90 m	8,034 m	C7–B1 48 km	C7–B1 4.1 m/km
B 2	3,090 m 3,100 m	55 m 100 m	6,180 m 9,270 m	16 km	6.8 m/km
B 3	2,711 m	30 m	3,708 m	B1–B4 23 km	B1–B4 27.8 m/km
B 4	2,752 m	50 m	7,716 m	15 km	2.7 m/km
Average	3,001 m	65 m	6,981 m		

of the Kodiak-Bowie chain. In the heavily glaciated northern Gulf of Alaska southeast of Middleton Island, compressional ridges forming on the slope of this subduction margin have created a trellised slope-canyon drainage system for the delivery of the glacial detritus to the deep sea. Southeast of Middleton Island the transpressive nature of the plate interaction has resulted in the erosion into the continental slope of small dendritic-patterned canyon systems. Farther southeast along-margin, the slope is intensively eroded by gullies that extend directly downslope. The absence of large canyons on the continental slope is likely a reflection of the relative youthfulness of this margin. The extensive slope gullies have not had time to coalesce into large canyons. The numerous gullies do, however, coalesce on the rise to feed large volumes of glacially derived sediment to large deep-sea channels. The Surveyor and Chirikof channel systems are active pathways for carrying Yakutat terrane glacial marine sediment to the abyssal plain and, in the case of Surveyor Channel, to the Aleutian Trench. These channels have acted as conveyors for Gulf of Alaska continental margin sediment at least throughout the Pleistocene.

Acknowledgments

We appreciate the assistance provided by the officers and crew of the R/V *Farnella*, the marine logistics and support groups, the technicians and engineers, the navigation and processing groups who all provided vital services necessary to collect and process the large amount of data used in preparation of this chapter. We are especially thankful to Greg Gable and Doug Rearic (deceased) for their expertise and care in processing of the GLORIA imagery included in this chapter. Steven Eittreim and Stephanie Ross provided constructive criticism of the manuscript.

References

Atwood, T. A., Bruns, T. R., Carlson, P. R., Molnia, B. F., and Plafker, G. 1981. Bathymetric map of the eastern Gulf of Alaska. *U.S. Geological Survey Miscellaneous Field Study Map MF-859*, 3 sheets, scale 1:250,000. Reston, Va.: USGS.

Bering Sea EEZ-SCAN Scientific Staff 1991. Atlas of the Exclusive Economic Zone, Bering Sea, Alaska. *U.S. Geological*

Survey Miscellaneous Investigations Series 1-2053, scale 1:500,000. Reston, Va.: USGS.

Bruns, T. R. 1983. A model for the origin of the Yakutat block, an accreting terrane in the northern Gulf of Alaska. *Geology* 11: 718–21.

Bruns, T. R., and Schwab, W. C. 1983. Structure maps and seismic stratigraphy of the Yakataga segment of the continental margin, northern Gulf of Alaska. *U.S. Geological Survey Miscellaneous Field Studies Map MF-1424*, 20 p., 4 sheets, scale 1:250,000. Reston, Va.: USGS.

Bruns, T. R. 1985. Tectonics of the Yakutat block, an allochthonous terrane in the northern Gulf of Alaska. *U. S. Geological Survey Open-File Report 85-13*, 112 p. Reston, Va.: USGS.

Bruns, T. R., and Carlson, P. R. 1987. Geology and petroleum potential of the southeast Alaska continental margin. In *Geology and Resource Potential of the Continental Margin of Western North America and Adjacent Ocean Basins – Beaufort Sea to Baja California*, eds. D. W. Scholl, A. Grantz, and J. G. Vedder, pp. 269–82. Circum-Pacific Council for Energy and Mineral Resources Earth Science Series, Vol. 6. Houston, Tex.: American Association of Petroleum Geologists.

Bruns, T. R., Carlson, P. R., Stevenson, A. J., Fisher, M. A., Ryan, H. F., Mann, D. M., Dobson, M., Huggett, Q., Parson, L., and Fannin, N. G. T. 1989. GLORIA images from the Gulf of Alaska and British Columbia: Subduction zones, transforms and channels. *EOS, Am. Geophys. Union Trans.* 70: 1338.

Bruns, T. R., Stevenson, A. J., and Dobson, M. R. 1992. GLORIA investigation of the Exclusive Economic Zone in the Gulf of Alaska and off southeast Alaska: M/V Farnella cruise F7-89-GA, June 14–July 13, 1989. *U.S. Geological Survey Open-File Report 92-317*, 18 p. Reston, Va.: USGS.

Carlson, P. R., Molnia, B. F., Kittleson, S. C., and Hampson, J. C. 1977. Distribution of bottom sediments on the continental shelf, northern Gulf of Alaska. *U.S. Geological Survey Miscellaneous Field Study Map MF-876*, 13 p., 2 sheets, scale 1:500,000. Reston, Va.: USGS.

Carlson, P. R., Bruns, T. R., Molina, B. F., and Schwab, W. C. 1982. Submarine valleys in the northeastern Gulf of Alaska: Characteristics and probable origin. *Mar. Geol.* 47: 217–42.

Carlson, P. R. 1989. Seismic reflection characteristics of glacial and glacimarine sediment in the Gulf of Alaska and adjacent fjords. *Modern Glacimarine Environments: Glacial and Marine Controls on Modern Lithofacies and Biofacies*, eds. R. D. Powell and A. Elverhoi, pp. 391–416. *Mar. Geol.* 85: 391–416.

Carlson, P. R., Bruns, T. R., and Fisher, M. A. 1990. Development of slope valleys in the glacimarine environments of a complex subduction zone, Northern Gulf of Alaska. In *Glacimarine Environments: Processes and Sediments*. eds. J. A. Dowdeswell and J. D. Scourse, pp. 139–53. Geological Society, Special Publication, No. 53. London.

Eyles, C. H., Eyles, N., and Miall, A. D. 1985. Models of glacimarine sedimentation and their application to the interpretation of ancient glacial sequences. *Paleogeog. Palaeoclim. Palaeocol.* 51: 15–84.

Eyles, C. H. 1987. Glacially influenced submarine channel sedimentation in the Yakataga Formation, Middleton Island, Alaska. *J. Sed. Petrol.* 57: 1004–17.

Hampton, M. A. 1983. Geology of the Kodiak shelf, Alaska: Environmental considerations for resource development. *Cont. Shelf Res.* 1: 253–81.

Hampton, M. A., Carlson, P. R., Lee, H. J., and Feely, R. A. 1986. Geomorphology, sediment, and sedimentary processes. In

The Gulf of Alaska: Physical Environment and Biological Resources, eds. D. W. Hood and S. T. Zimmerman, pp. 93–143. Washington D.C.: National Oceanic and Atmospheric Administration.

Huggett, Q. J., Cooper, A. K., Somers, M. L., and Stubbs, A. R. 1992. Interference fringes on GLORIA side-scan sonar images from the Bering Sea and their implications. *Mar. Geophys. Res.* 14: 47–63.

Johnson, K. H. 1986. *Recent Sedimentation in the Central and Western Aleutian Trench*. M.S. Thesis, California State University. Hayward, Calif.

Jones, D. L., Silberling, N. J., Coney, P. J., and Plafker, G. 1986. Lithotectonic terrane map of Alaska (west of 141st meridan). *U.S. Geological Survey Miscellaneous Field Studies Map MF-1874*, scale 1:2,500,000. Reston, Va.: USGS.

Kulm, L. D., Von Huene, R., et al., 1973. *Initial Reports of the Deep Sea Drilling Project*, Vol. 18. Washington, D.C.: U.S. Government Printing Office, 1077 p.

Menard, H. W. 1964. *Marine Geology of the Pacific*. New York: McGraw-Hill, 271 p.

Molnia, B. F., and Carlson, P. R. 1978. Surface sedimentary units of the northern Gulf of Alaska continental shelf. *Am. Assoc. Petrol. Geol. Bull.* 62: 633–43.

Molnia, B. F., and Carlson, P. R. 1980. Quaternary sedimentary facies on the continental shelf of northeast Gulf of Alaska. In *Quaternary Depositional Environments of the Pacific Coast, Pacific Coast Paleogeography Symposium 4*, eds. M. E. Field, A. H. Bouma, I. P. Colburn, R. G. Douglas, and J. C. Ingle, pp. 157–68. Los Angeles, Calif.: Society of Economic Paleontologists and Mineralogists.

Ness, G. E., and Kulm, L. D. 1973. Origin and development of Surveyor deep-sea channel. *Geolog. Soc. Am. Bull.* 84: 3339–54.

Plafker, G., and Addicott, W. O. 1976. Glaciomarine deposits of Miocene through Holocene age in the Yakataga Formation along the Gulf of Alaska margin, Alaska. In *Recent and Ancient Sedimentary Environments in Alaska*, ed. T. P. Miller, pp. Q1–Q23. *Proceedings of Alaska Geological Society Symposium*, Anchorage, Ala.: Alaska Geological Society.

Plafker, G. 1987. Regional geology and petroleum potential of the northern Gulf of Alaska continental margin. In *Geology and Resource Potential of the Continental Margin of Western North America and Adjacent Ocean Basins – Beaufort Sea to Baja California*, eds. D. W. Scholl, A. Grantz, and J. G. Vedder, pp. 229–68. Circum-Pacific Council for Energy and Mineral Resources Earth Science Series, vol. 6. Houston, Tex.: American Association of Petroleum Geologists.

Rea, D. K, Basov, I. A., Janecek, T. R., and Leg 145 Shipboard Scientific Party 1993. Cenozoic paleoceanography of the North Pacific Ocean: Results of ODP Leg 145, the North Pacific transect. *EOS, Program and Abstracts, Am. Geophys. Union Trans.* 76,(16 Suppl.): 173.

Sharma, G. D. 1979. *The Alaskan Shelf: Hydrographic, Sedimentary, and Geochemical Environment*. New York: Springer-Verlag, 498 p.

Somers, M. L., Carson, R. M., Revie, J. A., Edge, R. H., Barrow, B. J., and Andrews, A. G. 1978. GLORIA II – an improved long range sidescan sonar. In *Proceedings of the Institute of Electrical Engineering on Offshore Instrumentation and Communication, Oceanology International Technical Session J*, pp. 16–24. London: BPS Publications.

Stevenson, A. J., and Embley, R. 1987. Deep-sea fan bodies, terrigenous turbidites sedimentation, and petroleum geology, Gulf of Alaska. In *Geology and Resource Potential of the Continental Margin of Western North America and Adja-*

cent Ocean Basins – Beaufort Sea to Baja California, eds. D. W. Scholl, A. Grantz, and J. G. Vedder, pp. 503–22. Circum-Pacific Council for Energy and Mineral Resources Earth Science Series, vol. 6. Houston, Tex.: American Association of Petroleum Geologists.

Swinbanks, D. 1986. New GLORIA in record time. *Nature* 320: 568.

Thrasher, G. P. 1979. Geologic map of the Kodiak outer continental shelf, western Gulf of Alaska. *U.S. Geological Survey Open-File Report 79-1267*, 2 sheets, scale 1:250,000. Reston, Va.: USGS.

Underwood, M. B. 1986. Transverse infilling of the central Aleutian Trench by unconfined turbidity currents. *Geo-Mar. Ltrs.* 6: 7–13.

16 Sedimentation along the fore-arc region of the Aleutian Island Arc, Alaska

Max R. Dobson,[1] Herman A. Karl,[1] and Tracy L. Vallier[2]

[1]Institute of Earth Studies, University of Wales, Aberystwyth, United Kingdom

[2]U.S. Geological Survey, Menlo Park, California

Abstract

A geophysical survey (including GLORIA) of the southern margin of the Aleutian Island Arc, north Pacific, from Umnak Island to Stalemate Ridge allowed fore-arc sedimentation patterns to be analyzed. Direct and indirect variables govern the availability, volume, and composition of sediment supplied to the fore-arc. Direct variables include the changing thermal aspect and convergence angle of the subducting plate and associated volcanism. Convergence along the Aleutian Arc ranges from orthogonal through oblique to parallel. Tectonics impact on both the magmatic arc and the fore-arc and include folding and block faulting in the fore-arc basin, plate decoupling, fragmentation and block rotation of the magmatic arc, and large-scale strike-slip with translation. Indirect variables include sea-level changes, global climate, ocean circulation, and biogenic productivity.

The high-latitude location of the arc has ensured that glacial pulses since the Pliocene exerted a marked influence on the patterns of sediment deposition along the slope and subsequent redeposition into the fore-arc basin. Four tectonically dominated depositional models are recognized. In an Unalaska Type A model orthogonal convergence and fore-arc compression result in slope aprons and fans feeding fore-arc depocenters by mass failure and debris flows, which translate downslope into channel-levee systems. In a Yunaska-Atka Type B model, convergence is slightly oblique, giving a mixed compressional and extensional tectonic pattern. Slope aprons are prominent sediment sources, but basin inversion with compensation cycles results in complex sequences. The emergence of discrete shear zones exceeding 20 km in width is a special feature. In an Adak Type C model convergence is oblique. Here the magmatic arc and fore-arc basin have been fragmented into blocks such that faults provide conduits for sediment from the shelf, thus bypassing the slope. Basins are elongate whereas strike-slip tectonics obscure distinctions between the outer fore-arc basin and the accretionary prism zones. An Attu Type D occurs where convergence approaches pure strike-slip and translation dominates such that sections of the fore-arc are moved laterally. Canyons transport sediment direct from a narrow magmatic arc to a fault-dominated fore-arc region.

Introduction

Modern deep water (>400 m) depositional systems associated with intraoceanic convergent margins, unlike readily observable fluviatile environments, are difficult to analyze in part because the geophysical methods available to investigate them limit the opportunity for understanding sedimentary processes. Indeed, the dependence on geophysical methodology allied to the problems of scale frequently make even direct comparisons with the rock record infeasible (Mutti and Normark 1991). Moreover, unlike the relatively uniform depositional conditions found along most passive margins, depositional systems associated with convergent margins are dominated by subduction-controlled tectonics and volcanism. The varied impact of tectonic activity on the rate of arc erosion, the processes and patterns of sediment dispersal, and the sites for sediment deposition make modeling and the development of recognition criteria for this environment daunting. Nevertheless, a wide-ranging series of sedimentation models, derived from studies of modern convergent margins, have evolved over the last twenty years; for collected papers see Leggett (1982) and Watkins and Drake (1982); for an extensive review with case studies see Pickering, Hiscott, and Hein (1989).

Early models of modern fore-arcs, as developed by Dickinson and Seely (1979), were related, at least in part, to frontal accreting subduction complexes. Subsequent studies made use of multichannel seismic methods to refine these models (Moberly, Shepherd, and Coulbourn 1982; Lu and McMillen 1983). But in order better to quantify the mechanical parameters and physical properties of these complexes, several DSDP (and now ODP) programs were designed specifically to address convergent margin problems. The extensive use of bathymetric and sidescan sonar (Sea-MARC), seismic and submersible data, in many cases directly linked to drilling programs, ensured that significant advances in understanding sedimentary processes along these margins were made despite the dimension of variability present (von Huene, Aubouin et al. 1985; Cadet et al. 1987). Understanding the processes of sedimentation has been constrained because in a tectonic context, where there is a continuum of response from orthogonal subduction to transpression and strike-slip, very little is known about variations in fore-arc structural fabric as a function of variable convergence (Le Pichon 1987).

Sumatra and New Zealand are significant in a tectonic context because they contain partially exposed margins that display increasing convergent obliquity. With respect to the former, part of the fore-arc sequence is exposed in the Nias Islands allowing comparisons to be made with geophysical data obtained offshore. This margin has been especially valuable in appreciating the complex interaction between sedimentation and tectonics, particularly repeated fore-arc basin tilting (Moore, Curray, and Emmel 1982; Stevens and Moore 1985). Subsequently, SeaMARC was used to survey part of the Sunda Arc where three structural styles were recorded (Breen, Silver, and Hussong 1986). In New Zealand (van der Lingen 1982), slope instabilities due to dextral strike-slip tectonics have generated slides and sheet-like deposits. Sedimentation patterns in intraoceanic settings have been examined along the Tonga-Kermadec margin (Gnibidenko et al. 1985) where slow sedimentation rates have allowed tectonic processes to be better appreciated, and in the area of the Lesser Antilles (Brown and Westbrook 1987). In the latter, fore-arc basins like the Tobago Basin have received such an abundance of sediment that they have overstepped the accretionary prism. Review papers concerned specifically with convergent margin sedimentation models (Underwood and Bachman 1982; Underwood 1991) emphasize the complexity of the tectonic environment, the importance of trench fill, and the associated formation of an imbricate wedge, but especially the interplay of the several main controlling variables on the distribution of sediment in time and space.

The north Pacific convergent margin, which extends westward from the Gulf of Alaska Transform Margin in the east to the Kuril Trench in the far west, ranges from Andean type off the Alaskan Peninsula (Bruns et al. 1987) to intraoceanic volcanic arc (Scholl, Vallier, and Stevenson 1987). Dickinson and Seely (1979) classified the fore-arc region of the Aleutian Arc as terraced and ponded with a filled trench. Initial studies conducted along this margin concentrated on elucidating the bathymetry, tectonics, and geological history (Nilsen and Zuffa 1982; Schoff, Vallier, and Stevenson 1982; Scholl et al. 1987; and Vallier et al. 1994 for a recent review). Sedimentological studies have included (1) provenance and transport processes recorded in the Aleutian Trench sediments (Underwood 1986a, 1986b), which have shown that much of the trench material has an external source; (2) the mapping of canyons (Lewis, Ladd, and Bruns 1988; Bruns et al. 1989); and (3) the provenance and dis-

persal of muds within the fore-arc (Underwood and Hathon 1989).

In 1988 and 1989 a comprehensive GLORIA sidescan sonar survey was made of the Aleutian margin between the longitudes of 164 W and 168 E. Supporting seismic reflection profiles were also collected. This chapter presents the results of an analysis of these geophysical data with emphasis on the processes and patterns of sedimentation associated mainly with the youngest (Pliocene to Holocene) rocks. A series of sedimentation models are presented that reflect the changing structural regimes recorded along the arc. The structural regimes present along the arc are related to the orientation of convergence between the Pacific Plate and the North American Plate. Convergence is orthogonal at 164° W but becomes progressively more oblique towards the west, such that at 172° E convergence between the plates is essentially parallel (Engebretson, Cox, and Gordon 1984).

Controls on sedimentation in intraoceanic fore-arc settings

Patterns of sedimentation in both time and space in intraoceanic fore-arc basin settings are governed by direct (subduction and volcanism) and indirect (sea-level changes, oceanic circulation, global climate, and biogenic productivity) variables (Dobson, Scholl, and Stevenson 1991; Winsemann and Seyfried 1991). Subduction is the most important direct influence as it conditions both the level of magmatic activity and intensity of arc-wide tectonics. It is very much the age, thermal aspect, and velocity of a down-going plate that influences both the scale of volcanic activity and elevation of a magmatic arc (Marlow et al. 1983; Delong, Fox, and Mcdowell 1978; Scholl et al. 1987). Explosive volcanism, in particular, can be a source of large volumes of sediment supplied to a fore-arc basin dominantly in the form of ash falls. Whilst the elevation of a magmatic arc, relative to sea level, has a direct bearing on the volume of sediment supply to a fore-arc basin, earthquakes, associated with subduction-induced uplift and subsidence, can trigger the mass movement downslope of previously deposited sediment. Faulting, particularly strike-slip motion, and folding in a fore-arc basin are strongly linked to the obliquity of the down-going plate relative to the convergence of the overriding plate (Harding, Vierbuchen, and Christie-Blick, 1985; Stevens and Moore 1985).

Oblique convergence between tectonic plates promotes disruption and fragmentation of both the active magmatic arc and sediment dominated fore-arc basin. The controlling factor appears to be the degree of interplate shear stress or plate coupling (Goldfinger et al. 1992). Recognizable structural features indicative of interplate shear stress include fragmentation of the magmatic arc into discrete blocks that may or may not be rotated; lateral translation of these separate blocks; and strike-slip faulting in the magmatic arc, fore-arc basin, and accretionary prism regions (Scholl et al. 1987).

Arc-wide variations in the scale of fragmentation are functions of obliquity of subduction, internal composition of an arc, and the distribution of strike-slip shear zones. Overall, these disruptions can create depressions along a magmatic arc, summit basins, and depocenters in a fore-arc basin, as well as linear depressions within an accretionary prism. Basins may form as a direct result of the rotation of individual blocks. The broad effect is to generate a series of structural domains and associated depocenters that vary in form and type according to their position along an arc (Scholl et al. 1987).

Subduction generates underthrusting or a synthetic thrust belt, which, when oblique to the convergence vector of the obducting plate, promotes shortening or folding in the cover and rotation of fractured elements (Harding 1983; Aydin and Page 1984; Harding et al. 1985; Goldfinger et al. 1992). There can also be a decoupling of orthogonal and along-strike components with the added possibility of every variation in between. The siting of major transverse fractures in a magmatic arc and fore-arc basin may be influenced in part by the structural fabric, age, and convergent rate of the down-going oceanic crust as well as the internal fabric of the magmatic arc, especially the location of major plutonic masses. Large topographic irregularities on the subducting plate, like ridges and seamounts (Delong et al. 1978; Lonsdale 1988; Vallier et al. this volume), that fail to be subdued by normal faulting activity along an outer trench high can deform a fore-arc basin. This deformation will involve both the accretionary prism and basinal zones of a fore-arc, if present (Cadet et al. [1987] and Pickering et al. [1989] for a review of this topic).

Indirect variables, particularly global climatic cycles, are significant in sediment supply terms, especially for those island arcs located in high latitudes. During phases of glacial maxima, low sea level provides the conditions to rework sediment previously stored on the continental shelf, and material supplied by glaciers serves to ensure high rates of sedimentation in a fore-arc basin (Underwood 1991). Excepting most Andean type margins, during highstands and climatic amelioration, opportunities for supplying sediment to a fore-arc basin from the magmatic arc may be reduced whilst provenances external to an arc, including the ocean domain with its biogenic potential, are enhanced (Scholl et al. 1987; Hathon and Underwood 1991). The absence of glacial overprint means that depositional systems are more directly related to the morphotectonic evolution of the magmatic arc (Dobson et al. 1991; Winsemann and Seyfried 1991).

Data sources and methods

Technical details of the GLORIA system are available elsewhere (Somers et al. 1978; Laughton 1981). It should be emphasized that the shipboard sonar mosaic, which was routinely used for initial interpretation, has a spatial resolution of 125 m along track and 50 m across track. Therefore, most sedimentary features will lack coherence. Because variable backscatter is a function of seafloor roughness, some structures at the centimeter scale (Kidd, Hunter, and Simon 1987) may be recorded. Analysis of GLORIA data was further complicated as a result of the very steep slopes traversed. Tracklines were oriented parallel to the trend of the island arc and therefore orthogonal to the fore-arc slope, which meant that signal response levels from the upslope side, compared to the downslope side, had a tendency to be higher. A further complication, in terms of subsequent interpretation, is that since subbottom penetration targets may record on GLORIA images, as a result of the operating frequency employed (6.3 to 6.7 kHz), any brightness of the backscatter may be a function of a buried surface (Gardner et al. 1991). The digital GLORIA sonar images were corrected for geometry and radiometry (Chavez 1986) and used to improve the interpretation.

A 3.5-kHz seismic profiler with a calculated vertical resolution of 0.8 m and a 160 m^3 airgun system with a vertical resolution of 25 m were routinely deployed. Interpretation methods involved making echocharacter maps based on 3.5-kHz profiles (Pratson and Laine 1989). A modified version of the Pratson and Laine classification, developed for the New England margin and incorporating the work of Kidd et al. (1987) and O'Leary and Dobson (1992), has been used here particularly to delimit the reflective forms on GLORIA. Employment of an established classification was necessary because, apart from two DSDP sites (Scholl and Creager 1973), a few dredge samples (Scholl et al. 1987), and a set of ninety-eight piston and gravity cores (Underwood and Hathon 1989), the region has not been systematically sampled.

Regional physiography, stratigraphic and structural setting

The curvilinear Aleutian Island Arc, extending westward from Unimak Island at the tip of the Alaskan Peninsula toward the Russian mainland of Kamchatka, is 2,200 km long (Figure 16–1). A volcanic-crested cordillera, about 100 km wide, is fronted to the south by a fore-arc basin and trench slope between 60 and 110 km wide. Total relief from volcanic crest to trench floor exceeds 10,000 m. Today the subaerial magmatic arc has a subdued topography because of the level of erosion it received, probably in the late Tertiary (Scholl et al. 1987). Nineteen active volcanoes rise above the beveled platform.

The Aleutian Shelf, for the purposes of this study, extends from the present shoreline to water depths of 400 m. The 400-m depth is employed here because it conforms to the operating depth limit for GLORIA. Water depths from the

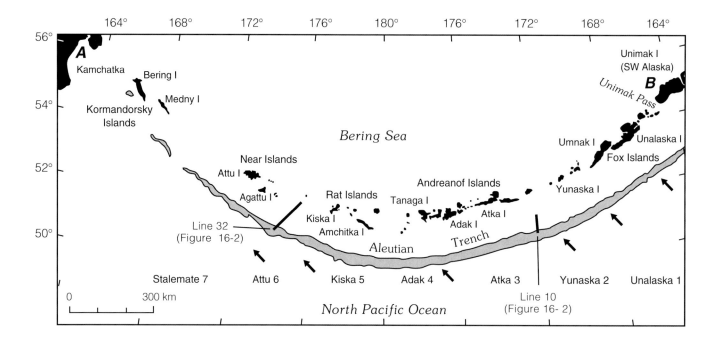

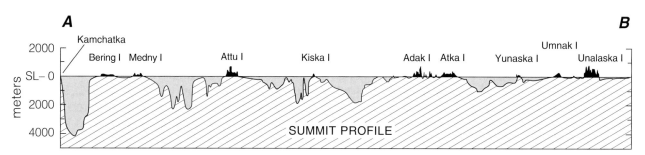

Figure 16–1. Regional physiography, including bathymetry, principal islands, and plots of latitude and longitude to indicate the extent of the named and numbered sectors used to describe the arc. Convergence vectors are indicated by black arrows. The trench is indicated by shaded area. A summit profile is included to show the degree of arc peneplanation. Sunday Basin is located immediately east of Kiska Island (see Figure 16–12). The positions of the two multichannel seismic profiles illustrated in Figure 16–2 are shown by solid lines.

400-m shelf edge to the middle of the upper slope range to 2,000 m, with a downslope width of between 15 and 20 km that produces a slope angle of between five and six degrees. From the middle of the upper slope to the fore-arc basin terrace, water depths increase from 2,000 to 4,600 meters, with a variable width averaging 20 km, which gives a slope angle of between four and six degrees. The fore-arc basin terrace is located in water depths between 4,600 and 4,800 m, while the width, although variable, particularly in the west, averages about 40 km. Downslope angles thereby vary from 1 degree to 0.3 degrees. Depending on the height of the outer terrace antiformal high (Figure 11 in Scholl et al. 1982) or trench-slope break, a reverse slope of one degree arcward occurs. The presence of a reverse slope is related to the scale of movement along the margin of the accretionary prism and the level of sediment fill in the fore-arc basin terrace region. The accretionary prism, which has been examined only in

reconnaissance fashion (Scholl et al. 1987), usually occurs as three poorly defined ridges separated by troughs; the maximum width of this whole complex is about 30 km. The water depth increases from 4,800 m along the inner ridge to more than 7,000 m, at least, along the Aleutian Trench floor, which provides for a general slope of four to six degrees. The flat-floored trench varies in width from 11 to 16 km except in the Stalemate Ridge collision zone where the width is less than 5 km. Its depth ranges from a relatively shallow 4,500 m near Kodiak Island in the east to 7,200 m in the central and western part of the Aleutian Arc.

Arc inception is calculated to have been at 56 Ma based on an increase in Kula-Pacific Plate velocity (Scholl, Buffington, and Marlow 1975; Engebretson et al. 1984) associated with a jump in the locus of subduction from along the Beringian margin to the present position. Three prominent rock series informally termed lower, middle, and upper are separated by major unconformities recognized on seismic profiles (Scholl et al. 1987; Vallier et al. 1994). Rocks of the lower series range in age from arc inception to 37 Ma based on dates obtained from volcanics on several islands (Scholl, Greene, and Marlow 1970; Vallier et al. 1994). Pillow lavas and volcaniclastic rocks dominate the basal part with mixed clastic and volcaniclastics contributing to the upper part, although laminates of chert with silt and occasional limestones and tuffaceous greywackes were also deposited. The lower series can be traced acoustically from the shelf seaward, where it is characterized by coarse and laterally discontinuous layering.

Middle series rocks of mostly Oligocene and Miocene ages form dip-slope blanket sequences between 1 and 3 km thick. Material was derived from an emergent arc that experienced protracted erosion. Sedimentation ranged from nonmarine and shallow shelf to an extensive fore-arc apron, whilst the facies recorded include fluviatile deltaic slope turbidites and an increasingly prominent pelagic component composed mainly of diatoms. According to Scholl and Creager (1973) the Aleutian Ridge was uplifted in the mid to late Miocene; this resulted in high rates of fore-arc sedimentation recorded at DSDP Site 186.

The upper series, including the Pliocene to recent sequences (younger than 5.3 Ma), unlike the slope setting of the two previous series, largely accumulated in fore-arc and summit basins. Transition of the fore-arc region from a slope to perched basinal setting may have occurred approximately six million years ago when large volumes of sediment flooded the hitherto sediment-starved trench floor (Figure 9 in Scholl et al. 1987). Compressional subduction-driven deformation generated an accretionary prism initially where the trench fill was thickest, notably between 172° W and 176° W, where even today the trench sediment thickness in places exceeds 3 km (Scholl et al. 1982). Development of the accretionary prism or subduction complex is thought to have accelerated with the introduction of turbidite deposits along the trench as a result of glaciation of continental Alaska that commenced in the late Miocene (McCarthy et al. 1984; McCarthy and Scholl 1985). Permanent ice cover in the northern hemisphere was established about 4.5 Ma ago (Rea, Basov, and Janecek 1993). Alaskan ice fields nourished valley glaciers capable of carrying large volumes of sediment to the Gulf of Alaska (Bruns et al. 1989). Immense channel-levee systems that developed on the floor of the Gulf of Alaska became conduits for the transfer of material as turbidity flows into the Aleutian Trench possibly as far as longitude 180°. This fine-grained sediment, transported up to 2,000 km along the trench, provided material for the accretionary prism, although it is known (Underwood 1986a) that trench sand of Holocene age did not come from continental sources. Thus, the marked angular unconformity that separates the middle from the upper series is considered to have been caused by the formation of a subduction complex (Figure 16–2).

Upper series sediments consist of bioturbated fine sands and silts with vitric ash and a significant diatom percentage (Scholl and Creager 1973; Sancetta 1979). Clayey diatomaceous ooze is especially common, with silty sands occurring as discrete layers attributed to turbidity-current processes. Apparently, biogenic productivity has fluctuated areally and temporally with a nonuniform dilution of the diatom content by arc-derived material. Extensive turbidite deposition occurred in the late Pliocene and earliest Pleistocene owing to erosion of the continental shelf during glacial maxima (von Huene, Larson, and Crouch 1973; Underwood 1991). Indeed, the growth of massive stratovolcanoes in the late Pliocene, with concomitant uplift, promoted the formation of Alpine glaciers. Increasing diatom content since the late Upper Miocene is proposed to have been the result of intensification of ocean circulation and upwelling of nutrient waters in the north Pacific (Scholl et al. 1987).

According to Jarrard (1986), the Aleutian Arc is difficult to classify in terms of strain. Strain ranges from mildly compressional, with the most recent folding being early Miocene to late Pliocene in age, to mildly tensional and associated with Quaternary faulting and uplift (Ryan and Scholl 1989). Locally, compressional deformation of the fore-arc basin manifests as antiformal structures, the axes of which are generally parallel to the trench axis. On a large scale, the entire Aleutian Arc is interpreted to be segmented into partially rotated blocks bound by transverse strike-slip faults (Geist, Childs, and Scholl 1987). Arc fragmentation history may be very old and related to the 43 Ma reorganization of north Pacific plates and subsequent plate motions (Scholl et al. 1982, 1983a, 1983b; Engebretson et al. 1984; Lonsdale 1988).

Arc fragmentation determines much of the location, form, and scale of the sedimentary depocenters. Fragmentation rates are difficult to quantify, but using paleomagnetic data from the U.S. Pacific Northwest, a block rotation rate of between sixteen and twenty-two degrees in 12 to 15 Ma has been calculated (Wells et al. 1984). Fragmentation is best seen in the western Adak sector; here broad submarine canyons, oblique to the curve of the magmatic arc, extend across the fore-arc basin (Gibson and Nichols 1953; Gates and Gibson 1956; Perry and Nichols 1965; Anderson 1971; Spence 1977; LaForge and Engdahl 1979; Geist et al. 1987; Scholl et al. 1987). Tectonically controlled canyons of this type are frequently the surface expression of deep sliver or wedge-shaped depocenters which, although prominent in the western arc, occur to a limited extent in the east (Figure 16–3). In the east, the downslope courses of canyons are occasionally diverted by upstanding small fault blocks located in the fore-arc region.

Accelerated distension, related to block rotation in the late Cenozoic, has created a series of summit basins that contain up to 5 km of sediment fill (Geist et al. 1987). The largest summit basin, the Sunday Basin, occurs in the Kiska sector (Scholl et al. 1987). The scale and distribution of these crestal basins are partly a function of the internal fab-

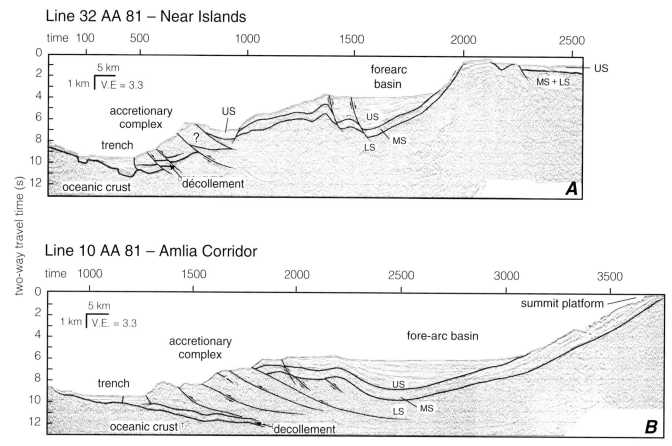

Line 32 AA 81 – Near Islands

Line 10 AA 81 – Amlia Corridor

Figure 16–2. Interpreted multichannel seismic profiles (from Vallier et al. 1994; for location see Figure 16–1). Upper profile (line 32) is located at the eastern end of Attu, sector 6. The lower profile (line 10) transects the Atka, sector 3. Both profiles illustrate uplift of the middle series caused by the emplacement of an accretionary prism (subduction complex), which together have resulted in the creation of a fore-arc basin. Note the site of the outer terrace high. Horizontal and vertical scales are provided.

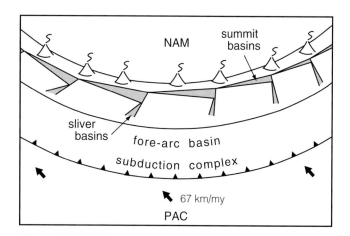

Figure 16–3. A simple model to show arc fragmentation by block rotation caused by increasing obliquity of plate convergence. The fault patterns produced by rotation determine the sites for summit basins and sliver basins (from Scholl, Vallier, and Stevenson 1987). Note the significance of the R' faults (Aydin and Page 1984) that define the margins of the sliver basins. These are referred to in sector 4. The line with black triangles, located close to the outer terrace high, marks a prominent thrust (see Scholl et al. 1987, Figure 6). NAM refers to the North American Plate and PAC is the Pacific Plate.

ric of the magmatic arc and partly due to the angle of convergence. These basins, because of the large volumes of sediment they can accommodate, must influence the quantity of sediment subsequently available for deposition on the lower slopes of the arc. However, since most summit basins occur on the north side of the arc platform, it is the volume of sediment supplied to the Bering Sea that is most affected.

In the near central section, at 176° W, the oblique component of convergence is more than thirty degrees (Vallier et al. 1994); here the Pacific Plate is subducting beneath the North American Plate at less than the orthogonal rate of 8 cm/yr (probably 6 cm/yr). This oblique convergence promotes the development of strike-slip motion, initially in the accretionary prism and the fore-arc region, but as the obliquity increases this tectonic activity is extended to the magmatic arc (Newbury, Laclair, and Fujita 1986). Strike-slip tectonics in the accretionary prism generates linear depressions. Ryan and Scholl (1989) recognized that the Hawley Ridge Shear Zone, located in the fore-arc basin in the Atka sector, was due to strike-slip motion, whereas towards the west, large offsets in the crestal zone of the magmatic arc,

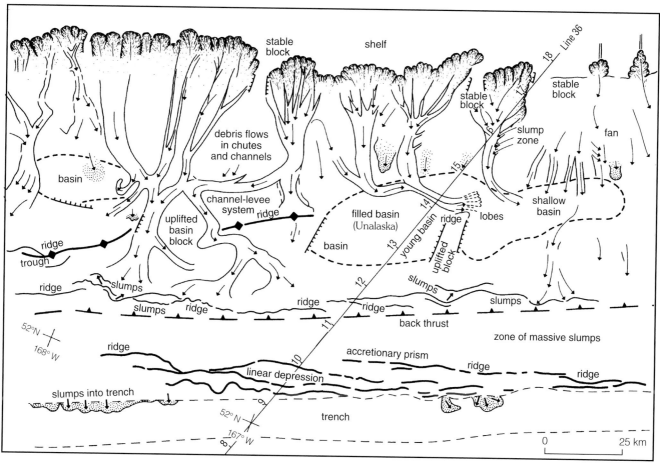

Figure 16–4. Diagram of the patterns of sedimentation in the Unalaska, sector 1, 164°–168° W. Arrows indicate the direction of sediment transport; dashed lines refer to basinal areas. A prominent thrust seen on multichannel seismic profiles (Scholl et al. 1987) is shown as a discontinuous line with teeth pointing north. Slumps, where imaged on GLORIA, are shown as dotted areas (see Figure 16–6, which is a computer-enhanced mosaic of this sector of the fore-arc basin). Line 36, a seismic profile, is shown in Figure 16–5. Numbers along the line are time fixes to aid location. The sediments contained in the slope apron are reworked downslope as debris flows that become channel-levee systems. Sediment is transported both across and along the fore-arc basin.

imaged on aerial photographs, are similarly interpreted as due to strike-slip motion.

Results of the GLORIA and geophysical surveys

We informally and arbitrarily divide the Aleutian Island Arc into seven geographic sectors (Figure 16–1), each extending over four degrees of longitude. The results of the geophysical survey follow for each.

Unalaska, sector 1, 164°–168° W

The Unalaska sector contains the largest group of volcanic islands found along the entire arc; included are most of the Fox Islands, notably Unalaska as well as Akutan, Akun, and part of Unimak. Because wide shelves provide sediment holding sites during eustatic highstands, it is pertinent to

note that the shelf in this sector at 168° W is nearly 40 km wide while at 164° W it is 100 km wide (Figure 16–1), probably signaling a change in the underlying basement (Nilsen and Zuffa 1982). The transition from continental accreted terranes to intraoceanic island arc lies close to 164° W (Bruns et al. 1987).

Plate convergence along this entire sector appears to be orthogonal with no evidence from the available geophysical data of tectonic fragmentation or significant strike-slip faulting, although Unimak Basin may have developed through oblique extension (Bruns et al. 1987). Instead, several subparallel compression ridges are developed in the fore-arc basin to the west of Unalaska Basin (Figures 16–4 and 16–5). Along the eastern half of the sector, between 164° and 165° W, the shelf edge is straight with no indentations caused by canyon-head erosion. GLORIA images (Figure 16–6) reveal the presence of a series of about ten small chutes that have directed sediment downslope, along a 20-km-wide section, towards the filled Unimak Basin in the

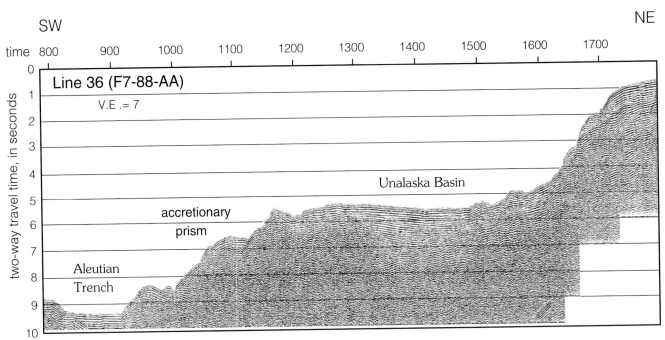

Figure 16–5. Part of seismic line 36 (cruise F7-88; processed records on file at U.S. Geological Survey). Time fixes along the top of the figure are included on Figure 16–4 to aid location and orientation. Fore-arc basin sediment fill exceeds 1,500 m. Contained within the fore-arc basin fill are several very low-angle unconformities, produced as a result of tilting towards the magmatic arc as a result of compression.

fore-arc terrace region. On the basis of poor GLORIA signal return this section of the fore-arc basin is probably veneered with fine-grained hemipelagic sediments. Close to 165°30′ W, GLORIA images, recorded at the extreme shallow-water range of the survey, show that two canyons originate on the adjacent shelf. Both canyons have deeply incised the shelf edge, cutting into the underlying middle series, and due to slope failure have produced an extensive amphitheater-type morphology (Figures 16–4 and 16–6). They extend into the eastern half of Unalaska Basin. A fan-shaped body of sediment, 25 km wide at the base of the slope, occurs at 166° W (see GLORIA mosaic Figure 16–6). Its fan-shaped appearance is enhanced by (1) a series of debris-choked linear channels or chutes that tend to radiate outward as they extend downslope toward Unalaska Basin and (2) a convex morphology with a relief between 100 and 200 m above the level of the adjacent prograding slope aprons. In view of the gradients involved, many of these canyons probably discharge sediment avalanches (Prior and Bornholm 1989).

Along the western half of the Unalaska sector, between 166° and 168° W (Figure 16–4), the upper slope is thickly draped with a slope apron that reveals evidence of failure. Amphitheater-shaped canyon heads have eroded back the shelf edge providing a source of sediment to north-south – oriented fore-arc basin feeder channel systems that are broad (6 km wide), shallow (~500 m), and short (~15 km) with steep gradients (~5 degrees). These feeder channel systems are partially choked with debris flows. Combined GLORIA

and seismic-profile data indicate the presence, downslope from the feeder channels, of a channel-levee system with an overall relief of ~200 m. The system is oriented east-west along the axis of the fore-arc basin with overbanking extending up to 12 km both to the north and south. This channel-levee complex terminates as a series of 5-km-long lobes that encroach into a bathymetric low (Figure 16–4) located in the western half of the Unalaska Basin. Many channel-levee systems show bank stability and a marked vertical growth component. This aggradation is clearly seen on the seismic profiles although high-amplitude reflectors (HAR) are absent, which suggests the deposits are dominantly fine grained (Weimer 1991).

In the region of 168° W a complex of several narrow channels extends downslope where, in the vicinity of the fore-arc basin zone, they merge to form two broad channels. These two 200-m-deep channels cross the sediment-filled fore-arc basin and a compression ridge and flow around an uplifted block of middle series sediments before terminating against the outer terrace antiformal high or trench-slope break. Discharged material is ponded along the line of the trench-slope break and locally spills into linear depressions formed in the accretionary prism, probably as unconfined turbidity flows.

Seismic-reflection profiles (Figure 16–5) reveal the overall style and sequence of sediment fill in the fore-arc basin along this sector. Although not seen on seismic line 36, numerous seaward-dipping normal faults recognized as

Figure 16–6. A computer-enhanced GLORIA sidescan sonar image (cruise F7-88) covering virtually the whole of Unalaska, sector 1. Light areas refer to high backscatter. The shelf edge is marked by a series of embayments with an associated pinnate pattern interpreted as sites of slope failure that converge to form broad channels or canyons. Large fan-shaped sediment bodies may also be seen. Only two canyons appear to have supplied sediment from the inner shelf direct to the slope apron. Irregular channel-levee systems extend into the fore-arc basin, located in the lower half of the mosaic. See Figure 16–4 for an interpretation of part of this mosaic. Width of the mosaic is about 100 km.

gravity-slide faults have been reported to occur along the slope (Bruns et al. 1987). Emplacement of an accretionary prism created a wide (~40 km) fore-arc depression that became flat-floored as deposition proceeded. Based on the evidence of single-channel seismic profiles, sedimentation processes in the fore-arc basin involved stacked channel-levee systems each up to 20 km wide (Figure 16–4). Accumulation continued until the basin had filled to the level of the outer terrace high or trench-slope break, at which point an uninterrupted regional slope, extending from the middle upper slope to the accretionary prism, allowed unconfined sediment flows to overspill down across the inner trench slope.

The 30-km-wide accretionary prism consists of several prominent ridges and slope basins oriented about fourteen degrees to the line of the trench. GLORIA images (Figure 16–6) indicate that both the uplifted middle series along the trench-slope break and the inner accretionary prism sequences are actively slumping. Mass movement occurs both northwards into the fore-arc basin and southwards, in some cases as far as the trench. Slumping is especially prominent where a ridge intersects with the trench. Based on seismic profiles, the slump masses that extend into the trench form mounds more than 7 km wide.

The trench at this sector is from 11 to 16 km wide and filled with as much as 2 km of sediment. On the evidence of the available seismic profiles, the fill consists of about 200 m of parallel-bedded sediments, below which a further series of evenly bedded sediments are separated by a very low angle unconformity that dips northwards. Small debris flows extend south from accretionary prism across the trench floor for up to 4 km.

Yunaska, sector 2, 168°–172° W

In this sector the magmatic arc is 60 km wide and the width of the shelf averages 20 km. Apart from the large island of

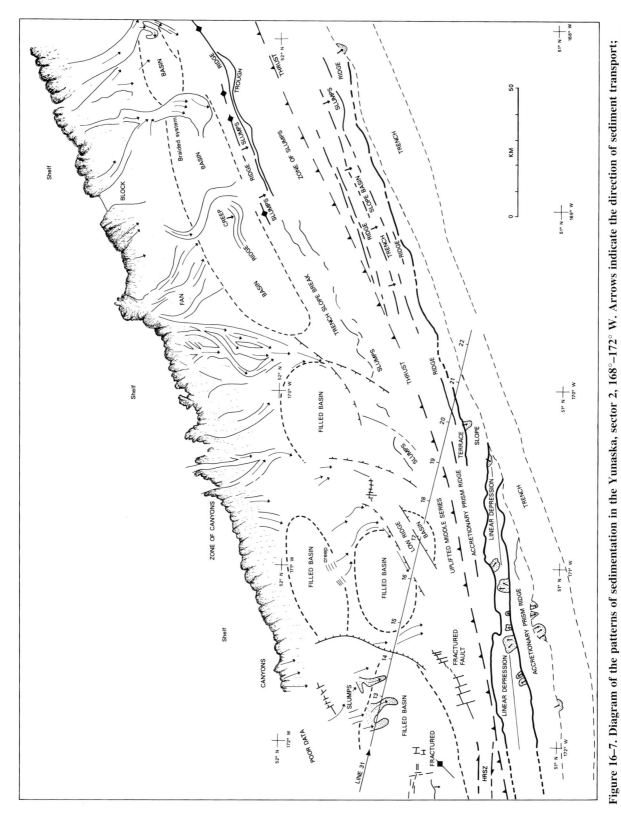

Figure 16–7. Diagram of the patterns of sedimentation in the Yunaska, sector 2, 168°–172° W. Arrows indicate the direction of sediment transport; dashed lines refer to the basinal areas. A thrust is shown as a discontinuous line with teeth pointing north. Slumps are shown as dotted areas. Line 31 refers to a seismic profile (Figure 16–8). Numbers along the line are time fixes to aid location. A combination of filled basins and horst-graben faulting determine the sediment transport dispersal patterns. Note especially the scale of the linear depressions located in the accretionary prism and the associated slumping.

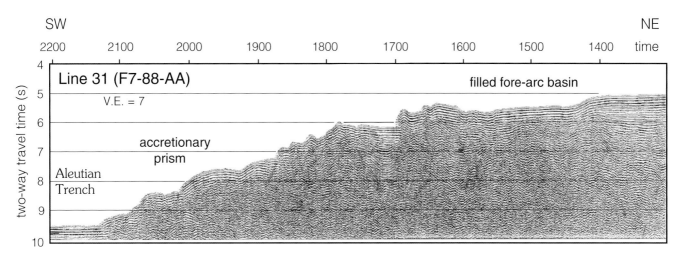

SW   NE

2200   2100   2000   1900   1800   1700   1600   1500   1400   time

Line 31 (F7-88-AA)   filled fore-arc basin

V.E. = 7

accretionary
prism

Aleutian
Trench

two-way travel time (s)

Figure 16–8. Single-channel seismic profile (cruise F7-88) located in Yunaska, sector 2 (Figure 16–7), at 169° W. The profile extends diagonally across the fore-arc and records a series of normal faults that control the location of the several depocenters present in the fore-arc basin.

Umnak only small insular volcanoes such as the Islands of the Four Mountains are subaerially exposed. Convergence, relative to the magmatic arc, is slightly oblique (~10 degrees).

Seismic profiles indicate that west of 169°30' W the fore-arc has been faulted. Normal faults affect the fore-arc basin and have generated a horst-graben regime, largely within the middle series (Figures 16–7 and 16–8). Although the fault patterns are poorly imaged on GLORIA it would appear they are curvilinear in plan. Amukta Summit Basin is bounded by an orthogonal set of normal faults striking parallel to the trend of the arc (Scholl et al. 1987; Geist et al. 1987).

Except for volcanic ash and pelagic material, and one shelf-sourced canyon, sediment supplied to the fore-arc basin in this sector emanates from sites of mass failure at or just below the shelf break. A slope apron is present along the eastern half of this sector and appears to have been well nourished with sediment, although, judging from seismic profiler evidence and GLORIA images, the volume of sediment being transported toward the fore-arc basin is reduced in comparison with sector 1. Between 168° W and 169° W a broad, shallow, sinuous braided channel system (Belderson et al. 1984) extends across the fore-arc toward the outer terrace high. Systems exhibiting low bank stability usually indicate a sand-rich character. At 169° a 30-km-long by 10-km-wide zone, consisting of a series of sinuous ridges and associated hollows (Figure 16–7), may be the result of large-scale sediment creep. Certainly the structure appears sufficiently topographically prominent to divert a channel system. Between 169° and 171° W the upper slope is etched by a series of canyons that direct confined sediment flows toward the fore-arc basin. A 45-km-wide fan-shaped sediment body with a convex upper surface extends out across the slope. Narrow linear channels radiate out across the fan

and transfer sediment to the fore-arc basin. From the hummocky evidence of the seismic profiles (Pratson and Laine 1989), it appears many of these channels are filled with debris flows. Several debris flows are more than 20 km long and, where unconfined, 3 km wide. Seismic profiles indicate that these debris flows are covering earlier sequences consisting of stacked and abandoned channel-levee systems.

Between 170° and 172° W seismic profiles show four oval depocenters, containing on average about 1.5 km of sediment fill. These fore-arc subbasins are between 40 and 60 km long and 25 to 30 km wide and appear to be fault-controlled by a pattern of horst and graben structures (Figures 16–7 and 16–8). Most of the larger subbasins exhibit a convex upper surface that is raised 50 to 75 m above the surrounding seabed. This positive seafloor relief and the fault patterns influence sediment dispersal. Channel thalwegs are directed both around the flanks of the inverted basins and along the trend of the bordering faults (see fix 16 on line 31, Figure 16–8). Between 171° and 172° W slope failure in the form of canyon-head erosion is largely absent, implying that sediment build-out across the slope has been on a small scale. Certainly subaerial exposure along this section of the magmatic arc is very much less (Figure 16–1). This anomaly may signal long-term sediment abandonment. Moreover, individual seismic profile reflectors may be traced without interruption from the filled fore-arc basin across the outer terrace high towards the slope basins, suggesting that sediment overspill has occurred.

Beyond the fore-arc subbasins, uplifted middle series sediments show evidence of arcward downslope failure. Failure zones, seen on GLORIA images as narrow curvilinear belts of strong reflectivity, are up to 20 km wide. Between the subdued trench-slope break and the Aleutian Trench the accretionary prism is both narrow (20 km) and structurally

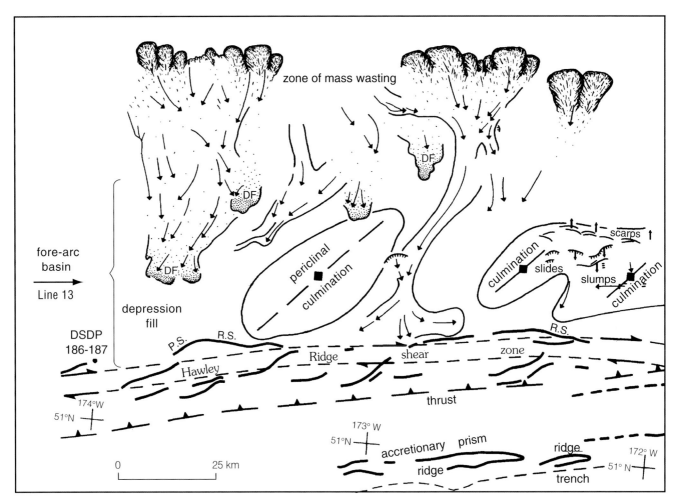

Figure 16–9. Diagram of the patterns of sedimentation in the Atka, sector 3, 172°–176° W. Arrows indicate the direction of sediment transport; solid lines refer to the periclinally folded basinal areas. A thrust is drawn as a discontinuous line with teeth pointing north. Debris flows (DF) are shown by dotted areas with flow arrows. The Hawley Ridge Shear Zone (HRSZ) is indicated by dashed lines. R.S. refers to Riedel Shears and P.S. to P Shears. Line 13 refers to a seismic profile (Figure 16–10). Culminations are shown using an anticline symbol. Deep-Sea Drilling Project Sites 186 and 187 are also shown.

complex. As in sector 1, both ridges and slope basins are oriented about ten degrees to the line of the trench. The existence of a linear depression more than 130 km long and 20 km wide that may be described as an immature slope basin (Underwood and Bachman 1982) adds to the complexity of the prism. The basin contains less than 100 m of sediment fill. Accretionary prism ridges contribute sediment to the depression through slumping and the development of debris flows. Sediment is transported south into the trench as a series of slumps. As in sector 1, where a ridge intersects the line of the trench, extensive slumping occurs. All the available evidence suggests that these slumps form a minor component of the overall sediment budget of the trench along this sector of the arc. Sediment overspill, as unconfined turbidity flows, across the outer terrace high could be the source of the magmatic arc material identified in trench samples (Underwood 1986b).

Atka, sector 3, 172°–176° W

Near continuous crestal exposure consisting of the Andreanof Islands of Atka and Amlia characterize this sector, which is also fronted by a 30-km-wide shelf (Figure 16–9). Seismic profiles indicate that the Amlia Summit Basin, lying close to Atka Island, contains more than 2 km of upper series sediments (Scholl et al. 1987). Pacific plate convergence is oblique and ranges from fifteen degrees at about 172° W to more than twenty degrees at 176° W.

Multichannel seismic-reflection profiles collected on a cruise of the R/V *S.P. Lee* (Scholl et al. 1987) provide a valuable additional database for understanding the tectonic complexity of this sector of the arc. Four profiles oriented approximately north-south include, from east to west, L-5-80, trackline 13 (172°40′ W); L-9-81, line 12 (173°30′ W); L-9-81, lines 9 and 10 (174° W′) on which DSDP Sites 187

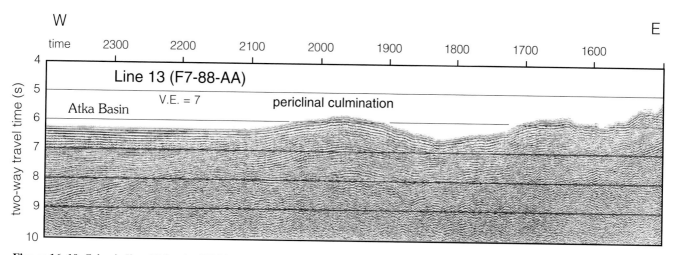

W

E

time

2300 2200 2100 2000 1900 1800 1700 1600

two-way travel time (s)

Line 13 (F7-88-AA)

V.E. = 7 periclinal culmination

Atka Basin

Figure 16–10. Seismic line 13 (cruise F7-88) shows the pattern of sedimentation described in the text as compensation cycles.

and 189 are located; and L-9-81, line 6 (175°30′ W). These data allowed the Hawley Ridge Shear Zone (HRSZ) to be identified (Ryan and Scholl 1985; Scholl et al. 1987). Along the line of the HRSZ, Riedel and P shears (Aydin and Page 1984) can be recognized on the GLORIA images (Figure 16–9). Several phases of folding, each with an elongate periclinal form, have been mapped based on the presence of unconformities, clearly recorded on seismic profiles (see Figures 16–9 and 16–10, line 13). Although there is some evidence for reactivation of earlier established periclines, each phase of compression generated a fresh series of elongated en echelon periclines. The southern margin of the fore-arc region has been affected by extensive shearing that has caused significant inversion of earlier depocenters resulting in the formation of the Hawley Ridge. This ridge is also extensively developed in the Adak sector 4 (Ryan and Scholl 1989).

Seismic profiles confirm that along the eastern half of this sector a well-nourished slope apron is present; by contrast, toward the west the slope apron appears to be less well developed. Despite these apparent variations both GLORIA images and seismic profiles indicate that the fore-arc basin along the whole sector has been well supplied with sediment. In particular, the western half of this sector is dominated by a large sediment-filled, flat-floored area that incorporates the Atka and Hawley Basins. Seismic profiles exhibit uniform laterally continuous reflectors that compare strongly with areally extensive turbidite sheet systems recognized by Moore et al. (1982) in the Sunda fore-arc.

Patterns of transport and sedimentation are strongly influenced by the distribution of the elongate domes, which have 60-km wavelengths and amplitudes of 300 m. Intervening depressions become sites of subsequent sedimentation; demarcation is seen as faintly perceptible very low angle unconformities. Discharge of sediment down conduits and poorly defined channels to lower slope depositional sites

is associated with increasing overbanking that eventually results in the filling of the tectonically formed depressions. More pronounced elongate domes develop patterns of slope failure. Failure in the form of slumping is directed both into the attenuated fore-arc basin to the north as well as to the south. A troughlike depression north of one of these domes is recorded on line 13 (Figure 16–10). These domes involve both middle and upper series sediments. Domes, which rise to more than 300 m above bathymetric background, promote transport deflection with lateral buildups; if the relief of a dome is 500 m or more above background, slumping and failure patterns develop (Figure 16–10). Subdued folds of 100 m or less appear to be readily overwhelmed by invading sediments. This type of shifting sediment fill may be seen as large-scale tectonically driven compensation cycles with depocenters arranged in a lateral en echelon pattern. These tectonically influenced compensation cycles should not be confused with those cycles that are the result of sediment buildups associated with submarine fan lobes (Mutti and Normark 1991).

DSDP Sites 186 and 187 (Scholl and Creager 1973), twenty-four dredge stations and seven piston and gravity cores (Underwood 1986b; Underwood and Hathon 1989), provide a valuable source of detail concerning the nature of the fore-arc basin sediment fill (Scholl et al. 1987). Many of the dredge sites are located on the upper to lower slope south of Amlia and Atka Islands in water depths ranging from 200 m to 4,200 m. In particular, sediments that fill Atka Basin contain volcaniclastic sands derived from the magmatic arc. They were deposited from mud-rich turbidite flows. At DSDP Site 187 the Quaternary sequence is about 320 m thick with an unconformity separating it from Upper Pliocene deposits, sediment thicknesses of this order representing the last 1.8 my suggest rates of sedimentation of between 17 and 20 cm per 1,000 years. Sediment packets delineated by bounding unconformities on the seismic profiles

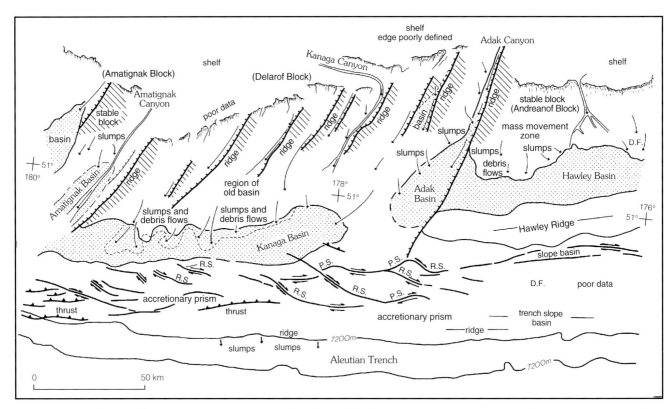

Figure 16–11. Diagram of the patterns and processes of sedimentation in the Adak, sector 4, 176°–180° W. Arrows indicate direction of sediment transport, which is dominated by slumps and debris flows. Canyons are structurally controlled and carry sediment from the shelf directly to the fore-arc basinal region. Upstanding margins of the tilted and rotated blocks are shown with diagonal lines and the term *ridge*. Normal faults have ticks on the downthrown side, but note these faults are also left lateral R faults. See discussion in text under sector 4. Basins are dotted to distinguish them from inverted basins like Hawley Ridge. Thrusts are shown with teeth to the direction of movement. R.S. and P.S. are as listed on Figure 16–9.

range up to 300 m in thickness, which might suggest tectonic reactivation episodes occur at intervals of about 1.5 my, equivalent to about 90 km of plate subduction. Most middle series sediment recovered consists of sandstones and siltstones; lower series material consists of sandstones and lava (Scholl et al. 1987).

Between 172° to 176° W, a distance of more than 300 km, the accretionary prism is only from 6 to 15 km wide. It is narrow, even compared to the equivalent development in the adjacent eastern sector, and is characterized by a sediment-starved linear trough or immature slope basin located near the top of the prism and a laterally inpersistent ridge. Both are oriented at a low angle to the line of the Aleutian Trench. The trench contains a component of arc-derived sediment suggesting transverse supply across the outer terrace high (Underwood 1986b).

Adak, sector 4, 176°–180° W

Structurally the sector is composed of the Delarof Block flanked by the Andreanof Block to the east and the Amatignak Block to the west (Figure 16–11). Three large

islands – Adak, Kanaga, and Tanaga – form the subaerial expression of the arc in the eastern half of the Adak sector. To the west only small scattered islands of the Delarof group, composed of uplifted older rocks, are present. Adjacent to 180° longitude lies a prominent gap in the arc, Amchitka Pass. The shelf, where it is developed, is very narrow (<10 km). The angular discordance between the subduction vector associated with the Pacific Plate and the orientation or alignment of the Aleutian Arc exceeds forty degrees. All available geophysical data confirm that the Delarof Block, which is about 170 km long, is fragmented into eight smaller blocks that vary in width between 10 and 20 km. Each of these small blocks has been rotated clockwise and tilted downslope in a southeasterly direction. It is possible that the faults margining the individual blocks are R′ (secondary) shears (Aydin and Page 1984) within an overall dextral shear couple driven by oblique convergence as proposed by Wells and Coe (1985) for the Cascadia subduction zone and subsequently discussed by Goldfinger et al. (1992).

As a consequence of these structural motions, a series of south-widening wedge-shaped basins or sliver basins have been generated that range from 2 to more than 8 km in width at their widest point. Adak Basin at 177° W is the best developed of the sliver basins. The angle of rotation involved in each case is between eight and ten degrees. Because of the cumulative effects of separate block rotation along the arc, the orientation of the fault that defines the margin of each sliver basin increases by about eight degrees towards the west relative to true north. At 177° W the margin fault is aligned N 20 E; at 178° W the fault that defines Kanaga Canyon is oriented N 32 E. At 179° W close to Amatignak Canyon, the fault orientation has reached N 40 E. With the exception of Kanaga Canyon (Figure 16–11), GLORIA failed to image the east-west – oriented tear faults that lie along the northern margin of the faulted blocks, partially because water depths are less than the operating depth of 400 m. Nevertheless, several scarps were imaged that might be candidates for such structures.

Sediment supply routes to the fore-arc basin reflect the impact of tectonsim. Because fault-controlled canyons transfer material directly from the inner shelf to the fore-arc basin, much of the upper slope along this sector is starved of sediment at present. Thus fragmentation of both the outer shelf and slope limits the potential for forming slope aprons. Sediment bypassing of the shelf edge and upper slope would explain the paucity of headward erosion features along this sector. Unlike sectors to the east, headward erosion of the shelf edge appears to be limited to small amphitheaters and minor canyon systems. As very little unconsolidated material has accumulated along the shelf break, the potential for mass movement is reduced even during lowstands of the sea. Indeed, GLORIA images of the northern half of the Delarof block are poorly reflective, which might indicate that it is veneered only with hemipelagic muds. By contrast, the southern halves of the rotated blocks are obscured by sediment. Even several of the upthrown block margins are buried beneath slumps. Yet, it is apparent from seismic profiles that the faults, which define the blocks, extend into the fore-arc basin.

The sliver basins are filled with more than 300 m of sediment at their widest points. Material forming the fill includes that derived from the shelf together with slumped material sourced from the angled slope of the adjacent blocks. Most sliver basins contain evidence of at least one angular unconformity, suggesting that movement of the blocks has been episodic.

To the east, the Andreanof block presents a different picture of arc evolution. This block, which is more than 80 km long, appears to have resisted the forces that fragmented the Delarof block. The Andreanof block remains massive, has no history of erosion by mass movement along the shelf edge, and has only one poorly defined canyon emanating from the shelf.

The fore-arc region is dominated by structures 80 km long and 20 km wide. These structures are either bathymetric lows, like the Kanaga Basin and Hawley Basin (Figure 16–11), or bathymetric highs such as Hawley Ridge. Based on evidence from seismic profiles, sediment is transported into the Kanaga and Hawley Basins as debris flows and slumps that extend from the sliver basins. Smaller debris flows emanate from the southern margin of the Andreanof block. Channel systems are rare to absent. Based on an analysis of the available seismic profiles (Pratson and Laine 1989), the basin infill consists of stacked slumps and debris flows. Hawley Ridge, which is at least 500 m thick, is faulted and has been influenced by a series of small thrusts and strike-slip faults (Scholl et al. 1982, Figure 11). It may be an example of a pressure ridge. The flanks of the ridge are marked by small slumps and debris flows directed to both the north and south. Hawley Ridge is separated from the accretionary prism by the Hawley Ridge shear zone.

At about 179° W, evidence obtained from seismic profiles indicates that an earlier basin containing more than 1,000 m of sediment is preserved between the sliver basins to the north and Kanaga Basin to the south. This earlier basin is presently being buried by recent slumps derived from the sliver basins. The infill of this buried basin complex thickens towards the east, suggesting that there was probably an earlier phase of arc fragmentation and block rotation.

The northern limit of the accretionary prism in this sector is difficult to identify because of the tectonic overprint. Thus only an estimate of the width is possible. To the east data are poor; to the west the width is between 15 and 20 km. It has been affected by strike-slip faulting with associated transfer faults and thrusting. En echelon conjugate Reidel shears and P shears have developed duplexes that individually are about 25 km long by 15 km wide. They are imaged by GLORIA along 100 km of the accretionary prism (Figure 16–11). As a result of a mixture of initial emplacement and the later impact of oblique subduction tectonics, both the surface expression of the prism recorded by GLORIA and the internal structure, seen on seismic profiles, are complex.

Kiska, sector 5, 180°–176° E

The sector is named after Kiska Island, although this part of the Aleutian Arc is frequently termed the Rat block, after Rat Island. Amchitka Island, in the eastern part of the sector (Figure 16–12), is the only other large island. Kiska sector extends a distance of 200 km from Amchitka Pass, a major break in the arc at 180°, to beyond Kiska Island where another break in the arc occurs. A large summit basin, the Sunday Basin, is located on the northeastern flank of Amchitka Island in 1,000 m of water.

Plate convergence in this sector has an obliquity of between sixty and eighty degrees. The effect of such a high angle of obliquity can be seen in the magmatic arc where the whole Rat block has been rotated and fractured by a series of normal faults oriented approximately 210°. Each fractured segment of the Rat block is tilted towards the south-

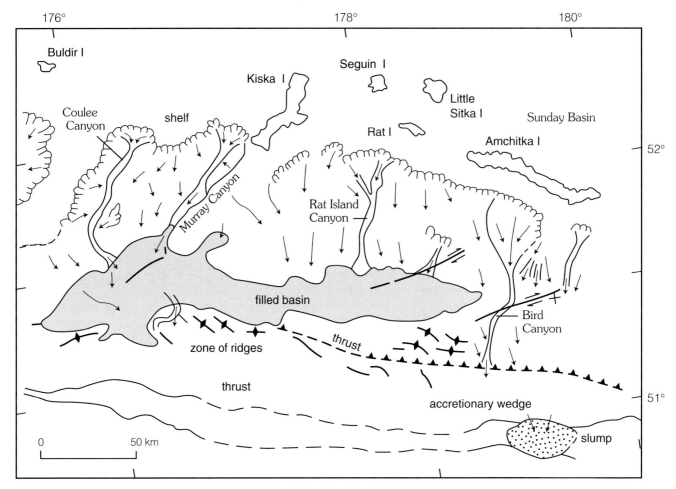

Figure 16–12. Diagram of the patterns of sedimentation in Kiska, sector 5, 180°–176° E. Arrows indicate direction of sediment transport. Basins are dotted. R.S. and P.S. are as listed on Figure 16–9. The canyons are tectonically controlled and locally extend to the accretionary prism or wedge. Note the slump that effectively blocks sediment transport along the trench.

east. Block rotation has also created tear faults in the magmatic arc behind the Rat block. Sunday Basin occupies one of these tear fault sites. South of the magmatic arc, the fore-arc basin is about 40 km wide.

The principal sediment dispersal routes are, from east to west, the Bird, Rat, Murray, and Coulee Canyons (Figure 16–12). Bird Canyon cuts across the fore-arc basin and extends to the accretionary prism. Part of the course of this canyon is controlled by a fault oriented east-west. GLORIA images are too poor to judge whether Bird Canyon empties sediment directly into the trench. The three western canyons all discharge sediment into a large basin, here termed informally the Murray-Rat Basin (Figure 16–12). Rat Island Canyon lies at the center of a large crescent-shaped headwall scarp or amphitheater that defines the shelf edge between Amchitka and Kiska Islands. It discharges sediment directly into the eastern half of the Murray-Rat Basin. Murray Canyon, aligned at 210 degrees, is fault controlled and deeply incised. Sediment is fed from this canyon into the western half of the Murray-Rat Basin. Coulee Canyon also

discharges sediment into the western end of the basin. A large number of small canyons have also been identified; their downslope orientation is determined by the tilt angle of the several blocks. They discharge sediment as confined flows into Murray-Rat Basin.

The accretionary prism zone in this sector is between 25 and 30 km wide; slumping is widespread. Compressional ridges, recognized both north and south of the trench-slope break, may be surface expressions of strike-slip movement. As in the previous sector, Reidel shears and P shears have developed. They occur along a 15-km-wide zone from the large slump located in the Aleutian Trench to the "zone of ridges" (Figure 16–12). Trench slope basins are restricted, narrow, and flat-floored. The trench between 180° and 176° E cannot be clearly defined from GLORIA images; as a consequence, the limits are inferred (Figure 16–12). Despite poor GLORIA images, the width of the trench is judged to vary from 20 km to as narrow as 5 km. Trench definition has been obscured by the effects of slumping from both the prism and the Pacific floor or outer trench high to the south.

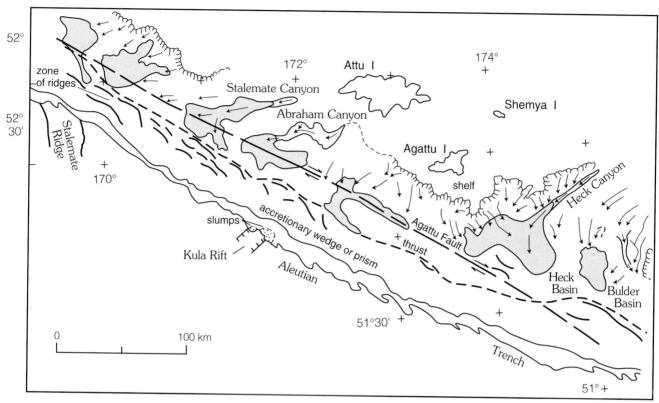

Figure 16–13. Diagram of patterns of sedimentation in Attu and Stalemate, sectors 6 and 7 respectively. Arrows indicate direction of sediment transport. Basin areas are dotted. The canyons are fault-controlled and supply sediment direct from the shelf to the fore-arc basin. The accretionary prism or wedge is narrow and poorly defined as is the trench. Stalemate Ridge in the far west effectively blocks the trench.

At 179° E the trench is partially filled by a large slump (Figure 16–12) that appears to have been derived from the accretionary prism.

Attu, sector 6, 176°–172° E and Stalemate, sector 7, 172°–168° E

Only two islands of any significance occur in these sectors; they are Agattu and Attu (Figure 16–13). Plate convergence between 176° and 168° E is almost entirely strike-slip. Obliquity ranges from seventy-five degrees in the east to about ninety degrees in the far west. The magmatic arc is fractured along the whole length of the two sectors, and the fore-arc basin is similarly splintered. A major dextral fault, here termed the Agattu Fault (Figure 16–13) can be recognized on multichannel seismic profiles by a large flower structure; the fault cuts through the fore-arc basin from 176° E to beyond the longitude of Stalemate Ridge. Stalemate Ridge is a prominent aseismic ridge (Lonsdale 1988) located on the Pacific Plate and clearly imaged on GLORIA (Vallier et al. this volume). Agattu Fault (Figure 16–13) is recognized as a major structural break within the fore-arc basin. Evidence from seismic profiles and GLORIA images suggests that south of Agattu Fault the fore-arc basin is being

driven westward at about 6 cm/yr by the relative motion between the Pacific and North American Plates (see Vallier et al. this volume). The westward transport of part of the fore-arc basin under the influence of the motion, particularly of the Pacific Plate, is described as translation. One of the surface expressions of translation is the occurrence of large linear depressions. The fore-arc basin, which averages only 15 km in width, is significantly narrower in these two sectors compared with the rest of the arc (Figure 16–13).

From east to west the pattern of sedimentation in the fore-arc basin is largely controlled by the position of the Agattu Fault, the only exceptions being a small basin informally termed Buldir Basin and Heck Basin (informal name) at the eastern end of the Attu sector. A multichannel seismic line obtained across the Buldir Basin (Vallier et al. 1994) indicates it is filled with about 2.4 km of upper series sediment despite having an area of only 500 km². These upper series sediments appear to overlie lower series rocks. Buldir Basin may be linked structurally to the larger Heck Basin further west.

Heck Basin is fed by sediments transported down the thalweg of the fault-controlled Heck Canyon and abundant smaller canyons that indent the western wall of Heck Canyon (Figure 16–13). Sedimentation rates in Heck Basin may be high since backfilling of Heck Canyon is occurring where it forms a flat floor. The area of Heck Basin is approximately 1000 km², and the basin has a maximum thickness of about 3 km. An elongate unnamed basin lies between, and south of, Agattu and Attu Islands (Figure 16–13). Agattu Fault determines the position of the basin axis. No major canyon directs sediment into the basin; rather, several small drainage systems bring sediment down from the shelf and upper slope regions. The basin, as mapped, has an area of about 350 km² and a sediment fill thickness of a little less than one kilometer.

Abraham and Stalemate Canyons both drain into basins informally named Abraham and Stalemate Basins, respectively. Abraham Basin has an area of about 550 km² and Stalemate Basin an area of about 750 km². Sediment thickness in both basins is only about one kilometer. Agattu Fault forms the southern side of both basins and through intermittent dextral movement has lifted the sediment fill into irregular folds. Abraham Canyon is relatively wide; its channel apparently follows a large fault that was mapped by Gates et al. (1971) on the island of Attu. Stalemate Canyon also follows a fault. The westernmost part of these sectors has two basins that lie athwart Agattu Fault. These unnamed basins have a combined area of about 1,000 km². Sediment fills (upper series) are as much as 2 km, indicating that abundant sediment has been deposited into the basin.

In these sectors the accretionary prism, like the fore-arc basin, is narrow and poorly developed. Trench-slope basins tend to be small, isolated, and flat-floored. Both Reidel shears and P shears are prominent in the Stalemate sector. West of longitude 174° E the prism narrows from about 30 km to less than 20 km at the longitude of Stalemate Ridge. Obduction of Stalemate Ridge has not only contributed to a reduction in size of the prism, but also to its partial destruction. Whereas the scale and width of the accretionary prism along much of the arc has prevented direct sediment supply to the trench from the magmatic arc, in these two sectors there is evidence for sediment dispersal pathways from the fore-arc basin to the trench floor. The trench is poorly imaged on GLORIA in the Attu and Stalemate sectors, partly as a result of the extensive slumping that has occurred from the prism and outer trench high. As a consequence it is irregular in plan, locally very narrow (<5 km), and at a shallower depth compared to sectors towards the east. No channels are discernible on the trench floor. However, at its eastern end the trench is open to input of sediment from the Alaskan Peninsula, and turbidity currents generated in the Gulf of Alaska are free to flow westward along the trench floor axis. In contrast, sediment is restricted from entering the trench longitudinally at the western end of the Aleutian Chain because Stalemate Ridge intersects the trench at 169° E, blocking sediment input from the west. Moreover, the trench floor slopes from east to west from the Alaskan Peninsula to longitude 180° where the slope direction of the axial gradient reverses (Scholl 1974).

Discussion

It is clear from analysis of the GLORIA image data that direct variables, particularly subduction and associated strike-slip tectonics, strongly influence the internal facies architecture of the upper series in the Aleutian fore-arc. It is also clear that indirect variables have had a significant impact. In this regard, global climate changes since the late Miocene have been especially important because of the greater area available for subaerial erosion during sea level lowering, the opportunity to transport sediment direct to the fore-arc basin, and the impact of glaciation.

Sediment feeder systems to the fore-arc basin

Sectors well nourished with sediment possess a thick slope apron; particularly well developed in the three eastern sectors, they extend from the shelf edge to the fore-arc basin. Expansion of slope aprons by progradation is dependent partly on indirect variables. During glacial maxima, the Aleutian Arc Shelf would have been widely exposed and crossed by small piedmont-type valley glaciers carrying debris towards the slope. Meltwater streams probably transported material to the shoreline where wave activity would have further reworked the sediment across the shelf and onto the slope. In the eastern part of the arc close to the Alaskan Peninsula, where land masses are large, valley glaciers probably extended to the shelf edge and discharged material directly down the slope (Bruns et al. 1989). The fan-shaped bodies identified in the first two sectors may have formed through this process. In sector 1, canyons occur on the shelf suggesting that some valley glaciers extended only partway across the wide shelf. Indeed, to judge from their form they appear to be rare examples of point-sourced supply. Those sectors affected by direct glacial activity would rapidly prograde. By contrast, during interglacial highstands, as at present, both the outer shelf and slope are starved of sediment.

Most slope aprons along the arc are between 15 and 20 km wide with gradients of about five degrees. Reworking of the slope-apron sediment through failure and mass movement is common. Although failure is aided by the five-degree depositional dip angle of the surficial layer, it is also conditional on a large supply of material to the shelf edge. Canyon-head mass failure is probably triggered by seismic activity (Carlson, Karl, and Edwards 1982) or an increase in hydrostatic pressure with piping of water along discrete porous layers (O'Leary and Dobson 1992). The shelf head zone of each such site of failure consists of broad concave amphitheater-shaped depressions. Canyons and the channel systems that emanate from them appear choked with debris and chaotic slide material. Repeated failure and mass flow result in headward retreat that deeply etches the shelf break to extend onto the shelf. Nevertheless, many such canyon-sourced systems become leveed fan channels carrying sed-

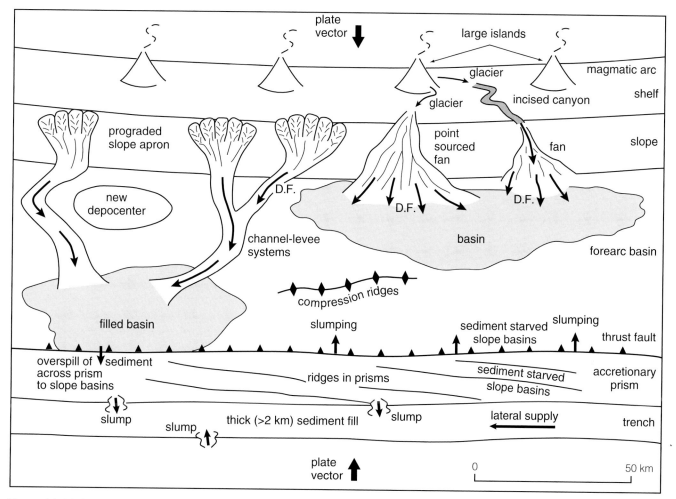

Figure 16–14. Sediment processes model for an orthogonal convergent setting during an episode of low sea level and glacial maxima, the Unalaska Type A. Extensive subaerial exposure of the magmatic arc supports valley glaciation. Both glaciers and meltwater rivers transport sediment to the shelf edge. Point sourced fans may be supplied with sediment direct from tidewater glaciers or from canyons cut across the shelf. Slope aprons will prograde, slump, and generate debris flows (DF) that may translate into high-density turbidity flows forming channel-levee systems. These processes are independent of the position of sea level. At low latitudes rivers and delta growth will replace glacier supply systems. Tectonic controls on sedimentation involve tilting of the fore-arc basin and the creation of antiformal highs. Tilting contributes to the rejuvenation of the slope apron which promotes further failure. The accretionary prism consists of ridges and intervening troughs oriented at a low angle to the line of the trench. Slope basins are sediment starved, particularly if fore-arc tilting is active. Slumping along the outer terrace high (close to the thrust fault) is extensive.

iment as turbid flows to the middle upper slope and fore-arc basins. Other systems carry material as debris flows to the middle upper slope where broad fans form. Unconfined turbidity flows emanating from the slope apron may deposit areally extensive sheet sands.

Large sections of the shelf edge and upper slope along the arc appear sediment starved. Sediment starvation can be attributed to three factors: the level of subaerial exposure of the magmatic arc at lowstands; bypassing of the slope; and the varying geology of the arc. Subaerial exposure levels during periods of lowstand broadly decrease along the arc from east to west. Judging from bathymetric data Unalaska sector 1 and Atka sector 3 have the greatest potential for

high levels of sediment supply. By contrast, the transport of material from the magmatic arc direct to the fore-arc region along established incised canyons or channels is rare in the eastern arc; however, it becomes increasingly common towards the west. Fragmentation, a common feature of the western arc, has created fault-controlled linear depressions along which sediment transport tends to be directed, effectively bypassing the slope. Unlike the Izu-Bonin fore-arc (Klaus and Taylor 1991), canyons that extend from the shelf to the outer terrace high are rare along the Aleutian fore-arc. Only Bird Canyon in the Kiska sector reaches the accretionary prism. Overspill from the fore-arc terrace across the outer terrace high is intermittent and controlled by the degree of tilting and sedimentation in the fore-arc basin. Seismic profiles suggest that draping of the accretionary prism with sediment derived from unconfined turbidity flows has occurred.

Sedimentation processes recorded for the Aleutian Arc are described below in terms of generalized models that may have wider application.

Tectonically dominated high-latitude depositional models

UNALASKA TYPE A, INVOLVING ORTHOGONAL SUBDUCTION AND FORE-ARC BASIN COMPRESSION (FIGURE 16–14), LOCATED BETWEEN 164° AND 169°30′ W

In the absence of large-scale tectonic fracturing of the magmatic arc, available eroded material will tend to accumulate as prograding slope aprons. Because the rapid buildup of these slope aprons probably occurs during eustatic lowstands, the pattern of activity would involve a succession of progradations, slope failures, and resedimentation followed by highstand quiescences (Winsemann and Seyfried 1991). Where large glaciers are able to extend across the shelf during lowstands, they will discharge material directly down the slope creating a large fan of unsorted sediment. Sedimentation patterns along a high-latitude intraoceanic arc before any glacial phase would have been different in style, for whereas sea-level oscillations could influence the availability of sediment supplied to the slope, only rivers would be available to transport material to the shelf edge. Such point-sourced supply would tend to favor the formation of submarine fan systems.

Slope aprons located in areas prone to earthquakes are susceptible to slope failure in the form of slumps and debris flows, a process that can result in the emplacement of a middle upper slope debris wedge. Debris flows can also transform into turbid flows with the creation of channel-levee systems that are capable of transferring large volumes of sediment into the fore-arc basin. Mass movement therefore cannot be judged as an accessory mechanism in this environment; rather, it is recognized as a major agent for initiating transfer of sediment to the fore-arc basin. Turbidites are known to be a significant component of the sediment fill of the fore-arc basin (Stewart 1978) although their origin may not be confined to channel-levee systems. Many may result from unconfined sheet flows.

The formation of an accretionary prism can generate elongate ridges and depressions seen as slope basins that may be 40 km wide and capable of accommodating more than 2 km of sediment. They may not parallel the line of the Aleutian Trench. In the case of the Aleutian Arc, prism formation causes the uplift of older, predominantly middle series rocks that lie arcward of the trench-slope break. Expansion of the accretionary prism causes further uplift that tilts the fore-arc basin and imposes a compressional regime. The

former promotes low-angle unconformities (Figure 16–5), whereas the latter introduces ridges. The impact of compression ridges on sediment transport patterns in the fore-arc region will be to redirect flow along the basin axis rather than across it.

With the rise and expansion of an accretionary prism, sediment failure develops along the line of the outer terrace high. Slumping of accretionary prism sediment into the fore-arc basin can be volumetrically significant. Moreover, because the material has a trench provenance, stratigraphic affinities across the fore-arc basin will prove complex. This secondary sediment source may increase in importance where abandonment of supply to the slope apron occurs. Abandonment across the fore-arc basin is recorded as a hemipelagic sediment veneer with diatomaceous rich layers and ash bands. Where such abandonment has not occurred, eventual filling of the fore-arc basin allows sediment to be transported from the slope apron across the trench-slope break and down the lower trench slope. Onlap and offlap relationships, documented for the Sunda fore-arc (Stevens and Moore 1985), are expected to occur in this setting. This overspill of unconfined turbidity flows may involve flow stripping (Underwood and Norville 1986) although, because of the presence of linear depressions in the accretionary prism, sediment may fail to reach the trench.

The changing patterns of sediment transport recorded for fore-arc basin sequences in this Unalaskan-type setting are a function of compression tectonics and associated tilting of the basin floor due in part to the emplacement of an expanding prism. Basin sequences might, therefore, record complex sediment dispersal patterns. Except at points of deflection, flows will radiate across the fore-arc with a major component of transverse flow. Deflection will align flows partially at right-angles to the line of convergence. Subsequent tectonic activity might impose an angular unconformity to be followed by deposits that show flow directions orthogonal to those in the underlying sediments.

YUNASKA TO ATKA TYPE B, SLIGHTLY OBLIQUE WITH MIXED COMPRESSIONAL AND EXTENSIONAL ASPECTS (FIGURE 16–15), LOCATED BETWEEN 169°30 W, AND 176° W

Sediment supply mechanisms will be similar to those described for the Unalaska type. In those sectors of an arc where the obliquity of convergence ranges from less than ten degrees to more than twenty degrees, the tectonic patterns will include both extensional and strike-slip components. Moreover, a rapidly expanding accretionary prism will impose an additional compressive regime on a fore-arc basin. Partial block rotation in the magmatic arc and fore-arc basin, observed as horst-graben extensional tectonics where the obliquity of convergence is about ten degrees, largely determine the position of depocenters. In this context, it has been recognized (Howell et al. 1980) that horst-graben structures form in oblique convergent settings be-

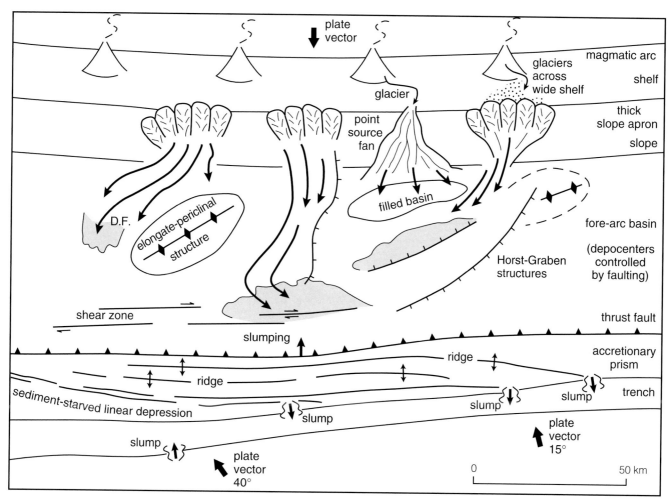

Figure 16–15. Sediment processes model for slightly oblique to strongly oblique settings during an episode of low sea level and glacial maxima, the Yunaska to Atka Type B. Sediment supply closely follows that described for the A type model. Sites for sedimentation are increasingly controlled by tectonic activity affecting the fore-arc basin. Horst-graben extensional structures are replaced along the fore-arc by elongate periclines and strike-slip tectonics that produce compensation cycles and significant realignment of sediment transport directions. Accretionary prism conditions are as for Figure 16–6 in the east except that sediment supply from the fore-arc basin to the trench slope basins is rare. Sediment is transported laterally along the trench.

tween parallel strands of bordering wrench zones; where wrench faults diverge, adjacent blocks move away from each other. As obliquity approaches twenty degrees, elongate periclines or domal structures evolve in response to strike-slip tectonics. Oval depocenters are developed between the elongate periclines and can exceed 1,000 square km in area.

Slope-apron progradation and associated mass failure are major features of sedimentary basins developed in this type of tectonic setting. Basin inversion, seen as areas of positive seafloor relief and incipient slumping, influences sediment dispersal paths such that channel systems are sinuous and locally fault controlled. Inversion tectonics also has an influence on facies distribution, observed as compensation cycles, where depositional sites shift to occupy the depressions between the positive areas. Very low angle hiatuses

and the infilling of slump scars are a feature of depositional site switching. Submarine relief that reflects tectonic movement in the form of growing antiforms and adjacent depressions has been recognized in an ancient fore-arc basin sequence in central Japan (Tokuhashi 1989).

The emergence of discrete shear zones marks this type of arc environment where convergence obliquity exceeds twenty degrees. Shear zones may be greater than 20 km in width, particularly if the associated Reidel and P shear faults are well developed. In the Yunaska to Atka sectors of the Aleutian Arc, shear zones extend from the accretionary prism in the east to the outer margin of the fore-arc basin in the west. As in sector 1, overspill of sediment across the outer terrace high will occur although shear zones could provide conduits for lateral along-arc dispersal.

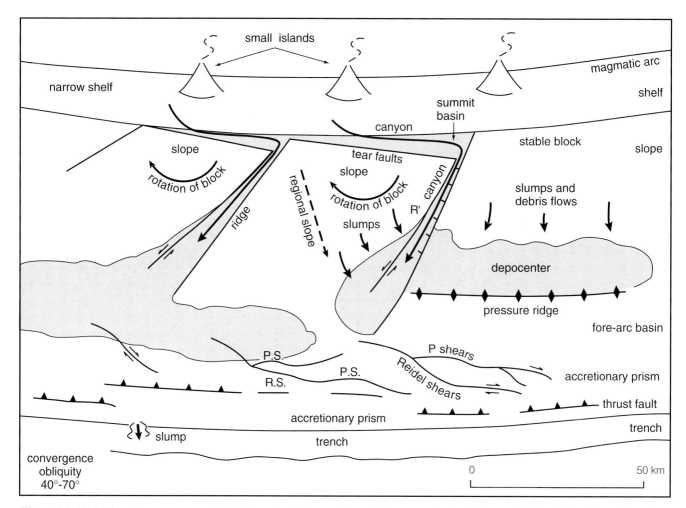

Figure 16–16. Sediment processes model for strongly oblique settings during an episode of low sea level and glacial maxima, the Adak Type C. The magmatic arc consists of small islands and a narrow shelf such that the volume of sediment available for the fore-arc is reduced compared to those sectors where convergence is orthogonal. Fragmentation of the arc and associated block rotation produce fault-controlled canyon routes and basins. Slope aprons fail to develop, except along the flanks of stable blocks, due to bypassing by canyons. Depocenters become elongate and may be constrained by pressure ridges produced by strike-slip tectonics. Tilting of the fore-arc basin does not occur. Strike-slip tectonics extends throughout the accretionary prism and into the fore-arc basin. As a consequence, the ridge-trough morphology has been largely destroyed as has the outer terrace high.

ADAK TYPE C, STRONGLY OBLIQUE WITH STRIKE-SLIP (FIGURE 16–16), LOCATED BETWEEN 176° W AND 176° E

In this model, oblique convergence ranges from about forty degrees to seventy-five degrees. Both the magmatic arc and fore-arc basin are fragmented into blocks that are on average about 50 km wide. Each block is bounded by listric-shaped tear faults and listric normal faults. Whilst sediment sources are similar to those described for the previous two model types, actual sedimentation will occur in the summit basin on the magmatic arc and in sliver basins, both of which contain channel systems that transport sediment to the fore-arc basin. Basin geometry is distinctive because both types of depocenter have, in plan and section, a wedge shape. Although the two types of basins invariably are tectonically linked, they are oriented orthogonal to each other; moreover, both have a half-graben form. Bathymetric evidence from the Adak sector indicates that there is an abrupt drop in the seafloor between summit and sliver basins, where the former type passes into the latter.

Because structurally controlled canyons direct sediment flows to the fore-arc basin, most of the upper slope is sediment starved, resulting in a general absence of slope aprons. Only where large blocks have resisted fragmentation will slope aprons form and mass movement be a feature of downslope sediment transport. Episodic phases of block rotation will be responsible for angular unconformities in basin sequences. Slump and debris flow sequences will be common whereas the rates of sedimentation, certainly as far as summit basins are concerned, will be strongly influenced by

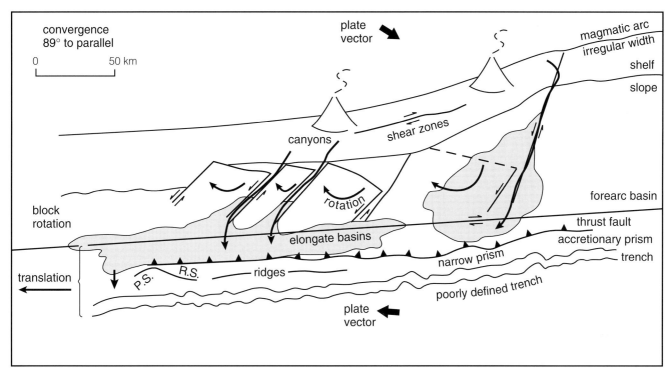

Figure 16–17. Sediment processes model for strongly oblique to dominantly strike-slip settings, the Attu Type D. The pattern is similar to that shown in Figure 16–16 except that the shelf is narrow and irregular in width with strike-slip faults. Block rotation locally exceeds forty-five degrees such that the summit basin form is lost whilst canyons direct sediment into the fore-arc basin. Fore-arc basins are narrow and linear and subject to translation. Both the accretionary prism and the trench are poorly defined as a result of laterally extensive slumping.

episodes of regression. The infill of both basins will have a similar composition except that, whereas the overall sediment grain size in the sliver basin will be finer, being downslope from the principal source area, greater slope angles in the sliver basin may influence the style of sediment transport.

Depocenters in the fore-arc region will tend to be asymmetric in plan. Slumps and debris flow sequences will dominate the fore-arc basin infill except where canyons occur, and particularly if they emanate from the shelf, turbidites will dominate the basin infill. Compression ridges will constrain the oceanward margin of the basins. Convergence obliquity may be the cause of any faulting observed in the accretionary prism. Strike-slip motion seen as Reidel shears and P shears, together with thrust faults, will tend to fragment the accretionary prism. The consequence of this development on sediment patterns could involve the formation of thick linear tectonic melanges that may be superficially indistinguishable from some types of debrites.

ATTU TYPE D, DOMINANTLY STRIKE-SLIP (FIGURE 16–17), LOCATED FROM 176° E TO BEYOND 170° E

In this type convergence ranges from about seventy-five degrees to pure along-arc translation. Block fragmentation is

similar in form to that in the Adak type except that rotation of individual blocks exceeds forty-five degrees, whereas strike-slip faults occur throughout the arc. Extreme elongation and associated narrowing of the fore-arc basins results from translation. Because of the small scale of the magmatic arc, sediment supply is reduced and may not be comparable with the other model types. Fault-controlled canyons will transport sediment to the fore-arc basin, bypassing the slope. Summit basins will fail to develop because of the extent of block rotation and associated tilting, but small sliver basins will persist. The accretionary prism will be narrow, poorly defined, and together with an attenuated trench, be involved in lateral translation. Where convergence approaches pure strike-slip, as in the westernmost part of the Aleutian Arc, translation will dominate and be seen as transform motion (Scholl et al. 1987), effectively transporting whole sections of the arc (terranes?) laterally.

Conclusions

High-latitude curvilinear intraoceanic convergent margins are important sites of sedimentation. Sediment accumulation patterns reflect the impact of convergence tectonics. In the Aleutians the full convergent range from orthogonal to pure strike-slip and transform motion results in at least four distinct depositional patterns. Where convergence is dominantly orthogonal, compression ridges in the fore-arc affect sediment transport paths. Back tilting of the fore-arc terrace repeatedly shifts the depocenter arcward, produces low-angle unconformities, and realigns transport paths. Well-nourished slope aprons are the prime source of sediment for the fore-arc basin, initially through mass movement processes that may translate to turbidity currents. Point-sourced slope fans provide additional material, but sediment supplied to the fore-arc by shelf-sourced canyons that bypass the slope are rare. Channel-levee systems, which are common in the eastern sectors, tend not to migrate but shift abruptly due to supply changes. Migration is usually a response to tectonics.

Where convergence becomes oblique the fore-arc basin is affected by faulting, including incipient strike-slip motion and folding. All these tectonic structures affect sediment deposition rather than patterns of supply. By contrast, in sectors affected by marked convergent obliquity, arc fragmentation seen as block rotation creates fault-controlled conduits that transfer sediment from the magmatic arc direct to the fore-arc basin. Slope aprons are not widely developed. Depocenters are of two types, fault-controlled wedge-shaped sliver basins and narrow elongate basins affected by strike-slip tectonics. Shear zones and the lateral translation of large sections of the fore-arc are a special feature of sectors where near-parallel convergence occurs. Both sample and seismic evidence indicate that the sediments filling the Aleutian fore-arc are mud dominated.

The accretionary prism is wide and well developed in the east but narrow and poorly defined in the west. Ridges and linear depressions are aligned at a low angle to the margin of the trench in the east where slumping is common. Trench slope basins are immature and sediment starved. The trench fill is considered to have been derived predominantly from lateral sources including the Surveyor Channel, continental Alaska (Underwood 1986a), and the reworking of the Zodiac Fan (Stevenson and Embley 1987).

Bypassing of the fore-arc basin through the transfer of sediment as confined flows along canyons from the magmatic arc to the trench has not been observed along the Aleutian Arc, but there is some evidence that unconfined turbidity flows overspill from the fore-arc, cross the trench-slope break, and extend down the lower trench slope. Because of the presence of sediment-starved linear depressions in the accretionary prism, it seems unlikely that material derived through overspilling actually reaches the trench (Underwood 1986b).

References

Anderson, R. E. 1971. Tectonic setting of Amchitka Island, Alaska: *U.S. Geological Survey Open-File Report,* 46 p. Reston, Va.: USGS.

Aydin, A., and Page, B. M. 1984. Diverse Pliocene-Quaternary tectonics in a transform environment. San Francisco Bay regions, California. *Bull. Geol. Soc. Am.* 95: 1303–17.

Belderson, H. H., Kenyon, N. H., Stride, A. H., and Pelton, C. D. 1984. A 'braided' distributary system on the Orinoco deep-sea fan. *Mar. Geol.* 56: 195–206.

Breen, N. A., Silver, E. A., and Hussong, D. M. 1986. Structural styles of an accretionary wedge south of the island of Sumba, Indonesia, revealed by SeaMARC II side scanner. *Bull. Geol. Soc. Am.* 97: 1250–61.

Brown, K. M., and Westbrook, G. K. 1987. The tectonic fabric of the Barbados Ridge accretionary complex. *Mar. Petrol. Geol.* 47: 71–81.

Bruns, T. R., von Huene, R., Culotta, R. C., Lewis, S. D., and Ladd, J. W. 1987. Geology and Petroleum Potential of the Shumagin Margin, Alaska. In *Geology and Resource Potential of the Continental Margins of Western North America and Adjacent Ocean Basins – Beaufort Sea to Baja California,* eds. D. W. Scholl, A. Grantz and J. G. Vedder, pp. 157–190. Circum-Pacific Council for Energy and Mineral Resources, Earth Sciences Series, vol. 6. Houston, Tex. American Association of Petroleum Geologists.

Bruns, T. R., Carlson, P. R., Stevenson, A. J., Fisher, M. A., Ryan, H. F., Mann, D. M., Dobson, M. R., Huggett, Q., Parsons, L., and Fannin, N. G. T. 1989. GLORIA images from the Gulf of Alaska and British Columbia: Subduction zones, transforms and channels. *Am. Geophys. Union Trans.* 70: 1338.

Cadet, J. P., Kobayashi, K., Jolivet, L., Aubouin, J., Boulegue, J., Dubois, J., Hotta, H., Ishii, T., Konishi, K., Nitsuma, N., and Shimamura, H. 1987. Deep scientific dives in the Japan and Kuril Trenches. *Earth Plan. Sci. Ltrs.* 83: 313–28.

Carlson, P. R., Karl, H. A., and Edwards, B. D. 1982. Puzzling mass movement features in the Navarinsky Canyon Head, Bering Sea. *Geo-Mar. Ltrs.* 2: 123–7.

Chavez, P. S. 1986. Processing techniques for digital sonar images from GLORIA. *Photogram. Engrg. Rem. Sens.* 52: 1135–45.

Delong, S. E., Fox, P. J. and Mcdowell, F. W. 1978. Subduction of the Kula Ridge at the Aleutian Trench. *Geol. Soc. Am. Bull.* 89: 83–95.

Dickinson, W. R., and Seely, D. R. 1979. Structure and Stratigraphy of Forearc Regions. *Am. Assoc. Petrol. Geol. Bull.* 63: 2–31.

Dobson, M. R., Scholl, D. W., and Stevenson, A. J. 1991. Interplay between arc tectonics and sea-level changes as revealed by sedimentation patterns in the Aleutians. In *Sedimentation Tectonics and Eustacy,* ed. D. I. M. Macdonald, pp. 151–64. Special Publication No. 12, International Association of Sedimentologists. Oxford: Blackwell.

Engebretson, D. C., Cox, A., and Gordon, R. G. 1984. Relative motions between ocean plates of the Pacific Basin. *J. Geophys. Res.* 89: 291–310.

Gardner, J. V., Field, M. E., Lee, H., Edwards, B. E., Masson, D. G., Kenyon, N., and Kidd, R. B. 1991. Ground-truthing 6.5-kHz sidescan sonographs: What are we really imaging? *Geophys Abs.* pp. 7–8. (Supplement to EOS, Jan., 1991).

Gates, O., and Gibson, W. 1956. Interpretation of the configuration of the Aleutian Ridge. *Geol. Soc. Am. Bull.* 67: 127–46.

Gates, O., Powers, H. A., Wilcox, R. E., and Schafer, J. P. 1971. Geology of the Near Islands, Alaska. *U.S. Geological Survey Bulletin 1028-U,* pp. 709–822. Reston, Va.: USGS.

Geist, E. L., Childs, J. R., and Scholl, D. W. 1987. Evolution and petroleum geology of Amlia and Amukta intra-arc summit basins, Aleutian Ridge. *Mar. Petrol. Geol.* 4: 334–52.

Gibson, W., and Nichols, H. 1953. Configuration of the Aleutian Ridge Rat Islands – Semisopochnoi Island to west of Buldir Island. *Geol. Soc. Am. Bull.* 64: 1173–86.

Gnibidenko, H. S., Anosov, G. A., Argentov, V. V., and Pushchin, I. K. 1985. Tectonics of the Tonga-Kermadec Trench and Osbourn Seamount junction area. *Tectonophysics* 112: 357–83.

Goldfinger, C., Kulm, L. D., Yeats, R. S., Applegate, B., Mackay, M. E., and Moore, G. F. 1992. Transverse structural trends along the Oregon convergent margin: Implications for Cascadia earthquake potential and crustal rotations. *Geology* 20: 141–4.

Harding, T. P. 1983. Divergent wrench fault and negative flower structure, Andaman Sea. In *Seismic Expression of Structural Styles,* ed. A. W. Bally, pp. 4.2-1–4.2-8. American Association of Petroleum Geology Studies in Geology Series 15. Tulsa, OK: AAPG.

Harding, T. P., Vierbuchen, R. C., and Christie-Blick, N. 1985. Structural styles, plate tectonic settings, and hydrocarbon traps of divergent (transtensional) wrench faults. In *Strike-Slip Deformation, Basin Formation, and Sedimentation.* eds. K. T. Biddle and N. Christie-Blick, pp. 51–77. Society of Economic Paleontologists and Mineralogists Special Publication 37. Los Angeles, Calif.: SEPM.

Hathon, E. G., and Underwood, M. B. 1991. Clay mineralogy and chemistry as indicators of hemipelagic sediment dispersal south of the Aleutian arc. *Mar. Geol.* 97: 145–66.

Howell, D. G., Crouch, J. K., Greene, H. G., McCulloch, D. S., and Vedder, J. G. 1980. Basin development along the late Mesozoic and Cenozoic Californian margin: A plate tectonic margin of subduction, oblique subduction and transform tectonics. In *Sedimentation in Oblique-Slip Mobile Zones,* eds. P. F. Ballance, and H. G. Reading, pp. 43–62. Special Publication No. 4, International Association of Sedimentologists. Oxford: Blackwell.

Jarrard, R. D. 1986. Relations among subduction parameters. *Rev. Geophys.* 24: 217–84.

Kidd, R. B., Hunter, P. M., and Simon, R. W. 1987. Turbidity current and debris flow pathways to the Cape Verde Basin. In *Geology and Geochemistry of Abyssal Plains,* eds. P. P. E. Weaver and J. Thomson, pp. 33–48. Geological Society Special Publication No. 31. Oxford: Blackwell.

Klaus, A., and Taylor, B. 1991. Submarine canyon development in the Izu-Bonin forearc: A SeaMARC II and seismic survey of Aoga Shima canyon. *Mar. Geophys. Res.* 13: 131–52.

LaForge, R., and Engdahl, E. R. 1979. Tectonic implications of seismicity in the Adak canyon region, central Aleutians. *Seismol. Soc. Am. Bull.* 69: 1515–32.

Laughton, A. S. 1981. The first decade of GLORIA. *J. Geophys. Res.* 86: B11,511–34.

Leggett, J., ed. 1982. *Trench-Forearc Geology.* Geological Society Special Publication No. 10. 576 p. Oxford: Blackwell.

Le Pichon, X. 1987. *Report of the Second Conference on Scientific Ocean Drilling (COSOD II).* Strasbourg, France: European Science Foundation, 142 p.

Lewis, S. D., Ladd, J. W., and Bruns, T. R. 1988. Structural development of an accretionary prism by thrust and strike-slip faulting: Shumagin region, Aleutian Trench. *Geol. Soc. Am. Bull.* 100: 767.

Lonsdale, P. 1988. Paleocene history of the Kula plate: Offshore evidence and onshore implications. *Geol. Soc. Am. Bull.* 100: 733–54.

Lu, R. S., and McMillen, K. J. 1983. Multichannel seismic survey of the Columbia Basin and adjacent margins. In *Studies in Continental Margin Geology,* eds. J. S. Watkins, and C. L. Drake, pp. 395–410. AAPG Memoir 34. Tulsa, Okla.: American Association of Petroleum Geologists.

Marlow, M. S., Scholl, D. W., Buffington, E. C., and Alpha, T. R. 1973. Tectonic history of the central Aleutian Arc. *Geol. Soc. Am. Bull.* 84: 1555–74.

McCarthy, J., Stevenson, A. J., Scholl, D. W., and Vallier, T. L. 1984. Speculations on the petroleum geology of the accretionary body: An example from the central Aleutians. *Mar. Petrol. Geol.* 1: 151–67.

McCarthy, J., and Scholl, D. W. 1985. Mechanisms of subduction accretion along the central Aleutian trench. *Geol. Soc. Am. Bull.* 96: 691–701.

Moberly, R., Shepherd, G. L., and Coulbourn, W. T. 1982. Forearc and other basins, continental margin of northern and southern Peru and adjacent Ecuador and Chile. In *Trench-Forearc Geology,* ed. J. Leggett, pp. 171–90. Geological Society Special Publication No. 10. Oxford: Blackwell.

Moore, G. F., Curray, J. R., and Emmel, F. 1982. Sedimentation in the Sunda Trench and fore-arc region. In *Trench-Forearc Geology,* ed. J. Leggett, pp. 245–58. Geological Society Special Publication No. 10. Oxford: Blackwell.

Mutti, E., and Normark, W. R. 1991. An integrated approach to the study of turbidite systems. In *Seismic Facies and Sedimentary Processes of Submarine Fans and Turbidite Systems,* eds. P. Weimer and M. H. Link, 75–106. Frontiers in Sedimentary Geology. New York: Springer-Verlag.

Newbury, J. T., Laclair, D. L., and Fujita, K. 1986. Seismicity and tectonics of the far western Aleutian Islands. *J. Geodyn.* 6: 13–32.

Nilsen, T. H., and Zuffa, G. G. 1982. The Chugach Terrane, a Cretaceous trench-fill deposit, southern Alaska. In *Trench-Forearc Geology,* ed. J. K. Leggett, 213–27. Geological Society (London) Special Publication No. 10. Oxford: Blackwell Scientific.

O'Leary, D. W., and Dobson, M. R. 1992. Southeastern New England continental rise: origin and history of slide complexes. In *Geologic Evolution of Atlantic Continental Rises,* eds. C. W. Poag and P. C. de Graciansky, pp. 214–65. New York: Van Nostrand Reinhold.

Perry, R. B., and Nichols, H. 1965. Bathymetry of Adak Canyon Aleutian Arc, Alaska. *Geol. Soc. Am. Bull.* 76: 365–70.

Pickering, K. T., Hiscott, R. N., and Hein, F. J. 1989. *Deep-Marine Environments,* 416 p. London: Unwin Hyman.

Pratson, L. F., and Laine, E. P. 1989. The relative importance of gravity induced versus current controlled sedimentation during the Quaternary along the middle U.S. continental margin revealed by 3.5 kHz echo character. *Mar. Geol.* 89: 87–126.

Prior, D. B., and Bornholm, B. D. 1989. Submarine morphology and processes of fjord fan deltas and related high-gradient systems: Modern examples from British Columbia. In *Fan Deltas: Sedimentology and Tectonic Settings,* eds. W. Nemec and R. J. Steel, pp. 125–43. Glasgow: Blackie and Sons.

Rea, D., Basov, I. A., and Janecek, T. 1993. North Pacific transect science operator report Leg 145. *JOIDES J.* 19: 21–8.

Ryan, H., and Scholl, D. W. 1985. Formation of outer arc structural highs, central Aleutian Ridge: *EOS Abs. Am. Geophys. Union Trans.* 66: 1105.

Ryan, H., and Scholl, D. W. 1989. The evolution of forearc structures along an oblique convergent margin, Central Aleutian Arc. *Tectonics* 8: 497–516.

Sancetta, C. 1979. Oceanography of the north Pacific during the last 18,000 years: Evidence from fossil diatoms. *Mar. Micropaleont.* 4: 103–23.

Scholl, D. W., Greene, H. G., and Marlow, M. S. 1970. Eocene age of the Adak Paleozoic (?) rocks, Aleutian Islands, Alaska: *Geol. Soc. Am. Bull.* 81: 3583–92.

Scholl, D. W., and Creager, J. S. 1973. Geologic synthesis of Leg 19 (DSDP) results: Far north Pacific, Aleutian Ridge, and Bering Sea. eds. J. C. Creager, and D. W. Scholl. In *Initial Reports of the Deep Sea Drilling Project, Leg 19.* Washington, D. C.: U.S. Government Printing Office.

Scholl, D. W. 1974. Sedimentary sequences in North Pacific Trenches. In *The Geology of Continental Margins,* eds. C. A. Burk and C. L. Drake, pp. 493–504. New York: Springer-Verlag.

Scholl, D. W., Buffington, E. C., and Marlow, M. S. 1975. Plate tectonics and the structural evolution of the Aleutian-Bering Sea region. In *Contributions to the Geology of the Bering Sea Basin and Adjacent Regions,* ed. H. B. Forbes, pp. 1–31. GSA Special Paper 131. Boulder, Colo.: Geological Society of America.

Scholl, D. W., Vallier, T. L., and Stevenson, A. J. 1982. Sedimentation and deformation in the Amlia fracture zone sector of the Aleutian trench. *Mar. Geol.* 48: 105–34.

Scholl, D. W., Vallier, T. L., and Stevenson, A. J. 1983a. Arc forearc and trench sedimentation and tectonics; Amlia corridor of the Aleutian Ridge. In *Studies in Continental Margin Geology,* eds. J. S. Watkins and C. L. Drake, pp. 413–39. AAPG Memoir 34. Tulsa, Okla.: American Association of Petroleum Geologists.

Scholl, D. W., Vallier, T. L., and Stevenson, A. J. 1983b. Geologic evolution of the Aleutian Ridge – Implications for petroleum resources. *J. Alask. Geol. Soc.* 3: 33–46.

Scholl, D. W., Vallier, T. L., and Stevenson, A. J. 1987. Geologic evolution and petroleum geology of the Aleutian Ridge. In *Geology and resource potential of the continental margins of western North America and adjacent ocean basins – Beaufort Sea to Baja California,* eds. D. W. Scholl, A. Grantz, and J. G. Vedder, pp. 123–56. Circum-Pacific Council for Energy and Mineral Resources, Earth Sciences Series, vol. 6. Houston, Tex.: American Association of Petroleum Geologists.

Somers, M. L., Carson, R. M., Revie, J. A., Edge, R.H., Barrow, B. J., and Andrews, A. G. 1978. GLORIA II – an improved long range sidescan sonar. *Ocean Int.* 78: 16–24.

Spence, W. 1977. The Aleutian Arc, tectonic blocks, episodic subduction, strain diffusion and magma generation. *J. Geophys. Res.* 82: 213–30.

Stevens, S. H., and Moore, G. F. 1985. Deformational and sedimentary processes in trench slope basins of the western Sunda Arc, Indonesia. *Mar. Geol.* 69: 93–112.

Stevenson, A. J., and Embley, R. 1987. Deep-sea fan bodies, terrigenous turbidite sedimentation, and petroleum geology, Gulf of Alaska. In *Geology and Resource Potential of the Continental Margins of Western North America and Adjacent Ocean Basins – Beaufort Sea to Baja California,* eds. D. W. Scholl, A. Grantz, and J. G. Vedder, pp. 503–22. Circum Pacific Council for Energy and Mineral Resources, Earth Sciences Series, Vol. 6. Houston, Tex.: American Association of Petroleum Geologists.

Stewart, R. J. 1978. Neogene volcaniclastic sediments from Atka Basin, Aleutian Ridge: *Am. Assoc. Petrol. Geol.* 62: 87–97.

Tokuhashi, S. 1989. Two stages of submarine fan sedimentation in an ancient forearc basin, central Japan. In *Sedimentary Facies in Active Plate Margins,* eds. A. Taira and F. Masuda, pp. 439–68. Tokyo: Terra Science.

Underwood, M. B., and Bachman, S. B. 1982. Sedimentary facies associations with subduction complexes. In *Trench-Forearc Geology,* ed. J. K. Leggett, pp. 213–27. Geological Society

(London) Special Publication No. 10. Oxford: Blackwell Scientific.

Underwood, M. B. 1986a. Sediment provenance within subduction complexes – an example from the Aleutian forearc. *Sed. Geol.* 51: 57–73.

Underwood, M. B. 1986b. Transverse infilling of the central Aleutian Trench by unconfined turbidity currents. *Geo-Mar. Ltrs.* 6: 7–13.

Underwood, M. B., and Norville, C. E. 1986. Deposition of sand in a trench-slope basin by unconfined turbidity currents. *Mar. Geol.* 71: 383–92.

Underwood, M. B., and Hathon, E. G. 1989. Provenance and dispersal of muds south of the Aleutian Arc, North Pacific. *Geo Mar. Ltrs.* 9: 67–75.

Underwood, M. B. 1991. Submarine canyons, unconfined turbidity currents and sedimentary bypassing of forearc regions. *Rev. Aquat. Sci.* 4(2-3): 149–200.

Vallier, T. L., Scholl, D. W., Fisher, M. A., Bruns, T. R., Wilson, F. H., von Huene, R., and Stevenson, A. J. 1994. Geologic framework of the Aleutian arc, Alaska. In *The Geology of North America,* vol. G-1, *The Geology of Alaska,* eds. G. Plafker and H. C. Berg. Boulder, Colo.: Geological Society of America. p. 367–388.

Van der Lingen, G. J. 1982. Development of the North Island subduction system, New Zealand. In *Trench-Forearc Geology,* ed. J. K. Leggett, pp. 259–72. Geological Society (London) Special Publication No. 10. Oxford: Blackwell.

Von Huene, R. J., Aubouin, J., et al. 1985. *Initial Reports Deep Sea Drilling Project 84.* Washington, D.C.: U.S. Government Printing Office.

Von Huene, R., Larson, E., and Crouch, J. 1973. Preliminary study of ice-rafted erratics as indicators of glacial advances in the Gulf of Alaska. In *Initial Reports of the Deep Sea Drilling Project, 18,* eds. L. D. Kulm, R. von Huene, Washington, D.C.: U.S. Government Printing Office. p. 835–842.

Watkins, J. S., and Drake, C. L. 1982. *Studies in Continental Margin Geology.* AAPG Memoir 34, 801 p. Tulsa, Okla.: American Association of Petroleum Geologists.

Weimer, P. 1991. Seismic facies, characteristics, and variations in channel evolution, Mississippi Fan (Plio-Pleistocene), Gulf of Mexico. In *Seismic Facies and Sedimentary Processes of Submarine Fans and Turbidite Systems,* eds. P. Weimer and M. H. Link, pp. 323–347. Frontiers in Sedimentary Geology. New York: Springer-Verlag.

Wells, R. E., Engebretson, D. C., Snavely, P. D., and Coe, R. S. 1984. Cenozoic plate motions and volcano-tectonic evolution of western Oregon and Washington. *Tectonics* 3: 275–94.

Wells, R. E., and Coe, R. S. 1985. Paleomagnetism and geology of Eocene volcanic rocks of southwest Washington; Implications for mechanisms of tectonic rotation. *J. Geophys. Res.* 90: 1925–47.

Winsemann, J., and Seyfried, H. 1991. Response of deep-water fore-arc systems to sea-level changes, tectonic activity and volcaniclastic input in Central America. In *Sedimentation Tectonics and Eustacy,* Special Publication No. 12, ed., D. I. M. Macdonald, pp. 273–92. International Association of Sedimentologists. Oxford: Blackwell.

17 Aleutian Basin of the Bering Sea: Styles of sedimentation and canyon development

Herman A. Karl, Paul R. Carlson, and James V. Gardner

U.S. Geological Survey, Menlo Park, California

Abstract

The present-day physiography and Quaternary sediments of the Aleutian Basin of the Bering Sea are the culmination of an evolution most recently dominated by mass wasting. These processes have modified the structural framework produced by plate convergence during early Tertiary time. The products of mass sediment transport are so pervasive as to make the Aleutian Basin possibly the world's best example of a basin and margin modified by these processes.

Three provinces, fan, debris flow, and basin plain, characterize the upper 50 m of sediment in the Aleutian Basin. Sediment reaches the floor of the Aleutian Basin through numerous canyons and gullies of the insular and continental slopes and through three main canyon-channel systems: Bering, Umnak, and Pochnoi channels. Bering Channel, morphologically the most impressive of the three, extends for more than 500 km and terminates in a presumably young submarine fan. Bering Fan, lacking the sloping wedge geometry and the upper-, middle-, and lower-fan subdivisions that are commonly used to describe submarine fans, resembles a single depositional lobe. Sediment shed from the Aleutian Island Arc and Bowers and Shirshov Ridges has been transported down the flanks of these features resulting in debris flow deposits that have accumulated around the perimeter of the Aleutian Basin. Two debris flows that originated from the Aleutian Arc (Umnak and Pochnoi) are dominant among the debris flows that occupy the perimeter of the basin. Umnak debris flow is the youngest of the debris flow deposits and overlies the margins of Bering Fan. The basin plain province, areally the most extensive and volumetrically the most important of the provinces, is constructed of turbidites interbedded with pelagic diatom ooze, ice-rafted debris, and volcanic ash.

Evidence of downslope mass movement is ubiquitous around the margins of the Aleutian Basin. Mass wasting processes not only distribute sediment to the basin plain of the Aleutian Basin but also are the principal control on the morphology of the basin margins. The most striking examples of margin mass wasting are the seven large canyons that incise the Beringian margin and the displaced massive slide blocks that rest on the floor of Aleutian Basin. Zhemchug Canyon, one of the largest submarine canyons in the world, appears to have originated by the collapse of the margin through a combination of massive slides and multiple retrograde failures.

Although some of the largest submarine canyons in the world incise the Beringian margin, only Bering Canyon has a submarine fan associated with it; Bering Fan is morphologically atypical, and possibly is in an early stage of development. If fans were built at the mouths of the other canyons, they are now buried at the base of the slope. The thick base-of-slope deposits represent debris from the collapse of the outer shelf and slope derived from a combination of multiple retrograde failures and singular massive slumps and slides, and to a lesser extent, by debris flows and turbidity currents.

Introduction

Background and purpose

Some of the largest submarine canyons in the world incise the Beringian continental margin. Scholl et al. (1970) and Carlson and Karl (1988) used evidence from seismic-reflection profiles to show the importance of mass transport of sediment in and out of the Beringian Canyons. During 1986 and 1987, the U.S. Geological Survey mapped the entire United States portion of the Aleutian Basin and Beringian margin using GLORIA sidescan sonar and seismic-reflection profiling (Bering Sea EEZ-SCAN Scientific Staff 1991). The GLORIA images show that mass wasting is the dominant recent erosional process on the continental slope (Carlson, Karl, and Edwards 1991).

The GLORIA mosaic enabled us to identify the principal routes of sediment transport to the basin and to contrast the sedimentation processes that have recently operated in the

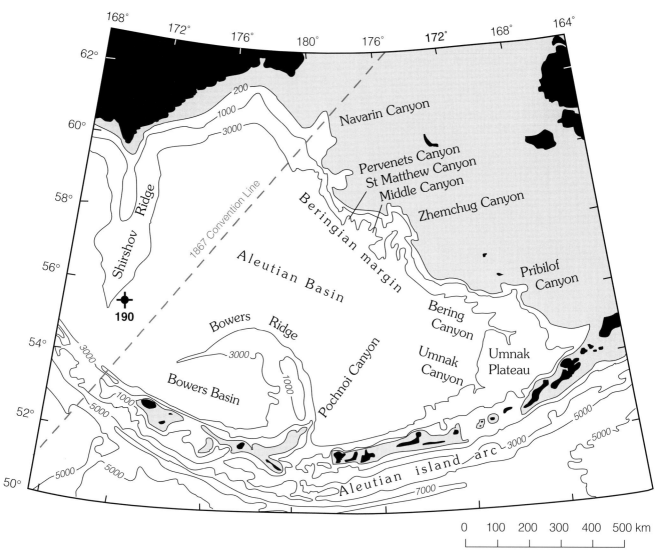

Figure 17–1. Location map of Bering Sea showing major physiographic features and bathymetry, in meters.

Aleutian Basin. The purpose of this chapter is to describe the various styles of sedimentation and mass failures in and around the Aleutian Basin and to contrast these with one another and to speculate on the development of these features.

Morphologic features of the Aleutian Basin

The Aleutian Basin is surrounded by the topographically steep Beringian continental margin on the northeast and by the steep insular margin of the Aleutian Island Arc on the south (Figure 17–1). We were unable to map west of the 1867 Convention Line and could not image Shirshov Ridge. However, this morphologic structure occurs less than 50 km west of our survey and has had some influence on the sedimentation in Aleutian Basin. Nine major and two minor bathymetric features complicate the otherwise monotonously featureless Aleutian Basin. The major features include the

large canyons of the Beringian margin (Navarin, Pervenets, St. Matthew, Middle, Zhemchug, Pribilof, and Bering Canyons), Umnak Plateau, and Bowers Ridge and the minor but significant features of Umnak and Pochnoi canyon-channel complexes (Figure 17–1). Each of the major and minor bathymetric features has had its own pronounced effect on the sedimentation in the Aleutian Basin.

Methods

The data used for this study include complete coverage with processed GLORIA imagery and more than 40,000 line kilometers of systematically collected two-channel 160-in³ airgun seismic-reflection profiles and 3.5-kHz high-resolution seismic-reflection profiles collected at nadir (Bering Sea EEZ-SCAN Scientific Staff 1991) (Figure 17–2). Navigation for the four cruises was by a combination of GPS,

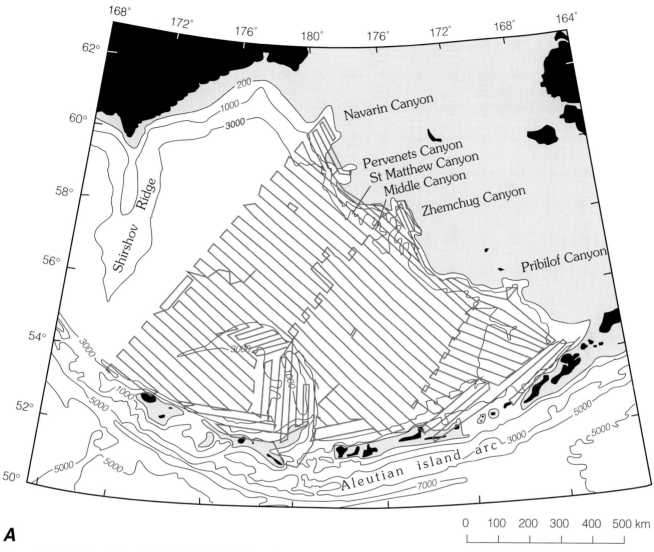

A

Figure 17–2. A, Trackline chart of EEZ-SCAN 86 cruises.

Loran C, and NNSS transit satellite. Positional accuracy of the smoothed navigation tracks was typically better than ±100 m.

When several different acoustic systems are used to investigate an area, the disparate vertical resolutions of the various systems are commonly not appreciated. Unfortunately, the differences do not provide a continuous spectrum of spatial resolutions, but rather occupy small sections of a wide spectrum, but with all sections separated from one another (see discussion by Mutti and Normark 1987). For example, GLORIA imagery has a theoretical resolution of the 6.5-kHz frequency of about 20 cm, but the signal penetrates as much as 10 m. The complicating factor with these data is that the backscatter signature of each pixel is the integrated resultant of reverberations from within a volume with horizontal dimensions of ∼120 m × 50 m and a variable vertical dimension of <10 m (Gardner et al. 1991; Lyons,

Anderson, and Dwan 1994; Somers this volume). Consequently, the 20-cm vertical resolution is convolved into the sum of all the backscattered intensities from throughout that volume. An additional complication is that the acoustic energy intersects the seafloor at a wide range of grazing angles. In contrast, the 3.5-kHz acoustic energy intersects the seafloor at vertical to near-vertical incidence and the frequency has a theoretical vertical resolution of ∼40 cm. However, the 3.5-kHz signal is the resultant of the integrated reflections from an area on the seafloor subtended by a cone with an apex angle of 20°. At water depths typical of the Aleutian Basin, each 3.5-kHz ping insonifies between 0.9 and 1.5 km² of seafloor. The 3.5-kHz system used for these surveys typically penetrated between ∼25 to ∼50 m, and thus, the data represent the sum of constructive and destructive reflection interferences over a considerable, and variable, volume. Even less resolution was obtained by the

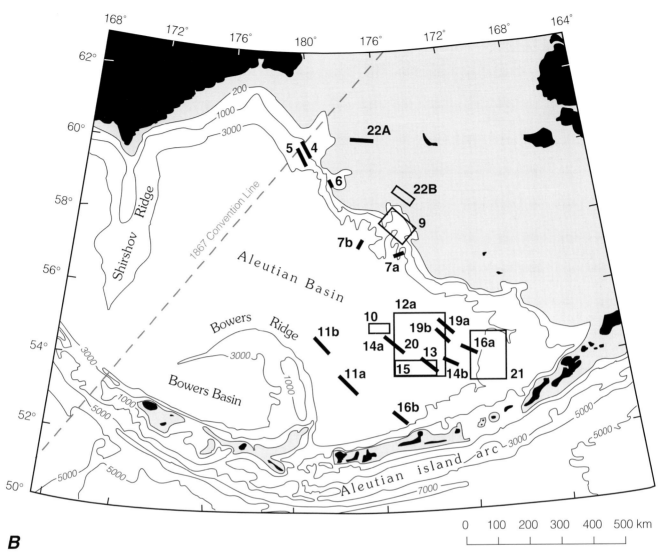

B

Figure 17–2. *Continued*. **B, Locations of seismic (bold lines) and sidescan (box) figures. Numbers refer to other text figures.**

airgun's received signal that was band-pass filtered between 40 and 80 Hz, providing a range of theoretical vertical resolutions on the records of between ~ 20 to ~ 40 m. The geometry of the omnidirectional airgun source and receivers allows this system to show only the regional characteristics of the sediment section on a rather coarse scale. These are the constraints of the data used for the following analyses of the sedimentation patterns of the Aleutian Basin.

Morphologic setting

The bathymetry of the Bering Sea is divided by the Beringian continental slope, which extends about 1400 km from the Aleutian Island Arc to the Russian Far East margin (Figure 17–1). This rugged steep slope separates the shallow (<150 m), flat (0.02°) Bering continental shelf and the deep (>3600 m), flat (0.02°) Aleutian Basin. The Bering

Shelf south of the Bering Strait is about 600 km wide from shoreline to the 150-m-deep shelf break. The Beringian continental slope ranges in width from 30 to 50 km and descends from the shelf break to an average depth of about 3,000 m. The gradient of the slope averages 5° and ranges from 3° to 8°. The slope is characterized by numerous submarine canyons and gullies. The base-of-slope, at about 3,600 m depth, is a narrow province that grades into the abyssal plain of Aleutian Basin. The base-of-slope is cut by several large channels that are continuations of submarine canyons but do not terminate on submarine fans, with the exception of Bering Channel. The channels are located at depths of between 3,000 and 3,600 m, with widths that range from 25 to 100 km and gradients from 0.5° to 2.3°.

The Aleutian Arc insular slope, extending about 2,200 km long, varies in width from 20 to 90 km from the shelf break at about 200 m to the base of the slope at about 3,000 m.

Table 17–1. *Dimensions of canyons that incise the north side of the Aleutian Arc*

Canyon	Head (depth m)	Bottom (m)	Length (km)	Gradient (degree)
Pochnoi	500	3,500	124	1.4
Tanaga	600	3,100	74	1.9
Atka	700	3,200	43	3.3
Amlia	1,100	3,400	83	1.6
Seguam	200	3,400	122	1.5
Amukta	600	3,200	130	1.2
Umnak	200	3,200	209	0.8
Bering	120	3,200	495	0.4

Table 17–2. *Volumes of the largest Beringian canyons compared to several other well-known canyons (adapted from Carlson and Karl, 1988)*

Rank	Canyon volume (km^3)	Canyon
1	5,800	Zhemchug
2	5,400	Navarin
3	4,300	Bering
4	2,950	Swatch of No Ground
5	2,300	Congo
6	1,800	Middle
7	1,700	Pervenets
8	1,300	Pribilof
9	830	Kamchatsky
10	740	St. Matthew
11	450	Monterey
12	425	Astoria
13	300	Hudson
14	70	Wilmington
15	10	La Jolla

The gradient of the slope averages about 6° and ranges from 2° to 8°. This slope is also extensively dissected by submarine gullies and canyons of widely varying dimensions (Table 17–1). The base-of-slope is cut by canyon-channel systems that extend about 10 km seaward of the slope to the 3,600-m isobath and grade into the abyssal basin (Table 17–1). Umnak and Pochnoi Channels are 5 km and 6 km wide, respectively, with gradients of 0.8° and 1.4°. The Aleutian Basin abyssal plain is extremely flat (1:4,000 to 1:7,000) and encompasses an area of about 105,000 km^2.

Beringian margin canyons

Characteristic morphology

Seven submarine canyons incise the Beringian margin. These canyons are named from north to south respectively, Navarin, Pervenets, St. Matthew, Middle, Zhemchug, Pribilof, and Bering (Figure 17–1). These canyons rank among the largest in the world when measured by volume of material removed (Table 17–2). Zhemchug Canyon, the largest of these canyons, has a volume of at least 5,800 km^3, more than an order of magnitude larger than the more familiar Monterey Canyon on the California continental margin (450 km^3) or Hudson Canyon on the New York margin (300 km^3) (Carlson and Karl 1988). Even St. Matthew, the smallest of the Beringian canyons, with a volume of 740 km^3, is larger than most well-known canyons (Table 17–2). The Beringian canyons are unusual not only because of their enormous size, but also because of their distinctive shapes (Figure 17–3). The seven Beringian canyons can be grouped into four general morphologic categories based principally on the geometry of the canyon where it breaches the outer continental shelf and upper slope. These groupings consist of (1) Navarin and Pervenets Canyons, (2) St. Matthew and Middle Canyons, (3) Zhemchug and Pribilof Canyons, and (4)

Bering Canyon. A summary of the topography of the canyons follows; detailed descriptions can be found in Carlson and Karl (1988).

NAVARIN AND PERVENETS CANYONS

Navarin and Pervenets Canyons resemble amphitheaters (gently sloping outer-shelf, upper-slope platforms that are hemispherical in plan view) cut by a system of primary and tributary channels (Figure 17–3). Navarin Canyon forms a more or less trapezoidal depression defined by the 150-m isobath on the continental shelf. The canyon is about 150 km across at its widest point on the shelf and has been eroded back about 100 km from the original shelf break at a depth of approximately 175 m. The canyon gently slopes down from the 150-m isobath to the shelf break where it narrows to a width of 100 km and is incised 1150 m. Two main channels in the upper reaches of the canyon merge at a depth of about 3100 m to form a single thalweg that continues to the abyssal plain.

Pervenets Canyon incises the margin about 75 km south of Navarin Canyon. Like Navarin Canyon, the head of Pervenets Canyon occupies a trapezoidal depression as defined by the 150-m isobath that extends about 100 km landward from the regional shelf break. The canyon is 70 km across at its widest point on the shelf and narrows to 40 km at the shelf break where it is incised 850 m. Two main channels characterize the upper part of Pervenets Canyon and coalesce at a depth of about 600 m to form a single channel that merges with Navarin Canyon at a depth of about 3500 m. Pervenets Canyon has a volume of 1700 km^3, about one-third the size of Navarin Canyon.

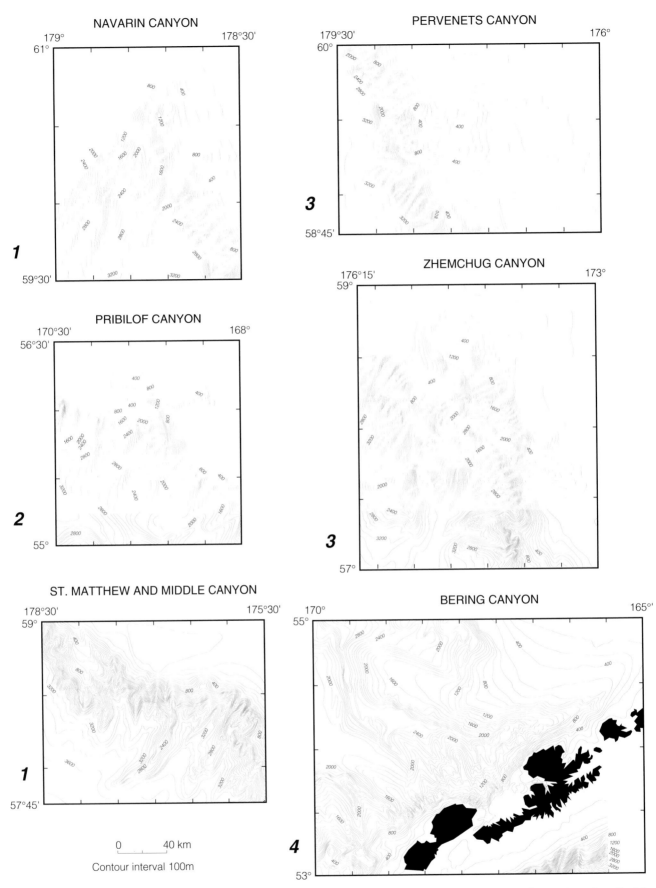

Figure 17–3. Bathymetric maps of the seven large Beringian canyons illustrating the four morphological groupings described in the text.

ST. MATTHEW AND MIDDLE CANYONS

St. Matthew and Middle Canyons barely indent the shelf break and have only minor morphological expression on the shelf (Figure 17–3). The canyons consist of systems of channels confined within poorly defined entrants or drainage patterns on the upper slope and outermost shelf. The drainage systems are separated by an elongate ridge or divide perpendicular to the trend of the shelf break. Both canyons have pronounced east and west branches with numerous smaller channels that incise the slope between the branches. These smaller channels are tributaries of the major channels (Carlson and Karl 1984). The east and west branches of St. Matthew Canyon join to form a single channel on the lower slope, and in a similar fashion, the two branches of Middle Canyon converge. The two resultant channels continue to the abyssal plain. St. Matthew Canyon is more prominent and better defined on the outer shelf than Middle Canyon, which consists of a series of small gullies at its head that barely indent the shelf. St. Matthew Canyon originates about 5 km landward of the shelf break at a depth of 145 m. The more-prominent west branch of St. Matthew Canyon is V-shaped, 6 km wide, and has a relief of 30 m at the shelf edge.

ZHEMCHUG AND PRIBILOF CANYONS

Zhemchug and Pribilof Canyons, in contrast to the more or less trapezoidal shape of Navarin and Pervenets Canyons, are very elongate parallel to the shelf edge at the outer shelf and upper slope (Figure 17–3). The rugged and steep-walled Zhemchug and Pribilof Canyons lack the broad, relatively gently sloping platforms that characterize the heads of Navarin and Pervenets Canyons. The rectangular gorges are divided into east and west wings by a ridge that occurs near the midpoint of the gorge and extends seaward normal to the trend of the shelf edge. Each wing of these canyons is characterized by a well-defined main channel with numerous tributaries. The main channels are oriented subparallel to the trend of the shelf break and merge downslope to form a single channel that exits onto the abyssal plain. The sidewalls of each of the canyons are dissected by numerous gullies.

Zhemchug Canyon occupies a 160-km-long, 30-km-wide steep-walled incision into the shelf that is oriented northwest-southeast, roughly parallel to the shelf-slope break. The two branches merge just seaward of a breach in Pribilof Ridge, an outer-shelf structural high (Marlow et al. 1976). The axial profiles of both branches show steplike steepening. The canyon has cut a gorge 100 km wide and 2,600 m deep at the regional shelf break. Seaward of the shelf break, the walls are still steep and have as much as 2,550 m of relief, but the floor becomes flat and is as much as 10 km wide. The gradient of the canyon decreases, and the canyon is not bathymetrically confined on the continental rise, resulting in a transformation of the canyon to a leveed deep-sea channel that winds southward at least another 100 km. Transverse profiles of the canyon are steep-walled and V-shaped landward of the shelf break.

Pribilof Canyon is 45-km wide at the shelf break and has a relief of 1,600 m. The two branches of Pribilof Canyon merge at a depth of about 1,500 m, at which point the canyon turns seaward 90° and passes through the outer-shelf bedrock high eventually to merge with Bering Canyon (Figure 17–3).

BERING CANYON

Bering Canyon is a V-shaped, sinuous gully that has a morphology typical of most of the world's submarine canyons. The canyon begins as a narrow wedge-shaped indentation at the shelf break and gradually widens downslope (Figure 17–3). Bering Canyon extends for 400 km across the outer shelf and slope to the continental rise where, at a depth of 3,200 m, it evolves into the Bering channel and fan system described in subsequent sections. In terms of length, it may be the longest canyon in the world. Bering Canyon is the only Beringian margin canyon that consists of a single main channel demonstrating a classical canyon morphology. The other canyons, in contrast, are best described as canyon systems that consist of an hierarchical arrangement of channels and tributaries within a discrete depression that varies in prominence, size, and shape.

Aleutian Basin sediment provinces

The floor of Aleutian Basin is a vast flat plain overlying a thick sediment section. Three major sediment bodies can be delineated on GLORIA sidescan imagery and 3.5-kHz seismic profiles, each implying a different process. We make a distinction between backscatter facies (GLORIA imagery) and reflection facies (3.5-kHz profiles) because of the fundamentally different physics involved in the interactions of the two acoustic signals with the sediment column. GLORIA acoustics insonify the seafloor at a wide range of grazing angles and record only a very small fraction of reverberated energy (Urick 1983; Gardner et al. 1991; Lyons et al. 1994). By contrast, much more of the 3.5-kHz energy is reflected back from the signal's near-vertical incidence, in the classic Zoeppritz formulation (Sheriff 1976). The consequence of these differences is that backscattered energy from the GLORIA sidescan can have characteristics that are very different from reflected energy from the 3.5-kHz seismic system.

GLORIA imagery, representing the top 10 m or less of the section, shows three backscatter patterns, termed here basin plain, fan, and debris flow backscatter provinces. Each province can be directly related to a different process, as their names imply. The three major backscatter provinces can, in turn, be subdivided into a total of seven localities: Umnak debris flows, Pochnoi debris flows, Bowers debris

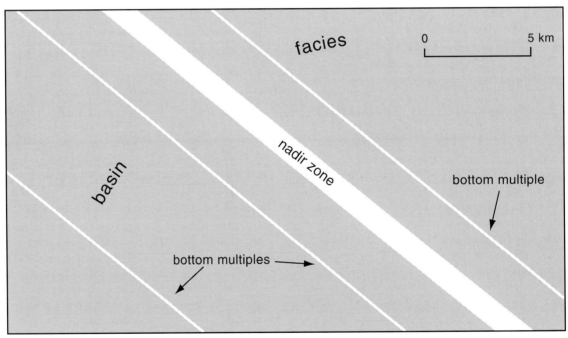

Figure 17–4. GLORIA mosaic and interpretation illustrating the characteristics of the basin backscatter facies.

flows, Shirshov debris flows, Bering margin debris flows, Bering Fan, and Aleutian basin plain.

The reflection facies of the 3.5-kHz records suggests that the top 50 m or less of the section can be divided into the same three primary provinces found in the GLORIA imagery. As will be shown, the airgun profiles show virtually no differences in acoustic character when the various sedimentological boundaries defined on GLORIA and 3.5-kHz imagery are crossed. Consequently, most of the following discussion is based on the 3.5-kHz profiles and GLORIA imagery.

Basin-plain province

The surface of the basin-plain province is very flat and featureless with relief less than 2 m in the center of Aleutian

NW

SE

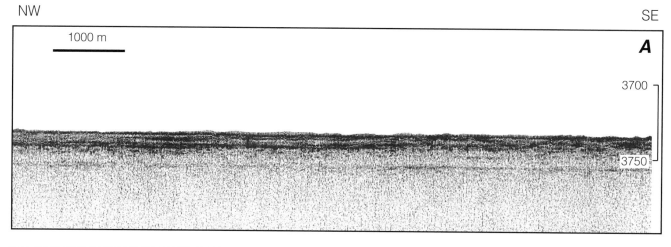

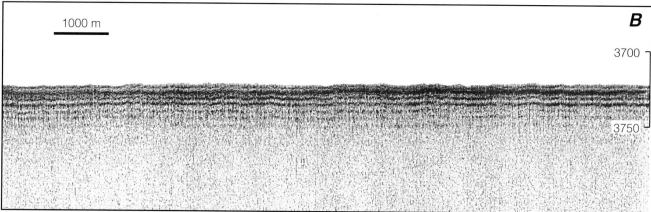

Figure 17–5. High-resolution seismic-reflection (3.5-kHz) profiles of the basin facies. Water-depth scale on right, in meters.

Basin and only up to 5 m of relief close to the basin margins. The slightly more irregular surface is probably the result of very thin (<2-m-thick) distal mass flow deposits, although these are not resolved on the 3.5-kHz records. GLORIA images of the basin-plain province have uniformly low backscatter with very little variation (Figure 17–4), but often with intense interference fringes (Huggett et al. 1992) and bottom multiples (Figure 17–4). The 3.5-kHz profiles in the basin-plain province are characterized by a relatively high amplitude surface reflection followed by a series of parallel and laterally continuous high-amplitude subbottom reflectors that lose amplitude with depth (Figure 17–5). This facies exhibits the classic character typical of basin-plain facies (e.g., Damuth 1980). The 3.5-kHz penetration in this province sometimes exceeds 70 m.

Bering Fan province

The backscatter facies in the Bering Fan province has a ropy alternating high- and low-backscatter pattern that appears to splay out from a point source, Bering Channel (Figure 17–6). The relatively high backscatter, lack of apparent interfer-

ence fringes, and outward-splaying flow lines within the fan backscatter facies makes mapping this facies on the sonographs relatively straightforward, especially when it is juxtaposed against the two adjoining backscatter facies. The reflection facies of this province is characterized by a very high amplitude surface reflection underlaid by a shallow zone of subparallel discontinuous high-amplitude subbottom reflectors with pinch-and-swales of various thicknesses (Figure 17–7). Seismic-reflection airgun profiles show no change in character when crossing from the fan to the basin-plain facies (Figure 17–8A). The lack of definition of the fan reflection facies on low-resolution seismic profiles suggests that the fan beds are somewhat less than 50 m thick.

Debris flow province

The GLORIA backscatter facies of the debris flow province is characterized as a mixed, alternating high- and low-backscatter pattern, but with large areas of high backscatter, and the entire facies has a somewhat braided appearance (Figure 17–9). The flow lines of this facies appear subparallel to convergent to one another downslope and appear to

Figure 17–6. A, GLORIA mosaic of Bering Fan.

coalesce from a line source. Although the debris flow and fan backscatter characteristics are somewhat similar, backscatter from the debris flows differs from the fan in that the debris flows do not appear ropy and generally have a much higher backscatter intensity than backscatter from the fan facies. The 3.5-kHz reflection characteristics of the debris flow province is one of very low amplitude surface reflection followed by an acoustically transparent layer, often followed in turn by a high-amplitude subbottom reflector and then more acoustically transparent section (Figure 17–10). The surface reflection is irregular to smooth with irregularities that range from <5 m to rough hyperbolated surfaces with >10 m of relief (Figure 17–10A). As with the reflection facies of the fan province, the reflection facies of the debris flow province is not resolved on the airgun pro-

files when crossing from the basin plain to the debris flow province (Figure 17–8B).

Areal distribution of facies

Basin-plain province

The centers of Aleutian and Bowers Basins are floored by the basin-plain facies (Figure 17–11). This province covers more than 100,000 km^2 and overlaps the debris flow province around the periphery of Bowers Basin, suggesting that the major episodes of debris flow events off Bowers Ridge are not recent. The basin-plain facies in the eastern region of Aleutian Basin is overlapped by debris flow fa-

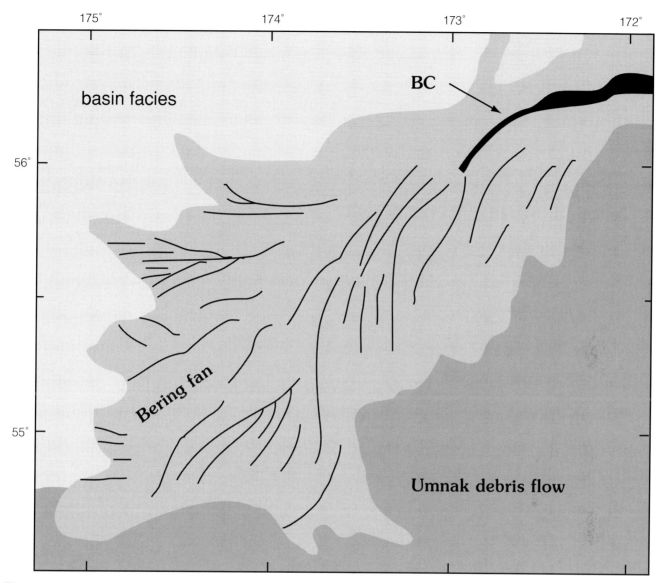

Figure 17–6. *Continued.* **B, Interpretation of Bering Fan and adjacent regions. Black zone labeled BC is Bering Channel. Lines on Bering Fan represent flow lines but do not correlate directly to distributary channels.**

cies and Bering Fan facies. These relationships suggest that sediment mass movements are relatively young, and most likely of Quaternary age. The basin-plain facies appears to have partially buried the large displaced blocks along the Beringian base-of-slope, suggesting that the collapse of the margin is older than the upper ~ 50 m of basin facies.

Bering Fan province

Bering Fan is composed of a canyon-channel complex, the continuation of Bering Canyon, and the submarine fan proper. Bering canyon-channel complex heads as a canyon at the junction of the Beringian continental slope, the Aleutian Island Arc, and Umnak Plateau (Scholl et al. 1970) (Fig-

ure 17–1). The canyon descends from <200 m at its head to 3500 m where, after joining with Pribilof canyon-channel, it evolves into a channel (Figure 17–12). The channel floor narrows from 25 km at its transformation from a canyon to a channel to less than 10 km wide where the channel looses its bathymetric relief approximately 50 km down-channel. The channel relief varies from 45 m at the canyon mouth to less than 10 m at its distal end. The channel is not well developed (Figure 17–13), and the levee developed on the north side of the channel never exceeds 25 m in height. After the channel traverses across about 100 km of fan, it evolves into a series of distributaries that splay out over a depositional lobe. The distributaries, resolved as small discrete channels on both the GLORIA imagery (Figure 17–6) and the 3.5-kHz profiles (Figure 17–14), are typically in-

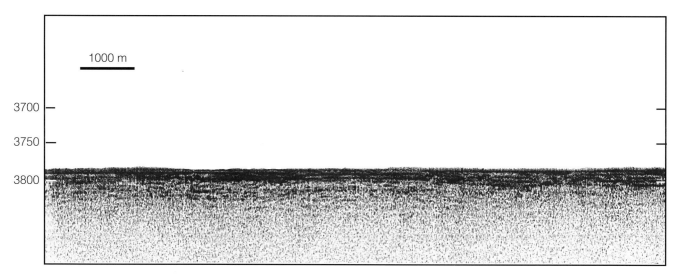

Figure 17–7. High-resolution seismic-reflection (3.5-kHz) profile of fan facies. Water-depth scale on left margin, in meters.

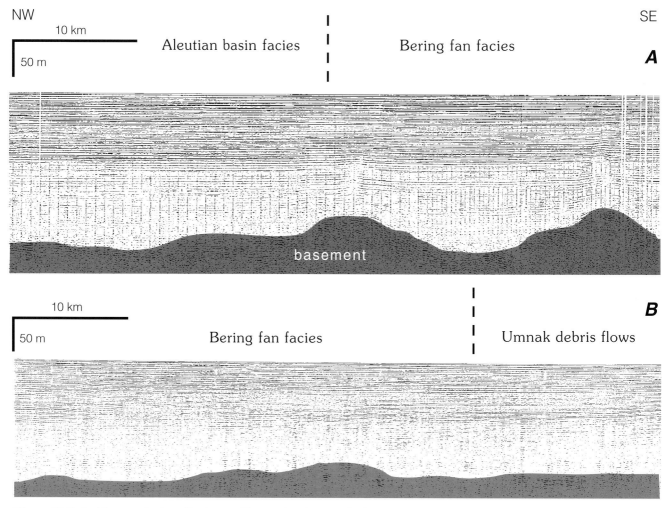

Figure 17–8. A, Airgun seismic-reflection profile crossing from basin facies to fan facies. B, Airgun seismic-reflection profile crossing from fan facies to debris flow facies. Note that no change occurs in the seismic records across these facies changes.

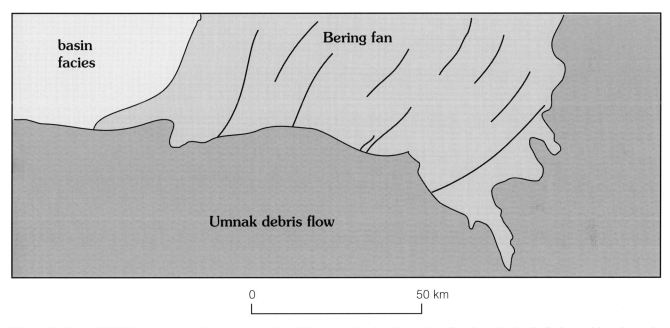

Figure 17–9. A, GLORIA mosaic and B, interpretation of the lower Bering Fan where it onlaps the basin facies and is onlapped by Umnak debris flows.

cised less than 10 m deep into the fan surface and can be followed for 10s of km.

Although Bering Fan covers more than 20,000 km², it is not easily divided into the conventional upper-, middle-, and lower-fan subdivisions often used to describe fans (e.g., Normark 1970; Damuth and Flood 1985). A longitudinal profile of the fan surface shows very little change throughout its length (Figure 17–12) and does not have the geometry of a sloping wedge of sediment. The distributaries splay out across the entire fan surface rather than being confined to only one province of the fan, and discrete depositional lobes cannot be identified at the termination of the distributaries. It is possible that depositional lobes have been buried by

Umnak debris flows, but we are unable to see this in our data. Bering Fan appears most like a single depositional lobe, very similar to one of the depositional lobes of Mississippi Fan described by Twichell et al. (1991, see their Figure 19.9)

Because the fan sediments cannot be distinguished on air-gun profiles from the overlapping debris flow and basin-plain facies, and because of the lack of a pronounced channel, the lack of a well-developed levee for much of its length on the basin floor, and the apparent lack of more than one depositional lobe, we believe that Bering Fan is a relatively young, possibly late Pliocene and Quaternary, feature that has not been in existence for most of the history of Aleut-

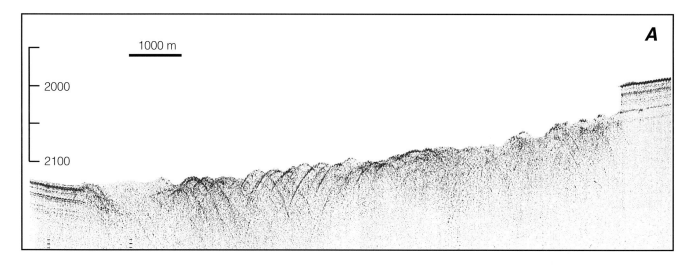

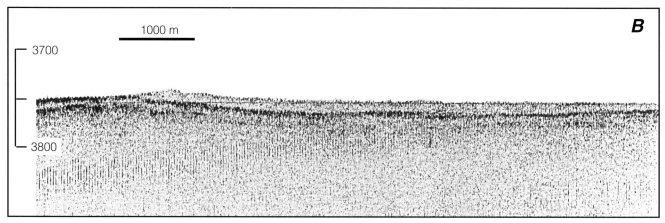

Figure 17–10. High-resolution seismic-reflection (3.5-kHz) profiles across the debris flow facies. Water-depth scale on left margin, in meters.

ian Basin. Contrary to the suggestion by Scholl et al. (1970), Bering Fan is the only submarine fan found in Aleutian Basin. It is curious that fans have not developed, or at least are so small that they cannot be identified on our 3.5-kHz and airgun profiles, along the base-of-slope of the Beringian margin.

Debris flow province

The debris flow facies displays two distinctly different characteristics, which are geographically separated. The submarine slides that evolved into the debris flows that descend down the Beringian margin carried large blocks, some kilometers in length (Figure 17–15). However, the Umnak, Pochnoi, Bowers, and Shirshov debris flows did not carry large blocks of intact sediment out onto the floor of Aleutian Basin, but rather spread relatively thin, presumably relatively fine grained material onto the basin. The zone of Beringian slides and debris flows on the basin floor extends less than 50 km beyond the base of slope, whereas the Pochnoi, Umnak, and Shirshov debris flows extend as much

as 200 km onto Aleutian Basin. Pochnoi debris flow deposits may merge with Bowers debris flows on the southeastern end of Bowers Ridge but the two debris flow units cannot be discriminated in our data. Debris flow deposits cover vast areas of Aleutian Basin; Umnak debris flows cover more than 55,000 km^2, Pochnoi debris flows cover more than 54,000 km^2, and Shirshov debris flows are considerably larger than the 25,000 km^2 that we could map. The only other area of debris flows that shows detached blocks of slide material within the debris flows is the upper west-facing edge of Umnak Plateau (Figure 17–15). Here slide blocks and debris flows have been displaced downslope, removing large sections of the rim of the plateau.

Discussion

Causes of differences in basin sedimentation

The Deep Sea Drilling Project (DSDP) drilled only one site, DSDP Site 190, in the Aleutian Basin (Figure 17–1), and

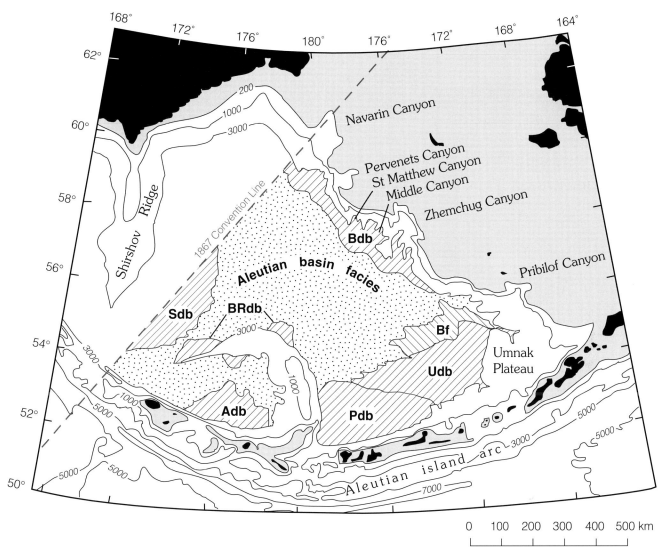

Figure 17–11. Compilation of the areal distribution of the three sediment facies. Bdb is Beringian debris flows, Udb is Umnak debris flows, Pdb is Pochnoi debris flows, Adb is Aleutian debris flows, BRdb is Bowers Ridge debris flows, SRdb is Shirshov Ridge debris flows, BF is Bering Fan, and the light gray area is the Aleutian basin facies.

this site is outside our study area and was not continuously cored (Creager et al. 1973). The other DSDP sites in the Bering Sea were drilled on Umnak Plateau and the flanks of Shirshov and Bowers Ridges and in the Russian sector. DSDP Site 190 recovered a complex Pliocene and Pleistocene section of diatomaceous silty clays with thin, very fine grained turbidites and thin horizons of ash (Creager et al. 1971). The sedimentation rates for the late Miocene to Holocene are only 7 cm/10[3] yr, a rate that may be representative only of the region of DSDP Site 190, a location far removed from the major canyons. Although many short cores have been collected in Aleutian Basin, especially by Lamont-Doherty Earth Observatory, few of these cores have been analyzed and no age dates have been published. One study (Horn, Ewing, and Ewing 1972) compiled grain-size

descriptions of Aleutian Basin cores and classified the region as one covered by very fine grained sandy and silty terrigenous turbidites. A subsequent study of cores from the Aleutian Basin (Gardner et al. 1982) demonstrated that large areas of the Aleutian Basin do not contain turbidites in the upper two meters of the section. The basin-plain facies, like most basin-plain deposits (Damuth 1980), is likely composed of a complex stratigraphy of hemipelagic muds, thin-bedded, fine-grained turbidites that evolved from the distal zones of the debris flow deposits, interbedded with pelagic diatom ooze, ice-rafted debris, and volcanic ash.

The basin-plain facies is overlapped by the fan facies of Bering Fan, but the contact between the two is not distinct even on the 3.5-kHz records. GLORIA imagery clearly shows that Umnak debris flow deposits overlap both the

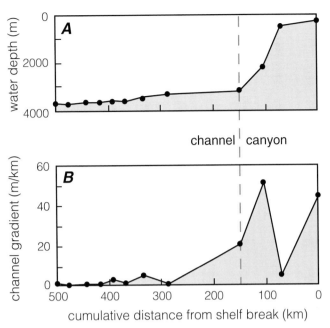

channel | canyon

Figure 17–12. A, Profile and (B), gradient of Bering Channel.

Bering Fan and Pochnoi debris flows (Figure 17–6). The basin-plain facies overlaps the Bowers debris flow deposits, but the Shirshov debris flow deposits appear to blanket the basin-plain facies. Consequently, it appears that the basin-plain province has been encroached upon, first by Bering Fan and then later by the debris flow deposits originating from the Aleutian Island Arc and the ridges.

The focus of sedimentation appears to be the eastern portion of the Aleutian Basin bounded by the Beringian continental margin, Umnak Plateau, and the eastern Aleutian Island Arc. There is a regional slope to the Aleutian Basin with the center of the basin being about 200 m deeper than the regional average for the perimeter. The tectonically active Aleutian Island Arc is the obvious source for the Pochnoi debris flows and some of the Umnak debris flows. The collapse and retrograde erosion of the Beringian margin is the source for the Beringian slide blocks and debris flows. Shirshov and Bowers Ridges are clearly the sources for their respective debris flows. Bering Fan appears to be supplied by sediment shed off the Alaska Peninsula as well as the southeastern Beringian margin and the Aleutian Arc.

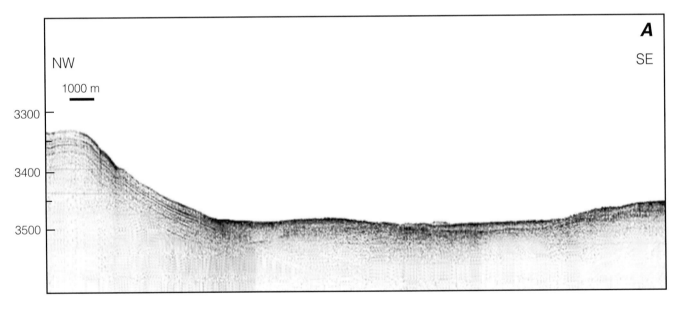

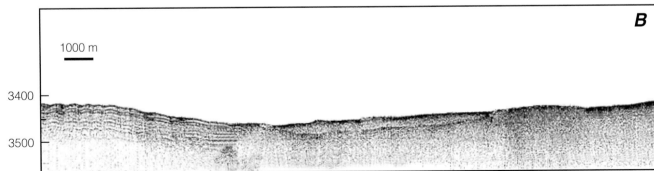

Figure 17–13. High-resolution seismic-reflection (3.5-kHz) profiles across Bering Channel. Water-depth scale on left margin, in meters.

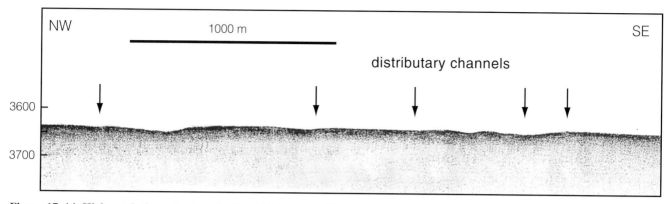

Figure 17–14. High-resolution seismic-reflection (3.5-kHz) profile across Bering Fan showing distributary channels. Water-depth scale on left margin, in meters.

Milliman and Meade (1983) give the combined sediment yield for the major southern Alaskan rivers (Kuskokwim and Yukon Rivers) of 440 tons/km^2/yr but suggest that southern Alaskan yields may have been an order of magnitude greater during glacial periods. The pattern of sedimentation suggests that a significant amount of this late Pleistocene and Holocene southern Alaskan river sediment yield was funneled into Bering Canyon, then down onto Bering Fan, and ultimately onto Aleutian Basin. This scenario would account for the regional gradient for the Aleutian Basin, for the formation of Bering Fan, and the lack of modern submarine fans elsewhere in the Aleutian Basin.

The large debris flows and detached slide blocks shed off the Aleutian Island Arc suggest that large-magnitude tectonic activity on the island arc may have triggered gravity flows off the island arc. Although Shirshov and Bowers Ridges appear not to be seismically active in modern times (U.S. Coast and Geodetic Survey 1970), they may have been seismically active at other times during the Quaternary to have produced such massive debris flows. The summits of both ridges are near 1,000 m deep and much too deep to be affected by storm wave base, a potential cause for sediment instability. Likewise, the presence of numerous recently active faults on the outer Beringian Shelf that parallel the Beringian margin (Abers et al. 1993) suggests earthquakes may have played a dominant role in generating mass movement of sediment (Carlson et al. 1991). Cooper et al. (1986) suggest that the failure along the western margin of Umnak Plateau was caused by the emergence of diapirs that ring the perimeter of the plateau.

Although correlating pulses of basin sedimentation to fluctuating eustatic sea levels is very much in vogue, this mechanism does not appear to offer much promise to explain the triggering mechanism for the debris flow deposits in the Aleutian Basin because the mass-flow deposits were generated well below eustatically lowest sea levels. For instance, the surface of Umnak Plateau is at about 900 m water depth, so that a drop in sea level of 120 m would have

little or no effect on the sediments there. The same argument holds for Bowers and Shirshov Ridges. Although the Aleutian Island Arc would have felt some influence from eustatically fluctuating sea levels, the steep insular slopes have very little in the way of shallow terraces for periodically accumulating and then offloading sediment. In addition, the Bering Sea is known to have been frozen over during the last glacial maximum (Hopkins 1979; Gardner et al. 1982; Sancetta and Robinson 1983) and presumably was frozen during most if not all of the late Neogene and Pleistocene glaciations. A frozen glacial-age Bering Sea should have inhibited significant amounts of sediment moving seaward of the frozen shoreline as littoral drift.

The basin province, predominantly composed of hemipelagic diatomaceous silty clays and turbidites, may well have been modulated by global climate cycles. Diatomaceous sediment and thin turbidites were produced during interglacial conditions when the Bering Sea was seasonally ice free, and very little sediment was deposited during glacial conditions when the Bering Sea was ice covered.

The apparent youth of Bering Fan and the various debris flow provinces, as well as the sudden appearance of terrigenous sediments in the Pliocene section at DSDP Site 190 (Creager et al. 1973), suggest a major change in sedimentation at some threshold in the glaciation of Alaska. Although initial glaciation in Alaska has been dated as old as middle Miocene (Lagoe et al. 1993), full-scale glaciation of the northern hemisphere did not occur until about 2.4 Ma (Raymo et al. 1989). Once full-scale northern-hemisphere glaciation commenced, it may well have been some time before the effects of Alaskan glaciation were felt out into the Aleutian Basin, some 1500 km distant from the Alaska Range and more than 400 km away from the Kuskokwim and Yukon Rivers. It appears to us more likely that the youth of the sedimentation events surrounding the Aleutian Basin is directly related to diapirism on Umnak Plateau and to neotectonics that have occurred, and are occurring, along the Aleutian Island Arc and the Beringian margin.

Figure 17–15. A, GLORIA mosaic of west edge of Umnak Plateau.

Origin of the Beringian canyons

The origin and early history of the Beringian canyons are not completely understood. Because of the area's former history as a convergent margin, the compositions and structures

of the slope may have contributed to its relative weakness to later erosional processes. Two major events pervasively influenced the evolution of the canyons. The first event was the shift of plate convergence from the Beringian margin to the Aleutian Trench, and the second event was Cenozoic

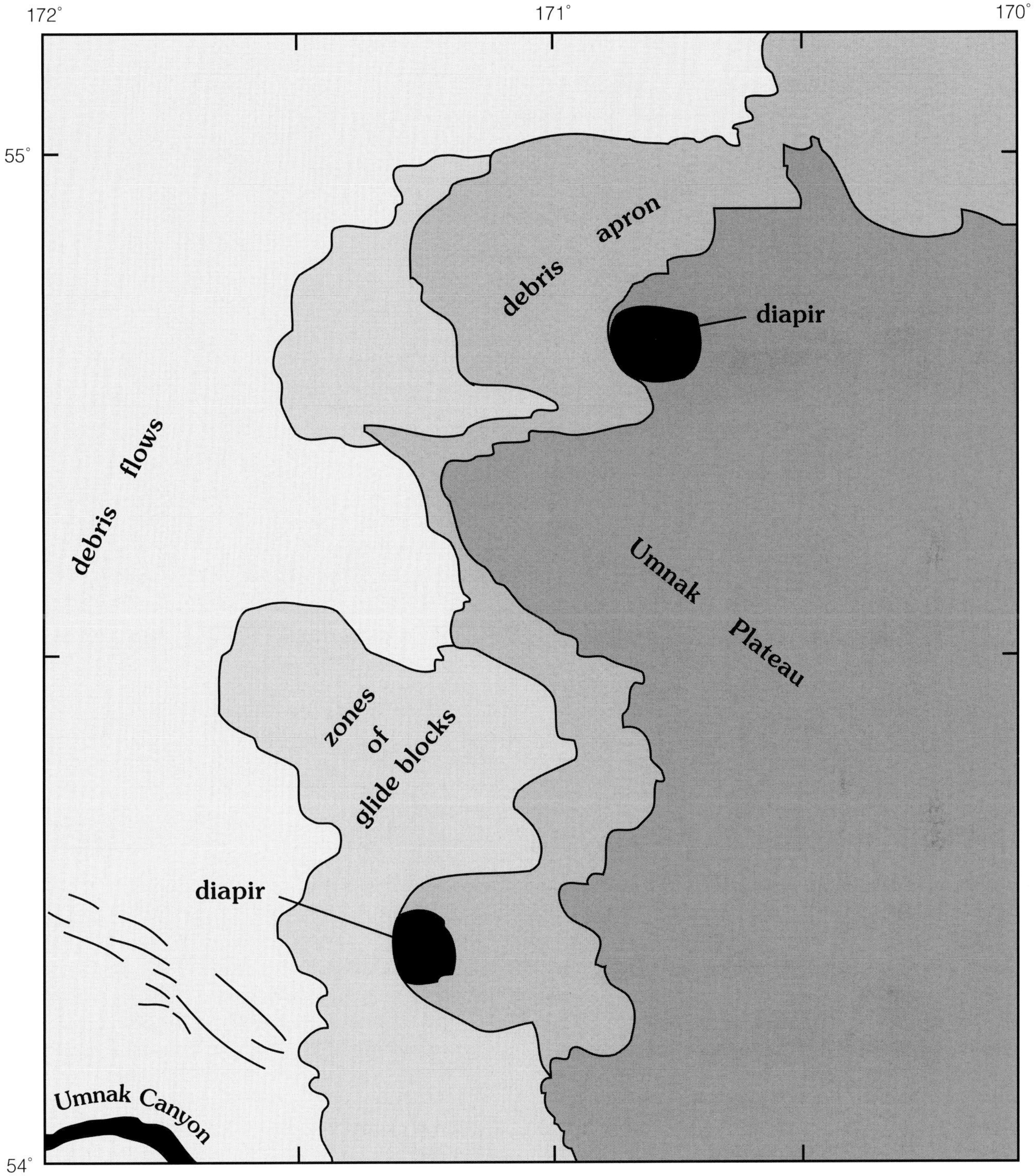

Figure 17–15. *Continued*x. B, Interpretation of extensive failure zone and the proximal debris flows from Umnak Plateau.

glaciation. The Beringian margin is the site of a former sub- duction zone where the Pacific (Kula) Plate was subducted beneath the North American Plate (Scholl, Buffington, and Marlow 1975). The Aleutian Island Arc began to form prob- ably during the early Tertiary as subduction shifted from the Beringian margin to the Aleutian Trench (Scholl et al. 1975).

The evolution of the canyons, particularly in the Ceno- zoic, exhibits intermittent stages of channel development and abandonment (Carlson and Karl 1988). Seismic- reflection profiles show many buried canyons and chan- nels. Buried channels, as wide as 15 km and filled with as much as 400 m of sediment, cross the continental rise ad-

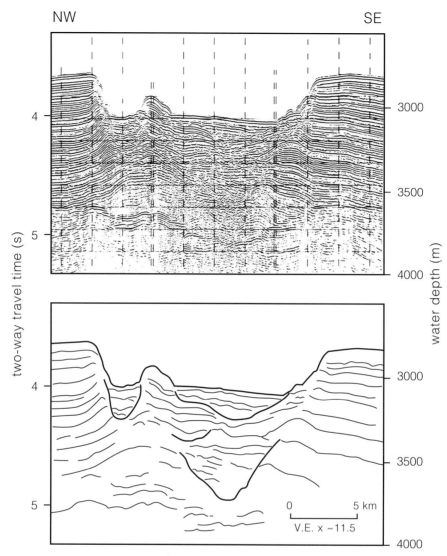

Figure 17–16. Seismic profile showing three sequences of cutting and filling across lower Navarin Canyon distributary system.

jacent to the mouths of several of the large canyons (Figure 17–16). Some active channels on the continental rise are of comparable size. Levees with relief of several tens of meters are associated with some of these channels. Flat-lying reflectors, probably the result of overbank deposition from turbidity currents, distinguish the interchannel areas.

Shape and size are the two parameters that distinguish the Beringian canyons from other well-known canyons (such as Monterey, Hudson, etc.). In fact, several of the Beringian canyons have shapes and sizes that more closely resemble large embayments in the margin that contain two or more canyons. The four morphological categories into which the seven canyons are grouped are a function of structure and basement rock type. Previous investigators have demonstrated that the shape, size, and location of the canyons on the margin are principally controlled by tec-

tonic structures (Scholl et al. 1970; Carlson and Karl 1988; Carlson et al. 1991). Profiles across the canyons show them bounded by and associated with faults (Scholl et al. 1970; Carlson and Karl 1988). Navarin, Zhemchug, and Pribilof Basins are large structural basins that underlie Navarin, Pervenets, Zhemchug, and Pribilof Canyons. Based on estimates of slumps and slides on seismic-reflection profiles and GLORIA imagery, over seventy percent of the Beringian margin is being eroded by mass movement mechanisms. Seismic-reflection profiles, GLORIA imagery, and sediment-core data demonstrate that these canyons have been excavated and developed primarily by processes of mass movement that include mud flows, debris flows, slumps, and massive slides (Carlson and Karl 1988; Carlson et al. 1991). Figures 17–17, 17–18, and 17–19 illustrate examples of these various mass movement products.

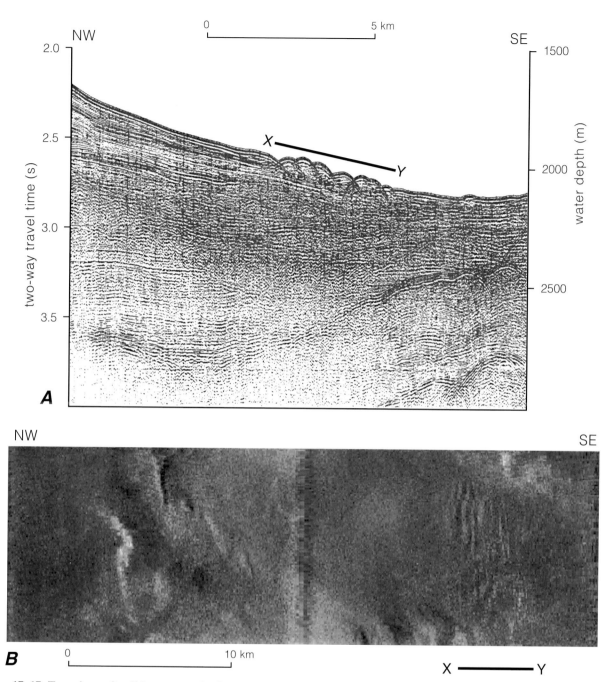

Figure 17–17. Two views of a slide mass on the floor of Navarin Canyon. A, Seismic-reflection profile showing hummocky sediment comprising the toe of the slide. B, GLORIA image of the same slide. The width of the slide, about 10 km.

ZHEMCHUG AND PRIBILOF CANYONS

Because Zhemchug Canyon is the largest, and in our opinion, the most representative of development by mass wasting processes of the Beringian canyons, we first discuss it and its morphologic companion, Pribilof Canyon. Zhemchug and Pribilof Canyons represent a mature stage of canyon development. Scholl et al. (1970) identified a series of faults on the outer shelf that parallel the shelf edge. Zhemchug Basin occupies the graben between the two outermost

faults (Scholl and Hopkins 1969). The structural geometry of Pribilof Basin, which underlies Pribilof Canyon, is similar to that of Zhemchug Basin (Marlow et al. 1976); consequently, the developmental history of Zhemchug Canyon serves as a model for both canyons. Zhemchug Canyon, which makes the largest incision in the Beringian margin, has breached Zhemchug Basin, which underlies the Bering Shelf. The canyon is eroding into the basin fill, and the shape of the basins and bounding faults (Scholl et al. 1970) con-

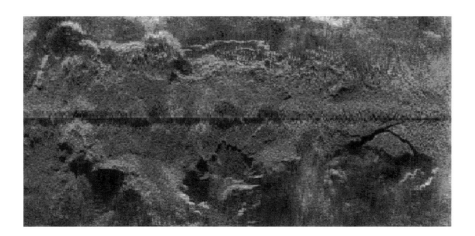

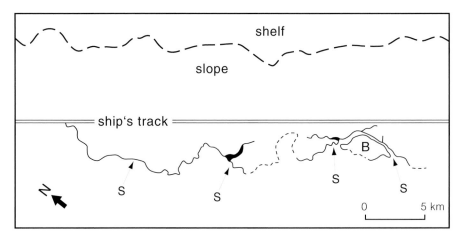

Figure 17–18. GLORIA image and interpretation of upper slope between Navarin and Pervenets Canyons that shows a 4-km-long, 2-km-wide block *(B)* **pulling away from long, irregular slump scarps** *(S)*.

trol the configuration of the developing canyon heads (Figure 17–20).

We speculate that Zhemchug Canyon was formed by headward erosion, principally by removal of large block slides and also to a lesser degree by other canyon-erosion processes such as turbidity currents, until the outermost normal fault was breached and the canyon began to erode the structural basin. A large sediment mass at the base-of-slope near Zhemchug Canyon may have been one of the large blocks that was dislodged from the margin and slide downslope during the initial stage of canyon development (Figures 17–19 and 17–20). After the canyon breached the outer bounding fault, sediment in the structural basin between the basin-bounding faults was preferentially eroded to the northwest and southeast parallel to the faults producing the winged-shape Zhemchug Canyon. As masses of sediment became detached from the walls of the canyon, the resulting blocks, slumps, and slides clogged major sediment dispersal pathways in the growing gorge and formed a labyrinth of channels. Subsequent mass movement on a smaller scale

and other secondary processes have produced the intensely gullied walls evident on GLORIA imagery (Figure 17–21).

NAVARIN AND PERVENETS CANYONS

Navarin and Pervenets Canyons are, perhaps, a less mature stage of canyon development than Zhemchug and Pribilof Canyons or are eroding into a different structural and lithologic framework. Both Navarin and Pervenets Canyons are underlaid by Navarin Basin, but neither of these canyons has yet breached the basin. A broad wave-cut platform underlies the head of Navarin Canyon (Marlow et al. 1982). This terrace, and possibly a similar surface under Pervenets Canyon, has influenced the amphitheater shape of these canyons. When Navarin and Pervenets Canyons eventually erode through this platform into the basin fill, they may develop shapes similar to Zhemchug and Pribilof Canyons.

ST. MATTHEW AND MIDDLE CANYONS

The reason for the location of St. Matthew and Middle Canyons is not as clear as that of the other five. The west-

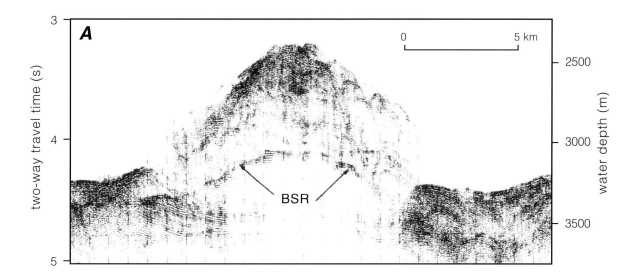

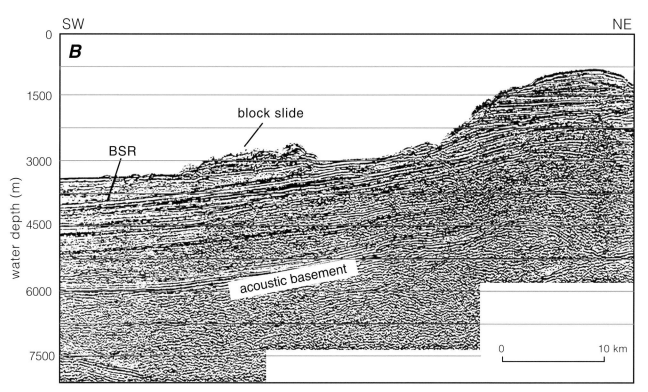

Figure 17–19. A, Seismic-reflection profile of a large mound of debris near the mouth of Zhemchug Canyon. B, Seismic-reflection profile of a large slide block at the base of Middle Canyon.

ern branch of St. Matthew Canyon is fault controlled, and the location of the other major channels of St. Matthew and Middle Canyons may be influenced by outcrops of basalt. Both canyons have developed in rock consisting of burrowed, moderately indurated, uniformly hard mudstone and basalt (Carlson and Karl 1984). Erosion of these canyons is predominantly headward with very little lateral cutting. Perhaps the reason that St. Matthew and Middle Canyons have not developed well-defined depressions on the outer shelf is that they represent an early stage of canyon development by mass movement processes or, alternatively, headward erosion is restricted by basaltic ridges on the upper slope (Carlson and Karl 1988).

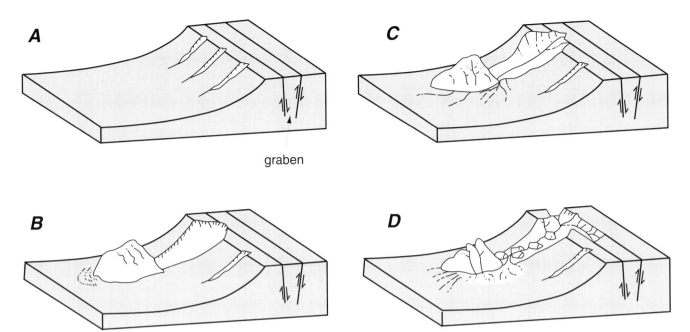

Figure 17–20. Model of the development of Zhemchug Canyon. A, Gullies erode into the upper slope. Outer shelf is cut by normal faults that parallel the shelf edge, and a graben formed. B, Large blocks detached from the upper slope initiating breaching of the shelf-slope break. C, Continued mass movement resulted in headward and lateral canyon excavation of the continental margin. Here, the headward growth of the canyon reached the shelf-edge faults. D, Excavation of the canyon progressed laterally along the fault scarps, and the canyon developed its present-day morphology.

BERING CANYON

The origin and location of Bering Canyon, a classic V-shaped canyon, is probably related to and controlled by faulting (Scholl et al. 1970). The location and course of the canyon are further constrained by Umnak Plateau and volcanic Bogoslov Island (Carlson and Karl 1988). Slides and slumps are clearly evident within Bering Canyon. The V-shape of Bering Canyon suggests that turbidity currents have been a relatively more important mechanism for its development than for the other Beringian canyons where mass wasting was and is the dominant mechanism. Like St. Matthew and Middle Canyons, erosion of Bering Canyon is principally headward with little lateral cutting.

Influence of canyons on basin sedimentation

Although a thick wedge of sediment exists at the base of the continental slope of the Beringian margin (Cooper et al. 1986), we have not recognized any classic submarine fans. One would expect large, discrete fans at the base of the enormous Beringian canyons, but, as previously discussed, only Bering Canyon has a modern fan associated with it, and Bering Fan is very different morphologically from other well-studied fans. Why is a recognizable fan associated only with Bering Canyon and none of the other six canyons?

During the low stands of sea level throughout the late Neogene and Quaternary, large amounts of sediment eroded from the highlands and volcanic mountains of the Alaskan

Peninsula and Aleutian Arc immediately adjacent to Bering Canyon would have been funneled into and down the canyon to be deposited in the Aleutian Basin (Gardner, Dean, and Vallier 1980) and form the Bering Fan. Although no large buried channels have been identified on the continental shelf extending from the Alaskan mainland to the head of Bering Canyon, it is possible that sediment from the mainland also reached the canyon through drainage systems developed on the exposed shelf. It is unknown if any of the Beringian canyons are active at the present time.

The sedimentation history of the other six Beringian canyons is very different from that of Bering Canyon. We offer three alternatives as to why no fans are associated with these canyons. The simplest explanation is that not enough sediment was delivered to the canyons to produce fans and that the thick accumulations of sediment at the base of the slope (as thick as 11 km) and in the basin (as thick as 2 km) are primarily the result of mass wasting processes and pelagic and hemipelagic sedimentation.

A more complex variation of the first explanation is that even during the lowest stands of eustatic sea level, all the canyons except Bering Canyon were never close enough to a highland to have received large volumes of sediment. When eustatic sea level reached a minimum, the heads of Navarin, Pervenets, Zhemchug, and Pribilof Canyons would have been large shallow embayments along the glacial-age coastline. As for Bering Canyon, no large buried channels have been identified extending from the heads of these

Figure 17–21. GLORIA mosaic of Zhemchug Canyon illustrating the intense gullying that characterizes the present-day canyon.

canyons to either the Alaskan or Siberian mainland. However, an extensive system of well-developed small channels has been mapped from very high resolution seismic-reflection profiles near the head of Navarin Canyon (Figure 17–22A; Carlson and Karl 1988). Small channels in the subsurface also are seen on high-resolution profiles collected on the shelf adjacent to Pervenets and Zhemchug Canyons (Figure 17–22B). This finding suggests that a system of small periglacial streams crossed the emergent shelf and carried sediment to the glacial-age coastline (Karl and Carlson 1982; Karl, Cacchione, and Carlson 1986; Carlson and Karl 1988). If the canyon morphology at that time was similar to what it is today, the canyon heads would

have been large bays. These bays could have been sinks for much of the sediment transported to the shore by the periglacial streams. Accumulations of cross-stratified sediment deposits as thick as 120 m found in the head of Navarin Canyon support this hypothesis (Karl et al. 1986). The quantity of debris reaching the canyon heads would have been greatly reduced and the rate and intensity of turbidity-current generation associated with canyon development would have been greatly mitigated, thereby reducing the amount of material delivered to the basin floor. Instead of sediment more or less constantly being funneled down the canyons, large deposits of sediment probably accumulated in the bays. These accumulations periodically became

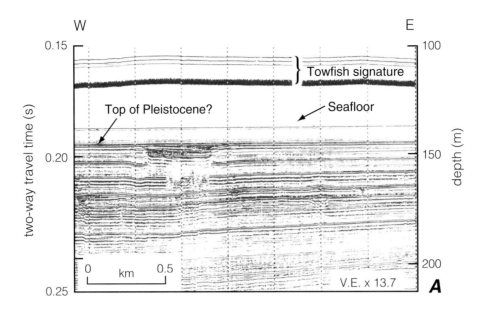

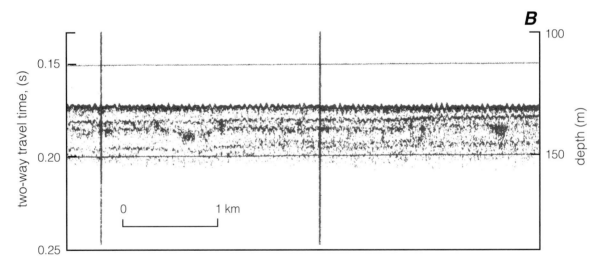

Figure 17–22. A, High-resolution seismic-reflection profile (Huntec deep-tow system, 0.5 to 10.0 kHz) showing small buried channel near the head of Navarin Canyon. This channel and the smaller channels on the profile are examples of an extensive system of channels identified on the Huntec records. B, High-resolution seismic-reflection profile (3.5 kHz) showing small buried channels on the shelf at the head of Zhemchug Canyon. This profile subparallels the shelf edge and extends from the head of Navarin Canyon to Zhemchug Canyon. Small channels have been identified along the entire profile.

unstable and moved downslope during mass transport events as slumps and slides.

The material filled the paleo-trench at the base of the slope and helped to produce the base-of-slope deposits. Although turbidity currents probably helped erode the canyon, the principal agent of canyon development was mass transport of sediment. This scenario is analogous to the Holocene transgression whereby many present-day shelf-edge canyons associated with rivers (e.g., Columbia River/Astoria Canyon; Hudson River/Hudson Canyon) were cut off from their principal source of sediment and isolated the canyons from

the shoreline because of the rise of sea level. Coarse sediment carried by these rivers is presently trapped in estuaries and prevented from being directly funneled into the canyons. The decrease in sediment supply slowed the development and growth of many canyon/fan systems during eustatic high stands.

A third possible explanation is that because mass wasting was so pervasive along the Beringian margin, formation of submarine fans was inhibited, overwhelmed, disrupted, or masked by the processes and products of mass movement of sediment. That is, fan morphology in the classic sense may be masked by the large slump and slide blocks at the base of the canyons and slope. If discrete classical fans did form when terrigenous debris was carried to the shelf edge during the numerous eustatic fluctuations of sea level since the Pliocene, these fans are now buried within the base-of-slope sediment. Whereas we have not been able to recognize complete and unequivocal fan complexes, seismic-reflection profiles illustrated in Carlson and Karl (1988) show evidence of buried channel-levee deposits that may represent the remnants of classic submarine fans. If fans morphologically similar to the modern Bering Fan formed at the base of the slope, then it is possible that such a non-classical fan facies would not be recognizable in the subsurface because we cannot distinguish the facies of Bering Fan from debris flow facies and basin facies on seismic profiles. Bering Fan does not exhibit the classic morphology of well-studied fans such as Astoria, Navy, and Monterey Fans and is principally resolved by backscatter differences on the GLORIA imagery (in fact, it is doubtful that, without backscatter data, Bering Fan even would have been discovered). Even though we speculate that, because of its proximity to highlands, Bering Canyon was a conduit for enormous amounts of debris during the late Pleistocene and early Holocene, no large, discrete, wedge-shaped deposit of sediment (the classic fan) has developed at its mouth on the basin floor.

Summary of styles of sedimentation and canyon development

The morphological development and depositional history of the Aleutian Basin have been controlled primarily by three styles of despositional processes: (1) mass wasting processes associated with growth and development of submarine canyons, (2) sediment gravity-transport processes responsible for debris flow and fan deposits on the basin floor, and (3) pelagic and hemipelagic processes that contribute to continuous sediment accumulation in the basin. These depositional processes have produced three recognizable depositional facies (basin plain, debris flow, and submarine fan) in the basin and sculpted the basin margins.

Mass wasting processes are clearly the dominant style of sedimentation along the continental margins (i.e., slope and rise) of the Aleutian Arc and the Beringian continental shelf. These processes not only distribute sediment to the basin floor but also are the principal control on the morphology of the basin margins. The gigantic canyons (Navarin, Pervenets, St. Matthew, Middle, Zhemchug, Pribilof, and Bering) of the Beringian margin primarily are the result of mass wasting (the V-shaped Bering Canyon less so than the others). Although the enormous solitary sedimentary blocks that rest on the basin floor at the mouths of several of the Beringian canyons (Figure 17–19) suggest the dominant mechanism for excavating the canyons has been large, in-

frequent slumps and slides, it is more probable that multiple relatively frequent retrograde failures quantitatively are more important for canyon development.

The next most important style of sedimentation is sediment gravity flow, which includes debris flows, mud flows, and turbidity currents. We have identified five major debris flow regions (Umnak, Pochnoi, Bowers, Shirshov, and Bering margin) in the Aleutian Basin (Figure 17–11). Bering Fan Province is the result of sediment gravity flows generated by the large quantity of sediment shed from the Aleutian Arc and Alaska Peninsula that was funneled into the Bering Canyon during eustatic low stands. Bering Fan does not resemble the traditional wedge-shaped submarine fan composed of an amalgamation of depositional lobes. Instead, based on the available acoustic data, it appears most like a single depositional lobe.

The basin-plain province, although not as impressive a style of sedimentation as the styles of mass wasting, is areally extensive and volumetrically important. This province is composed of a complex stratigraphy of hemipelagic turbidites interbedded with pelagic diatom ooze, ice-rafted debris, and volcanic ash.

Tectonic activity is probably the principal mechanism that generated the mass movement of sediment along the margins of the Aleutian Basin. Fluctuating eustatic sea level was not likely a dominant factor in triggering mass transport events, although the basin-plain province may have been affected by changes in eustatic sea level.

Acknowledgments

We appreciate the review of a draft of this chapter by C. Hans Nelson, T. L. Vallier, and an anonymous reviewer. Their comments improved the clarity and organization of the final manuscript.

References

Abers, G. A., Eckstrom, Marlow, G., and Geist, E. L. 1993. Bering Sea earthquake of February 21, 1991: Active faulting along the Bering shelf edge. *J. Geophys. Res.* 98(B2): 2155–65.

Bering Sea EEZ-SCAN Scientific Staff 1991. Atlas of the U.S. Exclusive Economic Zone, Bering Sea. *U.S. Geological Survey Miscellaneous Investigations Series I-2053*, 152 p. Reston, Va.: USGS.

Bouma, A. H., Normark, W. R., and Barnes, N. E., eds. 1985. *Submarine Fans and Related Turbidite Systems*. New York: Springer-Verlag, 351. p.

Carlson, P. R., and Karl, H. A. 1984. Discovery of two new large submarine canyons in the Bering Sea. *Mar. Geol.* 56: 159–79.

Carlson, P. R., and Karl, H. A. 1988. Development of large submarine canyons in the Bering Sea, indicated by morphologic, seismic, and sedimentologic characteristics. *Geol. Soc. Am. Bull.* 100: 1594–615.

Carlson, P. R., Karl, H. A., and Edwards, B. D. 1991. Mass sediment failure and transport features revealed by acoustic techniques, Beringian margin, Bering Sea, Alaska. *Mar. Geotechnol.* 10: 35–51.

Cooper, A. K., Stevenson, A., Kenyon, N., and Bishop, D. 1986. GLORIA study of the Exclusive Economic Zone off Alaska – southern Bering Sea: Initial report for cruise F4-86-BS. *U.S. Geological Survey Open-File Report 86-596.* Reston, Va.: USGS.

Creager, J. S., Scholl, D. W., et al. 1973. Initial Reports of the Deep Sea Drilling Project, Vol. 19, Washington, D.C.: U.S. Govt. Printing Office, 913 p.

Damuth, J. E. 1980. Use of high-frequency (3.5-12 kHz) echograms in the study of near-bottom sedimentation processes in the deep sea: A review. *Mar. Geol.* 38: 51–75.

Damuth, J. E., and Flood, R. D. 1985. Amazon Fan, Atlantic Ocean. In *Submarine fans and related turbidite systems.* eds. A. H. Bouma, W. R. Normark, and N. E. Barnes, pp. 97–106. New York: Springer-Verlag.

EEZ-SCAN 85 Scientific Staff 1987. Atlas of the U.S. Exclusive Economic Zone, Eastern Caribbean area. *U.S. Geological Survey Miscellaneous Investigations Series I-1864-B.* 58 p. Reston, Va.: USGS.

EEZ-SCAN 84 Scientific Staff 1988. Physiography of the western United States Exclusive Economic Zone. *Geology* 16: 131–4.

Gardner, J. V., Dean, W. E., and Vallier, T. V. 1980. Sedimentology and geochemistry of surface sediments, outer continental shelf, southern Bering Sea. *Mar. Geol.* 35: 299–329.

Gardner, J. V., Dean, W. E., Klise, D. H., and Baldauf, J. G. 1982. A climate-related oxidizing event in deep-sea sediment from the Bering Sea. *Quat. Res.* 18: 91–107.

Gardner, J. V., Field, M. E., Lee, H., Edwards, B. E., Masson, D. G., and Kenyon, N. 1991. Ground-truthing 6.5-kHz sidescan sonographs: What are we really imaging? *J. Geophys. Res.* 96: 5955–74.

Hopkins, D. M. 1979. Landscape and Climate in Beringia during late Pleistocene and Holocene time. In *The First Americans: Origins of Affinities and Adaptions,* eds. W. S. Laughlin and A. B. Harper, pp. 15–41. New York: Gustav Fisher.

Horn, D. R., Ewing, J. I., and Ewing, M. 1972. Graded-bed sequences emplaced by turbidity currents north of 20°N in the Pacific, Atlantic, and Mediterranean. *Sedimentology* 18: 247–75.

Huggett, Q. J., Cooper, A. K., Somers, M. L. and Stubbs, A. R. 1992. Interference fringes on GLORIA side-scan sonar images from the Bering Sea and their implications. *Mar. Geophys. Res.,* 14: 47–63.

Karl, H. A., and Carlson, P. R. 1982. Large sand waves in Navarin Canyon head, Bering Sea. *Geo-Mar. Ltrs.* 2: 157–62.

Karl, H. A., Cacchione, D. A., and Carlson, P. R. 1986. Internal-wave currents as a mechanism to account for large sand waves in Navarin Canyon head, Bering Sea, *J. Sed. Petrol.* 56: 706–14.

Lagoe, M. B., Eyles, C. H., Eyles, N., and Hale, C. 1993. Timing of Late Cenozoic tidewater glaciation in the far North Pacific. *Geol. Soc. Am. Bull.* 105: 1542–60.

Lyons, A. P., Anderson, A. L., and Dwan, F. S. 1994. Acoustic scattering from the seafloor: Modeling and data comparisons. *J. Acoust. Soc. Am.* 95: 2441–51.

Marlow, M. S., Scholl, D. S., Cooper, A. K., and Buffington, E. C. 1976. Structure and evolution of Bering Sea shelf south of St. Lawrence Island. *Bull. Am. Assoc. Petrol. Geol.* 60: j161–83.

Milliman, J., and Meade, R. H. 1983. World-wide delivery of river sediment to the oceans. *J. Geol* 9: 1–21.

Mutti, E., and Normark, W. R. 1987. Comparing examples of modern and ancient turbidite systems: Problems and concepts. In *Marine Clastic Sedimentology,* eds. J. K. Leggett and G. G. Zuffa, pp. 1–38. London: Graham and Trotman.

Normark, W. R. 1970. Growth patterns of deep-sea fans. *Bull. Am. Assoc. Petrol. Geol.* 54: 2170–95.

Raymo, M. E., Ruddiman, W. F., Backman, J., Clement, B. M., and Martinson, D. G. 1989. Late Pliocene variation in northern hemisphere ice sheets and North Atlantic deep water circulation. *Paleoceanography* 4: 413–46.

Sancetta, C., and Robinson, S. W. 1983. Diatom evidence on Wisconsin and Holocene events in the Bering Sea. *Quat. Res.* 20: 232–45.

Scholl, D. W., and Hopkins, D. M. 1969. Newly discovered Cenozoic Basins, Bering Sea shelf, Alaska. *Bull Am. Assoc. Petrol. Geol.* 53: 2067–78.

Scholl, D. W., Buffington, E. C., Hopkins, D. M., and Alpha, T. R. 1970. The structure and origin of the large submarine canyons of the Bering Sea. *Mar. Geol.* 8: 187–210.

Scholl, D. W., Buffington, E. C., and Marlow, M. S. 1975. Plate tectonics and the structural evolution of the Aleutian-Bering Sea region. In *Contributions to the Geology of the Bering Sea Basin and Adjacent Regions,* ed. R. B. Forbes, pp. 1–32. Geological Society of America Special Paper 151. Boulder, Colo.: GSA.

Sheriff, R. E. 1976. *Encyclopedic Dictionary of Exploration Geophysics.* Tulsa, Okla.: Society of Exploration Geophysicists, 266 p.

Twichell, D. C., Kenyon, N. H., Parson, L. M., and McGregor, B. A. 1991. Depositional patterns of the Mississippi Fan surface: Evidence from GLORIA II and high-resolution seismic profiles. In *Seismic Facies and Sedimentary Processes of Submarine Fans and Turbidite Systems,* eds. A. Weimer and M. H. Link, pp. 349–63. New York: Springer-Verlag.

Urick, R. J. 1983. *Principles of Underwater Sound,* Ed. 3. New York: McGraw-Hill, 423 p.

U.S. Coast and Geodetic Survey 1970. *Seismicity of Alaska – NEIC Map 3011.* Washington, D.C., 1 sheet.

18 Geology of the Kula Paleo-Plate, North Pacific Ocean

Tracy L. Vallier,[1] Carlos A. Mortera-Gutierrez,[2] Herman A. Karl,[1]
Douglas G. Masson,[3] Libby Prueher,[4] and Thomas E. Chase[1]

[1]U.S. Geological Survey, Menlo Park, California
[2]Texas A&M University, College Station, Texas
[3]Institute of Oceanographic Sciences, Southampton, United Kingdom
[4]University of Michigan, Ann Arbor, Michigan

Abstract

The only recognized fragment of the Kula Plate occurs south
of the western Aleutian Islands in the North Pacific Ocean.
The paleo-plate remnant is now trapped within the Pacific
Plate and bounded by Stalemate Ridge on the south and
west, by the inner wall of the Aleutian Trench on the north,
and by the Kula Rift (central graben of the Kula-Pacific
Ridge, an extinct spreading center), on the east. The paleo-
plate, presently about 7,600 km^2 in area, is sliding past the
arc at about 80 km/my and is currently being deformed; subduction is occurring, at least as far west as 169° E longitude. In the region where the Kula Paleo-Plate and the Stalemate Ridge are colliding with the Aleutian Ridge (KUSAR
collision zone), the trench floor is raised about 1,400 m and
the fore-arc region of the Aleutian Ridge is greatly deformed. The collision left a tectonic indentation (Attu Bight)
as the bathymetrically high Stalemate Ridge scraped along
the inner trench wall during its westward migration on the
Pacific Plate.

The presence of magnetic anomalies 18R to 21N indicates
a maximum age range for the Kula Paleo-Plate of about 47.5
to 40.5 Ma. Seafloor spreading rates of the Kula Paleo-Plate
were slower than rates on the Pacific Plate during final stages
of spreading along the Kula-Pacific Ridge. Cessation of
spreading apparently occurred after the plate motion change
recorded by the Hawaii-Emperor Bend.

The oceanic plate between 169° E and 174° E longitudes,
which includes the Kula Paleo-Plate, is tectonically fragmented and seismically active as it obliquely converges and
underthrusts the Aleutian fore-arc region. West of 174° E,
fault patterns on the outer trench high form acute angles with
the trench axis. Fault trends compared to the trench axis in
the central part are nearly 6° greater than faults to the east.
Further west (at about 171° E to 172° E), two fault trends
are evident: one fault trend pattern forms highly acute angles (about 15°) with the trench axis and other faults strike
nearly parallel to the trench. Near Stalemate Ridge, fault angles are nearly parallel to the trench axis. The two dominant
fault trends in the outer trench slope and their apparent rotation as the oceanic plate migrates westward stimulate questions about the nature of brittle deformation in this highly
oblique convergent plate boundary. We propose two possible reasons for the unique fault orientations in this region.
The first possible reason is closely related to the tectonic activity within the region itself. In this scenario, faulting of
the oceanic plate near the trench initially resulted from plate
bending related to the subduction process. Subsequent
changes in fault trends would thereby be related to later rotation of the stress field in a highly oblique convergent
regime as the plate moved westward, which would be a re-
orientation of an extensional stress direction due to right-
lateral shear. Alternatively, the faults may have rotated since
they initially formed farther east where convergence is less
oblique; subsequently, they were transported westward
where northwest-directed shear became the dominant stress
regime.

Relatively large and shallow earthquakes that have occurred in the region between Stalemate Ridge and the Aleutian Ridge, and their focal mechanisms, indicate that the upper structures of the oceanic plate near the trench and within
the outer fore-arc region are in stress regimes that result from
strong coupling caused by the extreme oblique convergence.
The overall focal mechanisms for earthquakes within the
oceanic plate near the trench (dominant extensional-stress
regime) differ from the overall mechanisms in the fore-arc
region (dominant compressional-stress regime). First-
motion focal mechanisms of earthquakes in the oceanic plate
indicate predominantly normal faults with a very minor
strike-slip component. In the Aleutian Ridge, however, first-
motion focal mechanisms show that thrust faults are dominant. Their solutions have larger strike-slip components than
the earthquakes in the oceanic plate. The amounts of shallow seismic energy released in both plates are approximately
the same.

The Kula Paleo-Plate remnant will continue to be subducted and carried westward if the relative plate motions do

333

not change significantly. As it migrates westward, the paleo-plate may be entirely consumed beneath the Aleutian Ridge. Alternatively, it may collide with the Kamchatka Peninsula in 11 to 12 million years.

Introduction

GLORIA (Geological LOng-Range Inclined Asdic) sonar images of the northern Pacific seafloor, combined with seismic-reflection, gravity, bathymetric, and magnetic data, were used to map the Kula Paleo-Plate. During late Cretaceous time, the Kula Plate probably formed most of the northern Pacific Ocean floor. Pitman and Hayes (1968) first recognized northward-younging E-W magnetic anomalies in the North Pacific Ocean. Their studies implied that a former oceanic plate had been subducted. Grow and Atwater (1970) named this former plate Kula, meaning "all gone" in an Athabaskan Indian dialect. However, the Kula Plate is not "all gone." A fragment trapped within the Pacific Plate remains today south of the Near Islands segment of the Aleutian Ridge (Lonsdale 1988). Present-day Pacific Plate motion indicates that the Kula Paleo-Plate is obliquely converging with, and being subducted beneath, the North American Plate.

This chapter describes the remnant of the Kula Paleo-Plate and the surrounding area in the northwest Pacific Ocean Basin (Figure 18–1). The recent collection (Karl et al. 1987) of GLORIA, seismic, gravity, bathymetric, and magnetic data by the U.S. Geological Survey (USGS) and the U.K. Institute of Oceanographic Sciences (IOS), combined with preexisting seismic, bathymetric, and magnetic data (Nichols and Perry 1966; Buffington 1973; Lonsdale 1988), provide new insights into the geologic history of the region and further evidence for tectonic fragmentation of both the oceanic plate and the fore-arc region of the western Aleutian Ridge. Our study complements the insightful work of Lonsdale (1988), who documented the presence, approximate size, and probable age of the Kula Plate and reconstructed the Paleogene Kula Plate and Pacific Plate motion histories in the North Pacific Ocean. This information in combination with our present knowledge of the Aleutian Ridge (Scholl, Vallier, and Stevenson 1983, 1988; Lonsdale 1988; Stone 1988; Vallier et al. 1994), provides a better understanding of the North Pacific Ocean floor and the Aleutian Island Arc.

We concentrated our studies on the area where the Kula Paleo-Plate and Stalemate Ridge are colliding with the Aleutian Ridge (KUSAR collision zone). This collision, combined with the essentially dextral movement between the oceanic and continental plates, has greatly affected the structural development of the region and the associated sedimentary processes (Dobson, Karl, and Vallier this volume). The present Kula Paleo-Plate (Figure 18–2) is shaped somewhat like a skewed isosceles triangle with its apex near the intersection of Stalemate Ridge and the Aleutian Trench and

its base along the axis of the fossil Kula-Pacific spreading center (Kula-Pacific Ridge). The plate has a maximum 150 km length and 75 km width. Its area is approximately 7,600 km^2 (2,200 nautical mi^2), about 1.5 times larger than the state of Delaware in the United States. We have no knowledge of the maximum size of the original Kula Plate, but plate reconstructions of the North Pacific (Rea and Dixon 1983; Engebretson, Cox, and Gordon 1984, 1985; Rea and Duncan 1986; Lonsdale 1988) suggest that the Kula was a major oceanic plate during the evolution of the late Cretaceous and Paleogene Pacific Ocean floor.

Bathymetry and physiography

The Kula Paleo-Plate remnant is bounded by Stalemate Ridge on the south and west, the inner wall of the Aleutian Trench on the north, and Kula Rift (graben along the axis of the Kula-Pacific Ridge) on the east (Figures 18–2, 18–3, and 18–4). The paleo-plate, exclusive of the raised parts along Stalemate Ridge, ranges in depth from about 4,100 m on a sediment-covered seamount near the Kula Rift to about 7,200 m in the trench.

Stalemate Ridge, a fossil fracture zone (Lonsdale 1988), is the dominant bathymetric feature of the region (Figure 18–3). The ridge rises to a depth of less than 2,600 m just east of its intersection with Kula-Pacific Ridge and to about 2,900 m where the ridge bends northward. The depth to the top of the ridge increases as it approaches the Aleutian Trench where it is subjected to uplift along the outer trench slope. Stalemate Ridge has a maximum local relief of about 1,400 m on the east to 2,000 m on the west (Figure 18–3). A bathymetric low that parallels the north side of the ridge is a probable transform valley, referred to as the Stalemate transform fault by Lonsdale (1988). Because of the probable difference in age of more than 50 million years between the south (and west) side of Stalemate Ridge (mid-Cretaceous, about 100 million years old; Renkin and Sclater 1988) and north (and east) side of Stalemate Ridge (Eocene, about 47 to 41 million years old; Lonsdale 1988, and this study), the depth of igneous crust (basement) is greater on the south. Stalemate Ridge is being transported northwest along the Aleutian fore-arc region at about seventy-eight mm/yr and is colliding obliquely with, rather than being completely subducted beneath, the Aleutian Ridge. The forceful collision of Stalemate Ridge with the Aleutian Ridge not only has deformed the Aleutian Ridge fore-arc region, but also may have contributed to the extreme strike-slip faulting in the western part of the Aleutian Arc that is described by Geist, Scholl, and Vallier (1991, 1994).

At the intersection of Stalemate Ridge and the Aleutian fore-arc region, the flat floor of the Aleutian Trench is raised as much as 1,400 m (Figures 18–3 and 18–5). West of Stalemate Ridge, the Aleutian Trench is present as a bathymetric depression (Figure 18–4). Available earthquake and seismic-reflection data are not sufficient to show whether

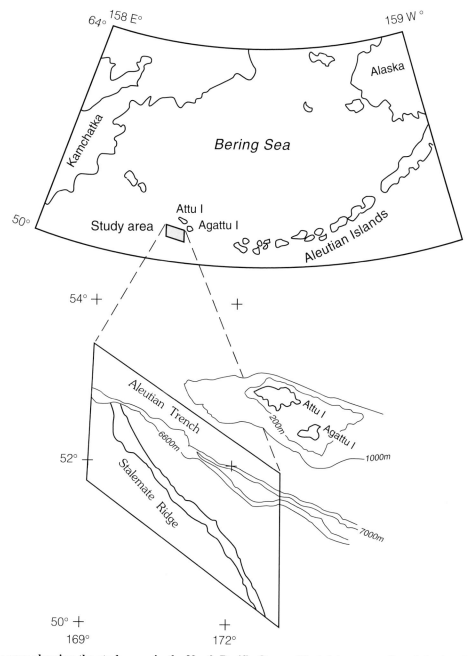

Figure 18–1. Index map showing the study area in the North Pacific Ocean. Most data were collected during the combined U.S. Geological Survey and U.K. Institute of Oceanographic Sciences GLORIA cruise F2-87-AA on the R/V *Farnella* (Karl et al. 1987).

subduction is occurring or if most of the relative motion is taken up along strike-slip faults parallel and subparallel to the trench. In Figure 18–5 thrusting is evident in the seismic reflection profiles that cross Stalemate Ridge perpendicular to the trench, which suggests that subduction is occurring as far west as those profiles. The trench west of Stalemate Ridge, however, may be merely a wide graben that marks a zone of strike-slip faults. However, we suspect that because it would be difficult to maintain a deep trench without subduction, a small amount of subduction is occurring west of the collision point.

East of Stalemate Ridge, the Aleutian Trench floor is more than 7,200 m deep along the length of the Kula Paleo-Plate. In places, faults oriented obliquely to the trench cut the outer wall of the trench and extend partway across the sediment floor (Figures 18–2 and 18–4). A landward indentation (bight) of the inner trench wall is present near the east side of the KUSAR collision zone (Figures 18–3 and 18–4). The indentation probably formed by tectonic erosion that was induced by the interaction of Stalemate Ridge with the inner trench wall, followed by subduction of the eroded fragments. This indentation, informally called Attu Bight, has a length

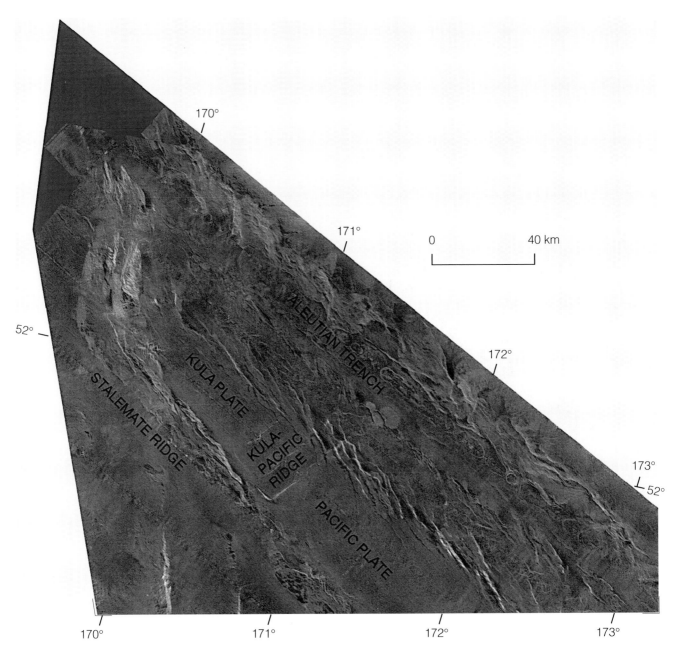

Figure 18–2. GLORIA mosaic of the remnant of the Kula Paleo-Plate and surrounding area. The GLORIA image was processed from data collected on cruise F2-87-AA (Karl et al. 1987). Methods used for constructing the GLORIA image, including data collection and image processing, are described by Somers (this volume) and Chavez et al. (this volume).

of about 40 km and offsets the inner trench wall a maximum of 10 km. The oceanic plate is presently moving northwestward at about seventy-eight km/my, which indicates that Stalemate Ridge interacted with the Aleutian inner trench wall to form the Attu Bight between about 300 and 800 thousand years ago. Bathymetry and interpreted structural features in the proximity of the KUSAR collision zone (Figures 18–2 to 18–5) suggest that another bight may be forming at the present time. Stalemate Ridge probably is too large (and high) for uniform and uncomplicated subduction.

Magnetics

Kula Paleo-Plate magnetic anomalies (Figures 18–4, 18–6, and 18–7) are mapped as 18R through 21N (R, reversed polarity chron; N, normal polarity chron). This interpretation differs from magnetic anomalies 18R through 22N as proposed by Lonsdale (1988). Because of the trackline spacing during our 1987 cruise, we obtained a more comprehensive database that included magnetic profiles in the KUSAR collision zone. We were not able to map anomalies 21R and

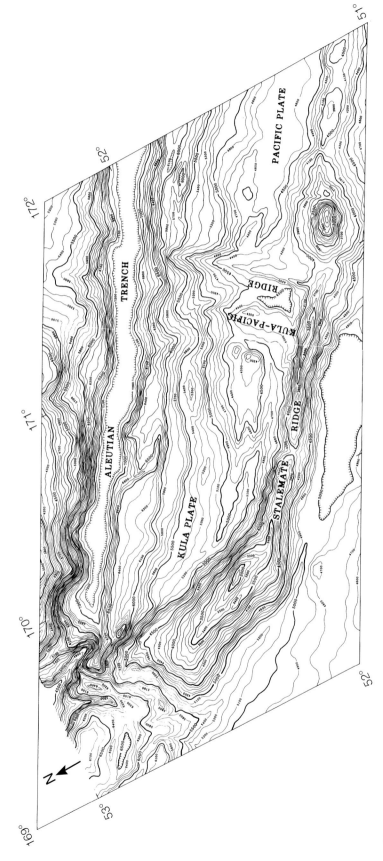

Figure 18–3. Bathymetric map of the Kula Paleo-Plate and surrounding region. Note the seamounts near the intersection of Stalemate Ridge and the Kula Rift.

337

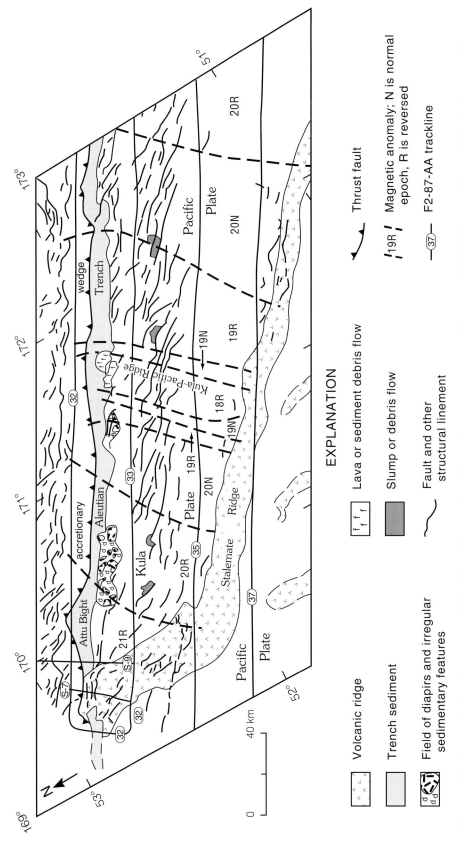

Figure 18–4. Generalized geologic map of Kula Paleo-Plate region. The geology was interpreted from the GLORIA image (Figure 18–2), bathymetry (Figure 18–3), and seismic profiles (tracklines S-7, S-9, 33, and 35).

EXPLANATION

Volcanic ridge

Trench sediment

Field of diapirs and irregular
sedimentary features

Lava or sediment debris flow

Slump or debris flow

Fault and other
structural linement

Thrust fault

19R Magnetic anomaly; N is normal
epoch, R is reversed

37 F2-87-AA trackline

0   40 km

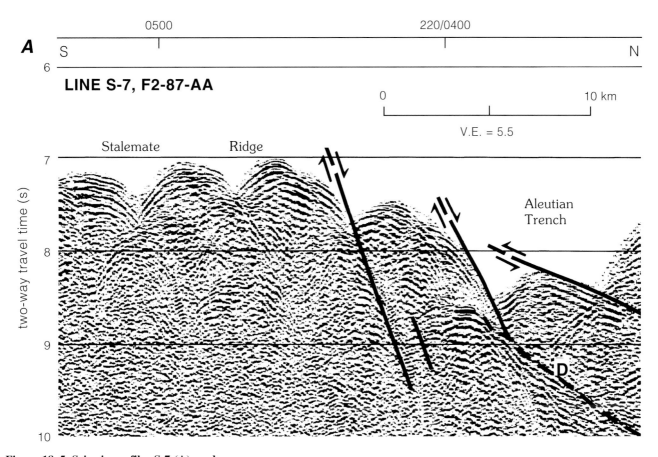

Figure 18–5. Seismic profiles S-7 (A), and

22N near Stalemate Ridge. Anomaly 18R occurs along the Kula-Pacific Rift, and anomaly 21N is partially subducted in the KUSAR collision zone (Figure 18–6). Although some of anomalies 21N and 18R apparently are missing, we estimate that the age interval represented by the remnant Kula Paleo-Plate is about 47.5 to 40.5 million years (geomagnetic polarity time scale of Cande and Kent 1992). The small fragment of Kula Paleo-Plate thereby represents about seven million years of spreading history.

Lonsdale (1988) stated that spreading along the Kula-Pacific Ridge was remarkably asymmetric during the Eocene. We confirmed his interpretation by measuring widths of magnetic anomalies in a direction perpendicular to the Kula-Pacific Ridge and calculating approximate half spreading rates (Figure 18–8) for anomalies 20R (54 mm/yr, Pacific; 21 mm/yr, Kula), 20N (40 mm/yr, Pacific; 30 mm/yr, Kula), 19R (43.5 mm/yr, Pacific; 7 mm/yr, Kula), and 19N (22 mm/yr, Pacific; 18.5 mm/yr, Kula). Spreading on the Pacific Plate limb decreased somewhat regularly in time from 20R through 19N, whereas spreading rates on the Kula Plate limb were more sporadic. An abrupt slowing of Kula Plate spreading occurred during anomaly 19R (7

mm/yr), between about 43 and 42 Ma, but the rate increased again during 19N time (about 18.5 mm/yr). The estimated half spreading rates of both oceanic plate limbs and the rates estimated by Cande and Kent (1992) for CK92 of the North Pacific Basin are compared in Figure 18–8.

Other features stand out on the magnetic map (Figure 18–6). For example, magnetic highs on anomaly 20N mark volcanoes that apparently formed near the intersection of the Kula-Pacific Ridge and the Stalemate transform fault. The growth of three edifices on the east side (Pacific limb) of Kula-Pacific Ridge compared with only two on the west side (Kula limb) probably was influenced by the asymmetry of spreading rates. The fact that the magnetic signatures of these probable volcanoes are the same as that of the adjacent seafloor suggests that the edifices were built during short time intervals within the same magnetic chron in which the adjacent seafloor was formed. The magma would thereby be produced beneath the Kula-Pacific spreading ridge. We suspect that these volcanoes are magmatic products of a leaky transform fault zone (Stalemate) when it was a plate boundary before about 41 Ma.

North-trending magnetic anomalies (Figure 18–6) on the

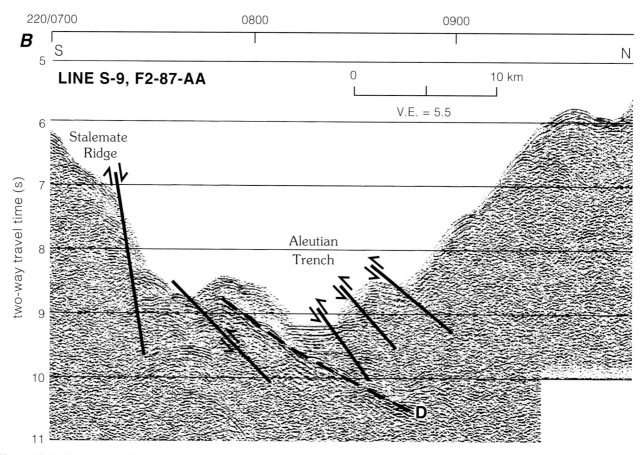

Figure 18–5. *Continued.* **S-9 (B) of cruise F2-87-AA showing the intersection of Stalemate Ridge and the inner wall of the Aleutian Trench. The thick line marked D on profile S-7 shows that Stalemate Ridge is probably being subducted under the Aleutian Ridge. The seismic data was processed by Ray Sliter at the USGS seismic laboratory.**

south side of Stalemate Ridge are associated with bathymetric ridges (Figures 18–2 and 18–3). These ridges parallel several other ridges that occur on the Pacific Plate farther east and south of Stalemate Ridge. It is possible that they formed during the Cretaceous magnetic quiet superchron, but their relative magnetic amplitudes and rugged relief suggest that they formed later, possibly during the latest Cretaceous or early Tertiary. Some may have formed as a response to regional strain related to the spectacular change in spreading direction that occurred in the early Tertiary (Rea and Duncan 1986) particularly during the formation of anomalies 23 and 24 (Lonsdale 1988), about 51 to 55 Ma.

Sedimentary features on the Kula Paleo-Plate

The original acoustic basement relief of the Kula Paleo-Plate is mostly blanketed and smoothed by a thick sediment cover. Seismic profiles along trackline 33 (Figure 18–9A) and trackline 35 (Figure 18–9B) show that two major acoustic

units overlie an irregular igneous acoustic basement. The lower unit along line 35 (Figure 18–9B) is of probable Eocene and Oligocene age (referred to as Kula Paleo-Plate Acoustic Unit 1); it may contain sediments as young as middle Miocene in age. This unit, therefore, would be correlative with the middle and lower units of Scholl et al. (1977), based on their studies of the Mejii sediment tongue and underlying sedimentary acoustic units. It also may be correlated with the lower unit of Eocene to lower Miocene sediments drilled during Leg 145 on Detroit Seamount, just south of our studied area (Scientific Party 1993; Rea et al. 1995). The upper acoustic unit, exclusive of the trench fill, ranges from 0.2 sec. to 1.0 sec. (two-way travel time) in thickness and is of probable Miocene to recent age. Using acoustic rms (root mean square) velocities in the range from 1.6 km/s to 2.1 km/s (equation 8b in Carlson, Gangi, and Snow 1986), the thickness of the upper acoustic unit ranges from about 160 m to 1050 m on the Kula Paleo-Plate. The lower unit has an acoustic thickness that ranges from 0.11 s to 0.51 s (two-way travel time). Assuming acoustic rms velocities from 1.7 km/s to 2.3 km/s for the sediments (equa-

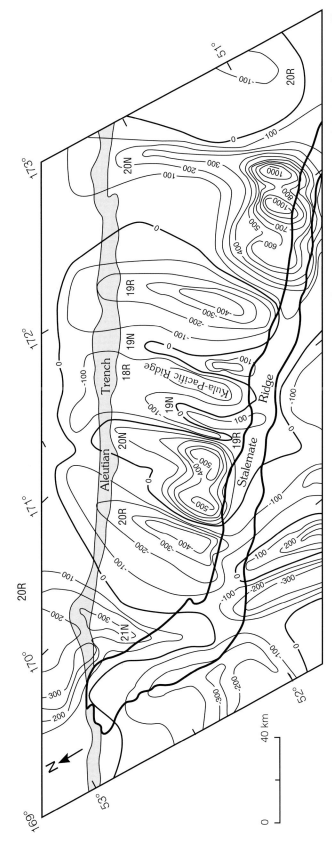

Figure 18–6. Magnetic map of the oceanic plate (including Kula Plate, western part of the Eocene Pacific Plate, and a small segment of the late Cretaceous (?) Pacific Plate south of Stalemate Ridge). The data were contoured at 100 nanoteslas. Notice magnetic field attenuation as the oceanic plate disappears under the Aleutian Ridge north of trench.

341

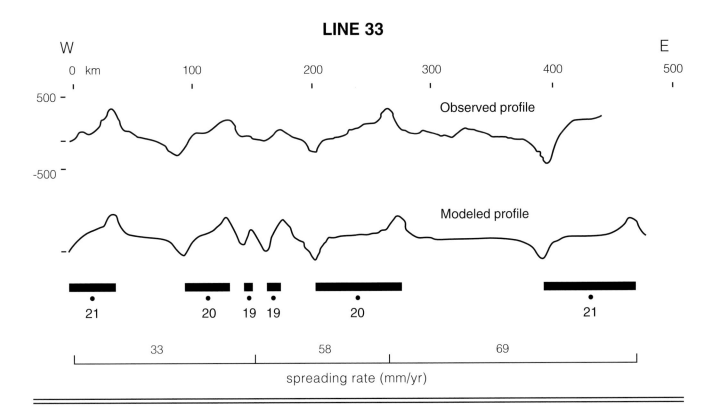

LINE 33

W   E

Observed profile

Modeled profile

21   20   19   19   20   21

33   58   69

spreading rate (mm/yr)

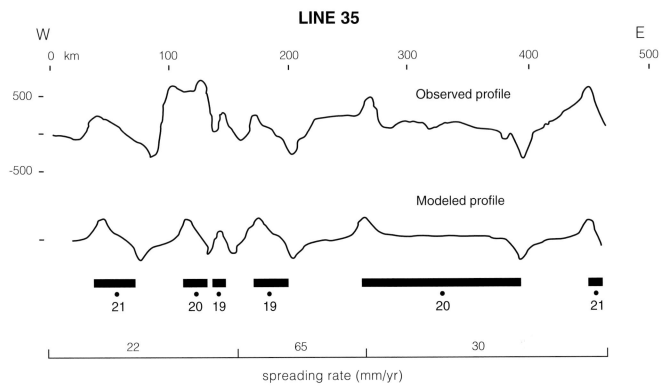

LINE 35

W   E

Observed profile

Modeled profile

21   20 19   19   20   21

22   65   30

spreading rate (mm/yr)

Figure 18–7. Synthetic and real magnetic profiles along lines 33 and 35 using the MACS (Magnetic Anomaly Construction Set) program developed for the Macintosh computer by Eric Rosencrantz of University of Texas, Austin. The following parameters were used in modeling: present inclination, 50°; remnant inclination, 30°; declination, 3°; lineation azimuth, 40°; depth to magnetic layer, 6.5 km; thickness of magnetic layer, 0.5 km; magnetization, 10 A/m; and phase shift of 72.98.

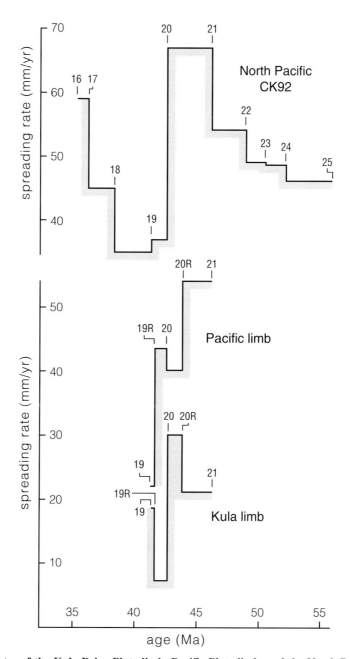

Figure 18–8. Half spreading rates of the Kula Paleo-Plate limb, Pacific Plate limb, and the North Pacific Plate composite (CK92, from Cande and Kent 1992). Distances between magnetic reversals in both limbs were constrained from the contouring in Figure 18–6. Note the large differences in spreading rates between anomaly 19R and 20R for both limbs; the large fluctuations are consistent with fluctuation observed in the CK92 North Pacific Plate composite profile between anomalies 17 and 20.

tion 11 in Carlson et al. 1986), thicknesses range from about 100 to 500 m.

The seismic profile along line 35 (Figure 18–9B) on the Pacific Plate west of Stalemate Ridge shows three acoustic units (rather than the two that occur on the Kula Paleo-Plate) overlying probable late Cretaceous igneous crust. The combined lower two acoustic units (1A and 1B) ranges from 0.10 to 0.39 sec. in two-way travel time (about 85 m to 450

m of sediment thickness using the rms velocities of Carlson et al. 1986) and the upper unit (2) ranges between 0.15 and 1.05 s (about 120 m to nearly 1,100 m of sediment thickness).

The seismic profile along line 33 (Figure 18–9A), which lies nearer the trench, shows the extensive faults within the oceanic plate. Faults have formed grabens and half grabens between uplifted blocks; near the trench, turbidites trans-

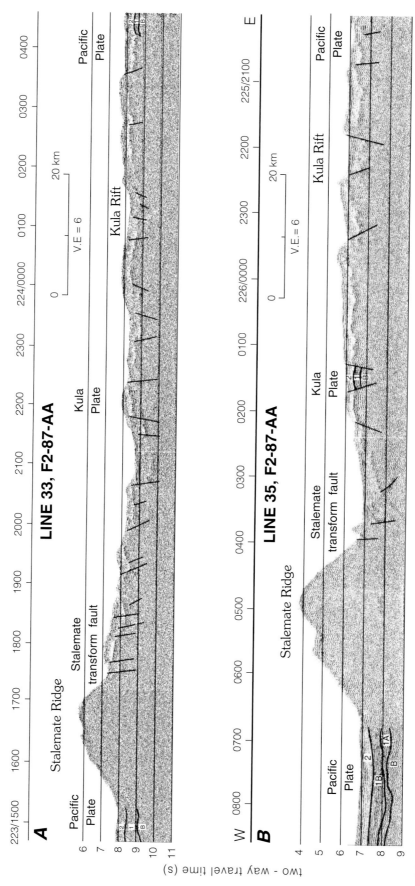

Figure 18–9. Seismic profiles (cruise F2-87-AA) along tracklines 33 (9A) and 35 (9B). Notice a significant difference between the acoustic stratigraphy on each side of Stalemate Ridge in both profiles. In the southern profile (9B), the Kula Rift basement shows the topography of an old spreading ridge as pointed out by Lonsdale (1988). In line 33 (9A), shot semiparallel to and near the trench, the seafloor of Kula limb is extensively faulted and the spreading ridge topography is subdued.

344

ported from source regions along the Aleutian Ridge have partially to totally filled these fault-bounded basins (Underwood et al. 1987).

Thickness of trench sediments in the western Aleutian region ranges between 0 km where the trench floor is elevated at the intersection of the trench and Stalemate Ridge to about 2 km. Thicknesses in the trench elsewhere along the Aleutian Arc are somewhat irregular, but in the central Aleutians near the Amlia Fracture Zone, between longitudes 173° W and 174° W, the thickness in one deep basin is about 4 km (Scholl, Vallier, and Stevenson 1982).

The GLORIA data exhibit some obvious sediment slumps and two fields of probable sediment diapirs (Figures 18–2 and 18–4). Slumps of igneous basement (bounded by normal faults) at the intersection of Stalemate Ridge with the Aleutian Trench (lines S-7 and S-9, Figure 18–5) probably formed as a result of faulting that was initiated by the plate-bending mechanism. Alternatively, the slumps may be related to the strong dextral shear that is suspected to occur where Stalemate Ridge is obliquely underthrusting the Aleutian fore-arc region. Circular and elliptical structures along the outer trench slope near the KUSAR collision zone (Figures 18–2 and 18–4) are probable diapir fields. Deformation of the Kula Paleo-Plate along the trailing edge of the KUSAR collision zone, combined with the faulting due to bending of the oceanic plate, presumably facilitated the release of fluids from the sediment units that led to the diapiric intrusions.

Enigmatic high-backscatter features in the Aleutian Trench, where it intersects the Kula-Pacific Rift, are probable sediment slumps. However, they may be lava flows. Two distinct and adjacent fields are evident (Figures 18–2 and 18–4). The simplest explanation is that the features are young sediment slumps, caused by abrupt oversteepening of the sediment-filled Kula Rift as it began descending into the trench. Strong nearby earthquakes also could trigger the slumps. The occurrence of high-backscatter sediment slumps in the Aleutian Trench as observed on GLORIA images, however, is rare. The features described here are different from other known slump deposits in the trench by having higher backscatter characteristics and by the absence of a large sediment source. They may be lava flows. The backscatter intensity is similar to that recorded on GLORIA images from young lava flows near the Hawaiian Arch (R. Holcomb, personal communication, 1994). The presence of young lava flows in a trench setting that is not associated with the subduction of an active spreading center has interesting tectonic implications. If these features are lava flows, then with time they may be accreted onto the inner trench wall during the subduction process. Lava extrusions synchronous with trench sedimentation could explain the enigmatic occurrence of oceanic igneous rocks in accretionary prisms that are the same age as, or younger than, the enclosing sedimentary rocks, such as those mapped on Kodiak Island (Moore et al. 1983).

Deformation of the fore-arc region at the KUSAR collision zone and tectonic fragmentation of Kula Paleo-Plate

The largest component of Pacific Plate – North American Plate motion is nearly parallel to the axis of the Aleutian Ridge in the Near Islands region (Minster and Jordan 1978). The crusts of both the overriding and the oceanic plate are extensively deformed as shown by the acoustic backscatter of multiple fault traces in the oceanic plate, the evidence for fault displacements in seismic-reflection profiles, and the extensive shallow seismicity. Evidently, two different styles of deformation are dominant: (1) a compressional style occurs in the Aleutian fore-arc region, and (2) an extensional style is dominant in the oceanic plate.

The deformational responses in both plates are primarily the result of shear produced by extreme oblique convergence. The tangential component of plate convergence is significantly larger than the normal convergent component (orthogonal to the trench axis); consequently the amount of slab subduction is reduced, and major parts of the oceanic plate, and probably of the Aleutian Ridge itself, migrate parallel to the trench. In the Aleutian fore-arc region, the largest component of motion is nearly parallel to the Aleutian Ridge. However, plate convergence with attendant dextral shearing has created a narrow accretionary wedge about 12 to 15 km wide. Southeast of Agattu Island a décollement was acoustically imaged (multichannel seismic-reflection profile) for about 10 km beneath the accretionary wedge (Vallier et al. 1994). A short décollement also was acoustically imaged (single-channel seismic-reflection profile) along trackline S-9 (Figure 18–5) where Stalemate Ridge underthrusts the inner trench wall.

Faults are by far the dominant structures in the region, occurring both in the Aleutian fore-arc region and on the ocean plate (Figures 18–2, 18–4, and 18–10). Figure 18–10 shows the fault fabric and distribution of significant seismicity (1957–1992). Overall, long linear faults in the outer fore-arc region parallel the inner trench wall, but near the KUSAR collision zone, lineaments in the outer fore-arc region are decidedly curved, suggesting that Stalemate Ridge deformed the strata during collision (Figure 18–4). The fore-arc region is undergoing extreme deformation as shown by the abundance of shallow earthquakes, particularly at distances from the trench of 30 km and greater (Figure 18–10). Similarly, the upper crust of the oceanic plate is undergoing extensive brittle deformation near the trench. Faults in the oceanic plate are particularly abundant along and near the trench and occur within a narrow zone less than about 50 km wide.

The KUSAR collision zone

The KUSAR collision zone, where Stalemate Ridge and the Kula Paleo-Plate are crashing into the Aleutian fore-arc re-

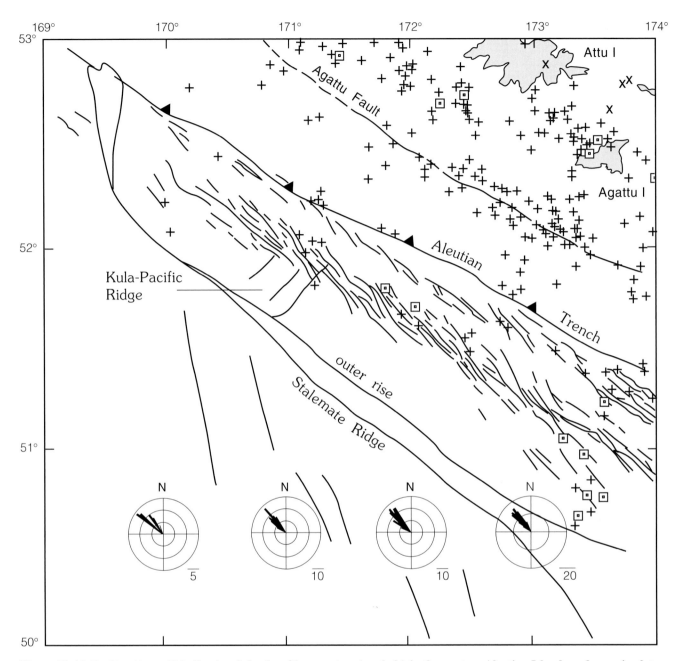

Figure 18–10. Fault patterns (thin lines) and depths of hypocenters (symbols) in the western Aleutian Islands and oceanic plate regions based on the recent relocations of seismic events in the Aleutian arc-trench system by Boyd et al. 1995). The faults and the trends of other structures in the region were digitized from the signal-processed GLORIA sonar images with overlapped swaths. Rose diagrams: the average fault trends estimated from measured orientations of digitized fault segments within one degree longitude intervals. The locations of fourteen large earthquakes from the catalog of Harvard University's Centroid Moment Tensor solutions (Dziewonski et al. 1990) are marked with square symbols. Shallow hypocenters: most earthquakes (plus symbols) occurred in the upper plate rather than in the subducting plate at depths shallower than 50 km; the moderate-depth events (cross symbols), ranging from 50 to 100 km, may have taken place at the interface between the descending slab and the upper plate. Agattu Fault trace is well correlated by a narrow band of shallow seismicity.

gion, is a dynamic area where deformation has greatly affected both the bathymetry and structure (Figures 18–2 through 18–5). Apparently, Stalemate Ridge is moving west on the oceanic plate at nearly 80 km/my and is aggressively eroding the inner wall of the trench and deforming strata in

the fore-arc region (Vallier et al. 1992). Long faults curve around, and trend approximately parallel to, the nose of Stalemate Ridge (Figures 18–2 and 18–5).

One consequence of the collision is especially evident at Attu Bight (Figure 18–4). We suspect that the inner trench wall was initially uplifted during the passage of Stalemate Ridge; a weakened inner trench wall subsequently collapsed into the trench, thereby forming an indentation or bight. Similar tectonic erosion during seamount (or ridge) subduction has occurred along the inner wall of the Tonga Trench where the Louisville Ridge is colliding (Ballance et al. 1989). Collot and Fisher (1989) described a somewhat similar phenomenon in the New Hebrides Island Arc where a seamount formed an indentation along the inner trench wall. Masson et al. (1990) described similar effects from seamount collisions with the inner wall of the Java Trench.

Stalemate Ridge may be deforming internally. In fact, some of the abrupt northward curvature of the ridge near the trench may be the result of tectonic stacking and ductile squeezing of the lower and middle crust as the ridge has been pushed along the fore-arc region. GLORIA images show evidence for surface deformation along and across the northern end of Stalemate Ridge, particularly between the western bend of the ridge and the trench. Three discontinuous backscatter (fault?) alignments trend northwest (about 319°) across the ridge and appear to bisect the entire ridge to form an en echelon shear zone (Figures 18–2 and 18–10) in the upper part of the oceanic plate. These fault trends form acute angles of less than 15° with the trench, and their strikes are nearly parallel to the direction (315°) of North American – Pacific Plate relative motion (Figure 18–11). The crust seemingly thickens behind (east of) Stalemate Ridge (Figure 18–9, tracklines 33 and 35); our seismic data, however, are not able to resolve deeper crustal structure.

No significant seismicity (greater than 3.5 Mw) has been recorded at the north end of Stalemate Ridge (Figure 18–10). Furthermore, regional seismic activity (Mw >3.5) decreases from east to west in the fore-arc region. We propose that the NW-trending faults along Stalemate Ridge near the trench are probably the result of en echelon shear failures caused by the westward movement of the oceanic plate relative to the Aleutian fore-arc region. This shear failure either creates very small scale undetectable (with present instrumentation) seismic activity, is not capable of accumulating more stress, or has not been active during the time of recording earthquake activity in the region.

We suspect that the KUSAR collision zone is where the North American and Pacific Plate margins initially changed tectonic stress regimes from one mostly produced by oblique plate convergence to one produced by a transform plate boundary. The absence of large earthquakes in the KUSAR region may be related to the very weak strength of the upper crust. The KUSAR collision zone thereby would be an accommodation zone of brittle deformation.

Tectonic fragmentation of the oceanic plate

Extensive surface brittle deformation of the oceanic plate on the trench outer rise (Figures 18–2, 18–4, and 18–9) resulted from the high obliquity of plate convergence combined with plate bending during the subduction process. GLORIA sidescan sonar images display the backscatter of numerous long fault escarpments that disrupt the ocean floor south of the Aleutian Trench (Figure 18–2). These escarpments form linear fault patterns, generally striking diagonal to the Aleutian Trench west of the Rat Fracture Zone, but some fault segments strike nearly parallel to the trench near the KUSAR collision zone. The diagonal fault trends in the KUSAR region, as pointed out by Masson (1991), significantly depart from the widely held concept that faults near the trench strike either semiparallel to the trench as a result of bending of the oceanic plate (Hilde and Sharman 1978; Jones et al. 1978); or oblique to the trench because their trends are affected by preexisting spreading fabrics (Warsi, Hilde, and Searle 1983; Masson 1991).

In the GLORIA sonar mosaic (Figure 18–2), the continuity of fault escarpments was not imaged across the tracklines. As explained by Gardner et al. (this volume), the lack of backscatter along the tracklines produced long bands of artifacts on the GLORIA images; the tracklines are shown as bands with no faults. However, seismic-reflection profiles recorded simultaneously along the GLORIA lines show direct evidence that faults disrupt the seafloor across the tracklines (Figures 18–5 and 18–9). Many of the faults, therefore, may be longer than shown on Figure 18–10. Some of these fault escarpments form elongated graben structures that extend out from the trench onto the outer rise. These structural relationships across the outer rise closely correspond to the gravity field in the region (Vallier, Karl, and Underwood 1987; Sandwell and Smith 1992).

Some faults can be traced continuously at least 50 km (Figures 18–2, 18–4, and 18–10). Masson (1991) measured the strike of distinguishable fault escarpments in the oceanic plate south of the Aleutian Trench from shipboard GLORIA mosaics; he integrated the fault strikes in one-degree longitude intervals. The measured trends, however, were not weighted to lengths. We have continued his work by making a comparative analysis of the fault trends with respect to the trends of the convergent plate motion vectors, studying the preexisting fault fabric and the nodal planes of earthquake mechanisms (Table 18–1), and remapping the fault trends directly from the processed GLORIA images from the region.

We studied fault trends within six one-degree longitude segments across the trench outer rise. Fault trends in the six segments range from 295° to 333°; average trends range from 311° to 317° (Table 18–1). Overall, the dominant fault trends within each segment form acute angles (between 9° and 25°) relative to the trench trend (Figure 18–10) and are approximately at right angles (between 85° and 92°) with respect to the preexisting fault fabric that had formed parallel to the midocean ridge and was a consequence of the original spreading dynamics (MacDonald 1982). We assume, therefore, that the preexisting faults formed shortly after the oceanic crust was magnetized during the process

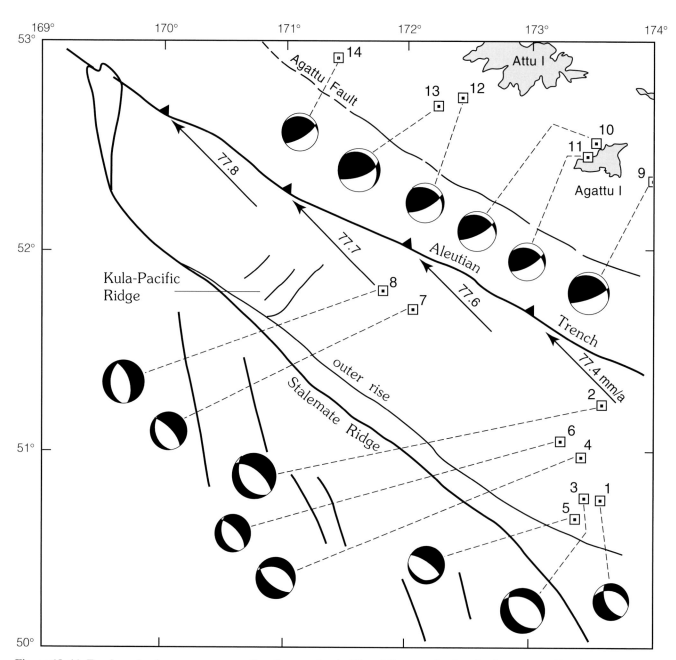

Figure 18–11. Focal mechanisms, convergence directions, and velocities of the oceanic plate in the western Aleutian Islands region. The relative plate motion of the oceanic plate with respect to North American Plate (based on NUVEL 1, DeMets et al. 1990) varies from 313° to 315° azimuths with rates from 77.9 to 77.3 mm/yr^{-1} at the trench axis between longitudes 169° E and 174° E. The best double-couple solutions of the focal mechanisms (beach balls, locations shown in squares) are listed in Table 18–2. Solutions are from Harvard's CMT solution catalog covering the years from 1978 to 1992. Sizes of beach balls are relative to their Mw magnitude.

of seafloor spreading. The average trends of these preexisting faults in the region are estimated from the seafloor magnetic anomalies lineations as mapped by Lonsdale (1988).

Normal faulting of the oceanic plate near a trench is generally attributed to bending stresses caused by the subduction process (Jones et al. 1978). However, the fault patterns

along this obliquely convergent margin apparently reflect two regimes of brittle deformation that control the preferential orientations. The fault patterns and orientations change from a densely fractured oceanic crust, with primary faults striking diagonally and at higher angles to the trench in the west, to a less fractured crust with faults striking semi-parallel to the trench in the east (Figure 18–10). This change

Table 18–1. *Preferential trends of the oceanic-plate structures*

Longitude	Trench	Pac-NA*		Faults†		Preexisting Fabric‡		Nodal Planes	
°E	Az.	Az.	Rate	Trend	Pairs	Trend	Magnetics	Trend	Range
174	293°	315°	77.3	312°	354	46°	20R—Pac.		
								312°	±21° (6)
173	296°	315°	77.4	316°	346	46°	20N—Pac.		
172	293°	314°	77.6	317°	293	46°	19R—Pac.		
								329°	±10° (2)
171	298°	314°	77.7	316°	235	45°	20N—Kula		
170	299°	313°	77.8	§324° and 306°	76	45°	21N—Kula		
169	306°	313°	77.9	§310° and 300°	17		Stalemate		

Azimuths (Az.) of the Aleutian Trench and their corresponding Pacific (Pac)–North America (NA) plate motion vectors are listed for six locations at the trench; in addition, the average estimates of the trends of the fault patterns, the preexisting seafloor-spreading fault fabric, and the compound nodal focal-plane directions seaward of the trench are listed for each longitudinal degree intervals.
*Rates (mm y^{-1}) and plate motion vectors at the Aleutian Trench are based on NUVEL1 Pac-NA closure-fitting Euler pole (48.3 °N, 77.0° W, $\Omega = -0.79$ deg. m.y.$^{-1}$, DeMets et al. 1990)
†The average fault trends are estimated by measuring the orientations of digitized fault segments (pairs of points) within one degree longitude interval and between the trench and outer rise, the reference longitude location at the trench is the mid-point of the interval.
‡The average trends of preexisting seafloor spreading fault fabric are determined from the configuration of magnetic-anomalies, conformed by Lonsdale (1988).
¶Compound nodal plane direction are estimated from the sum of the directions of fault plane solutions of large earthquakes within a region of radius distance of less than half of degree. Number of nodal plane solutions between parentheses.
§Bimodal distribution in the fault orientations.

in fault trends definitely is not related to faults that formed parallel to the preexisting spreading fabric; the change may instead be linked to tectonic processes that occur solely along the plate boundary. As the oceanic plate moves westward, the oceanic crust is first brittlely deformed by the plate-bending stress regime that operates mainly west of about 174° E; a strike-slip shear stress regime becomes more and more dominant in the oceanic crust as the plate migrates farther west into a zone where the plate convergence becomes highly oblique.

Average fault trends on the oceanic plate (about 314° to 317°; Figure 18–10) are nearly parallel to the azimuths of the relative plate motion vectors of the oceanic plate with respect to the North American Plate (Figure 18–11). Although these fault trends most likely are related to the high obliquity of convergence vectors, which apparently rotated the stress fields (Masson 1991), the faults may have formed farther east where convergence is less oblique, and subsequent plate motion carried them west. This scenario, however, would probably require large-scale strike-slip faulting along the trench inner wall and either small amounts of subduction or none at all. The change in fault orientation (40° clockwise) observed along the trench at 179° E, east of Rat Fracture Zone (Mortera-Gutierrez et al. 1992), coincides with the region where the trend of anomaly 23R forms a 45°

angle with the trend of anomaly 24N. This trend change may not be a coincidence, but rather the result of rock failure in the direction of the conjugate plane to the preexisting faults. This hypothesis is an attractive explanation for understanding rock failure in the presence of inherited fault fabric; a thorough discussion of this phenomenon is beyond the scope of this chapter.

Regional seismotectonics

Earlier studies (Cormier 1975; Newberry, LaClair, and Fujita 1986) discussed the diversity of seismicity and faulting in the region. Cormier (1975) pointed out that a significant number of earthquakes do not have strike-slip motion as predicted by the relative plate motions. In fact, the slip vectors of earthquakes deviate (20° and less) from those predicted by the RM2 regional convergence vector of Minster and Jordan (1978). Newberry, LaClair, and Fujita (1986) studied focal mechanisms in the Near Islands and Komandorsky Islands regions and concluded that there are several distinct tectonic zones. The seismicity data suggested to them that subduction terminates near 173° E. However, this conclusion conflicts with our seismic profiles, which clearly show partial underthrusting of the oceanic crust beneath the Aleut-

Table 18–2. *Best double couple solutions of earthquakes near the Kula Plate*

No.	Date yymmdd	Longitude °E	Latitude °N	Depth * km	MW†	P. Nodal Plane Strk	P. Nodal Plane Dip	P. Nodal Plane Rake	A. Nodal Plane Strk	A. Nodal Plane Dip	A. Nodal Plane Rake	Fault Motion Components
1	880204	173.6	50.7	15	5.0	153	66	−55	274	41	−142	N, ss, s
2	800503	173.6	51.1	15	5.8	302	57	−69	147	36	−69	N, ss, d
3	880207	173.5	50.8	15	6.3	147	47	−65	293	48	−114	N, ss, s
4	781004	173.0	51.0	15	5.3	133	48	−73	289	45	−108	N, ss, s
5	880213	173.4	50.7	15	5.2	318	62	−101	162	30	−69	N, ss, s
6	860409	173.2	51.0	15	5.3	318	62	−101	162	30	−69	N, ss, d
7	890525	172.1	51.7	15	5.2	319	66	−102	167	27	−65	N, ss, d
8	891009	171.8	51.8	15	5.7	339	64	−97	174	27	−76	N, ss, d
9	790923	174.0	52.3	33	5.8	281	23	122	67	71	78	T, ss, d
10	850809	174.1	52.5	41	5.4	293	27	137	63	72	69	T, ss, d
11	850731	173.4	52.4	29	5.6	292	20	139	62	77	75	T, ss, d
12	781119	172.4	52.7	38	5.3	297	24	135	68	74	72	T, ss, d
13	871020	172.2	52.7	33	6.0	312	27	150	69	77	67	T, ss, d
14	920321	171.4	52.9	18	5.5	257	18	92	74	72	89	T, ss, d

Focal mechanism solutions gathered from the Harvard's Centroid-Moment Tensor (CMT) Solutions Catalog of large earthquakes since 1978 (e.g. Dziewonski et al. 1983; 1989). The selection criteria of the dominant fault mechanism is based on Anderson (1951) experimental failure criteria. The focal solutions of each seismic event is defined by its conjugate nodal fault planes and the type of failure mechanism. Fault orientations of the P. (primary) and A. (auxiliary) nodal planes are defined by their strike (Strk), dip and rake. The dominant fault mechanism of each event are defined either as extension event (N, normal fault stress regime and the P. plane dip is greater than 45°) or compression event (T, thrust fault stress regime and the P. plane dips less than 45°). Some focal mechanisms show small component of strike-slip (ss) fault motion with either dextral (d) or sinistral (s) lateral motion.
*Focal depths are fixed to 15 km because of the limitations on the CMT method to determine earthquake depths shallower than 15 km (Dziewonski et al. 1981; Dziewonski and Woodhouse 1983).
†Mw (moment magnitude of seismic events) where Mw = 2/3 log M_o − 10.7, where M_o is the seismic moment in dyne-cm (Hanks and Kanamori 1979).

ian Ridge to at least the KUSAR zone near 169° E (Figure 18–5), and we suspect that some underthrusting is occurring even farther west.

Large earthquakes occurring in subduction zones reflect 95% of the seismic moment released at convergent plate boundaries (Dziewonski et al. 1990). The epicenters of earthquakes with magnitudes (Mw) greater than 3.5 that occurred since 1957 in this region were relocated by Boyd et al. (1995) and are shown here in Figure 18–10. Plate corrections were not applied to determine hypocenters because of the uncertainty remaining on the slab geometry within this region. The seismicity, mostly shallow in both the oceanic and the overriding plates, occurs at depths of less than 40 km in the oceanic plate and less than 60 km in the region between the trench and the highest parts of the island arc (Figure 18–10). Few earthquakes greater than Mw 3.5 occur near the toe of the accretionary prism. Small seismic events probably occur within the outer trench slope, but they have not been recorded by regional seismic networks. For example, a short-time deployment of ocean bottom seismometers recorded many small shallow events in the outer trench slope of the central Aleutian region (Frohlich et al. 1982).

Focal mechanism solutions of fifteen earthquakes (1977–1993), from both the oceanic plate and the Aleutian fore-arc region (Figure 18–11, Table 18–2), are compared with the relative plate motion vectors of the oceanic plate with respect to the North American Plate based on the NUVEL 1 global model (DeMets et al. 1990). On the basis of normal fault criteria (Anderson 1951), the azimuth of primary nodal planes dipping greater than 45° are close to the average fault trends and to plate motion vectors (Table 18–1). The focal mechanism solutions on the oceanic plate show very small dextral strike-slip components (Figure 18–11). Furthermore, azimuths of focal mechanism solutions in the oceanic plate show a westward change in the trend of nodal planes from 312° to 329° (between 173° E and 171° E).

In Figure 18–11, seismic events 1 to 8 took place in the oceanic plate near the Aleutian Trench; the rest of the events occurred in the Aleutian fore-arc region. For solutions 1 and 3, there was some ambiguity in selecting the primary nodal planes, because both conjugate nodal planes are suitable to fit the dominant seismic strain. The earthquake for event 4 occurred closer to the trench than other events in the oceanic plate, but its location is near the zone where fault patterns change trends. The solution to event 6 is significantly different from other events in the oceanic plate because it took place south of Stalemate Ridge where the tectonic stress regime is probably different from the events close to the trench (Zoback et al. 1989). For large earthquakes in the fore-arc region (numbers 9 to 15 on Figure 18–11), events occurred mostly south of the ridge axis at depths between 29 and 49 km. The solution of event 10 is significantly different (shallow depth and focal mechanism solution) from the rest of the fore-arc events because the earthquake occurred very near (or under) Agattu Island. Event 10 is the shallowest (about 18 km deep) recorded thrusting event. Event 15, in the far western part of the region, is significantly different from the events farther east. Its primary nodal plane dips northwest about 18° from the horizontal plane. We suspect that this earthquake is not well located, and that it most likely took place in the fore-arc region nearer the trench where thrust faults with that motion are more likely to occur. The total seismic moment released in fifteen years by large earthquakes (estimated from Table 18–2) in this region is 2.37×10^{23} dyne-cm yr^{-1} for the fore-arc region and 1.55×10^{23} dyne-cm/yr for the outer trench slope.

Evolution and fate of the Kula Paleo-Plate

The Kula Plate is just about "all gone"; most of it has been subducted. By backward reconstruction, it is possible to obtain some understanding of its age, original size, and past movement (Delong, Fox, and McDowell 1978; Byrne 1979; Rea and Dixon 1983; Engebretson et al. 1984, 1985; Lonsdale 1988), and we can predict its probable fate. As suggested by Engebretson et al. (1984) and strengthened by Lonsdale (1988), the Kula-Pacific Ridge stopped spreading at about the same time as the abrupt change in plate motion that is recorded by the Hawaii-Emperor Bend (Dalrymple and Clague 1976).

The slowdown and subsequent termination of spreading along the Kula-Pacific Ridge may have important consequences for interpretations of ancient Pacific Plate motion. It is probable that the cessation of activity along the Kula-Pacific spreading center is directly related to the change in Pacific Plate motion that is recorded by the Hawaii-Emperor Bend. Our data, however, suggest that the age of the bend might be somewhat younger. If the age of the Hawaii-Emperor Bend (Dalrymple and Clague 1976; Clague and Dalrymple 1987) of 43.1 ± 1.4 Ma (range 44.5 to 41.7 Ma) is accurate, then the significant slowdown of spreading rates at the beginning of anomaly 19R time (42.63 Ma; time scale of Cande and Kent 1992) in the western Aleutian region (Figures 18–6 and 18–7) could be related to that change in plate motion. However, cessation of spreading along the

Kula-Pacific Ridge occurred during 18R time (probably in the interval of about 41 to 40.5 Ma; the age range of anomaly 18R is 41.35 to 40.22 Ma), which is somewhat younger than the calculated minimum age of the bend. Either it took more than a million years for the effect of the change in plate motion to reach the Kula-Pacific Ridge or the Hawaii-Emperor Bend is somewhat younger than previously thought. We doubt if the shutdown of spreading along the Kula-Pacific Ridge would be synchronous with the plate motion change, but we do suggest that the age of the Hawaii-Emperor Bend may be closer to, or even less than, the minimum age of about 41.7 Ma (Dalrymple and Clague 1976).

It is probable that the igneous basement of the deep Bering Sea (Aleutian Basin), which lies north of the Aleutian Ridge, is part of the Kula Plate (Marlow and Cooper 1983). According to current interpretations, Kula Plate subduction jumped from the Alaskan and Russian continental margins to the Aleutian Arc about 55 to 56 Ma (Scholl et al. 1983, 1988), thereby trapping part of the Kula Plate behind the new arc.

The 56 to 55 Ma change in spreading patterns noted by Lonsdale (1988) greatly affected the tectonics along most margins of the North Pacific Ocean. The change probably was caused by an increase in slab pull along the western boundary of the North Pacific as relatively older and denser asthenosphere was subducted. This increase in slab pull may have accelerated the subduction process, forcing the Kula-Pacific Ridge to stop spreading. Thereby, the Kula and Pacific Plates were welded together to form the late Eocene to present-day Pacific Plate.

We can speculate on the fate of the Kula Paleo-Plate remnant south of the Aleutian Ridge. Our data show that subduction of the plate is continuing at least as far west as Stalemate Ridge, but we suspect that underthrusting is also occurring west of that ridge. If the present motion continues, the remaining fragment will be partially subducted under the Aleutian Ridge. The high component of arc-parallel plate motion, however, may result in collision (and subduction?) of a much smaller Kula Paleo-Plate with the Kamchatka margin in about eleven to twelve million years.

Acknowledgments

We thank the ship's crew of the R/V *Farnella* in addition to the USGS and IOS scientific staff during cruise F2-87-AA in the North Pacific Ocean. We are grateful for reviews of the manuscripts by Eric Geist, Colin Jacobs, David Rea, and Holly Ryan, who made many helpful comments and suggestions for improvements. Ray Sliter processed the seismic records, and Clint Steele and Carolyn Degnan assisted with magnetic and bathymetric data. We thank Susan Vath for help with the illustrations.

References

Anderson, E. M. 1951. *The Dynamics of Faulting*, Ed. 2, Edinburgh: Oliver and Boyd.

Ballance, P. F., Scholl, D. W., Vallier, T. L., Stevenson, A. J., Ryan, H. F., and Herzer, R. H. 1989. Subduction of a Late Cretaceous seamount of the Louisville Ridge at the Tonga Trench: A model of normal and accelerated tectonic erosion: *Tectonics* 8: 953–62.

Boyd, T. M., Engdahl, E. R., and Spence, W. 1995. Seismic cycles along the Aleution arc: Analysis of seismicity from 1957 through 1991. *J. Geophys. Res.* 100: 621–644.

Buffington, E. C. 1973. *The Aleutian-Kamchatka trench convergence.* Los Angeles, University of Southern California, *Ph.D. dissertation,* 363 p.

Byrne, T. 1979. Late Paleocene demise of the Kula-Pacific spreading center. *Geology* 7: 341–4.

Cande, S. C., and Kent, D. V. 1992. A new geomagnetic polarity time scale for the Late Cretaceous and Cenozoic: *J. Geophys. Res.* 97: 13,917–51.

Carlson, R. L., Gangi, A. F., and Snow, K. R. 1986. Empirical reflection travel time versus depth and velocity versus depth functions for the deep-sea sediment column. *J. Geophys. Res.* 91: 8,249–66.

Clague, D. A., and Dalrymple, G. B. 1987. The Hawaii-Emperor volcanic chain. In *Volcanism in Hawaii: U.S. Geological Survey Professional Paper 1350*, eds. R. W. Decker, T. L. Wright, and P. H. Stauffer, pp. 5–54. Reston, Va.: USGS.

Collot, J.-Y., and Fisher, M. A. 1989. Formation of fore-arc basins by collision between seamounts and accretionary wedges: An example from the New Hebrides subduction zone. *Geology* 17: 930–3.

Cormier, V. F. 1975. Tectonics near the junction of the Aleutian and Kurile-Kamchatka arcs and a mechanism for middle Tertiary magmatism in the Kamchatka Basin. *Geol. Soc. Am. Bull.* 86: 443–53.

Dalrymple, G. B., and Clague, D. A. 1976. Age of the Hawaii-Emperor Bend. *Earth Plan. Sci. Ltrs.* 31: 313–29.

Delong, S. E., Fox, P. J., and McDowell, F. W. 1978. Subduction of the Kula Ridge at the Aleutian Trench. *Geol. Soc. Am. Bull.* 89: 83–95.

DeMets, C., Gordon, R. G., Argus, D. F., and Stein, S. 1990. Current plate motions. *Geophys. J. Int.* 101: 425–78.

Dziewonski, A. M., and Woodhouse, J. H. 1983. An experiment in the systematic study of global seismicity: Centroid-moment tensor solutions for 201 moderate and large earthquakes of 1981. *J. Geophys. Res.* 88: 3,247–71.

Dziewonski, A. M., Chou, T.-A., and Woodhouse, J. H. 1981. Determination of earthquake source parameters from waveform data for studies of global and regional seismicity. *J. Geophys Res.* 86: 2,825–52.

Dziewonski, A. M., Friedman, A., Giardini, D., and Woodhouse, J. H. 1983. Global seismicity of 1982: Centroid-moment tensor solutions for 308 earthquakes. *Phys. Earth Plan. Int.* 33: 76–90.

Dziewonski, A. M., Ekström, G., Woodhouse, J. H., and Zwart, G. 1990. Centroid-moment tensor solutions for October-December, 1989. *Phys. Earth Plan. Int.* 62: 194–207.

Engebretson, D. C., Cox, A., and Gordon, R. G. 1984. Relative motions between oceanic plates of the Pacific Basin. *J. Geophys. Res.* 89: 10,291–310.

——— 1985. Relative motions between oceanic and continental plates in the Pacific Basin. *Geol. Soc. Am. Special Paper 206*, 59 p. Boulder, Colo.: GSA.

Frohlich, C., Billington, S., Engdahl, E. R., and Malahoff, A. 1982. Detection and location of earthquakes in the central Aleutian subduction zone using land and ocean-bottom seismograph stations. *J. Geophys. Res.* 87: 6,853–64.

Geist, E. L., Scholl, D. W., and Vallier, T. L. 1991. Collision of the Aleutian Island Arc with Kamchatka. *EOS Trans. Am. Geophys. Union* 72: 440.

Geist, E. L., Vallier, T. L., and Scholl, D. W. 1994. Origin, transport, and emplacement of an exotic island-arc terrace exposed in eastern Kamchatka, Russia. *Geol. Soc. Am. Bull.* v. 106, p. 1182–94.

Grow, J. A., and Atwater, T. 1970. Mid-Tertiary tectonic transition in the Aleutian arc. *Geol. Soc. Am. Bull.* 81: 3,715–22.

Hilde, T. W. C., and Sharman, G. F. 1978. Fault structure of the descending plate and its influence on the subduction process. *EOS Trans. Am. Geophys. Union* 59: 1182.

Jones, G. M., Hilde, T. W. C., Sharman, G. F., and Agnew, D. C. 1978. Fault patterns in the outer trench walls and their tectonic significance. *J. Phys. Earth* 26 Supplement: S85–S101.

Karl, H. A., Vallier, T. L., Masson, D. G., and Bishop, D. G. 1987. Cruise report for EEZ-SCAN cruise F2-87-AA, western Aleutian arc and adjacent North Pacific. *U.S. Geological Survey Open-File Report 87-643*, 6 p. Reston, Va.: USGS.

Lonsdale, P. 1988. Paleogene history of the Kula Plate: Offshore evidence and onshore implications. *Geol. Soc. Am. Bull.* 100: 733–54.

MacDonald, K. C. 1982. Mid-ocean ridges: Fine scale tectonic, volcanic and hydrothermal process within the plate boundary zone. *Ann. Rev. Earth Plan. Sci.* 10: 155–90.

Marlow, M. S., and Cooper, A. K. 1983. Wandering terranes in southern Alaska: The Aleutian microplate and implications for the Bering Sea. *J. Geophys. Res.* 88: 3,439–46.

Masson, D. G., Parson, L. M., Milsom, J., Nichols, G., Sikumbang, N., Dwiyanto, B., and Kallagher, H. 1990. Subduction of seamounts at the Java Trench – a view with long-range sidescan sonar. *Tectonophysics* 185: 51–65.

Masson, D. G. 1991. Fault patterns at outer trench walls. *Mar. Geophys. Res.* 13: 209–25.

Minster, J. B., and Jordan, T. H. 1978. Present-day plate motions. *J. Geophys. Res.* 83: 5331–54.

Moore, J. C., Byrne, T., Plumley, P. W., Reid, M., Gibbons, H., and Coe, R. S. 1983. Paleogene evolution of the Kodiak Islands, Alaska: Consequences of ridge-trench interaction in a more southerly latitude. *Tectonics* 2: 265–93.

Mortera-Gutierrez, C. A., Carlson, R. L., Scholl D. W., and Vallier, T. L. 1992. Azimuthal differences of fault fabric in the outer rise of the western Aleutian Trench. *EOS Trans. Am. Geophys. Union* 73: 527.

Newberry, J. T., LaClaire, D. L., and Fujita, K. 1986. Seismicity and tectonics of the far western Aleutian Islands. *J. Geodyn.* 6: 13–32.

Nichols, H., and Perry, R. B. 1966. Bathymetry of the Aleutian arc. *ESSA Monograph 3*, Rockville, Md.: U.S. Department of Commerce, scale 1:400,000.

Pitman, W. C., and Hayes, D. E. 1968. Seafloor spreading in the Gulf of Alaska. *J. Geophys. Res.* 73: 6,571–80.

Rea, D. K., and Dixon, J. M. 1983. Late Cretaceous and Paleogene tectonic evolution of the north Pacific Ocean. *Earth Plan. Sci. Ltrs.* 64: 67–73.

Rea, D. K., and Duncan, R. A. 1986. North Pacific plate convergence: A quantitative record of the past 140 m.y.. *Geology* 14: 373–6.

Rea, D. K., Basov, I. A., et al. 1995. Proc. ODP, Sci. Results, 145. College Station, Tx.: (Ocean Drilling Program).

Renkin, M. S., and Sclater, J. G. 1988. Depth and age in the North Pacific. *J. Geophys. Res.* 93: 2,919–35.

Sandwell, D. T., and Smith, W. H. F. 1992. Global marine gravity from ERS-1, Geosat and Seasat reveals new tectonic fabric. *EOS Trans. Am. Geophys. Union* 73: 133.

Scholl, D. W., Hein, J. S., Marlow, M. S., and Buffington, E. C. 1977. Mejii sediment tongue: North Pacific evidence for limited movement between the Pacific and North American plates. *Geol. Soc. Am. Bull.* 88: 1,567–76.

Scholl, D. W., Vallier, T. L., and Stevenson, A. J. 1982. Sedimentation and deformation in the Amlia fracture zone sector of the Aleutian Trench. *Mar. Geol.* 48: 105–34.

——— 1983. Arc, fore-arc, and trench sedimentation and tectonics: Amlia corridor of the Aleutian Ridge. In *Studies in Continental Margin Geology American Association of Petroleum Geologists, Memoire 34*, eds. J. Watkins and C. Drake, pp. 413–39. Tulsa, OK: AAPG.

——— 1988. Geologic evolution and petroleum geology of the Aleutian Ridge. In *Geology and Resource Potential of the Continental Margin of western North America and Adjacent Ocean Basins – Beaufort Sea to Baja California*, eds. D. W. Scholl, A. Grantz, and J. G. Vedder, pp. 123–56. *Circum-Pacific Council for Energy and Mineral Resources Earch Science Series*, vol. 6. Houston, Tex.: American Association of Petroleum Geologists.

Scientific Party, ODP Leg 145 1993. Paleoceanographic record of North Pacific quantified. *EOS Trans. Am. Geophys. Union* 74: 406–11.

Stone, D. B. 1988. Bering Sea-Aleutian Arc, Alaska. In *The Ocean Basins and Margins: The Pacific Ocean*, vol. 7B, eds. A. E. M. Nairn, F. G. Stehli, and S. Uyeda, pp. 1–84. New York: Plenum Press.

Underwood, M. B., Karl, H. A., Vallier, T. L., and Masson, D. G. 1987. Sedimentation within the Amchitka-Attu corridor of the Aleutian Trench. *EOS Trans. Am. Geophys. Union* 68: 1468.

Vallier, T. L., Karl, H. A., and Underwood, M. B. 1987. Geologic framework of the western Aleutian arc and adjacent North Pacific seafloor. *EOS Trans. Am. Geophys. Union* 68: 1486.

Vallier, T. L., Karl, H. A., Prueher, L. M., and Masson, D. G. 1992. Deformation in the western Aleutian fore-arc region caused by impingement of Stalemate Ridge on the trench inner wall. *Geol. Soc. Am. Abs. with Prog.* 24: 87.

Vallier, T. L., Scholl, D. W., Fisher, M., Bruns, T., Wilson, F., von Huene, R., and Stevenson, A. J. 1994. Geologic framework of the Aleutian arc, Alaska. In *Geology of Alaska: Geological Society of America DNAG*, vol. 10, eds. T. Plafker and H. Berg, pp. 367–88. Boulder, Colo.: GSA.

Warsi, W. E. K., Hilde, T. W. C., and Searle, R. 1983. Convergence structures of the Peru Trench between 10°S and 14°S. *Tectonophysics* 99: 313–29.

Zoback, M. L., Zoback, M. D., Adams, J., Assumpção, M., Bell, S., Bergman, E. A., Blümling, P., Brereton, N. R., Denham, D., Ding, J., Fuchs, K., Gay, N., Gregersen, S., Gupta, H. K., Gvishiani, A., Jacob, K., Klein, R., Knoll, P., Magee, M., Mercier, J. L., Müller, B. C., Paquin, C., Rajendran, K., Stephansson, O., Suarez, G., Suter, M., Udias, A., Xu, Z. H., and Zhizhin, M. 1989. Global patterns of tectonic stress. *Nature* 341: 291–8.

Index